COMPUTER SCIENCE AND STATISTICS:
Proceedings of the
Fifteenth Symposium on the Interface

COMPUTER SCIENCE AND STATISTICS:

PROCEEDINGS OF THE FIFTEENTH SYMPOSIUM ON THE INTERFACE

Houston, Texas, March 1983

edited by

JAMES E. GENTLE

Manager, Research and Design
IMSL, Inc.
Houston, Texas, U.S.A.

1983

NORTH-HOLLAND PUBLISHING COMPANY – AMSTERDAM · NEW YORK · OXFORD

© North-Holland Publishing Company, 1983

All rights reserved. No part of this publication may be reproduced, stored in a retrieval system, or transmitted, in any form or by any means, electronic, mechanical, photocopying, recording or otherwise, without the prior permission of the copyright owner.

ISBN: 0 444 86688 4

Published by:
NORTH-HOLLAND PUBLISHING COMPANY – AMSTERDAM · NEW YORK · OXFORD

Sole distributors for the U.S.A. and Canada:
ELSEVIER SCIENCE PUBLISHING COMPANY, INC.
52 VANDERBILT AVENUE
NEW YORK, N.Y. 10017

This work relates to the Department of the Navy Purchase Order N00014-83-M-0041 issued by the Office of Naval Research. The United States Government has a royalty-free license throughout the world in all copyrightable material contained herein.

Library of Congress Cataloging in Publication Data

Symposium on the Interface (15th : 1983 : Houston, Tex.)
  Computer science and statistics.

  1. Mathematical statistics--Data processing--
Congresses. I. Gentle, James E., 1943-
II. Title.
QA276.4.S95 1983   519.5'028'54   83-11462
ISBN 0-444-86688-4

PRINTED IN THE NETHERLANDS

Preface

The rapidly occurring advances in computing science are making a major impact on the activities of the applied statistician. Conversely, the needs of the statistical data analyst have supplied the computer scientist and the numerical analyst with interesting and challenging problems. This exciting area in the interface of the disciplines has been explored since 1967 in the Symposia on the Interface.

The sessions of the Fifteenth Symposium, held in Houston, generally covered the traditional topics of the interface, such as graphics, numerical algorithms, statistical program packages, and simulation techniques. Reflecting the impact of one of the most significant current trends were the two sessions on smaller computers. The current status of the available software for the desktop computers is rapidly changing. The generally negative reports on this software given at the Symposium did contain a few bright spots, but indicated a need for extensive development of system support software for the micros. There were sessions on computing problems in specific areas of statistics, such as time series and survival analysis, and on general techniques, such as optimization. Other sessions considered such problems as tools for program development and measurement of software complexity.

The format of the Fifteenth Symposium was similar to that of recent Interface Symposia. Following the keynote address in the morning of the first day, there were three concurrent sessions consisting primarily of presentations by invited speakers and audience discussion. Many of the presentations incorporated audio-visual effects that can not be made available through these proceedings; nevertheless the written versions of the papers contained in this volume generally summarize the presentations and perhaps make some points in a more considered manner than the corresponding live presentations. In addition to the audio-visual effects and the spontaneity of oral presentations, there are other aspects of the symposium that the proceedings cannot capture. Much of the real interchange of ideas and information occured in informal contact and in demonstrations. (As I was leaving the hotel around 12:30 a.m. Friday morning, I saw a group animatedly discussing a computing problem in survival analysis. I will not try to describe the problem here!) During the noon break on both days contributed papers were presented in poster sessions. Written versions of these papers also appear in this volume.

There are a number of people who contributed to the success of the Fifteenth Symposium on the Interface and I would like to convey my thanks to all of these. In particular, I wish to thank Doris Ann Moore for help with so many of the details. I would like to thank Cecile Blake, Karen Clark, Bill Evans, Joyce Gentle, Byron Howell, Don Kainer, Tim Leite, Lindsay Reed, Bill Sallas, and Jim Thompson for help with local arrangements. I would like to thank Tom Aird, Tom Boardman, John Dennis, H. E. Dunsmore, Rich Heiberger, Bill Kennedy, Peter Lewis, Bob McGill, Webb Miller, Joe Newton, Michael Steele, Mike Tarter, Harry Wong, and Sid Yakowitz for their help in organizing the program. Tom Glenn and several other members of the Operations Division at IMSL were very helpful in various other aspects of putting on the Symposium, and I wish to thank all of them for their assistance.

Though the Interface Symposium is not sponsored by any separate organization, a number of the professional societies cooperate with the Symposium. I would like to thank the American Statistical Association, particularly its Section on Statistical Computing and its Houston and Southeast Texas Chapters, the Association for Computing Machinery, and the International Association for Statistical Computing for their cooperation.

Financial support for the Symposium came from the Office of Naval Research, the National Science Foundation, and IMSL, Inc. To all of these I extend my thanks.

James E. Gentle
Houston

Financial Supporters of the Fifteenth Interface Symposium

U. S. Office of Naval Research, U. S. Dept. of the Navy

National Science Foundation

IMSL, Inc.

Cooperating Organizations

American Statistical Association
    Statistical Computing Section
    Southeast Texas Chapter
    Houston Chapter

Association for Computing Machinery

International Association for Statistical Computing

Contents

1 Keynote Address
3   The Future
    R. W. Hamming

7 Statistical Databases
9   Statistical Databases: Characteristics, Problems, and Some Solutions
    Arie Shoshani

24 Some Statistical Data Base Requirements for the Analysis of Large Data Sets
    Richard J. Littlefield and Paula J. Cowley

31 RAPID: A Statistical Database Management System
    Roy G. Hammond

35 A Graphical Query System for Complex Statistical Databases
    Harry K. T. Wong and Whei-Ling Yeh

51 Simulation Output Analysis/Graphics
53 SIMTBED: Overview
    P. A. W. Lewis, D. G. Linnebur, E. J. Orav, L. Uribe, and H. W. Drueg

56 Simulation and Graphics--From Single Purpose Applications to Generic Interfaces
    Joseph Polito

61 Smaller Computers: Software Trends
63 Statistical Packages on Microcomputers
    Wayne A. Woodward and Alan C. Elliott

71 What More is There to Statistics on a Micro than Precise Estimates of Regression Coefficients? or Mainframe Statistical Software in a Box
    MaryAnn Hill and Jerome Toporek

80 Machines and Metaphors
    Joseph Deken

85 Software For Survival Analysis
87 SURVAN A Nonparametric Survival Analysis Package
    Barry W. Brown, M. Elizabeth Rozell, and Eula Y. Webster

93 Data Analysis Considerations and Survival Analysis Programs
    Frank E. Harrell, Jr.

97 Fitting Survival Models Using GLIM
   John Whitehead

104 An Interactive Statistical Analysis System for Cancer Clinical Trials Data
   Anne-Therese Leney and Marcello Pagano

108 Software for Survival Analysis
   Richard M. Heiberger and Milton N. Parnes

111 COMPUTING FOR TIME SERIES

113 Computing for Autoregressions
   H. Joseph Newton and Marcello Pagano

119 Time Series ARMA Model Identification by Estimating Information
   Emanuel Parzen

125 What Should Your Time Series Analysis Program Do?
   H. Joseph Newton

131 NUMERICAL ALGORITHMS

133 A Generalization of the Proposed IEEE Standard for Floating-Point Arithmetic
   W. J. Cody

140 A Package for Solving Large Sparse Linear Least Squares Problems
   Alan George and Esmond Ng

148 Exact Computation with Order-N Farey Fractions
   R. T. Gregory

154 Recent Advances in Generating Observations from Discrete Random Variables
   Bruce Schmeiser

161 SMALLER COMPUTERS: WORKSTATIONS AND THE HUMAN INTERFACE

163 Hardware for Kinematic Statistical Graphics
   Jerome H. Friedman and Werner Stuetzle

171 PATTERN RECOGNITION AND DENSITY ESTIMATION

173 Probability Density Estimation in Higher Dimensions
   David W. Scott and James R. Thompson

180 Pattern Recognition Applications to River Flow Analysis
   S. J. Yakowitz, T. E. Unny, and A. Wong

186 Methods for Bandwidth Choice in Nonparametric Kernel Regression
   John Rice

191 On Curve Estimation and Pattern Recognition in the Presence of Incomplete Data
   Michael Tarter and William Freeman

199 NASA's Fundamental Research Program in MPRIA
   R. P. Heydorn

208 A Mathematical Experiment
   Ulf Grenander

211 STATISTICAL PACKAGES: IMPLEMENTATION TECHNIQUES

213 Design and Implementation of MINITAB
    Barbara Ryan and Thomas A. Ryan, Jr.

217 The Implementation of PROTRAN
    Thomas J. Aird and John R. Rice

223 The Conversion of SAS^{TM} Software Products to Minicomputers
    Richard D. Langston

227 Implementation of SPSS and SPSS-X
    Jonathan B. Fry

231 TOOLS FOR DEVELOPING STATISTICAL SOFTWARE

233 Portable Tools and Environments for Software Development
    Webb Miller

235 NONNUMERIC ALGORITHMS

237 Letting MACSYMA Help
    Gail Gong

245 Codata Tools: Portable Software for Self-Describing Data Files
    Deane Merrill and John L. McCarthy

253 NEW AND INNOVATIVE WAYS OF LOOKING AT DATA

255 Interactive Tools for Data Exploration
    Sara A. Bly

261 DEVELOPMENTS IN OPTIMIZATION

263 Constrained Maximum-Likelihood Estimation for Normal Mixtures
    Richard J. Hathaway

269 SOFTWARE METRICS AND EFFORT ESTIMATION

271 Defining Metrics for Ada Software Development Projects
    Sylvia B. Sheppard, John W. Bailey, and Elizabeth Kruesi

277 Model Evaluation in Software Metrics Research
    S. M. Thebaut

286 Investigation of Chunks in Complexity Measurement
    John S. Davis

293 CONTRIBUTED PAPERS

295 Color Anaglyph Stereo Scatterplots---Construction Details
    Daniel B. Carr and Richard J. Littlefield

300 IMSL Libraries for Micro-Computers
    Henry Darilek and David Kortendick

304 Resistant Lower Rank Approximation of Matrices
    K. Ruben Gabriel and Charles L. Odoroff

309 Interactive Tools for EDA
    Herman Gollwitzer

310 On the Computation of a Class of Maximum Penalized Likelihood Estimators of the Probability Density Function
   V. K. Klonias and Stephen G. Nash

315 Econometric Calculations Using APL
   Stephen D. Lewis

319 A Survey of the Current Status of Statistical Computing in Colleges
   John D. McKenzie, Jr. and David P. Kopcso

324 Generating Normal Variates by Summing Three Uniforms
   Eugene F. Schuster

328 GRAPHPAK: Interactive Graphics for Analysis of Multivariate Data
   Danny W. Turner and Keith A. Hall

333 LP/PROTRAN, A Problem Solving System for Linear Programming Problems
   Granville Sewell

339 Evaluation of the Doubly Noncentral t Cumulative Distribution Function
   Michele Boulanger Carey

344 Elementary Methods for Approximating the Cumulative Distribution of Beta, F, T, and Normal Variables
   Dean Fearn

347 A Procedure for Approximating and Optimizing Multiattribute Value Functions Subject to Linear Constraints
   Thomas R. Gulledge, Jr. and Jeffrey L. Ringuest

352 L_1 Norm Estimates Using a Descent Approach
   Lee Ann Josvanger and V. A. Sposito

355 Security Issues for Dynamic Statistical Databases
   Mary D. McLeish

361 On the Use of Cluster Analysis for Partitioning and Allocating Computational Objects in Distributed Computing Systems
   Lennart Pirktl

365 Exact Factorization of Nonnegative Polynomials and its Application
   Mohsen Pourahmadi

368 A Comparison of Economic Designs of Non-Normal Control Charts
   M. A. Rahim and R. S. Lashkari

373 The Design of Unitest, A Unified Fortran Program Testing System
   Vincent L. Tang

375 Exact Sampling Distributions of the Coefficient of Kurtosis Using a Computer
   Derrick S. Tracy and William C. Conley

Availability of Proceedings

13, 14th (1981,82)	Springer-Verlag New York, Inc. 175 Fifth Avenue New York, NY 10010
12th (1979)	Jane F. Gentlemen Dept. of Statistics University of Waterloo Waterloo, Ontario Canada N2L 3G1
11th (1978)	Institute of Statistics North Carolina State Univ. P. O. Box 5457 Raleigh, North Carolina 27650
10th (1977)	David Hogben Statistical Engineering Laboratory Applied Mathematics Division National Bureau of Standards U.S. Dept. of Commerce Washington, D.C. 20234
9th (1976)	Prindle, Weber, and Schmidt, Inc. 20 Newbury St. Boston, Massachusetts 02116
8th (1975)	Health Sciences Computing Facility, AV-111 Center for Health Sciences Univ. of California Los Angeles, California 90024
7th (1974)	Statistical Numerical Analysis and Data Processing Section 117 Snedecor Hall Iowa State Univ. Ames, Iowa 50010
4,5,6th (1971,1972, 1973)	Western Periodicals Company 13000 Raymer Street North Hollywood, California 91605

The Sixteenth Symposium on the Interface will be held March 14-16, 1984, in Atlanta, Georgia. The chair will be Lynne Billard of the University of Georgia.

KEYNOTE ADDRESS

The Future
R. W. Hamming

THE FUTURE

R. W. Hamming

Naval Postgraduate School
Monterey, California

This paper predicts that: the past exponential growth in people will not continue, but the publication rate will; computers will have significant effects on the presentation of results; and the field is being saturated by increasing detail. is becoming increasingly personal instead of objective.
Statistics is changing. From compression of data we are going to expansion; from objective results we are going to personal; from a broad concensus of common theories we are going to elaborate specialized theories; and finally the reputation of statistics will continue to fall. You should consider how to alter these predictions by your future behavior.

The purpose of this talk is to persuade you to think about, and plan for, the future, both your own and that of statistics. The future has a habit of coming. It seems to me that because most people do very little long range planning they are forced to accept the future that comes rather than being prepared for, and to some extent making, the future they want.

Such planning implies that: (1) you have some idea of what the future is likely to be; (2) you have some scenarios of the future; (3) you have some desires and goals for the future; and (4) you are willing to plan and work to see that some of these desires are realized.

These are difficult to do. From long observations I believe I am right when I say that most people do not make such extensive plans for the future; they more or less let the future happen. Among statisticians I do not need to refute the common claim that since the future is uncertain, you therefore cannot be sure that your plans will be realized - hence the futility of planning, so they claim.

Let us look at the future. We have all heard that 90% of the scientists who ever lived are now alive. We know that this is the basis of growth since about the time of Newton (say 1660). We have been doubling knowledge about every 17 years since then. A lot of other observations, from the classical doubling period of library holdings to the growth of the number of employees in Bell Labs, support this doubling period.

Can this growth go on indefinitely? The popular expression, "Science, the endless frontier", suggests that it can, but if it did then in 340 years there would be 20 doublings - that is a million fold growth!

How have we coped with the growth since Newton's time? Essentially by specialization. In my lifetime I have seen an enormous increase in specialization. If you accept the million-fold growth of knowledge, then you must expect something like a million fields of specialization for every field we now have. I doubt you can believe that. Hence, although the past trends in information growth and specialization indicate one thing, common sense indicates another, slower, growth of knowledge in the future.

But there is a closer deterrent. At present there are not the students in school to carry on. Thus the doubling of the number of people must slaken off fairly soon.

Let us next look at publications. So long as we have the pattern of "publish or perish" we will have the glut of publication. The situation has positive feedback! I have noted with interest that top management in the schools, because they are less and less able to judge the quality of a candidate for promotion, are insisting more and more on volume of publication. They are measuring quantity, not quality. But is there any other solution for them? I cannot see it coming soon. It is less apparent in industry, but as specialization increases they will have, in their turn, to depend on outside evidence, and this tends to be quantity not quality. We ourselves are guilty! We will seldom express our opinions clearly about the trash that appears monthly lest we in our turn be critized; we only complain. Since we will not, then I say management cannot do other than increasingly use quantity in place of quality.

As a curious aside, I note that JASA appears to reject every first submission of a paper. Apparently no one, including the reviewers themselves, can write a proper paper the first time. I look at the delays indicated, and wonder if they are worth the "improvements" that come from the recycling. A statistical study of the dates of first submission and final acceptance would reveal a number of startling things, and raise serious questions about editorial policy.

The tricks for building up a big bibliography are well known. You take an idea, publish it in three papers rather than digest it first (risking prior publication by someone else) and have one good paper. You need energy to resubmit the paper after the first refereeing, since apparently (see above JASA remarks) no reviewer feels he is doing his job unless he asks for some alterations. Hence you must plan to send it back promptly with the changes asked, (even if in your eyes and others the changes are bad); you wear the reviewer down. It is a matter of having more energy than he has! It is also well known that you publish the same thing under different appearing titles in different places - the excuse is that it is of wide importance and needs wide circulation! I think that almost anyone can build up a bibliography in these days of specialization by becoming the recognized expert in their pinheaded field of specialization.

With the falling off of the number of scientists, can the growth of publication continue? Yes! The computer is the solution. Merely using a word processor will greatly ease the burden of rewriting the paper to satisfy the reviewer. It is also a great aid in preparing alterate versions for publication in various journals.

Many of you think that the computer will save us from being swamped by publications. You think that, by suitable cross referencing the papers, the computer can find the relevant literature. I doubt that it will do a lot, although there are some possibilities in this direction. The real truth of the matter is that a major problem in publication is to find a problem that has not yet been solved! That is often more work than solving the problem when you have found it.

Some of you may think that the computer can be programmed to search the product space of subclassification labels to find an empty hole that has surrounding papers already published. Then the computer merely names the missing entry, prints out the relevant papers along the lines of cross references, and you have on your desk all the material ready to whip out the missing theorems and proofs!

It is too glib for me to believe, but then I am not an ardent fan of artificial intelligence. If you believe that it can be done then maybe you can get the machine to write much of the paper too. In principle the machine has already delivered all the relevant papers needed to find the proof. But if you believe that you can write a program that can separate important information from junk, then I suggest you connect it to a source of random noise, and let it find important information. After all, a random source will, sooner or later, deliver every finite piece of information, and you claim your program can filter out the unimportant results! If you doubt the practicality of this proposal, then you need to back down and begin to consider, as I have, the impracticality of the earlier wishful thinking that computers will make significant improvements in our ability to do important research - note the qualifying words!

Is there no limit in the growth of knowledge and the concomitant specialization? As I earlier said, I do not believe in the "Science - the endless frontier". I believe, instead, in diminishing returns. When, finally, you have a massive three volume work on Maximum Likelihood Estimators, written by a select committee of experts, will you want to continue in the field? I suggest that most humans will find further effort repulsive! And when you have a five volume work, each 1000 pages or more, on Maximum Entropy, will that not be enough? How far do you believe people will be willing to push this endless frontier? Not how far can it be pushed!

I have already indicated a limit on the growth of activity in science because of the lack of students now in school to carry on. I have just indicated that I believe that the economic theory of diminishing returns also applies to research - the same amount of effort no longer produces in your own eyes the same value. There is a third limiting factor. It is the sense of creativity. Creativity is not specialized as a talent. One has a general urge to create. And if the apprenticeship in one field requires many long years before you can do real creations of importance in your own eyes, then you are likely to turn to other fields where the frontier is closer and the hope of creative activity is much higher. The creative person tends to go where there is a chance to be creative soon - not after 20 years of postgraduate study to find out what is known and only then begin to create.

I believe that this stage of diminishing returns is now here in statistics. The period of most activity often occurs after the period of diminishing returns starts. The inertia, the "publish or perish" syndrome, and the working habits of the people already in the field, all tend to maintain the frantic explosion of papers. How many of you believe that great, important progress is now going on in Statistics? I doubt that many of you do. What is going on is more filling in of missing pieces, finding variations similar to earlier parts, rewriting of things more abstractly that in fact adds little new knowledge. So far as I can see, the important papers, per person working and publishing, seems to be way down from what it was in earlier years, and I predict a further falling off, percentagewise.

You should be wary of doing things because they can be done. Without a sound basis, in mathematics and probability, a sound superstructure is very hard to build. There seems currently to be a growth in "personal statistics". Not that personal statistics may not get better results (without defining what I mean) than can objective statistics, but it is hardly a way of convincing others to commit their actions to your personal vision. For better or worse, the results you find must be communicable to others and win their assent before they can be translated into large scale actions. By "personal statistics" I mean the following: give the same data to 10 competent statisticians and you could get up to 20 different answers! The essence of science used to be reproducibility.

No doubt the connection between statistics and computing will become closer, but the theory of computer science seems to me to be going in directions that have little interest to statisticians at present. Of course some of the fallout from computer science may benefit you, but don't expect too much. Like statistics, computer science has become rather inwardly directed, as judged by the publication (or perish!) record.

Back to my opening remarks. What will happen, in the long run, to statistics and yourself? As to statistics I hope that there will be some serious examination of the foundations, both theoretical, which is happening, and experimentally which is not. I have been preaching for many years that 90% of the time the next independent measurement will fall outside of the previously announced 90% error bounds. I came to the idea from Youden's paper Enduring Values, and have seen mountains of evidence since then which support the claim - approximately. So I wonder why all the fancy statistics is studied and taught when the simplest parts fly in the face of reality. If, statistically speaking, statistics gives wildly wrong predictions, then what is wrong, and why will statisticians not look? The usual expert syndrome, I suspect.

I have meditated on this 90% effect for many years, and have come up with one small factor to partially explain it. When the experimentalist finally gets the equipment working, then the final stages of adjustment of the equipment begins. What criterion is used? Reproducibility, of course! Run-to-run should give consistent answers. But what is reproducibility except low variance? The equipment is adjusted to get low variance, and it is exactly this data that the statistician, sitting in his ivory tower, uses to estimate the probable error of the data.

Now we come down to predicting your future in statistics. I cannot say what your desires should be, nor what you should want, but I can say that if you look carefully and wonder about where you will be, and what statistics will probably be at the time of your mid-and end-career, then you are in a position to plan to have happen what you wish would happen.

I have omitted the topic of dramatic new discoveries that completely change the field. I have given some thought about their possibilities, but the results are so meager that I am ashamed to tell you. Hence, I will take cover under the standard remark, "barring sudden and unexpected developments in the world."

I have talked about only one aspect of the effects of computers on statistics. We need to examine the topic in more detail in our projections of the future. Perhaps the most dramatic advance will be in the general area of display in three or more colors. It is bound to be widely available and cheap in the near future. The ability to look at a set of data, displayed in many ways, and from many angles, with a rich battery of transformations to apply on demand, will encourage the art of prospecting in the raw data for the nuggets of value. Furthermore, the cheapness of computing will encourage many different summarizations of the data - until the summarizations exceed the original data in volume. We are entering the era of the explosion of the data. And the automation of the collection of data will exasperbate the problem many-fold. Unless we demand a mathematical basis for the processing we will face a deluge of personal data summarizations and presentations with no "degrees of freedom" decrease to compensate for the selection mechanism. I still remember the internal memorandum at Bell Telephone Labs that had less than 5 pages of text and over 100 pages of the output graphs. There is almost no upper limit on the ability to take a small amount of data and blow it up to an overwhelming amount.

In summary, I see changes I do not like. I was raised on the definition that statistics was the boiling down of the mass of data; I predict the regular creation of a mass of output from a small amount of input data. I used to believe that the purpose of statistics was to gain insight; tomorrow I see a volume of output that by its very mass can only prevent insight. I used to think statistics strove for objectivity; it seems headed toward personal data exploration. I see the continuation of mathematical statistics (how else do you get publications and theses?) to the neglect of the obvious current defects in prediction. I see the elaboration of every detail without end until nausea sets in; I do not see the statisticians daring to express value judgements on what is worthwhile. I see the reputation of statisticians matching that, in the public eye, of lawyers - people totally unconcerned with truth, justice, etc., but instead devoted to winning the case. In the future it will still be said with justice, "There are liars, damned liars, and statistics." Presumably I am an old fogey. If you do not want this future, now is the time for you to do something about it!

STATISTICAL DATABASES

Organizer: *Harry K. T. Wong*

Invited Presentations:

    Statistical Databases: Characteristics, Problems, and Some Solutions, *Arie Shoshani*

    Some Statistical Data Base Requirements for the Analysis of Large Data Sets, *Richard J. Littlefield and Paula J. Cowley*

    RAPID: A Statistical Database Management System, *Roy G. Hammond*

    A Graphical Query System for Complex Statistical Databases, *Harry K. T. Wong and Whei-Ling Yeh*

STATISTICAL DATABASES : CHARACTERISTICS, PROBLEMS,

AND SOME SOLUTIONS*

Arie Shoshani

Lawrence Berkeley Laboratory
University of California
Berkeley, California 94720

Abstract

The purpose of this paper is to describe the nature of statistical data bases and the special problems associated with them. Since statistical data bases are common in a variety of application areas, the paper begins by describing several examples that emphasize the complexity, the size and the difficulties of dealing with such data bases. A description is then given of the characteristics of statistical data bases in terms of data structures and usage. The remaider of the paper describes a large collection of problems, and when appropriate some solutions or work in progress. The problems and solutions are organized into the following areas: physical organization, optimization, logical modeling, user interface, integrating statistical analysis and data management, and security.

Table of content

1. Introduction
2. Some example data bases
3. Characteristics
   3.1. Category and summary attributes
   3.2. Sparse data
   3.3. Summary sets
   3.4. Stability
   3.5. Proliferation of terms
4. Physical organization
   4.1. Category attributes
   4.2. Sparse data
   4.3. Transposed files
   4.4. Partial cross product
5. Optimization
   5.1. Selection of physical structures
   5.2. Response Assembly
6. Logical modeling
   6.1. Representation of category and summary attributes
   6.2. Graph representation
   6.3. Summary sets
   6.4. Aggregation/disaggregation
7. User Interface
8. Integrating statistical analysis and data management
9. Security
   9.1. Limiting the query set
   9.2. Limiting overlap of query sets
   9.3. Random sample queries
   9.4. Partitioning the data base
   9.5. Perturbing data values
10. Concluding remarks

---

* This paper appeared also in the Proceedings of the Eighth International Conference on Very Large Data Bases, Sept. 1982.

1. Introduction

Statistical data bases (SDBs) can be described in terms of the type of data they contain, and their use. SDBs are primarily collected for statistical analysis purposes. They typically contain both parameter data and measured data (or "variables") for these parameters. For example, parameter data consists of the different values for varying conditions in an experiment; the variables are the measurements taken in the experiment under these varying conditions. The data base is usually organized into "flat files" or tables.

The statistical analysis process involves the selection of records (or tuples) using selection conditions on the parameters, taking a random sample, or using a graphics device to point to the items desired. Several variables are then selected for analysis. The analysis may involve applying simple univariate statistical functions to the value sets of the variables (e.g. sum, mean, variance) or using more complex multivariate analysis tools (e.g. multiple regression, log-linear models).

The statistical analysis process may involve several steps. It includes phases of data checking, exploration, and confirmation. The purpose of data checking is to find probable error and unusual but valid values (called "outliers" by statisticians), by checking histograms or integrity constraints across attributes. The purpose of data exploration is to get an impression of the distribution of variables and the relationships between them. This phase involves taking samples of the data, selecting records, and creating temporary data sets for use in graphical display and preliminary analysis. In the conformation phase, the analyst tests hypothesized distributions (which are based on the observations made in the exploratory phase) against the data base, or relationships between variables (cross tabulations). This process may then iterate several times until satisfactory results are achieved. A more detailed description of the statistical analysis process can be found in [Boral et al 82]. A compact description of data manipulation capabilities that are important for SDBs can be found in [Bragg 81].

At first glance it appears that the necessary data management functions can be supported by existing generalized data management systems. For example, one can view flat files as relations in a relational data management system, and the generation of subsets for analysis by using relational operators, such as "join" and "project". However, practice has shown that these data management systems have not been used for SDBs. Instead, one finds that statistical packages are used or special purpose software is developed for the particular data base in hand, or for a collection of data bases with similar characteristics. A case in point is the Census Data Base, which is collected and processed by special purpose software, distributed as flat files whose descriptions are quite complex, and therefore requires special purpose software for subsequent querying. An example of such a system which was designed specifically to manage geographically-based data is the Social, Economic, Environmental, Demographic Information System (SEEDIS), developed at at Lawrence Berkeley Laboratory [McCarthy et al 82].

Another approach to managing SDBs is by using statistical packages such as SAS [SAS 79] or SPSS [Nie et al 75]. While these packages have some data management capabilities their primary purpose is to provide statistical analysis tools to the analyst.

There are two main reasons for the fact that commercial data management systems have not been widely used for SDBs. The first reason is the storage and access inefficiency of these systems for SDBs. As will be discussed later, many SDBs have a high degree of data redundancy that can benefit from sophisticated compression techniques. The organization of the data into records (or tuples) makes retrieval inefficient in those cases where only a few attributes are needed for the analysis. Other data organization methods, such as organizing the data by columns instead of rows (called "transposed files") are usually more efficient [Teitel 77, Turner et al 79]. Most existing data management systems are designed for high volume interactive transactions with the possibility of concurrent access to the data. The large overhead required for the support of concurrent access is not necessary for SDBs. Analysts work with their particular subset of the data, and are willing to put up with occasional sequential access to the original data bases. A short discussion of additional reasons can be found in [Cohen & Hay 81].

The second reason stems from the lack of functionality and ease of use. Statistical functions available in commercial data management systems are quite limited, usually to simple aggregate univariate functions such as sum, maximum, or average. Most systems do not have facilities for supporting additional user-defined functions, although some provide an ability to create predefined functions in libraries. In addition, some query languages are quite complex when it comes to specifying aggregate functions. This is in part because of insufficient modeling of the SDBs, as will be discussed later.

Ease of use considerations are much more pragmatic. In order to perform statistical analysis, an analyst must eventually rely on more sophisticated statistical tools such as those found in statistical packages. This means that in order to use a data management system the analyst will need to become familiar with two systems, and the methods used to pass data between them. Often, the analyst will choose to stay with the essential statistical tools provided by the statistical package, and manage with the limited data management tools provided by them. However, for many applications more sophisticated data structures, such as networks, matrices, or vectors, and the operations to manipulate them are required. In such cases, the choice is between limited capabilities of a single system, or having to learn to use and interface the two systems. In a later section the problems of interfacing data management systems and statistical packages are discussed.

The purpose of this paper is to describe the special characteristics and problems of SDBs. When appropriate, solutions that have been proposed in the literature are pointed out. Many of the references appeared in the Proceedings of the First LBL Workshop on Statistical Database Management. In section 2 a few representative examples are described. Section 3 contains a discussion of the special characteristics of SDBs. The remaining sections describe problem areas requiring special attention.

2. Some example data bases

In order to provide an overview of the types of data that comprise statistical data bases, a few representative examples are discussed here. SDBs exist in virtually any enterprise that collects data. Often, the data is collected for the purpose of statistical analysis, such as census data, environmental data, and scientific experiments. But SDBs are also generated as a result of continuing operations, such as the collection of mortality data in hospitals, or logs of financial transactions in banks. SDBs are prevalent in scientific and socio-economic applications. However, they also exist in any business enterprise that needs to perform data analysis of its performance, market projections, or inventory trends.

Example 1: A scientific experiment

The purpose of this example is to show that the quantity of data contained in SDBs can increase rapidly.

An experiment in atmospheric physics was carried out by the Atmospheric Science Department at Battelle Pacific Northwest Laboratories, Richland, Washington. The purpose of the experiment is to investigate trends in the build-up of different elements in the atmosphere. such as carbon dioxide, which can lead to the depletion of the ozone layer. As part of the experiment, measurements are taken of the auroras at night to determine the amount of night light due to photochemical effects. An automatic scanning photometer was used, which can take measurements at different azimuths and elevations.

Measurements are taken at six different elevations every six degrees (i.e. sixty times for a full circle). The experiment is repeated on the average four times an hour. There are five filters, so five measurements are made at each position. A quick calculation shows that the number of measurements taken in an eight hour night exceeds 50,000 (60 times/circle x 6 elevations x 4 times/hour x 8 hours/night x 5 filters = 57,600). The experiment takes place at ten stations, and was designed to run for five years. This brings the total number of measurements for the entire experiment to about one billion.

In addition to the measured data, the parameter data must be stored in some form. For each measurement there are parameter data for the azimuth, elevation, year, month, day, time, filter type, and station number. There may be more information on who ran the experiment, malfunction of devices, etc. The parameter data can increase the size of the data base by an order of magnitude, unless special techniques are applied for efficient storage. Several such known techniques are discussed in a later section.

Example 2: County business patterns

The purpose of this example is to show that some "flat files" require a fairly complex description of what the rows and columns represent.

The table in Figure 1 is taken from [Gey 81], and is an example drawn from the U.S. Bureau of the Census, County Business Patterns, 1972. The rows represent industrial categories within states, and the columns represent several variables about number of employees, payroll, and number of reporting units (i.e. businesses). Several interesting observations can be made.

First, the industrial categories are organized into hierarchies. That is, categories (such as "mining") are broken into sub-categories, and therefore categories values represent totals of the sub-categories values. It is a fairly common phenomena that parameters are organized into hierarchies when the number of categories is large. Sometimes the hierarchies can be five or six levels deep, such as the classification of energy sources into the different types of oil, gas, and coal.

Second, the rows do not represent only a single parameter (industrial categories) but another parameter (state). Because all the information has to be squeezed into two dimensions, it is not surprising that several parameters have to be combined in a single dimension. It is not uncommon to have several parameters represented in a single dimension (such as industry by state by year), in which case each row (or column) represents some combination from each of the parameters.

Third, the columns represent several measured attributes, but one of the attributes, the number of reporting units, is broken by categories of employment size class. This complexity can grow into additional levels. For example, if the data covered several years, columns could be organized so that those belonging to the same year were grouped together.

SIC code	Industry	Number of employees, mid-March pay period	Taxable payrolls, Jan.-Mar. ($1,000)	Total reporting units	Number of reporting units, by employment-size class							
					1 to 3	4 to 7	8 to 19	20 to 49	50 to 99	100 to 249	250 to 499	500 or more
	CALIFORNIA											
	TOTAL	5 589 019	11 303 554	349 096	167 906	77 138	59 558	27 720	9 246	5 043	1 478	1 007
...	AGRICULTURAL SERVICES, FORESTRY, FISHERIES.	29 824	45 503	4 261	2 268	1 050	688	187	49	16	3	-
07	AGRICULTURE SERVICES AND HUNTING. . . .	27 436	39 402	3 901	2 071	1 005	589	170	47	16	3	-
071	MISC. AGRICULTURAL SERVICES	8 340	11 144	581	245	130	114	55	24	12	1	-
072	ANIMAL HUSBANDRY SERVICES	8 257	11 184	1 597	803	488	258	44	4	-	-	-
073	HORTICULTURAL SERVICES.	10 737	16 977	1 692	1 000	382	214	71	19	4	2	-
08	FORESTRY.	524	603	56	24	11	10	11	-	-	-	-
09	FISHERIES	1 864	5 498	304	173	34	89	6	2	-	-	-
...	MINING.	29 087	86 090	988	333	176	196	153	61	46	18	5
10	METAL MINING.	1 228	3 659	48	25	9	5	4	2	1	2	-
101	IRON ORES	(D)	(D)	2	-	-	-	1	-	-	1	-
103	LEAD AND ZINC ORES.	(D)	(D)	4	-	1	1	1	-	1	-	-
104	GOLD AND SILVER ORES.	28	22	14	12	2	-	-	-	-	-	-
106	FERROALLOY ORES EXCEPT VANADIUM . . .	(D)	(D)	6	3	1	1	-	-	-	1	-
1064	TUNGSTEN ORES	(D)	(D)	5	2	1	1	-	-	-	1	-
109	MISCELLANEOUS METAL ORES.	261	623	15	6	3	3	1	2	-	-	-
1092	MERCURY ORES.	176	400	11	3	3	3	1	1	-	-	-
13	OIL AND GAS EXTRACTION.	16 386	48 250	526	229	117	114	86	41	28	9	2
131	CRUDE PETROLEUM AND NATURAL GAS . . .	9 079	27 477	284	117	50	49	23	22	14	7	2
132	NATURAL GAS LIQUIDS	699	2 336	17	5	1	4	4	1	1	1	-
138	OIL AND GAS FIELD SERVICES.	6 607	18 434	324	106	66	61	59	18	13	1	-
1381	DRILLING OIL AND GAS WELLS.	2 048	6 939	63	23	7	10	11	4	7	1	-
1382	OIL AND GAS EXPLORATION SERVICES. .	522	1 519	40	25	7	4	1	2	1	-	-
1389	OIL AND GAS FIELD SERVICES, N.E.C..	4 037	9 975	221	58	52	47	47	12	5	-	-
14	NONMETALLIC MINERALS, EXCEPT FUELS. . .	8 404	24 240	263	65	45	68	54	15	9	5	2
141	DIMENSION STONE	27	34	10	8	1	1	-	-	-	-	-
142	CRUSHED AND BROKEN STONE.	1 255	3 721	69	13	12	24	15	5	-	-	-
1422	CRUSHED AND BROKEN LIMESTONE. . . .	291	747	18	4	2	9	1	2	-	-	-
1423	CRUSHED AND BROKEN GRANITE.	244	940	8	-	1	2	3	2	-	-	-
1429	CRUSHED AND BROKEN STONE, N.E.C . .	387	1 097	27	5	6	9	7	-	-	-	-
144	SAND AND GRAVEL	3 183	9 948	116	28	24	22	30	4	6	2	-
1442	CONSTRUCTION SAND AND GRAVEL. . . .	2 709	8 355	100	22	22	20	27	3	4	2	-
145	CLAY AND RELATED MINERALS	282	624	16	4	1	7	1	3	-	-	-
1459	CLAY AND RELATED MINERALS, N.E.C. .	183	403	11	3	1	4	1	2	-	-	-
147	CHEMICAL AND FERTILIZER MINERALS. . .	2 235	6 224	10	1	-	3	1	1	-	2	2
1474	POTASH, SODA, AND BORATE MINERALS .	(D)	(D)	8	1	-	1	1	1	-	2	2
149	MISCELLANEOUS NONMETALLIC MINERALS. .	1 399	3 645	36	6	7	10	7	2	3	1	-
1492	GYPSUM.	(D)	(D)	5	1	2	-	1	-	1	-	-
1496	TALC, SOAPSTONE, AND PYROPHYLLITE .	109	337	7	2	1	2	2	-	-	-	-
1498	PEAT.	(D)	(D)	3	-	1	1	-	-	1	-	-
1499	NONMETALLIC MINERALS, N.E.C.. . . .	716	1 964	13	1	2	5	2	1	1	1	-
--	ADMINISTRATIVE AND AUXILIARY.	3 062	9 932	48	11	5	9	9	3	8	2	1
...	CONTRACT CONSTRUCTION	315 454	867 330	26 196	12 364	5 624	4 891	2 269	665	298	63	22
15	GENERAL BUILDING CONTRACTORS.	83 839	217 293	7 310	3 811	1 490	1 188	516	191	89	19	6
16	HEAVY CONSTRUCTION CONTRACTORS.	62 280	193 209	2 639	950	592	562	317	117	69	21	11
161	HIGHWAY AND STREET CONSTRUCTION . . .	22 004	62 866	1 159	398	282	252	141	45	29	10	2
162	HEAVY CONSTRUCTION, N.E.C..	40 197	130 190	1 468	547	309	304	176	72	40	11	9
17	SPECIAL TRADE CONTRACTORS	166 859	447 477	16 200	7 598	3 535	3 128	1 425	354	134	22	4
171	PLUMBING, HEATING, AIR CONDITIONING .	35 456	108 807	3 389	1 485	790	712	298	68	30	5	1
172	PAINTING, PAPER HANGING, DECORATING .	13 598	31 734	2 063	1 307	357	258	110	20	10	1	-
173	ELECTRICAL WORK	27 389	85 976	2 350	1 071	509	460	232	50	23	2	3
174	MASONRY, STONEWORK, AND PLASTERING. .	19 417	49 023	2 170	1 098	472	378	171	34	14	3	-
1741	MASONRY AND OTHER STONEWORK	6 771	15 969	937	516	199	142	66	12	2	-	-
1742	PLASTERING AND LATHING.	9 869	26 447	841	354	194	174	87	18	12	2	-
1743	TERRAZZO, TILE, MARBLE, MOSAIC WORK	2 768	6 604	390	227	78	62	18	4	-	1	-
175	CARPENTERING AND FLOORING	11 659	26 006	1 273	643	285	204	100	31	10	-	-
1751	CARPENTERING.	6 888	14 800	555	248	124	97	54	23	9	-	-
1752	FLOOR LAYING & FLOORWORK, N.E.C.. .	4 768	11 178	717	394	161	107	46	8	1	-	-
176	ROOFING AND SHEET METAL WORK.	10 498	24 627	1 146	450	315	265	93	17	6	-	-
177	CONCRETE WORK	13 656	30 031	1 122	454	269	237	110	41	8	3	-
178	WATER WELL DRILLING	859	1 517	152	91	26	27	7	1	-	-	-
179	MISC. SPECIAL TRADE CONTRACTORS . . .	33 761	88 680	2 443	952	492	568	298	92	33	8	-
1791	STRUCTURAL STEEL ERECTION	5 994	16 447	230	36	36	73	51	23	9	2	-
1792	ORNAMENTAL METAL WORK	771	1 774	84	34	20	23	4	3	-	-	-
1793	GLASS AND GLAZING WORK.	1 867	5 215	276	136	68	57	13	1	1	-	-
1794	EXCAVATING AND FOUNDATION WORK. . .	2 101	5 378	285	149	55	55	22	4	-	-	-
1795	WRECKING AND DEMOLITION WORK. . . .	825	1 696	63	24	14	14	9	1	1	-	-
1796	INSTALLING BLDG. EQUIP., N.E.C. . .	1 817	7 294	43	7	10	13	5	3	2	3	-
1799	SPECIAL TRADE CONTRACTORS, N.E.C. .	20 382	50 873	1 461	566	288	333	194	57	20	3	-
--	ADMINISTRATIVE AND AUXILIARY.	2 476	9 351	47	5	7	13	11	3	6	1	1

FIGURE 1: County business patterns

Fourth, checking carefully, one can see that totals do not always equal the sum of the components. This is because some of the data in the lower levels were suppressed for privacy disclosure reasons. This shows, that in such cases, the higher level totals have to be stored in the data base, because they cannot be computed from the sub-totals.

A similar example in which four parameters are represented on a page, is shown in Figure 2. The example is taken from a publication of the U.S. Department of Labor, entitled Employment and Earnings, States and Areas, 1939-71. The parameters are state, SMSA, industry, and year. A SMSA (Standard Metropolitan Statistical Area) typically represents a collection of several counties. The columns represent information about the number of employees and their average earnings. Notice how the four parameters were displayed with headings and sub-headings in the left-hand column. Note also that there are quite a few null values, and that they tend to be concentrated in columns. This point is referred to in the section on physical organization.

ILLINOIS

Year	All employees (thousands)	Production or nonsupervisory workers: Average weekly earnings (dollars)	Production or nonsupervisory workers: Average weekly hours	Production or nonsupervisory workers: Average hourly earnings (dollars)
CHICAGO-NORTHWESTERN INDIANA (Con.)				
WHOLESALE AND RETAIL TRADE				
1964	591.5	–	–	–
1965	619.4	–	–	–
1966	650.3	–	–	–
1967	670.9	–	–	–
1968	690.4	–	–	–
1969	704.6	–	–	–
1970	704.6	–	–	–
1971	705.1	–	–	–
WHOLESALE TRADE — SIC 50				
1964	190.0	–	–	–
1965	199.0	–	–	–
1966	209.7	–	–	–
1967	216.4	–	–	–
1968	221.8	–	–	–
1969	226.5	–	–	–
1970	228.4	–	–	–
1971	226.5	–	–	–
RETAIL TRADE — SIC 52-59				
1964	401.5	–	–	–
1965	420.5	–	–	–
1966	440.6	–	–	–
1967	454.5	–	–	–
1968	468.6	–	–	–
1969	478.1	–	–	–
1970	476.2	–	–	–
1971	478.6	–	–	–
FOOD STORES — SIC 54				
1969	77.9	–	–	–
1970	77.6	–	–	–
1971	76.4	–	–	–
FINANCE, INSURANCE, AND REAL ESTATE				
1964	161.4	–	–	–
1965	163.7	–	–	–
1966	166.5	–	–	–
1967	173.0	–	–	–
1968	178.9	–	–	–
1969	183.7	–	–	–
1970	188.7	–	–	–
1971	192.3	–	–	–
SERVICES				
1964	421.8	–	–	–
1965	441.5	–	–	–
1966	466.0	–	–	–
1967	489.6	–	–	–
1968	502.7	–	–	–
1969	518.8	–	–	–
1970	524.2	–	–	–
1971	528.2	–	–	–
MEDICAL AND OTHER HEALTH SERVICES — SIC 80				
1969	116.0	–	–	–
1970	123.3	–	–	–
1971	130.2	–	–	–
GOVERNMENT — SIC 91-93				
1964	289.7	–	–	–
1965	308.0	–	–	–
1966	327.1	–	–	–
1967	348.0	–	–	–
1968	362.4	–	–	–
1969	372.6	–	–	–
1970	384.4	–	–	–
1971	394.6	–	–	–
DAVENPORT-ROCK ISLAND-MOLINE				
TOTAL NONAGRICULTURAL EMPLOYMENT				
1960	104.9	–	–	–
1961	104.1	–	–	–
1962	107.0	–	–	–
1963	111.4	–	–	–
1964	117.4	–	–	–
1965	122.6	–	–	–
1966	129.5	–	–	–
1967	133.4	–	–	–
1968	133.1	–	–	–
1969	134.2	–	–	–
1970	133.0	–	–	–
1971	129.8	–	–	–
CONTRACT CONSTRUCTION AND MINING				
1960	5.6	–	–	–
1961	5.1	–	–	–
1962	5.2	–	–	–
1963	5.6	–	–	–
1964	5.5	–	–	–
1965	6.1	–	–	–
1966	5.7	–	–	–
1967	6.8	–	–	–
1968	7.0	–	–	–
1969	6.7	–	–	–
1970	6.5	–	–	–
1971	6.1	–	–	–
MANUFACTURING				
1960	38.6	–	–	–
1961	37.8	–	–	–
1962	38.6	113.15	40.3	2.81
1963	41.1	119.29	40.6	2.94
1964	44.8	126.35	41.3	3.06
1965	46.3	128.42	41.0	3.13
1966	49.5	140.08	41.8	3.35
1967	50.0	141.33	40.6	3.48
1968	46.1	145.79	39.8	3.66
1969	44.8	153.97	39.9	3.86
1970	42.7	162.57	39.7	4.10
1971	40.0	175.77	39.4	4.46
DURABLE GOODS				
1960	29.9	–	–	–
1961	28.8	–	–	–
1962	29.8	118.30	40.6	2.91
1963	32.4	124.74	40.9	3.05
1964	35.8	132.63	41.9	3.16
1965	37.3	134.60	41.5	3.24
1966	40.4	146.48	42.3	3.46
1967	40.3	145.86	40.8	3.57
1968	36.7	150.07	40.0	3.76
1969	35.8	158.36	39.9	3.97
1970	34.0	165.42	40.0	4.13
1971	31.2	179.12	39.6	4.52
LUMBER AND FURNITURE — SIC 24,25				
1960	1.0	–	–	–
1961	.9	–	–	–
1962	1.1	87.78	40.5	2.17
1963	1.1	89.96	40.4	2.23
1964	1.2	91.24	40.5	2.25
1965	1.2	89.78	38.7	2.32
1966	1.0	93.86	39.2	2.39
1967	.8	100.22	39.5	2.54
1968	.8	98.76	38.1	2.59
1969	1.0	104.94	38.0	2.76
1970	1.0	114.56	38.4	2.99
1971	1.0	120.43	38.3	3.14
PRIMARY METAL INDUSTRIES — SIC 33				
1960	5.2	–	–	–
1961	5.2	–	–	–
1962	5.4	129.16	41.6	3.10
1963	5.8	137.02	43.2	3.17
1964	6.2	143.39	44.2	3.25
1965	6.5	154.13	45.2	3.41
1966	6.9	161.26	45.2	3.57
1967	6.7	160.62	43.4	3.70
1968	5.9	167.09	43.6	3.84
1969	6.5	179.40	44.1	4.07
1970	6.4	178.77	42.7	4.19
1971	5.7	198.08	42.8	4.63
FABRICATED METAL PRODUCTS — SIC 34				
1960	1.6	–	–	–
1961	1.5	–	–	–
1962	1.6	96.65	39.2	2.47
1963	1.5	96.59	39.1	2.47
1964	1.6	98.01	39.7	2.47
1965	1.8	106.05	40.7	2.61
1966	2.1	111.46	41.1	2.71
1967	1.9	106.53	38.8	2.74
1968	1.8	110.41	38.2	2.89
1969	1.7	121.74	39.9	3.05
1970	1.7	131.80	39.9	3.30
1971	1.5	139.74	40.0	3.49
MACHINERY, EXCEPT ELECTRICAL — SIC 35				
1960	18.6	–	–	–
1961	18.1	–	–	–
1962	18.6	119.62	40.0	2.99
1963	20.9	126.86	40.3	3.15
1964	23.7	135.75	41.4	3.28
1965	24.4	135.10	40.5	3.33
1966	26.5	149.69	41.7	3.59
1967	26.8	148.51	40.1	3.70
1968	24.2	152.83	38.9	3.93
1969	22.6	159.56	38.3	4.17
1970	21.3	169.36	39.3	4.31
1971	19.6	183.25	38.7	4.73
NONDURABLE GOODS				
1960	8.7	–	–	–
1961	9.0	–	–	–
1962	8.8	96.08	39.3	2.45
1963	8.7	99.13	39.4	2.52
1964	9.0	102.13	39.0	2.62
1965	9.0	103.27	39.0	2.65
1966	9.1	111.86	39.7	2.82
1967	9.7	120.31	39.6	3.03
1968	9.3	130.71	39.3	3.32
1969	8.9	139.34	40.1	3.47
1970	8.7	152.21	38.4	3.96
1971	8.8	164.82	38.6	4.27
FOOD AND KINDRED PRODUCTS — SIC 20				
1960	4.4	–	–	–
1961	4.5	–	–	–
1962	4.2	117.35	40.3	2.91
1963	4.2	122.39	40.2	3.05
1964	4.3	125.36	39.5	3.17
1965	4.3	128.12	39.7	3.23
1966	4.3	139.88	41.2	3.39
1967	4.8	149.50	40.3	3.71
1968	4.8	157.20	39.7	3.96
1969	4.8	172.43	41.0	4.20
1970	4.7	186.64	38.6	4.84
1971	4.8	204.67	38.9	5.27
PRINTING AND PUBLISHING — SIC 27				
1960	1.5	–	–	–
1961	1.6	–	–	–
1962	1.6	98.91	40.5	2.44
1963	1.6	102.17	41.2	2.48
1964	1.5	102.48	40.6	2.52
1965	1.6	104.36	40.1	2.61
1966	1.7	108.40	39.8	2.72
1967	1.7	114.27	39.9	2.86
1968	1.7	118.33	40.2	2.94
1969	1.7	131.43	39.6	3.32
1970	1.7	137.38	39.0	3.52
1971	1.6	152.72	39.5	3.86
TRANSPORTATION AND PUBLIC UTILITIES				
1960	6.9	–	–	–
1961	6.7	–	–	–
1962	6.6	–	–	–
1963	6.5	–	–	–
1964	6.5	–	–	–
1965	6.7	–	–	–
1966	7.1	–	–	–
1967	7.2	–	–	–
1968	7.0	–	–	–
1969	7.2	–	–	–
1970	7.3	–	–	–
1971	7.1	–	–	–

FIGURE 2: Cross-product of four category attributes.

The reasons for treating such data as "flat files" are three fold. First, it is a convenient way of viewing the data on a printed page or another two-dimensional medium, without involving graphics. It is also a natural way of generating a report. Second, statisticians traditionally like to view data in matrix form. Indeed, most statistical packages support only "flat files". Third, it is an economical way of representing the parameter data, even though some of it may still repeat (for example, in Figure 1, the different industry categories will repeat for each state). This preference naturally carried over to the storage of the data in this form, but necessitates some way of describing the meanings of rows and columns. Unfortunately, such descriptions can usually be found only outside of the data base.

Example 3: Geographically-based data

This example illustrates the complexity that can exist within a hierarchy of parameters. This is especially significant in geographically-based data bases, since the partitioning of geography can change with the application. In addition, the classification of the parameters can change over time.

Several census data bases use the following geographical hierarchy for gathering information on population, income, etc:

States
   SMSA (Standard Metropolitan Statistical Area)
      County (or county equivalent)
         Place
            Census Tract

The realities are such that this is not a strict hierarchy, i.e. descendent elements are not always wholly contained within a single parent element. For example, SMSAs are not always contained within states, they can cross state boundaries. Counties are always contained in SMSAs, except in New England. Places, which are usually cities, can be split between counties. Furthermore, places do not cover all areas in the U.S., since very sparsely populated regions are not assigned place names. Finally, in 90% of the cases, census tracts are contained within places; the remainder cross place boundaries. On the other hand all census tracts are contained within county boundaries which are the next level up in the hierarchy.

There is no regularity to this hierarchical organization. Much of it came about in an evolutionary fashion over the years. Nevertheless, most statistical analysts have to deal with this situation, usually with no tools other than programming languages and statistical packages that require flat files. The situation is further complicated since the definitions of these regions change over time due to legislative action or political needs. For example, existing county boundaries may be redefined and new ones introduced. This is not unique to geography. Another example in which changes evolve over time and are difficult to follow is in the classification of diseases. Since there are a very large number of diseases, a book of about 650 pages is needed to describe them. Every ten years an international meeting is held to review the classifications, and make changes and additions if necessary. The description of the differences between the two sets of classifications requires a 100 page document [DHEW 75].

It is therefore an extremely tedious task to correlate data over time, or between data sets collected by different agencies. In order to analyze the relationships between such data sets, projects have to include specialists in the details of such data, who may also be required to write special software to correlate them. An example of such a project is the PAREP (Populations at Risk to Environmental Pollution) project [Merrill et al 79], whose goal is to find correlations between cancer mortality and air pollution. In addition to several statisticians the project has a specialist who is an expert on such data.

Example 4: World trade

This example serves to illustrate the fact that many statistical data bases can be quite sparse, i.e. contain a large proportion of null values. Although this is an extreme example of sparsity, it is not uncommon to find data bases that are 40-50% sparse.

Over the years the World Bank has collected a large amount of economic data, mainly as time series [Johansson & Shilling 81]. One such data base contains data on trade activities between countries. There are hundreds of possible trade categories which are organized into hierarchies. The number of countries involved is also large, since the bank is particularly interested in developing countries. This data was collected over a period of several years. Thus, the number of all possible combinations for the different trade categories, by exporting countries, by importing countries, by year, is quite large.

The exact number of possible combinations is not the important issue here. Rather, it should be noted that most of the potential combinations have null values for the corresponding trade data. This is because most countries produce only a small number of trade materials, and trade with a small number of countries. If the data base is stored as a "flat file" (or a matrix form), it will contain 80-90% null values. Even though it is advantageous to store the data in a matrix form to simplify the software for managing them, the storage overhead is unacceptable. In practice, this data base is stored compactly, with null values removed, but requires more complex and specialized software for its access and manipulation.

Example 5: Energy data

This is not an example of a single data base, but rather of a large collection of data bases that are intended to be used together. This example was chosen as a way to show the complexity level that statistical data bases can reach when the data is collected from a diversity of sources.

The mission of the Energy Information Administration (EIA) of the Department of Energy is to collect, validate and analyze the U.S. energy data, and distribute statistical and analytical reports. The data is collected from electric utilities, oil producers, petroleum product retailers, oil refineries, natural gas pipeline companies, and a variety of businesses in the U.S. The data is collected by using fairly complex forms. The information collected with each form (or sometimes a set of forms) is assembled into a data base, called an energy information system. There are 147 such energy systems that the EIA collects and analyzes, although if one considers other government agencies who deal with energy data, the total is estimated at about 230.

The data is organized into four major areas of energy sources: coal, petroleum and petroleum products, natural gas, and electricity. Each area consists of about 20-30 energy systems (or data bases), and produces about 10-15 publications (reports) a year, such as "The Statistics of Publicly Owned Electric Utilities" or "World Crude Oil Production". In addition, there are about 40 energy systems on a variety of other topics, such as solar, automobile classifications, and conservation.

It can be easily seen that the amount of different data elements is very large. Indeed, the number of data elements for natural gas and electric power utilities alone is about 3,600. In total, the number of data elements is around 10,000. The number of elements makes it difficult to remember what they are, let alone to remember the precise acronyms used. What makes it even more difficult is that several data elements have similar meanings, and that the difference may be in the methods used to measure them or the units used. There must be some place where such detail information is kept and managed, or the meaning of the data is eventually lost.

There is a great deal of complexity in validating such data. Validation can be as simple as range checking, or as complex as checking whether certain totals on consumption match (or are within bounds of) the corresponding figures on sales. But even simple range checking is not always straight forward, since the ranges change over time and have yearly trends. Validation across data elements from different data bases is often complicated by the need to perform unit conversion and to check that the parameters for the measured values coincide.

Another need is to guarantee that proprietary information will not be compromised. Much of the information reported by companies and businesses, such as prices, quantities sold, etc., could give a competitive edge to a competitor. Usually, information about an individual company is protected by releasing only global figures (i.e. summarized over a set of individual companies). However, as will be discussed later, a data base can be easily compromised if summarization techniques alone are used.

3. Characteristics

A host of problems can be envisioned from the previous examples, some of them quite formidable. The next sections attempt to describe these problems within categories, such a "modeling", "security", etc. But first some characteristics that are common to SDBs are identified in this section. These characteristics are both in terms of the structure and use of the data.

3.1. Category and summary attributes

It should be clear by now that most SDBs can be thought of as having two types of data: measured data on which statistical analysis is performed, and parameter data which describe the measured data. Why the distinction? In traditional data management, data is organized into record types, relations, or entities whose columns represent attributes. Both parameter data and measured data are described in terms of the attributes, and no distinction is made.

To illustrate the reasons for this distinction, Figure 3 shows a simple data base represented in a table (relation) form. The first five attributes (oil type, state, county, year, month) represent the parameter data, and the last two (consumption, production) represent measured data. The attributes for the parameter data have been referred to in the literature as "category" attributes, since they contain categories for the measured data. The attributes for the measured data are referred to as "summary" attributes, since they contain data on which statistical summaries (and analysis) are applied. There are several points to note in Figure 3.

First, note that a combination of the category attribute values is necessary for each of the values of each summary attribute. That is, the category attributes serve as a composite key for the summary attributes. This relationship between category and summary attributes is key to some modeling techniques, as discussed in a later section on logical modeling.

OIL TYPE	STATE	COUNTY	YEAR	MONTH	CONSUMPTION	PRODUCTION
Crude	Alabama	County 1	1977	Jan	500	800
Crude	Alabama	County 1	1977	Feb	700	300
"	"	"	"	.	1700	700
"	"	"	"	"	.	.
"	"	"	"	"	.	.
"	"	"	"	Dec	.	.
"	"	"	1978	Jan	.	.
"	"	"	"	"	.	.
"	"	"	"	"	.	.
"	"	"	"	Dec	.	.
"	"	"	1979	.	.	.
"	"	"	.	.	.	.
"	"	County 2	.	.	.	.
"	"	"	.	.	.	.
"	"	"	.	.	.	.
"	Alaska	.	.	.	.	.
"	"	.	.	.	.	.
"	"	.	.	.	.	.
"	"	.	.	.	.	.
Heating	.	.	.	.	.	.
.	.	.	.	.	.	.
.	.	.	.	.	.	.

FIGURE 3: An example of category and summary attributes.

Second, as can be readily seen from Figure 3, there is a great amount of redundancy in the values of the category attributes. In many data bases all possible combinations of the category attributes (i.e. the full cross product) exist. In such cases each value of a category attribute repeats as many times as the product of the cardinality of the remaining category attributes. This is the main reason for the organization of SDBs into matrix form, as was shown in Example 2. A matrix organization replaces the need to store the category values in the data base by representing them as positions of the columns and rows. Clearly, there is a need for efficient storage and access of category attributes. This issue is discussed later in the section on physical organization.

Third, the range of category attributes is usually small, from as little as two (e.g. "sex") to a few hundreds (e.g. oil type). In contrast, summary attributes often have large ranges since they usually represent numeric measures. Often, category attribute ranges are grouped together so as to have fewer categories, such as using "age groups" rather than "age". Also, category values are more descriptive in nature, and therefore tend to be character data (e.g. industrial classes), while summary values tend to be numeric. Often, coded versions of the text are assigned to long category values.

Of course, there are exceptions to the above observations. Indeed, it is not always obvious whether an attribute should be considered a category or a summary attribute. For example, in a population data base which contains the attributes: race, sex, age, income, and profession, it is impossible to tell *a priori* which are category attributes and and which are summary attributes. Statistics on age can be requested while the others are considered category attributes, or age can be treated as one of the categories for income statistics. However, in most cases the distinction can be made if the original purpose for collecting the data and its intended use is considered. Actually, as can be seen from the examples above, in most data bases there is no doubt as to which are category and which are summary attributes.

3.2. Sparse data

As can be seen from Example 4 on world trade statistical data bases can be quite sparse. This a direct consequence of the cross product of all possible values of category attributes. There are two options of dealing with sparse data. The first is to leave the null values (or zeros, or any other designated constants) in the data base and then squeeze them out using compression techniques. The second option is to remove entries that have only null values from the data base. For example, in Example 2 on county business patterns most states have products in only a few of all possible industrial categories. Thus under a given state only the industrial categories which have non-null values have an entry. By doing so the row position can no longer be used to indicate a category, and therefore an additional column is required to describe the category value for each entry.

The second option is primarily used in publicly available data. One of the reasons for this is that data is physically represented as it would appear in a report or a two dimensional display. Some data bases are actually published in reference books, and the data that are distributed on a tape have the same format as that presented in the book. In the next section on physical organization, storage techniques for use with sparse data that are independent of a two dimensional layout are described.

3.3. Summary sets

When statistical data bases are very large, it becomes too expensive to work directly with the original data set. Users extract smaller data sets that are of interest to them, apply the usual selection functions to limit the number of entries in the data set (such as only the western states), apply projection functions to limit the summary data they are interested in, and join data from different data sets (although tools for joining are not always available). But in addition, a very common operation is to reduce the number of category attributes by summarizing over them. In the example shown in Figure 3, a user can request total consumption by oil type, by state, by year, so that the consumption values are totaled over the appropriate counties and months.

It is this last summarizing operation that usually involves some aggregate function, that inspired the term "summary sets". In an active data base, a large number of summary sets may be generated, causing management problems which will be discussed later.

3.4. Stability

Fortunately, a large proportion of statistical data bases are very stable. Initial corrections may be required but very little updating is necessary afterwards. This stems from the primary purpose of SDBs, which is to collect data for future reference and analysis. Once the data is collected, there usually is no reason to change it unless it is for the correction of identified errors. Even in data bases that are usually associated with a high degree of updating, such as inventories, the transactions are actually recorded over time, if further analysis is desired. Actually, most businesses, such as banks, retail stores, etc., record all transactions as verification that the transaction has taken place, along with the time and person performing the transaction.

The stability of SDBs is a benefit since many of the problems that arise in multiple updates to data bases which require concurrency control algorithms can be avoided. It also simplifies the management of summary sets in that it is not necessary to keep track of the dates of their creation and their dynamic updating. However, the possibility of updating should not be discounted altogether, but rather label the stable data bases as such, so as to take advantage of this property.

There is another benefit to the stability of data bases which takes advantage of the trade-off between retrieval and update operations. If one assumes very little or no updating, it is possible to design more efficient retrieval algorithms on account of slow updating.

3.5. Proliferation of terms

This phenomenon is not unique to statistical data bases, but exists whenever a data base contains a large number of attributes. Consider the energy data base described in Example 5. How is one to remember the content of the data base, let alone the names and acronyms of the attributes? When a data base has hundreds (or even a few tens) of attributes, it is necessary that some tools be provided for dealing with such complexity.

In order to formulate a query, a user must remember the following things in addition to the details of the query language: the names of data sets (or relations) needed, the names or acronyms of the attributes needed, the possible and legal values for these attributes, and the formats of the values (e.g. the format for age groups, or whether to use capitals in names of cities). In addition, the codes or abbreviations that were assigned to values (e.g. codes for states and counties) must be remembered. It is not surprising that such data bases require specialists to access them.

These difficulties are even more serious in SDBs, for two reasons. First, many data bases have categories that change their definitions over time. An example of this was mentioned previously where counties change their boundaries but not their names. Also, the same terms are used with slightly different meanings. For example, the term "state" may include Guam and Puerto Rico in one data base, but not in another. The second reason stems from the summary sets. With every new summary set that is created, new names are introduced, or perhaps old names with new meanings. It is necessary to control this proliferation of terms, and to keep track of what exists in the system.

The next sections organize the discussion of problems into research areas. Whenever appropriate, some solutions that have appeared in the literature are mentioned. This is not intended to be a comprehensive list of solutions, but rather to pick some representative solutions that we are familiar with as possible approaches to the problems.

4. Physical Organization

Most of the problems discussed in this section stem from the need to compress the data in large SDBs, while permitting fast access. There are a large number of known compression techniques ranging from coding to intricate text compression. The purpose of this section is to highlight some representative techniques that are particularly applicable to SDBs.

4.1. Category attributes

Whenever several category attributes are used jointly to form a composite key, a large storage overhead results. This point was illustrated previously in Figure 3, where there is much repetition of values in the category attributes columns. Because the category attributes form a cross product, the storage requirements are multiplicative in nature. For each additional category attribute in the composite key, the amount of extra storage required is the size of the storage required for the previous category attributes times the number of distinct categories in the additional attribute. It is therefore quite important that some compression techniques be used.

One common technique to reduce this overhead, is to encode the category values, and to store only the codes with the data base. This can result in great savings, since some category values are descriptive text (for example, the industrial categories shown in Figure 1). Furthermore, the amount of storage needed for the category values depends on the number of distinct category values. Thus, only one bit is necessary to encode the two values of sex, and only four bits for the twelve values of months. Two example systems that were specifically designed to manage SDBs, use this technique: the RAPID system [Turner et al 79], and the ALDS system [Burnett & Thomas 81]. As was pointed out in [Gey 81] it is unfortunate that many systems which use encoding, do not provide software for the automatic translation between the original values and the encoded values, but rather leave the burden on the user to determine which codes to use before querying the data base.

The previous technique still requires that the encoded values be stored repeatedly. Another approach is to use the logical extension of the matrix storage form. One can store the list of distinct category values of each attribute once (perhaps in a dictionary). Then, each category attribute can be used to form one dimension of a multi-dimensional matrix. For each combination of values from the category attributes, one can compute the appropriate position in the multi-dimensional matrix. There is a well-known algorithm for such a mapping (called "array linearization"); its use for category attributes is explained in [Eggers & Shoshani 80]. It is worth noting that the mapping is a simple computation, and therefore random access is essentially achieved.

4.2. Sparse data

As was discussed in Example 4 on world trade, SDBs can be quite sparse. The greater the sparseness, the greater the chance that longer sequences of null values can be found in the data. But, in addition, experience suggests that in SDBs null values (or other designated constants) tend to cluster. To see the reason for this, refer back to figure 3. Suppose that a certain state does not consume a certain oil type. Then in the consumption column there would be zero (or null) values in consecutive positions for all the counties in that state, for all years, for all months. Of course, the order of the category attributes will change the length of the null sequences.

This brings up the following interesting problem: given a certain order of the category attributes and given the precise layout of the corresponding measured values, find an efficient algorithm for determining the best reordering of the category attributes such that the length of null sequences is maximized.

The length of sequences is very important since compression techniques can take advantage of them by essentially replacing a sequence with a count and a value. This technique (called run length encoding) can result in substantial reductions in the size of the data, depending on the sparseness of the data base. The main problem with this technique is the need to access the data sequentially once it is compressed. The ability of random access according to relative position is lost. In [Eggers & Shoshani 80] a technique was developed where logarithmic access can be achieved for data whose null sequences have been compressed. The technique, called "header compression", makes use of a header which contains the counts of both compressed and uncompressed sequences in the data stream. The counts are organized in such a way as to permit a logarithmic search over them. A B-tree is built on top of the header to achieve a high radix for the logarithmic access. In a later paper [Eggers et al 81] the technique was extended to sequences of multiple constant values.

This header compression technique is also used to compress sequences of values that vary in size requirements (i.e. one byte, two bytes, etc.). This can be useful in the case where the distribution of summary attribute values is skewed in such a way that the majority of the values are small. As an example, consider seismic activity measurements where most of the measurements consist of low level background noise.

4.3. Transposed files

The tendency for clustering of null values often occurs within a single column (representing a single summary attribute). This suggests that from a compression point of view, it is advantageous to transpose files, i.e. to store values by attribute, rather than as records or tuples. As discussed in [Teitel 77] and [Turner et al 79], there are other reasons to prefer transposed files (sometimes called "attribute partitioning" or "vertical partitioning") in SDBs. It is argued that in SDBs very few attributes are requested in a single query, and it is inefficient to access data organized as records, since it is necessary to read the data of the other attributes which are of no interest from secondary storage. Another approach is to cluster the attributes which are likely to be accessed together, but it is not a simple matter to determine the preferred clustering from a set of representative queries [Hammer & Niamir 79]. Fully transposed files (i.e. no clustering of attributes) are used in the RAPID and ALDS systems mentioned above, and in earlier systems such as IMPRESS [Meyers 69] and PICKLE [Baker 76].

4.4. Partial cross product

The problem of efficiently storing the cross product of category attributes was discussed above. However, there are situations where not every possible combination of the category attributes is valid, i.e. for the combinations that are not valid, the values for all the summary attributes are null. In such a case, the entire entry is missing from the data base. An example of this was shown in Figures 1 and 2 where for each state many industrial categories are missing altogether. This situation is referred to as the "partial cross product".

The problem is to determine whether there is a way to compress partial cross products. Clearly, the method of value encoding still works, but is there a way to further compress the combinations of category attributes which are valid? In [Svensson 79] a technique which involves the use of a tree is suggested, but some redundancy of values is still left. In [Eggers et al 81] another solution is suggested. It combines the array linearization technique used for full cross product, and the header compression technique for null sequences. Imagine a vector of "ones" and "zeros" that corresponds to valid and invalid entries in the partial cross product, respectively. The partial cross product is treated as if it was a full cross product, and array linearization is used to map into this imaginary vector. Then, the header compression mapping is used to map from the imaginary vector into the actual positions of the valid entries. The outcome of this combination is that just a header is necessary to perform the entire mapping and to achieve a logarithmic access time.

5. Optimization

5.1. Selection of physical structures

Because of their special nature, SDBs offer opportunities for storage and access savings that use uncommon physical structures and access methods. In addition to the more common techniques, such as a variety of indexing and hashing methods, the gains that can be achieved by using techniques such as attribute partitioning (either full or clustered partitioning), encoding, array linearization, and different compression techniques must be considered.

Given a statistical data base and a set of representative queries, the problem is to determine how to partition, compress, or encode the data, and which access methods to use to access the data. In general, this is a very difficult problem, even with a limited set of choices. However, there is still the challenge of solving this problem for SDBs, given a small set of storage and access methods which seem most likely to benefit the application. A representative example of such work is described in [Lehot 77], where the problem of determining the optimal order of the category attributes in order to maximize the access time of a summary attribute was considered. The paper shows that under certain assumptions about the distribution of values and query characteristics, an optimal order can be found. It also contains several references to related work.

Of course, the complementary problem to that of selecting data structures for an application, is that of processing queries efficiently. Given that certain choices were made for the physical organization of a SDB, how can queries be processed optimally. Query optimization is still an active area in conventional data base research. The question is whether the known techniques can be applied to the more uncommon structures that are useful for SDBs.

5.2. Response assembly

Attribute partitioning and compression introduce the problem of assembling the response to a query. Suppose that a response needs to be extracted from several attributes that were partitioned, and that the partitions are stored on a disk. In order to assemble tuples (records) for the response, a value from each partition could be read, and the appropriate values written a tuple at a time. But if the primary buffer space is limited, this would result in an excessive number of reads to the disk. The reason is that a single value from each page is read in order to assemble a tuple. If, in addition, the values in the attribute partition are compressed, there is an overhead to be paid for each access to the compressed data. Thus, given a limited buffer, the problem is to minimize disk reads and the overhead of accessing compressed data.

In a paper published in this proceedings, a suggestion is made that special hardware can be used for the purpose of tuple assembly and compression [Hawthorne 82]. It suggests the use of microprocessors that are organized in a two level hierarchy. The leaf microprocessors are responsible for writing and reading the compressed data and the top level microprocessor(s) responsible for tuple assembly. Thus, the data from the different partitions could be read in parallel, and passed on for tuple assembly. This work is still in the design stage.

6. Logical modeling

Can benefits be gained from modeling the semantics of statistical data bases? Is it worth adding to the complexity of the data model? There is a long standing controversy as to whether logical data models should be semantically simple (such as the relational model), or whether they should contain more semantics about the data structures (such as having generalization hierarchies or distinguishing between entities and relationships). In the case of SDBs, the question is whether to model data types such as "matrix" and "time series", and concepts such as category and summary attributes.

6.1. Representation of category and summary attributes

This section points out some of the work done in modeling of category and summary attributes, and the benefits achieved. It is worth noting that practitioners make a distinction between parameters (which correspond to category attributes) and variables (which correspond to summary attributes) because it provides a better understanding of the content of the data base and how it was established. For example, in a scientific experiment, the parameters that can be set by the experimenter are referred to as the "independent variables", and the measured data as the "dependent variables".

One of the main benefits of modeling the semantics of category and summary attributes is the capability of "automatic aggregation". It is the ability of the system to infer the subsets of values over which an aggregation (or statistical) function should be applied. For example, consider the following query when applied to the data base in Figure 3: "find heating oil consumption in Alabama during 1977". It is obvious that the result should be the total heating oil consumption over all counties in Alabama and over all months in 1977. Yet, without the explicit semantics of category and summary attributes the system would not be able to infer what is obvious to us. The benefit to the user is that it is not necessary to explicitly express which category attributes to summarize over. This can greatly simplify aggregation expressions in query languages.

An example of adding the above mentioned semantics to an existing model is described in [Johnson 81]. Using the framework of the Entity-Relationship model, an additional type of entity is allowed, called a summary set, which captures the semantics of category attributes. In addition, an attribute which is designated as a summary attribute, can have an aggregation function (e.g. sum, average) or any other desired function (defined as a program) associated with it.

6.2. Graph representation

Another possibility is to have these semantic concepts represented internally, so that they are invisible to the user. An example of a system that takes this approach is SUBJECT [Chan & Shoshani 81], in which these semantic concepts are represented as a graph. There are two kinds of nodes: a "cross product" node, and a "cluster node". The nodes can be connected by arcs to form a directed acyclic graph. To illustrate the meaning of these node types, we have represented the data base in Figure 1 as the graph shown in Figure 4. Note that nodes marked "x" are cross product nodes, and those marked "c" represent cluster nodes.

Cluster nodes represent collections of items. Thus, the node "metal mining" at the bottom of the figure, represents the collection of iron ores, lead & zinc ores, etc. The node "metal mining" itself is one of the items under the node "mining". As can be seen, cluster nodes are used to represent a hierarchy of parameters. This is a way of representing the complex category attribute "Industrial classes". Cluster nodes are also used to represent the collection of summary attributes under the node labeled "variables".

Cross product nodes are used to represent composite keys of category attributes. Such is the case with the node "state by industry". The semantics of this cross product node is such that each of its instances is made up of a pair of instances, one taken from the node "industry" and one from the node "states".

This graph structure is invisible to the user and is used to support a menu driven interface. The user does not need to know the types of nodes, but the system can make use of them to provide automatic aggregation. The graph can be either browsed by moving up and down the nodes, or can be searched directly with key-words. The sharing of nodes provides the capability to use the same clusters (e.g. state names) across data sets, and to avoid confusion of names. One of the main advantages of this representation is that the user can be shown the content of the data base by gradually revealing more detail when requested. The possibility of viewing hierarchical menus of details alleviates the need to remember names and acronyms.

6.3. Summary sets

Summary sets are simply data base views that are generated by using aggregate functions. The main problem is one of managing a large number of sets. With each summary set new summary attributes are generated. Obviously, the newly computed values of the summary attributes have to be stored if recomputing them every time they are needed is to be avoided. But, is there a way to avoid duplicating the category attribute values? Similarly, new names are likely to be used for the new summary attributes (e.g "total consumption" when we summarize over consumption), but is there a way of using the same names of category attributes in the summary set?

This is another situation where distinguishing between the type of attributes can be beneficial. If the category attributes are organized as lists of category values (say, in a dictionary), then it is possible for the category attributes of the summary sets to "point" to these lists, and to share the same names. It is easy to visualize this point in terms of the SUBJECT graphs described above. If a category attribute is used in its entirety in the summary set, then a pointer to the corresponding node is all that is necessary. If a selection of a certain category is made, e.g. "Alabama", then the pointer points directly to the node representing "Alabama". If a selection of a subset of the category values was made (e.g. several states), then a new node is created whose members are that subset.

This idea is complementary to the technique described in the section on physical organization above, where the lists of category attribute values are stored only once, and array linearization is used to map between them and the appropriate positions of the summary attributes.

6.4. Aggregation/disaggregation

In Example 3 above on geographical categories, the difficulties of correlating data from different data bases on the basis of a common but not identical category attribute were described. This problem is referred to as "aggregation/disaggregation" because, as shall be seen shortly, it involves both functions. For example, suppose that unemployment rates are to be correlated with median family income. The problem is that unemployment rates may be available by federal regions, but median family income may be available only by census regions. Both federal and census regions represent groups of states, but the groupings are different. In order to perform the correlation task, it is necessary to disaggregate one of the summary attributes (e.g. unemployment rates) to the state level using some proxy variable (e.g., population), and then aggregate back to the desired level.

A recent paper [Merrill 82] describes the technique used to perform aggregation/disaggregation in the SEEDIS system which was mentioned in the introduction. It uses tables to represent the mapping between any two sets of category attribute values of interest. It deals only with mappings for geographical categories, but the amount of detail is still large since there are so many different geographic categories used by federal and state government. In [Johnson 81a] a technique is proposed for modeling these mappings by associating them with relationships of the Entity-Relationship model.

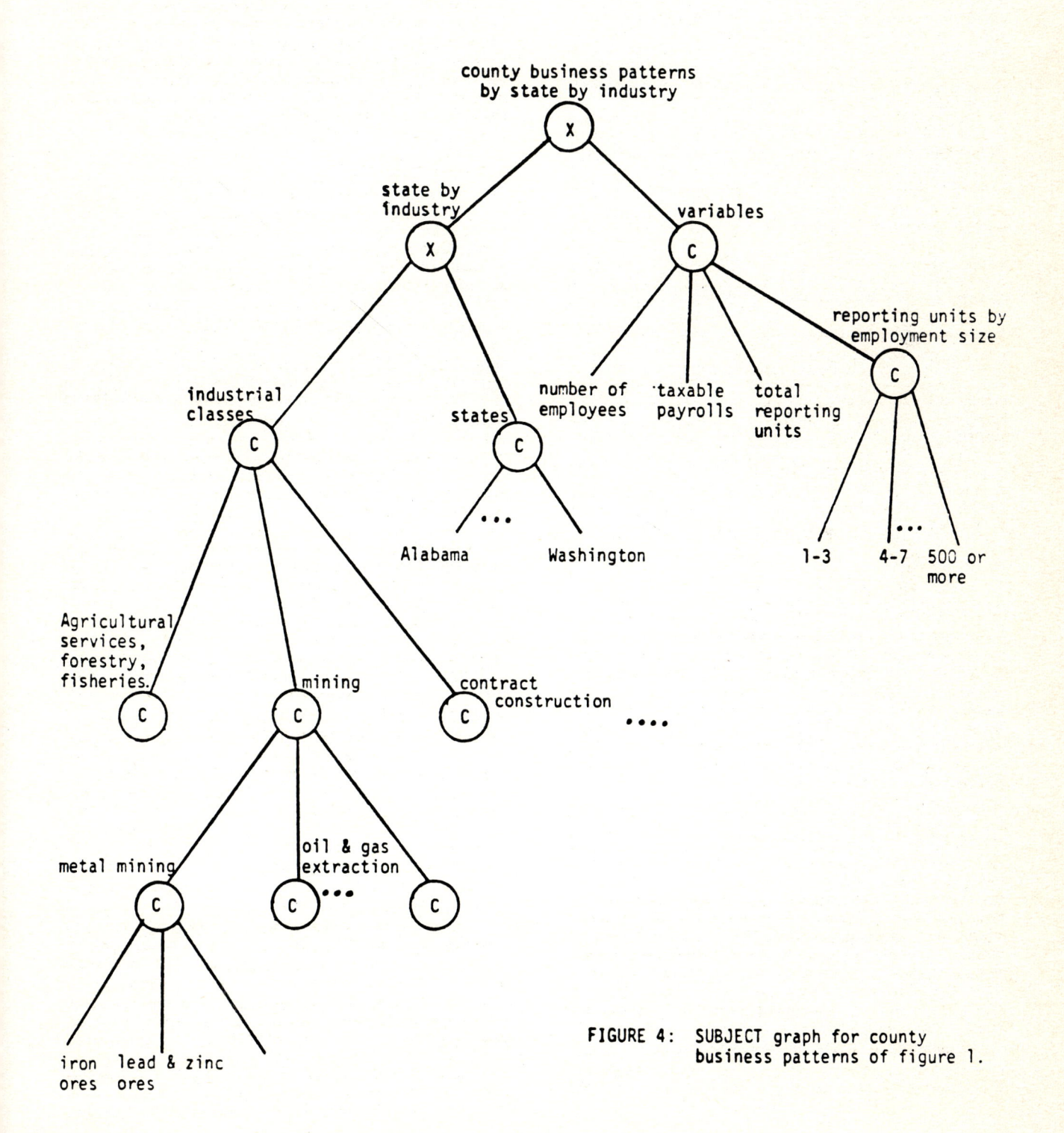

FIGURE 4: SUBJECT graph for county business patterns of figure 1.

7. User Interface

From the Examples section above, it is evident that one of the major problems for a user interfacing to large SDBs is to determine the content of the data base and the terms used for its attributes. Such problems are referred to as meta-data problems, since they deal with information about the data. Meta-data is much more complex than listing the record types (or relations) and the attribute names and types. It includes information such as missing data specification, data quality specification (to indicate how reliable the data could be considered), a history of data base creation and modifications, complex attribute structures (e.g. vectors to represent the boundaries of geographical regions), etc. In a paper published in this proceedings [McCarthy 82], a comprehensive list of requirements for meta-data is given, with special attention given to SDBs.

Whenever it is necessary to deal with the diversity and complexity of data bases such as the energy data of EIA described in Example 5, special techniques of classifying information may be needed. In fact, for the EIA data base, it was helpful to use a technique that is usually used in library systems, called "facet classification". Using this technique, summary attributes are described using facets. For example, some of the facets used for the energy data are: energy source (oil, coal, etc.), function (produced, shipped, etc.), units of measure, dates, etc. Each facet is a hierarchy of terms that can be quite deep, as is the case with energy source. A combination of the terms from the facet hierarchies is then used to describe a summary attribute (for example, heating oil refined in mid-western states during 1977). This technique can avoid conflicts in definition of similar attributes since they have to be defined using terms from predefined facets.

It is interesting to note that the SUBJECT graphs described previously are powerful enough to describe facets, since cluster nodes can represent a facet hierarchy. Indeed, SUBJECT is used to describe meta-data in a hierarchical manner, so that a user can start at a high level (e.g. population data, energy data, etc.), and gradually narrow down to the data set needed.

The distinction between meta-data and data is not always obvious. Information that is stored in the data base can sometimes be thought of as meta-data and vice versa. This is particularly true of the values of category attributes. For example, if a user is inquiring about the content of the data base in Figure 3, it is as natural to ask what are the summary attributes (e.g. consumption) as it is to ask what years are covered or what are the oil types. Again, this argues for associating the list of values of category attributes with a dictionary rather than to store them with the data.

What about query languages? Are there any special problems associated with SDBs? Aggregation is a predominant function that needs to be supported. However, it is perhaps the most awkward function to express in many query languages. In addition to enriching data models to support automatic aggregation, work is being done to simplify the expression of aggregate functions. For example, [Klug 81] proposes an extension to query-by-example in order to support aggregate functions in SDBs. Perhaps a combination of menu driven techniques (such as those used in SUBJECT), graphics techniques (such as described in [Wong & Kuo 82] in this proceedings), and simple command languages can bring about more convenient user interfaces.

8. Integrating statistical analysis and data management

In order to perform statistical analysis, an analyst needs both data management tools and statistical tools. Unfortunately, these tools are not usually integrated into a single system. Data management systems support only a limited number of statistical functions, and statistical packages have limited data management capabilities.

There are three possible approaches to this problem. The first is to enrich statistical packages with more general data base structures and more powerful data management functions. Evidence of this approach can be found in new releases of statistical packages, where many systems now support some kind of hierarchical or network data structure, while past versions supported only "flat files". However, they still lack many functions, such as joining two tables, or supporting summary sets (or views).

The second approach is to enrich existing data management systems with tools useful to an analyst, such as taking random samples, and a library of statistical operators. An example of this approach is described in [Ikeda & Kobayashi 81], where statistical facilities are added to a commercial data management system, Model 204.

The third approach involves interfacing statistical packages to data management systems. There are three variations to this approach. the first is to tightly couple each pair of systems. Usually a pair is selected for an application and is expected to last a long time. One such experience is described in [Weeks et al 81]. The second variation involves defining a standard data format that all systems accept. In the long run this is a more effective method to implement since each new system added is only required to communicate with the standard format in order to communicate to all systems. However, this technique may be less efficient in terms of processing time since two translators are involved, unless changes can be made to the software of the systems involved. This approach was taken in the SEEDIS project mentioned above, where a fairly simple standard, called CODATA was quite successful in integrating several components of the system. An essential feature of such a standard is that it is self describing, i.e. that data bases carry their own data definition. The third variation involves a monitor that takes care of interfacing the systems, but presents the user with the impression of a single system. an example of this variation is described in [Hollabaugh & Reinwald 81].

An important point to note is that regardless of the approach taken, it is quite essential that statistical operations should produce self-describing data structures that contain meta-data as well as data. Analysts have been burdened by having to keep hand-written documentation of the meta-data as they perform the analysis. As the analysis process progresses, it becomes increasingly difficult to keep track of these meta-data descriptions.

It is not clear which of the above approaches is the most successful. Perhaps future systems can be designed from the start to accommodate both statistical analysis and data management needs. The system S [Becker & Chambers 80] was designed with this goal in mind. It also uses a certain form of self-describing data structures.

9. Security

Security problems in statistical data bases arise from the wish to provide statistical information without compromising sensitive information about individuals. Providing the necessary security is also referred to as inference control since it is intended to prevent the inference of protected information from any collection of legitimate statistical queries.

Here again, it is convenient to distinguish between category and summary attributes. It is customary to consider the individual values of the summary attributes as the ones that should be protected. Consider, for example, a medical experiment in which individuals are treated with a certain drug and its effect are measured in terms of blood pressure, body temperature, etc. Other information about the individuals may be collected such as medical history, living conditions, income, etc. Clearly, some of the information is sensitive such as income or medical history. Suppose that the correlation between blood pressure and income is to be explored. A legitimate query might be "find the average blood pressure on a certain date for males whose income is over 50000". Date, sex and income are the "selecting" or category attributes, while blood pressure is the summary attribute over which a statistical operation is applied. Thus, "blood pressure" is the summary attribute that should be protected.

Unfortunately, it is not sufficient to protect only the summary attributes. Suppose, for example, that the blood pressure of some individual is known to an intruder. By repeated application of the above query at different income levels, he can infer information about the income of the individual, i.e. information about a category attribute can also be inferred. Furthermore, attributes can change their roles between category and summary. For example, in the query "find the average income of males whose blood pressure exceeds a certain level", income is the summary attribute, while blood pressure is the category attribute.

There is an extensive list of papers that discuss problems and propose protection mechanisms for statistical data bases. Several of these mechanisms are briefly discussed here. Interestingly, some of the important results are negative. The different techniques are applicable to different circumstances and needs.

9.1. Limiting the query set

One of the more obvious techniques is to limit the number of cases (individuals) that qualify as a result of a query, called a query set. If the query set is below a pre-specified threshold, then the statistical operation should not be applied over it and the query is refused. However, this technique is quite useless. Intuitively, several queries can be issued whose query sets overlap, until a desired individual value can be inferred. Indeed, this was not only shown to be the case [Schlorer 75, Denning et al 79], but an efficient procedure for finding the snooping queries (called a "tracker") has been devised [Denning & Schlorer 80].

9.2. Limiting intersection of query sets

To remedy this deficiency another approach has been suggested, where the system keeps track of previous queries and verifies that a new query request does not intersect excessively with previous queries [e.g. Dobkin et al 79]. Analyzing audit trails is not a simple matter, and techniques have been proposed for keeping track of the query sets of previous queries instead [Chin & Ozsoyoglu 81]. A new query is allowed as long as the intersection between its query set and those query sets previously answered do not fall below a pre-selected threshold. The main problem with this approach is that whether or not a certain query can be answered depends on the previous queries issued since the time that the data base was created. Also, for large data bases the number of previously accessed data sets may become quite large. This technique is thought to be advantageous for relatively small databases, such as medical experiments where the number of subjects is limited. In contrast, the technique described next is useful only when requested query sets are large.

9.3. Random sample queries

This technique, proposed by [Denning 80], applies the statistical operation on a set of values drawn randomly from the query set. This makes it impossible for users to control precisely the query set from which the responses to their queries are drawn. Clearly, for such an approach to be successful, the query sets requested must be large enough to allow responses based on random samples to be statistically meaningful. Thus, this approach is only useful with large data bases, and for applications whose typical query sets are large.

9.4. Partitioning of the data base

In many applications it is possible to pre-determine that groups of individuals (cases) should be always accessed together for statistical purposes [e.g. Yu and Chin 77]. The individuals within the groups cannot be accessed. In fact, a form of this technique, where the values for the groups are pre-aggregated, is probably the most widely used. Such is the case in census data bases, where information about individuals is prohibited by law. Only pre-aggregated data is available to users. One of the problems associated with such an approach is that sometimes the groups, as defined, may contain only a few individuals, and therefore one can infer information about individuals from the aggregated data. For example, if there are very few American Indian families living in a certain area, then aggregating over racial groups may reveal information about the individual families. This forces values to be suppressed from the data base as long as the number of individuals in a group is below a certain level.

9.5. Perturbing data values

There are two approaches to data perturbation: perturbing output values before presenting them to the user, and perturbing the actual values stored in the data base. Example papers of output perturbation are [Haq 77] and [Achugbue & Chin 78]. [Beck 80] discusses techniques for perturbing the stored values, and also covers previous work in the area.

The main difficulty with this approach is to insure that the error introduced is within acceptable bounds. There is a trade-off between the level of security that can be achieved and the variance of perturbation introduced. With a sufficient number of overlapping queries it is possible to narrow the range of values for an individual. The challenge is in developing techniques that can provide fairly accurate statistical responses, while keeping the inference of the range of individual values sufficiently large for security purposes.

To summarize, providing security in statistical data bases is a difficult problem. There seems to be no single general solution. Therefore, numerous techniques have been suggested that can be used for different applications and needs. Much of the research so far has been applied only with restricted assumptions, such as only SUM queries, or only a single case (record) for each element of the cross product of the category attributes [Kam & Ullman 77]. There is still active research in this area.

10. Concluding remarks

The purpose of this paper was to describe the characteristics and problems that exist in statistical data bases, and to highlight some of the interesting work now emerging in this area. It is inevitable that some work may have been overlooked, especially since this is an inter-disciplinary area.

New research efforts are now emerging in universities, such as the University of Florida and the University of Wisconsin. In addition, there are continuing efforts in laboratories and institutions, such as Lawrence Berkeley Laboratory, Lawrence Livermore Laboratory, Battelle Pacific Northwest Laboratory, Statistics Canada, Bureau of Labor Statistics, and Bell Laboratories.

The area of statistical data bases is quite important in that it encompasses a large variety of application areas that usually deal with large amount of data. Much of these data are now uselessly archived as there are no appropriate tools for their management and analysis. This area poses interesting, challenging, and real problems that should be addressed by data base researchers.

Acknowledgement

I would like to acknowledge the help received from Deanne Merrill and John McCarthy in selecting some of the examples. I am also grateful for the encouragement and support of many of my colleagues at Lawrence Berkeley Laboratory. This work was supported by the Applied Mathematical Sciences Research Program of the Office of Energy Research, U.S. Department of Energy under contract w-7045-eng-48.

References

[Achugbue & Chin 78] Achugbue, J.O., Chin, F.Y., Output Perturbation for Protection of Statistical Data Bases, Dep. Computing Science, University of Alberta, Canada, January, 1978.

[Baker 76] Baker, M., User's Guide to the Berkeley Transposed File Statistical System: PICKLE, Technical Report No.1, 2nd ed., University of California, Berkeley, Survey Research Center, 1976.

[Beck 80] Beck, L.L., A Security Mechanism for Statistical Databases, ACM Trans. Database Syst. 5, 3, (Sept. 1989), pp. 316-338.

[Becker & Chambers 80] S: A Language and System for Data Analysis, Bell Laboratories, July 1980.

[Boral et al 82] Boral, H., DeWitt, D.J., Bates D., A Framework for Research in Database Management for Statistical Analysis, Proceedings of the ACM SIGMOD International Conference on Management of Data, June 1982.

[Bragg 81] Data Manipulation Languages for Statistical Databases -- The Statistical Analysis System (SAS), Proceedings of the First LBL Workshop on Statistical Database Management, Dec. 1981, pp. 147-150.

[Burnett & Thomas 81] Burnett, R. A., and Thomas J. J., Data Management Support for Statistical Data Editing and Subset Selection, Proceedings of the First LBL Workshop on Statistical Database Management, Dec. 1981, pp. 88-102.

[Chan & Shoshani 81] Chan, P., Shoshani, A., Subject: A Directory driven System for Organizing and Accessing Large Statistical Databases, Proceedings of the International Conference on Very Large Data Base (VLDB), 1980, pp. 553-563.

[Chin & Ozsoyoglu 81] Chin, F.Y., Ozsoyoglu G., Auditing and Inference Control in Statistical Databases, March 1981, to appear in IEEE Transactions on Software Engineering.

[Cohen & Hay 81] Why Are Commercial Database Management Systems Rarely Used for Research Data? Proceedings of the First LBL Workshop on Statistical Database Management, Dec. 1981, pp. 132-133.

[Denning et al 79] Denning, D.E., Denning, P.J., Schwartz, M.D., The Tracker: A Threat to Statistical Database Security, ACM Trans. Database Syst. 4, 1, (March 1979), pp.76-96.

[Denning 80] Denning, D.E., Secure Statistical Databases with Random Sample Queries, ACM Trans. Database Syst, 5, 3, (Sep. 1980), pp. 291-315.

[Denning & Schlorer 80] Denning, D.E., Schlorer, J., A Fast Procedure for Finding a Tracker in a Statistical Database, ACM Trans. Database Syst. 5, 1, (March 1980), pp. 88-102.

[DHEW 75] U.S. Department of Health, Education, and Welfare, Comparability of Mortality Statistics for the Seventh and Eighth Revisions of the International Classification of Diseases, DHEW Publication 76-1340, Oct. 1975.

[Dobkin et al 79] Dobkin, D., Jones, A.K., Lipton, R.J., Secure Databases: Protection Against User Influence ACM Trans. Database Syst. 4, 1, (March 1979), pp. 97-106.

[Eggers & Shoshani 80] Eggers, S. J., Shoshani, A. "Efficient Access of Compressed Data," Proceedings of the International Conference on Very Large Databases, 6, 1980, pp. 205-211.

[Eggers et al 81] Eggers, S., Olken, F., Shoshani, A., A Compression Technique for Large Statistical Databases, Proceedings of the International Conference on Very Large Data Base (VLDB), 1981, pp 424-434.

[Gey 81] Gey, F.G., Data Definition for Statistical Summary Data or Appearances Can Be Deceiving, Proceedings of the First LBL Workshop on Statistical Database Management, Dec. 1981, pp. 3-18.

[Hammer & Niamir 79] Hammer, M., Niamir, B. "A Heuristic Approach to Attribute Partitioning," ACM SIGMOD Proceedings of the International Conference on Management of Data, Boston, 1979, pp. 93-101.

[Haq 77] Haq, M.I., On safeguarding Statistical Disclosure by Giving Approximate Answers to Queries, Int. Computing Symp., 1977, pp. 491-495.

[Hollabaugh & Reinwald 81] Hollabaugh L.A., Reinwald, L.T., GPI: A Statistical Package / Data base Interface, Proceedings of the First LBL Workshop on Statistical Database Management, Dec. 1981, pp. 78-87.

[Hawthorne 82] Hawthorne, P., Microprocessor Assisted tuple access, decompression and assembly for statistical database systems, Proceedings of the International Conference on Very Large Data Base (VLDB), 1982.

[Ikeda & Kobayashi 81] Ikeda H., Kobayashi, Y., Additional Facilities of a Conventional DBMS to Support Interactive Statistical Analysis, Proceedings of the First LBL Workshop on Statistical Database Management, Dec. 1981, pp. 25-36.

[Johansson & Shilling 81] Johansson, J.H., Shilling, J.D., Toward the Development of an Integrated Economic data Base at the World Bank, Proceedings of the First LBL Workshop on Statistical Database Management, Dec. 1981, pp. 39-40.

[Johnson 81] Johnson, R.R., Modelling Summary Data, Proceedings of the ACM SIGMOD International Conference on Management of Data, 1981, pp. 93-97.

[Johnson 81a] Johnson, R.R., A Data Model for Integrating Statistical Interpretations, Proceedings of the First LBL Workshop on Statistical Database Management, Dec. 1981, pp. 176-189.

[Kam & Ullman 77] Kam, J.B., Ullman, J.D., A Model of Statistical Databases and Their Security, ACM Trans. Database Syst. 2, 1, (March 1977), pp.1-10.

[Klug 81] Klug, A., Abe -- A Query Language for Constructing Aggregates-by-example, Proceedings of the First LBL Workshop on Statistical Database Management, Dec. 1981, pp. 190-205.

[Lehot 77] Lehot, P., Misuki, M., Rosenthal, A., Szabo, S., On the Optimal Attribute Ordering for an Indexed Sequential File Organization, IEEE Asilomar Conference on Computer Systems, 1977.

[McCarthy 82] McCarthy J., Meta data Management for Large Statistical Databases, Proceedings of the International Conference on Very Large Data Base (VLDB), 1982.

[McCarthy et al 82] McCarthy, J.L., Merrill, D.W., Marcus, A., Benson, W.H., Gey, F.C., Holmes, H., Quong, C., The SEEDIS Project: A Summary Overview of the Social, Economic, Environmental, Demographic Information System, Lawrence Berkeley Laboratory document PUB-424, April 1982.

[Merrill et al 79] Merrill, D., Levine, S., Sacks, S., Selvin, S., PAREP: Populations at Risk to Environmental Pollution, Lawrence Berkeley Laboratory Document LBL-9976, October 1979.

[Merrill 82] Merrill, D., Problems in Spatial Data Analysis, Proceedings of the Seventh Annual SAS Users Group International Conference, San Francisco, Feb. 1982.

[Meyers 69] Meyers, E.D. Jr., Project IMPRESS: Time Sharing in the Social Sciences, AFIPS Conference Proceedings of the Spring Joint Computer Conference, Vol. 34, 1969, pp. 673-680.

[Nie et al 75] Nie, N.H., et al, SPSS: Statistical Package for the Social Sciences, Second Edition, McGraw Hill, New York, 1975.

[SAS 79] SAS Institute, Inc., SAS User's Guide, 1979 Edition, Raleigh, North Carolina, 1979.

[Schlorer 75] Schlorer, J., Identification and Retrieval of Personal Records from a Statistical Data Bank, Methods Inform. in Medicine 14, 1, (Jan. 1975), pp. 7-13.

[Svensson 79] Svensson, P. On Search Performance for Conjunctive Queries in Compressed, Fully Transposed Ordered Files, Proceedings of the International Conference on Very Large Databases, 5, 1979, pp. 155-163.

[Teitel 77] Teitel, R.F., Relational Database Models and Social Science Computing, Proceedings of Computer Science and Statistics: Tenth Annual Symposium on the Interface, Gaithersburg, MD, National Bureau of Standards, April 1977, pp. 165-177.

[Turner et al 79] Turner, M. J., Hammond, R. and Cotton, F. A DBMS for Large Statistical Databases, Proceedings of the International Conference on Very Large Databases, 5, 1979, pp. 319-327.

[Weeks et al 81] Weeks, P., Weiss, S., Stevens, P., Flexible Techniques for Storage and Analysis of Large Continuing Surveys, Proceedings of the First LBL Workshop on Statistical Database Management, Dec. 1981, pp. 310-311.

[Wong & Kuo 82] A Graphical User Interface for Data Extraction, Proceedings of the International Conference on Very Large Data Base (VLDB), 1982.

[Yu and Chin 77] Yu, C.T., Chin, F.Y., A Study on the Protection of Statistical Databases, ACM SIGMOD Int. Conf. on Management of Data, 1977, pp. 169-181.

SOME STATISTICAL DATA BASE REQUIREMENTS FOR THE
ANALYSIS OF LARGE DATA SETS *

Richard J. Littlefield and Paula J. Cowley
Pacific Northwest Laboratory
Richland, Washington 99352

This paper defines a "statistical data base" as one that facilitates iterative analysis by storing both data and results. Five data management requirements are discussed that are important to the analysis of large data sets, where "large" means many cases, many variables, or great heterogeneity. Four of the requirements deal with the efficient storage of multiple shapes and sizes of data, storage of analysis results, storage of multiple versions of data, and storage of multiple subsets. The fifth requirement is for automatic tracking of analysis steps. Individual solutions to the first four requirements are identified in existing packages. No solution to the fifth requirement has yet been implemented, although a new concept of "data analysis environments" shows promise.

Keywords: Statistical data base; data management requirements; analysis of large data sets; data analysis environments.

* Work supported by the U.S. Department of Energy, Office of Basic Energy Sciences, Applied Mathematical Sciences program, under contract DE-AC-06-76RLO 1830.

1. INTRODUCTION

In contrast to other applications of data management, the analysis of large data sets requires an unusual combination of capabilities. In this paper, we will discuss five data management requirements that are important to the analysis of large data sets, but which have received relatively little attention from the data management community. We hope that, by calling attention to these requirements, we can stimulate development of techniques and software packages for meeting them. Several of the requirements affect the design of data management tools at their lowest level. It is therefore particularly important that these requirements be considered at the earliest stages of software design.

The paper is organized as follows. First, we will define our particular view of the term "statistical data base" and list the data management requirements that are appropriate to that view. Second, we will briefly describe the Analysis of Large Data Sets research program (ALDS) at the Pacific Northwest Laboratory (PNL). It is this program, with its unusual orientation and mix of statisticians and computer scientists, that has identified the requirements presented in this paper. Third, we will explain in more detail the individual data management requirements and the rationale behind them. Finally, we will summarize the requirements and suggest directions for further developments.

2. STATISTICAL DATA BASE REQUIREMENTS

There are several possible definitions for the term "statistical data base". One common interpretation is that a statistical data base contains data on which statistics are computed. This view is reflected, for example, by the SPSS analysis package [1], in which the "system file" contains essentially raw data on which various kinds of analyses can be performed. An alternative view is that a statistical data base contains the results of statistical analysis. For example, a data base full of U.S. census data summary tables might fall in this category.

Our definition, however, is the following:

A statistical data base is one that facilitates the iterative process of data analysis by storing and transferring data and partial results between manipulation, analysis, and display modules.

The data base can contain both data and analysis results, as well as certain ancillary information needed to assist and track the analysis process. Such a data base, particularly when applied to a "large" data set, has the following combination of requirements.

1. The data base must hold several shapes and sizes of data (vectors, matrices, and hierarchical structures containing both numbers and text).

2. Analysis routines interfaced to the data base must store their results back into the data base instead of merely listing them for the user's inspection.

3. The data base must provide efficient storage for multiple versions of the data, where the versions are algorithmically related or differ in only a small fraction of the cases.

4. The data base must provide for efficient storage of multiple overlapping subsets of the data.

5. The data base, in combination with whatever analysis routines are interfaced to it, must allow for automatic tracking of the various steps in the analysis process.

These requirements are discussed in more detail in succeeding sections.

3. THE ALDS RESEARCH PROGRAM

The Analysis of Large Data Sets research program (ALDS) at PNL is chartered to explore the opportunities and problems presented by large data sets. Staff includes both statisticians and computer scientists, and specific research areas include analysis techniques, exploratory graphics, and data management. The program is unusual in that it is not tied to any specific application or type of data. Instead, it has been free to consider a wide variety of applications and data organizations. As a result, the definition of "large" developed by the ALDS project has several independent aspects. The most important of these are:

1. Many cases. Some data sets have a simple structure, but contain so many cases that the sheer bulk of the data impedes managing and analyzing it. Data sets of this type are commonly produced by automated data gathering equipment.

2. Many variables. Some data sets contain a reasonable number of cases, but the number of variables per case is quite large. An example of this type is a medical case history. The major difficulty with many variables is the extremely large number of possible interactions between variables. Testing for and keeping track of these interactions can be a formidable bookkeeping task.

3. Heterogeneity. Some data sets, particularly those collected over a long period of time under varying conditions, exhibit considerable heterogeneity between data subsets. A major task here is identifying the homogeneous subsets and relating them to a common base.

Of course, these three aspects of large can also occur in combination. For example, the raw U.S. census data is certainly large in terms of number of cases. It is moderately large in terms of number of variables. In terms of homogeneity, the data set may or may not be large, depending on the success of the Census Bureau's efforts to assure uniform data collection practices.

Most of the ALDS project's efforts to date have been directed toward the first two aspects of large: many cases and many variables. Three particular problems were addressed in the area of data management.

The first problem was how to provide efficient access to individual variables across many cases. This type of access is required by most statistical analysis procedures. In response to this problem, a transposed file facility, called Self Describing Binary (SDB) [2], was designed and constructed. SDB files and their associated access routines provide a common data interface between ALDS software modules.

The second problem was how to provide efficient access to many variables within a case. This type of access is required by data verification and editing procedures. This type of access is provided within the SDB file framework by a buffering algorithm applied to the transposed data blocks.

The third problem was how to provide the analyst with efficient interactive access for data verification and editing. This problem was addressed by developing the ALDS Data Editor (ADE) [3,4]. ADE provides a variety of subsetting and editing operations.

These include: programmed or interactive modification of individual data elements; random, interactive, or algorithmic generation of data subsets; and selection by case or variable. A phrase structured command language with frequent feedback is used to minimize analyst error. In ALDS experience, data preparation is a major if not predominant task within the overall analysis process. ADE has proved to be a valuable tool in this regard.

The ADE data editor, SDB file facility, and various file format conversion utilities are the major data management components of the ALDS software system. The system also includes interfaces to the MINITAB [5] and RUMMAGE [6] statistical analysis packages, ARTHUR pattern recognition package [7], and a variety of more specialized tools, including projection pursuit and geographic mapping. The ALDS version of MINITAB has been especially heavily modified to serve as a testbed for high resolution color exploratory graphics techniques.

4.0 DISCUSSION OF THE REQUIREMENTS

4.1 Overview

Five major data management requirements for the analysis of large data sets were listed above. In this section, we shall discuss these requirements in more detail, paying particular attention to the reasons behind them.

There are basically four reasons behind the five requirements. The iterative nature of data analysis is central to all five. Iterative analysis by itself imposes the first two requirements: that the data base hold various shapes and sizes of data and that analysis packages store their results back into the data base. Iterative analysis applied to a data set with many cases imposes the third requirement, that the data base provide efficient storage for multiple versions of the data. Iteration applied to a data set with large heterogeneity imposes the fourth requirement, that the data base provide efficient storage for multiple overlapping subsets of the data. Finally, any complex iterative analysis imposes the fifth requirement, that the data base and analysis package provide for automatic tracking of the analysis steps. These relationships are described more fully in the following sections.

4.2 The Iterative Nature of Analysis

It is widely recognized that the analysis process is iterative, in the sense that results obtained at one stage of the analysis influence the procedures applied in subsequent steps. This view of iteration is shown in Figure 1. In this simplified view, the flow of data is essentially one-directional. That is, each step in the analysis accepts data as input and produces results as output, presumably for interpretation by the analyst. Certainly some analyses can be adequately described in this fashion. More commonly, however, some of the results from one analysis step are directly used as input to a subsequent step, serving as either data or control parameters. To describe this aspect, the more general diagram shown in Figure 2 is required.

The observation that some analysis results can serve as data for other analyses may not seem terribly profound. However, it has important implications for the design of statistical data bases. This is because even simple analyses produce results that have different shapes and sizes from their inputs. Figure 3 illustrates this idea with a linear regression. The regression inputs a matrix of independent variable values and a vector of dependent variable values. It outputs a short vector of regression

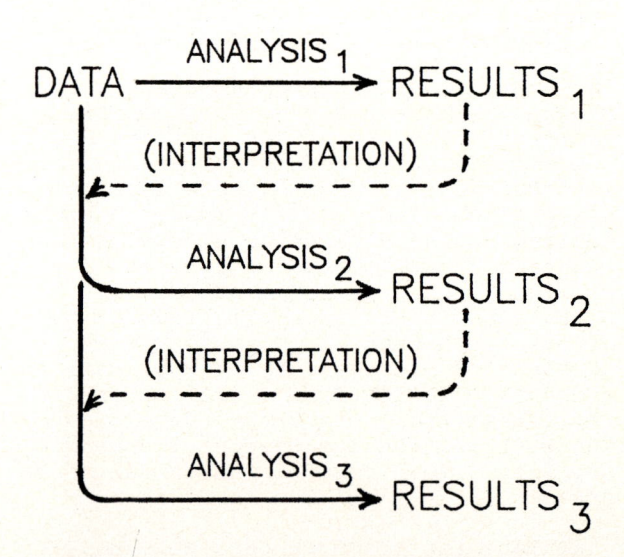

Figure 1. In this restricted view of iterative data analysis, the results of each analysis step affects subsequent steps indirectly, through interpretation by the analyst.

coefficients, a vector of residuals, a covariance matrix, and possibly a variety of regression diagnostics, such as the residual sum of squares and a vector of leverage values. Of all these results, only the vectors of residuals and leverage values could be considered to have the same size and shape as the input data.

Thus, we are led to the first requirement. The data base must be capable of holding multiple shapes and sizes of data. This requirement was largely ignored by early data management packages. For example, the SPSS "system file" can contain only data having a rectangular variable-by-case structure. More recent systems, such as S [8], RS/1 [9], the ALDS system, and even MINITAB, have overcome this deficiency. A variety of implementations have been used. For example, MINITAB uses memory-resident "worksheet" which can be written to and read from a single file. RS/1 takes the opposite approach, writing each of its rectangular "tables" to a separate file. In the ALDS system, multiple SDB files can be used in the same way. S takes an intermediate approach. All components of an S hierarchical structure reside on the same file, but different structures go on different files.

It is not sufficient, however, that a data management package support various sizes and shapes of data. The analysis packages interfaced to it must also use this capability. This aspect is commonly overlooked. Even when adequate data management capability is available, it is the rule rather than the exception that many analysis results are simply displayed for the analyst and cannot be saved. For example, the linear regression module in MINITAB will store the coefficient and residual vectors, but not the regression diagnostics. An outstanding exception to the rule is S, in which the system design specifies that each analysis module stores all its results.

4.3 Large Data Sets

The combination of large data sets and iterative analysis leads to several additional data management requirements. Efficient data storage and access is a major consideration. When the analysis is complex, automatic tracking of the analysis steps is also required to make efficient use of the analyst's time. In this section, we will discuss the data management requirements imposed by the three aspects of large: many cases, many variables, and heterogeneity.

As an analysis progresses, it is common to accumulate multiple versions of data that differ only slightly. With a data set containing many cases, this can produce a ponderous bulk of largely redundant data. The redundant data is usually produced in either one of two ways. The first can be termed "algorithmic". The most common example here is arithmetic transformation, such as scaling the data or taking exponential or square root functions. More complicated examples include functions of several variables and conditional operations. The second way can be termed "random". For example, this would include correction of a single erroneous observation, or an imputed value replacing missing data.

Different techniques are required to handle these two types of data modification. For algorithmic modification, the most reasonable approach is to store the original data and the modification rule. When access is required, the rule is re-evaluated. This produces a dramatic reduction in storage requirements at the cost of increased access time. This approach has been used by several packages. The SPSS *-type temporary data modification commands (*COMPUTE, *RECODE, etc.) are implemented in this way, although the rule is not actually stored as part of

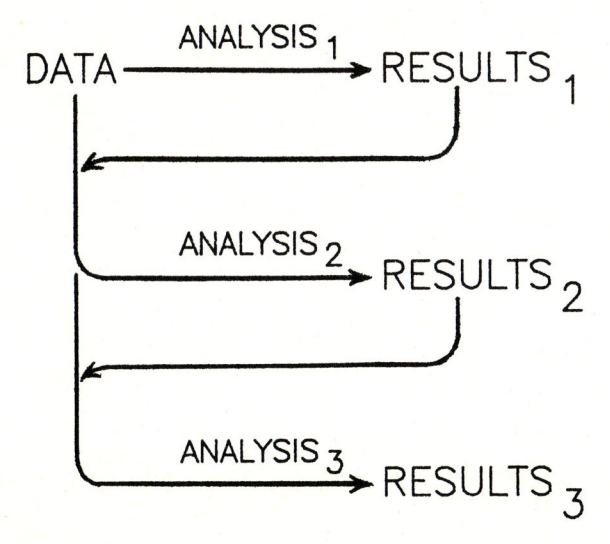

Figure 2. In this more general view of iterative data analysis, some results of each step can be used directly as input for subsequent steps, serving as data or parameters.

the data base. The ADE data editor has a similar XFORM command. Digital Equipment Corporation's DATATRIEVE package [10] supports a "COMPUTED BY" field specification that is stored as part of the data base. MINITAB, RS/1, and S appear to have no similar concepts.

For random data modifications, there simply is no rule to store. The most reasonable approach in this case would be to use differential files [11,12,13]. In this case, the data base would contain the original data plus a list of modified data elements. Access to the modified data would be provided by supplying values from the modified list if available, and the original data otherwise.

The differential file concept has been widely used in data base management systems to store updates to a data base and to provide backup and recovery capabilities. It is also used in systems for maintaining software source code, such as Control Data Corporation's UPDATE utility [14] and the UNIX SCCS package [15]. However, differential files apparently have never been applied as described here, for storing multiple versions of data during analysis.

Heterogeneous sets produce other requirements, notably the need for efficient storage of data subsets and for automatic tracking of the analysis process. To deal with heterogeneity, analysts frequently attempt to divide the data into relatively homogeneous subsets, and then either analyze the subsets independently or relate them to a common base. Several arrangements usually must be tried before an acceptable division is found. Thus, there is a need for efficient storage of these subsets.

As with data modification, several techniques can be used for storing subsets. If the subset comprises a large fraction of the total cases, then it may be most efficient to store the selection rule and re-evaluate it when access is required. If the subset is small, then it is more efficient to store a list of case numbers. However, if the subset is also widely distributed, this approach may require excessive input/output. In this case, the best tradeoff between storage space and access time may be achieved by making a physical copy of the subset.

All these approaches have been used by various systems. The SPSS command *SELECT IF and the ADE command COND represent the selection rule approach. The ADE "virtual subset" uses the case pointer approach, as do many other data management packages and bibliography retrieval systems, such as DATATRIEVE and Battelle's BASIS package [16]. Physical subsets, of course, can be generated by virtually any package. It is interesting to note that none of these packages can save the subset between sessions in any form except a physical copy.

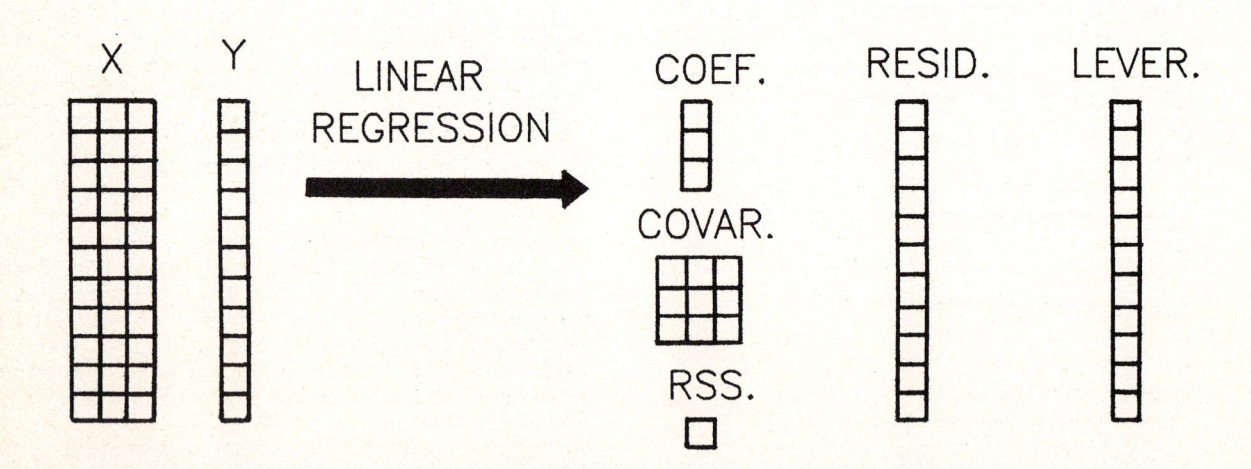

Figure 3. The results of many analysis procedures have different sizes and shapes from the input data. In this example, a linear regression inputs a matrix of independent variables and a vector of dependent variables, and outputs vectors of coefficients, residuals, and leverage values, a covariance matrix, and a scalar residual sum of squares.

Another effect of a data set with many variables or heterogeneity is to increase the complexity of the analysis. With any data set, exploratory analysis is likely to take the form of a tree, with most branches representing blind alleys that were evaluated and then rejected. The number of potential blind alleys increases rapidly with the number of variables. Even with a relatively small data set, using manual methods to track the analysis paths and results is an arduous bookkeeping task. With a complex data set, it can be nearly overwhelming. Thus, there is an urgent need for integrated analysis and data management packages that will automatically keep track of the analysis steps [17].

Existing analysis packages address this problem in only the most primitive ways. For example, both S and the current ALDS system provide the option of maintaining a command log. In combination with the original data set, a command log certainly in some sense captures the entire analysis process. There are problems with this approach, however. First, the command logs capture a large number of irrelevant commands, such as listing and plotting commands that merely refreshed the analyst's memory and did not constitute part of the actual analysis. Second, the analyst will frequently decide to return to a previous point in the analysis to try a different approach. Based solely on the original data and a copy of the command log, it can be difficult and time-consuming to restore an analysis to a previous state.

The ALDS project is currently developing the concept of "data analysis environment" [18] to avoid these problems. As currently conceived, a data analysis environment is a combination of the following:

1. The current state of the data set, including any modifications to the original data, new cases or variables, and analysis results that have been stored for subsequent use.

2. A listing of the sequence of analysis operations that produced this environment from a prior known environment.

3. Comments entered by the analyst to describe the environment and document the analysis process for future reference.

4. The analysis system status, including such analysis and user interface parameters as convergence criteria, default command options, plotting parameters, etc., which affect operation of the system but are not directly associated with the data.

The intent is for the analysis system to be capable of efficiently saving and restoring these analysis environments at any point. The analyst will be responsible for identifying when a useful new environment has been produced, since only he knows when this occurs. During the exploratory stages of an analysis, many of these environments may be created as various hypotheses are tested. The analyst can then temporarily set aside less promising environments, secure in the knowledge that he can return to them if more promising paths do not turn out well.

In most analyses, the difference between data sets in closely related environments will mostly take the form of varying data subsets and minor modifications. Thus, the concept of data analysis environment reinforces the requirements developed above for efficient storage of multiple versions and subsets of data.

5. CONCLUSIONS

In this paper we have discussed five data management requirements that are important to the analysis of large data sets. Four of these requirements deal with the efficient storage of multiple shapes and sizes of data, storage of analysis results, storage of multiple versions of data, and storage of multiple subsets. Each of these requirements has been individually met by various packages. However, no package has yet provided an integrated solution to all four requirements simultaneously. The fifth requirement is for automatic tracking of steps in the analysis process. Current implementations that address this requirement are inadequate at best. A new concept of "data analysis environments" being developed by the ALDS project shows promise, but has yet to be implemented. Its success will depend strongly on achieving an integrated solution to the first four requirements.

REFERENCES

[1] Nie, N. H., C. H. Hull, J. G. Jenkins, K. Steinbrenner, D. H. Bent. 1975. _SPSS: Statistical Package for the Social Sciences_. McGraw-Hill, Inc., New York.

[2] Burnett, R. A. 1981. "A Self-Describing Data File Structure for Large Data Sets." In _Computer Science and Statistics: Proceedings of the 13th Symposium on the Interface_, New York, NY, pp. 359-362.

[3] Thomas, J. J., R. A. Burnett, J. R. Lewis. 1981. "Data Editing on Large Data Sets." In _Computer Science and Statistics: Proceedings of the 13th Symposium on the Interface_, New York, NY, pp. 252-258.

[4] Burnett, R. A. and J. J. Thomas. 1982. "Data Management Support for Statistical Data Editing and Subset Selection." In _Proceedings of the First LBL Workshop on Statistical Database Management_, Menlo Park, CA, pp. 88-102.

[5] Ryan, T. A., B. L. Joiner, B. F. Ryan. 1976. _MINITAB Reference Manual_. Duxbury Press.

[6] Bryce, G. R. 1980. _Data Analysis in RUMMAGE - A User's Guide_. Brigham Young University, Provo, Utah.

[7] Duewer, D. L., J. R. Koskinen, and B. R. Kowalski. 1977. _ARTHUR_. Available from B. R. Kowalski, Laboratory for Chemometrics, Department of Chemistry, BG-10, University of Washington, Seattle, WA.

[8] Becker, R. A. and J. M. Chambers. 1981. _'S' - A Language and System for Data Analysis_. Bell Laboratories, Murray Hill, New Jersey.

[9] Bolt Beranek and Newman, Inc. 1981. _RS/1 User's Manual_. Cambridge, Massachusetts.

[10] Digital Equipment Corporation. 1980. _DATATRIEVE-11 V2.0 User's Guide_. Maynard, MA, July, 1980.

[11] Severance, D. G. and G. M. Lohman. 1976. "Differential Files: Their Application to the Maintenance of Large Databases." In _ACM Transactions on Database Systems_, Vol. 1, No. 3, September, 1976, pp. 256-267.

[12] Aghili, H. and D. G. Severance, "A Practical Guide to the Design of Differential Files for Recovery of On-Line Databases." In _ACM Transactions on Database Systems_, Vol. 7, No. 4, December, 1982, pp. 540-565.

[13] Verhofstad, J. S. M. 1978. "Recovery Techniques for Database Systems." In _ACM Computing Surveys_, Vol. 10, No. 2, June 1978, pp. 167-195.

[14] Control Data Corporation. 1978. _UPDATE 1 Reference Manual_. Control Data Corporation, Sunnyvale, CA.

[15] Bonanni, L. E. and A. L. Glasser. 1977. _SCCS/PWB User's Manual_. Bell Telephone Laboratories, Murray Hill, New Jersey.

[16] Battelle Development Corporation. 1981. _Basis User's Guide_. Columbus, Ohio.

[17] Denning, D., W. Nicholson, G. Sande, A. Shoshani. 1983. _National Research Council Panel Report on Statistical Database Management_, Washington D.C., February, 1983.

[18] Thomas, J. J. 1982. "A User Interaction Model for Manipulation of Large Data Sets." In _Proceedings of the Computer Science and Statistics 14th Symposium on the Interface_, Troy, New York, July, 1982.

RAPID: A Statistical Database Management System

Roy G. Hammond

EDP Planning and Support Division
Systems and Data Processing Branch
Statistics Canada
Ottawa, Canada K1A 0T6

Abstract
This paper will introduce the reader to RAPID - a Statistical DBMS developed at Statistics Canada. The differences between RAPID, conventional DBMS systems, and data management in the Statistical Packages are examined. Future directions in computer processing are contemplated, and in conclusion a case is made for a DBMS which supports informational and statistical queries in concert.

1. Introduction
Statistics Canada is charged with the responsibility to conduct and process all major surveys and censuses for the Canadian federal government. RAPID [RAPID, TURN79] is a DBMS developed at Statistics Canada for storing and accessing statistical data. RAPID is not available as a commercial product, although it is available to government agencies, educational institutions, and other non-profit-making organizations. It has been installed on machines with an IBM compatible architecture as far away as Japan, Chile, and Hungary.

The next section of this paper provides a basic description of RAPID and its uses. Section 3 will be used to compare some of RAPID's characteristics with those of commercial systems to demonstrate the differences between Statistical and Informational systems. In Section 4 we go back to RAPID (for computer scientists) to explain some of the technical approaches used in RAPID. Section 5 describes some directions for future statistical database development.

2. RAPID
This section provides an overview of RAPID's facilities, history, interfaces, and usage. Detailed documentation is available from Statistics Canada [RAPID].

2.1 RAPID - A simplified description
RAPID is based on the Relational Model [SAND81], forcing users to organize their data into simple arrays of rows (tuples) and columns (attributes). Each RAPID file (relation or array) is a self-contained operating system dataset. Metadata is stored as an integral part of each column. Rows and columns may be added and deleted - the default content of a newly created column is *null* for each currently allocated row, and the default content of a newly created row is *null* for each existing column. The deletion process frees space which will be reused automatically. Optionally, keys may be assigned to the rows for direct access. This gives four ways of accessing a data element: sequential, direct by any key, indirect by scanning other columns for a match on a query, and direct by relative row number.

All access to a RAPID database is performed through commands to the Host Language Interface (HLI). The RAPID Utilities are merely generalized applications programs.

There are many commands available for managing metadata and data. RAPID differs from most systems in the storing and retrieving of data in that the user must define *logical records* before accessing data. These definitions name and order the required columns, and specify data formats. This allows RAPID to access only the needed columns, and to present the application with *records* independent of other applications and internal data formats.

The basic HLI commands to open, close, store, and retrieve are very similar to the data management facilities of most programming languages and database management systems. Thus it is a simple task to convert most programs to/from RAPID.

RAPID has an integrated Data Dictionary system [HAMM82], backup, recovery, and a number of utilities not discussed in this paper. Here, the emphasis is on the commercial products with which RAPID has been interfaced.

2.2 Why was RAPID created?
At Statistics Canada, we did not originally want to build our own DBMS. The cost of building and supporting RAPID has been a strain on our resources, although it has paid for itself in processing cost reduction and increased facilities. The reasons for continuing to develop and maintain RAPID have changed over the years. In this subsection we will describe some of the events and their impacts.

The Census of Population, with its 33K geographic areas, 8M households, 25M persons, and thousands of published tables has provided a variety of data processing problems. For the 1971 Census an in-house tabulation system was written; its main feature being the use of transposed files containing micro-data. The days of having to maintain summary files to produce tabulations were over! Unfortunately, these transposed files were very difficult to maintain and for other applications to use.

By 1973, we had developed a transposed file extension to the commercially available TOTAL DBMS [TOTAL]. About this time our management decided that DBMS was the way of the future and committed the 1976 Census processing to new technology. The 1974 Census Test used the new processing software and our simple TOTAL extensions - surely a commercial system would appear to handle the 1976 volume.

By 1975, the Census was in trouble. Only one of the commercial systems could handle the volume, and it would consume several times the computing resources that would be available; not to mention needing a very difficult conversion. The selected alternative was a new system (RAPID) to replace TOTAL and the extensions with in-house software, requiring only minor changes to the Census programs.

In the late 70's many features of commercial DBMS systems were added to RAPID including logging, text handling, and integrated data dictionary. Concurrent update facilities were not added for reasons outlined in 3.2.

Today, hardware has improved, the Relational Model is being accepted, but there are still no commercial systems to handle either our volume or statistical queries.

2.3 Interfaces

Retrieval capability is the major criteria in the selection of an end user facility. In order to be competitive, commercial DBMS and statistical packages offer unique query facilities and languages. At Statistics Canada a number of competing tabulation and statistical packages are in use, and each has a dedicated user following. In developing RAPID we wanted to avoid creating yet another language, reach a broad user community, and minimize development and training expense. The solution was to provide interfaces to existing products such that users could share a database from any combination of their favorite packages. RAPID is currently interfaced with EASYTRIEVE [EASY], OPTIDATA [OPTI], SAS [SAS], SPSS [SPSS], STATPAK [STATPAK], and TPL [WEISS81]. The interfaces take on a variety of forms, depending on the particular package.

2.4 How is RAPID used?

The original intent of RAPID was for mass storage of micro-data, and our Census applications alone have about 3 giga-bytes of compressed data stored on RAPID files. Access to RAPID through the interfaces was another intent. Several South American installations use SPSS with RAPID instead of SPSS SAVE files to provide greater flexibility and disk space economy. One area of the U.S. Department of Labour uses a carefully chosen combination of SAS and application programs to maintain and access data on RAPID files [TRIM82].

Novel uses of RAPID include the Canadian Consumer Price Index processing [GOLD81] in which application programs fully exploit RAPID's host language interface and storage facilities. The Generalized Iterative Record Linkage System [GIRLS] uses sequential input files and eventually produces sequential output files; but uses RAPID in between for data, linkages, weights, rules and groupings.

3. Contrast with other systems

3.1 Query orientation

There are two diametric types of queries: *informational* ("Tell me all about John Smith.") and *statistical* ("What was the organization's average salary last month?"). They are diametric because an underlying storage structure optimized for one is basically the worst structure for the other. Conventional DBMS systems are aimed at the big market: the informational system. Indeed, a DBMS optimized for informational queries is essential to running a business today. A copy of the database, with a different structure [BATES82, COHEN81] is obviously needed for the statistical queries. RAPID is oriented to statistical queries.

3.2 Concurrency

A prime requirement for an information system is support for many concurrent users - both for read and update. The queuing and locking facilities, with their associated transaction backout and restart facilities, are the main resource consumers (cpu and I/O) in commercial systems. Statistical queries will cripple an information system. For example, in a banking system where each row represents an account, and one of the columns represents current balance, a statistical query to sum the balances must lock the portion of the column it has not yet accessed in case a transaction occurs to transfer funds between an account it has summed and one that it hasn't.

In a statistical DBMS, such as RAPID, we can take advantage of the characteristic that updates seldom occur. Updaters can be required to have exclusive control of a file, and readers are not encumbered with continuous concurrency overhead.

3.3 Metadata

An information database generally has a very limited number of views from the user perspective. A statistical database is quite the opposite [MCCART82]. Thus the metadata requirements for a statistical database go far beyond what is supported in commercial systems today. A discussion on metadata issues is beyond the scope of this paper. The RAPID developers have not solved the metadata problems, although some tools have been created [HAMM81].

3.4 Statistical packages

The developers of statistical packages have concentrated on functional facilities rather than data storage facilities. As a result, they provide neither the data compression (storage space utilization) nor the performance (access) that we expect [BATES82, TRIM82]. Perhaps, now that the statistical packages have extensive functional capability, the user community will tire of magnetic tape and press for better data management. As an example of the potential in this area, a STATPAK-RAPID cross-tabulation of two variables on a Census file will execute in five minutes on an Amdahl V6 computer (about 48 million data cells to be processed).

4. RAPID revisited

In the preceding sections, there have been general statements about RAPID's performance and organization. This section is for the computer scientist who is interested in how RAPID does its job.

It is RAPID's storage management, which is generally transparent to the user, that makes it a Statistical DBMS. As described in section 2.1, data is usually presented to the application record-at-a-time but, unlike most systems, it is not stored that way. Conceptually, RAPID is a transposed system in which each logical record on disk contains many elements from a single column. Elaborate buffering and access strategies perform the transformation to minimize the number of disk accesses and maximize data compression and the use of disk space. There are two types of storage managed by RAPID: *auxiliary* which is the disk space used to store data, and *main* which is used to buffer data between auxiliary storage and the application.

4.1 RAPID - Auxiliary storage

For each file (relation), the user gives RAPID a disk space allocation to work with. Utilities allow this space to be increased or decreased later without reorganization. RAPID formats this space into fixed length records whose size depends on parameters and device type. Each physical record is subdivided into two or more *segments*, which are the logical records on disk. Some of the physical records will be used for file metadata, a bit map of available segment slots, and a b-tree for direct access of segments.

The segment is the basic unit of RAPID storage, and each segment has a unique key which is either a column name or an internal code. When a column is created, its basic metadata is placed in a segment, keyed by column name and additional metadata may be internally assigned to other segments. When data is stored in a column, segments are allocated as needed to contain the data. The amount of data stored in a segment varies with the characteristics of the column and the (file defined) segment size. Generally, large segment sizes are used for large files. A one bit wide column with an 8K segment size would have 64K rows per segment, while a four byte wide column on the same file would have 2K rows per segment.

RAPID has an internally created and maintained one bit wide column to monitor rows in the file. When a row is allocated, a bit in this column is turned on, and when a row is deleted the bit is turned off. Data can be accessed for each active row and column intersection. If the segment that would contain a requested element has not been allocated, a segment of null values is materialized, and if it is updated it will eventually be allocated and stored. The algorithms for allocation, etc. are beyond the scope of this paper.

4.2 RAPID - Main storage

The previous section described segments as pieces of columns, and the way in which they are maintained on disk. In this section the strategy for movement of elements of data from segments to the application is introduced, explaining more of the reasons for RAPID's high speed under a variety of conditions.

When an application OPEN's a RAPID file, one of the parameters is the number of segment buffers to be used. The optimal number is dependent on the application logic and the amount of main storage available. Having at least the optimal number is very important to avoid thrashing, and there is no penalty for having an excess. These buffers are basically reused on a least recently used basis. There are other buffers and elaborate strategies used in accessing segments, but the pool of segment buffers is most important.

Example 1: The application has two sequential loops, each performing a sequential or skip sequential access to five columns. Five buffers is adequate since the first five columns will be finished before the second five start. The first logical READ involves five disk I/O operations to fill the buffers. Data elements are then converted as needed to the application's format and passed back as a record. Subsequent reads will cause an I/O only when associated segments are not in the buffer pool.

Example 2: Similar to Example 1, except that the two loops are interleaved. Ten buffers is optimal, otherwise five I/O operations would occur each time a different set of columns is accessed.

Example 3: The application requires random access to two columns. The optimal number of buffers would allow all segments of both columns to reside in the buffer pool together.

5. Future direction

The general strategy described in this section is a hypothesis on how our environment may change over the next few years. No one knows how the balance between desk-tops and mainframes will work out!

Statistical databases are likely to become increasingly important for business and government planning. The need is conveniently supported by hardware technology on three fronts: desk-top computers with the power of recent mini-computers, dramatic increases in the power and capacity of large computers (IBM has a machine in beta test with 2G bytes of main storage and appropriate cpu speed), and inexpensive, massive auxiliary storage (laser and disk). Our challenge is to develop the software that will take full advantage of this hardware.

The vendors of the Statistical Packages are poised to enter the desk-top environment, as demonstrated by their support on the mini-computers. This will support the consumers need for statistical software that is functionally identical in all environments. The desk-top will always lag behind the mainframe in terms of auxiliary storage, but will be ideal vehicles for data analysis of summary files. Since the desktops are personal work stations, they should basically be users of corporate data transferred to them through a communications network, and need little (if any) auxiliary storage and lots of main storage.

The mainstay of computing in most organizations is the massive mainframe complex. Its prime asset is its ability to provide centralized control over corporate data and communications. As more processing is distributed to the desk-tops, the mainframe's role can change towards that of a library. Its tasks would not be trivial: maintenance of vast amounts of metadata and data, physical and access security, and query processing to communicate packets of security cleared corporate data with the distributed processors.

One characteristic of hardware that does not appear to be keeping pace is the relative cost of transferring data from auxiliary storage to main storage. RAPID's approach to minimizing the amount to be transferred through its storage management seems to be a step in the right direction for static data. It is desireable that RAPID's function will be incorporated into statistical software on desk-top computers, and into database machines as part of the mainframe complex.

6. Conclusion

RAPID was once referenced as a maverick among Relational DBMS; perhaps that was because it is a Statistical DBMS, not just a Relational DBMS. The justification for a Statistical DBMS is clear, and it should reside in harmony with an Informational DBMS at the expense of data redundancy. The number of queries about (and requests for) RAPID suggest that a DBMS which supports both storage structures (transparent to the user except for performance), with appropriate metadata and DBA facilities, would be a viable commercial product. Perhaps that will be the evolution of RAPID.

References

BATES82 Bates, D., Boral, H., and DeWitt, D.J., *A Framework for Research in Database Management for Statistical Analysis,* Proceedings of 1982 ACM-Sigmod Conference, Orlando, pp.69-78

COHEN81 Cohen, E., Hay, R.A. Jr., *Why are commercial DBMS rarely used for research data?* Proceedings of the Workshop on Statistical Database Management, Menlo Park, CA., December 1981, pp. 132-133.

EASY *Easytrieve Reference Manual,* Pansophic Systems, Inc. 709 Enterprise Drive, Oak Brook, IL 60521

GIRLS *Generalized Iterative Record Linkage System,* EDP Planning and Support Division, Systems and Data Processing Branch, Statistics Canada, Ottawa, Canada K1A 0T6

GOLD81 Goldmann, G.J., *The Relational Model Applied to the Automation of the Canadian Consumer Price Index,* Proceedings of the Workshop on Statistical Database Management, Menlo Park, CA., December 1981, pp. 312-334.

HAMM81 Hammond, R.G., *Metadata in the RAPID DBMS,* Proceedings of the Workshop on Statistical Database Management, Menlo Park, CA., December 1981, pp. 123-131.

MCCART82 McCarthy, J.L., *Metadata Management for Large Statistical Databases,* Proceedings of 8th International Conference on Very Large Data Bases, Mexico, 1982.

OPTI *Extracto/Optidata,* Optipro, Inc., P.O. Box 9340, Ste-Foy, Quebec G1V 4B5.

RAPID *RAPID DBMS,* EDP Planning and Support Division, Systems and Data Processing Branch, Statistics Canada, Ottawa, Canada K1A 0T6

SAND81 Sandberg, G., *A Primer on Relational Data Base Concepts,* IBM System Journal, Vol 20-No 1-1981, pp. 23-40.

SAS *SAS User's Guide,* SAS Institute Inc., P.O. Box 8000, Cary, NC 27511

SPSS *Statistical Package for the Social Sciences,* McGraw Hill, New York.

STATPAK Podehl, W.M., *STATPAK - General Concepts and Facilities,* Systems Development Division, Systems and Data Processing Branch, Statistics Canada, Ottawa, Canada K1A 0T6

TOTAL *TOTAL Reference Manual,* Cincom Systems Inc., Cincinnati, Ohio.

TRIM82 Trimble, J.H. Jr., *Can SAS Replace Traditional Programming Languages in Large Systems Developments?* Proceedings of SAS International Users Group (SUGI82), pp. 335-337.

TURN79 Turner, M.J., Hammond, R.G., Cotton, P. *A DBMS for Large Statistical Databases,* Proceedings of 5th International Conference on Very Large Data Bases, Brazil, 1979, pp. 319-327.

WEISS81 Weiss, S.E., Weeks, P.L., Byrd, N.J., *Must We Navigate Through Databases?* Proceedings of the Workshop on Statistical Database Management, Menlo Park, CA., December 1981, pp. 111-122.

A Graphical Query System for Complex Statistical Databases

Harry K.T. Wong and Whei-Ling Yeh

Lawrence Berkeley Laboratory
University of California
Berkeley, California 94720

Abstract

This paper describes a system which uses graphics devices as tools to interface to complex databases. The motivation for this work comes from experience in using several high level database query languages. The difficulties associated with using query languages for large, complex databases such as some of the statistical databases (e.g., Census data, energy data) are examined. We will describe a system containing features such as subject directories, help messages, zooming facilities to the relevant part of the database schema, partial query formulation with intermediate results, aggregation facility, and graphics data display.

The system offers a graphics interface to the user. The database schema is displayed as a network of entity and relationship types. Queries can be expressed as traversal paths on this network. Partial queries (called "local queries") can be formulated and represented graphically and database retrieval results of any local query are available at any time. Local queries can be linked together to form larger queries and provide the basis for building queries in a piecemeal fashion. Parts of the schema can be selectively made visible or invisible and provide the basis of representing multiple levels of details of the schema. Flexible and powerful facilities for aggregation and graphics display of data are offered in an integrated fashion.

1. Introduction and Motivation

The main motivation of our work comes from experiences in using query languages from commercial Database Management Systems (DBMSs) and Statistical Packages. Even people with a computer science background (ourselves included) often have difficulty using the so-called "high level user friendly languages." Non-expert users may not have the patience, ability, or desire to learn and use these languages correctly. By non-expert users (as opposed to casual users), we mean non-computer science professionals such as social analysts, statisticians or accountants who have to deal with data regularly. The problem becomes much worse in an environment with very large databases that have very large and complex database definitions (schemas). Large statistical databases such as the Census database and energy database are examples of such an environment.

We believe that the following factors are the major reasons for the difficulty in using and understanding query languages.

The user has to remember too many things.

The names of the record types and attributes have to be remembered before the user can express a query. Lower level details such as the format and units of the attributes (e.g. $10,000, $10K, 10 or 10,000?) are only found by exploring the data, or looking the attribute definitions up in a dictionary (sometimes as parts of manuals). On the semantics side, the user has to determine the "meaning" of acronyms used to represent elements (record types and their attributes). All these problems are magnified when the database has a very complex schema (hundreds of record types, thousands of attributes).

Semantically poor data models.

Most high level query languages are based on mathematical concepts such as Predicate Calculus or Algebra or set theory. This sometimes leads to languages with a solid foundation; often, however, users don't relate to mathematical concepts such as range variables, join clauses, projections or complex aggregate functions. Often the problem comes from inadequate semantics in the underlying data model. An example is the join clause in the relational data model [1]. Join clauses are typically used for establishing relationships that exist implicitly between relations. The burden is put on the user to augment the semantics of the implicit relationships between relations and make them explicit. It is the responsibility of the user to restate these relationships *every* time the relations are referred to. We believe a more explicit model should be used to support the non-expert user interface to complex data.

No feedback during the query process.

The chances of formulating a complex query correctly on the first try (or even the first few trials) are slim. When using current languages, there is the fear and doubt as to whether the query is complete or whether some conditions are missing. Users could benefit from a facility that allows them to build a query in a piecemeal fashion with feedback

---

This work is supported by the Applied Mathematical Sciences Research Program of the Office of Energy Research, Department of Energy under Contract DE-AC03-76SF00098.

of partial results available to them at *any* time. Also, users of large statistical databases (statistical analysts) often do not have a set of fixed queries in mind. Rather their interaction with the databases is exploratory in nature and highly dependent on the intermediate results. Again a facility with feedback of partial results should be helpful in this kind of environment. To simulate this facility with current query systems requires stating separate queries. Query systems should be designed to encourage the experimental, exploratory nature of formulating queries in a piecemeal fashion.

Lack of levels of detail in schemas.

Users can be overwhelmed by a complex schema with thousands of elements in it. Most query systems have only two levels of detail at the schema levels: record (relation, set) level and attribute level. Even when the second level is suppressed, the user still has to locate the relevant record types. This is not easy if the schema has hundreds of record types in it. At the attribute level, there are applications where record types with up to several hundred attributes exist, each with its own code conventions. This is especially true for complex databases such as the Census database. It is not an easy matter to understand and select the relevant attributes for the query among these hundreds of attributes. What is needed is a mechanism to present the schema with varying amounts of detail controlled by the user.

Lack of meta-data browsing facility.

Complex databases often have a large number of data sets that deal with many diverse subject matters. For example, the Energy Information Administration (EIA) of the Department of Energy currently are maintaining 230 data sets of all aspects of energy such as sources, production, shipment, and consumptions. It is not an easy matter to even know what and where data is being kept. What is needed is a mechanism that can guide and encourage the user to explore and browse the meta-data to obtain a general view of the database and select subject matters that are of interest.

Limited and difficult-to-use aggregation facility.

It is difficult to express aggregation in most query languages. The syntax of embedding aggregate functions in a query is usually complex and the semantics is often confusing to the user [2]. Yet, aggregation is one of the most needed facilities for processing statistical databases in which the user can generate summaries by tabulating over individual data values. What is needed is a flexible and easy-to-use aggregation facility.

Lack of integrated, easy-to-use data display.

Most query systems provide report writers (some with graphics display facility). A common complaint is that the user has to learn another usually very different system, its syntax and user interface. Another complaint is the lack of integration between the query language and the report writer in that the user has to "get off" from one facility in order to get onto another. The latter is a serious shortcoming for data analysts since timely display of data is an important requirement of forming an impression of the data and detecting hidden relationships between variables. What is needed is an integrated, easy-to-use graphical data display facility which is invokable from any point of the querying stage on any data, not just on the final result of the query.

The approach being taken to address some of these problems is discussed in Section 2 of this paper. Section 3 will present an overview of a system called GUIDE (Graphical User Interface for Database Exploration) being implemented. In Section 4, an example session of using the system is described. Section 5 compares this system with some other systems with similar goals. Finally, Section 6 summarizes the paper, discusses some shortcomings of the system, and presents future plans.

2. Approach

The goal is to put together a set of facilities in an integrated system to try to address some of the above mentioned problems.

First, there are facilities in the system to remove the memory burden from the user. The facilities provide menus, examples, illustrations, and help messages at any stage of query formulation.

Second, a version of the Entity-Relationship (E-R) model [3] is used to represent relationships between entities explicitly. For experimental purposes, the graphics user facility interfaces to a subset of the query language CABLE [4] which is implemented on top of the database system DATATRIEVE [5].

Third, a simple graphics user interface is used for the following reasons:

-- The E-R model schema can be displayed as a network of objects, each object representing an entity or relationship type. This gives the user an overall view of the schema at all times.

-- Queries can be expressed as a traversal along the network of entities. Colors can be used to indicate the paths of the queries, and hence, pictorially indicate the scope and meaning of the queries.

-- Parts of the schema can be selectively made visible or invisible and provides the basis to the implementation of multiple levels of detail to aid in the understanding and use of the schema. (More details are discussed later.)

Fourth, the user can build up the query in a piecemeal fashion and have intermediate results of partial queries available at all times.

Fifth, to handle the problem of meta-data and the large number of the entities and their attributes, two kinds of directories are provided. The first is called a "hierarchical subject directory" (similar to that in [6] which can be used to organize the entities into logical groups hierarchically. The user is guided by the system through this directory to locate the relevant part of the schema for which queries can be expressed. This is also a useful facility to browse and explore the subject matters of the database. Figure 1 gives an example of the subject directory; more details will be presented in the Example Session section (Section 4). The second kind of directories are called "hierarchical attribute directories," and are used to organize attributes of entities (or relationships) into groups similar to the subject directory. Each entity or relationship type has an attribute directory. An example of an attribute directory is presented in Figure 2. Both kinds of directories are implemented as menus.

Sixth, a facility is made available to "rank" objects according to their "relevancy" to a particular group of users. The entity and relationship types are ordered and classified into groups according to the users' interests and the frequency of access in queries. Different groups of users may have different classifications. The first group of objects (with rank 1) is considered the most important objects or focal points of the schema. This is followed by the second group (with rank 2) of objects which together with objects from the first group gives the next, more detailed, level of the schema. Currently, up to 5 groups are allowed in the system. In the example presented in Section 4, the entity types PERSON, GEO-LOCATION, OCCUPATION and INDUSTRY were assigned with ranking 1, FAMILY, HOUSEHOLD are given ranking 2, and so on by a labor force analyst according to the experience in using the database. "Focus" can be specified in the schema so that the selected object will be placed in the center of the screen. Also, there are commands to move the picture around the screen, to zoom-in and zoom-out on the selected part of the picture. The idea is to present the right level of detail and the right part of the schema.

Seventh, an interactive aggregation facility is available in which the user is allowed to select those attributes from any entities as "parameter" attributes of the aggregation (called *category attributes* [7]). Also, the user can select a set of "measured" attributes from any entities (called *summary attributes*), and for each such attribute, a simple descriptive statistics function such as sum, average, etc. to apply.

Eighth, a graphical data display facility is made available for the display of the aggregation result. Data display formats such as tabular, pie, bar, and plot are implemented. The important property is that the interface is completely menu-driven, and integrated with the rest of the GUIDE system in that it can be invoked from any point in GUIDE without having to get off GUIDE itself.

All facilities are offered to the user through graphics menus.

3. Overview of System

There are five stages of query formulation in GUIDE: schema definition, schema exploration, query expression, aggregation, and graphics output display.

Data definition stage.

The Data Base Administrator (DBA) provides information about the schema during the definition stage. In addition to information on entities, relationships, and their attributes, there are examples and explanations of these objects. The graphical layout of the schema is fed into the system in this stage. Facilities are provided to the DBA to do the following:
-- specify a graphical layout of the schema;
-- build a hierarchical subject directory for the schema objects;
-- build a hierarchical attribute directory for each entity and relationship type;
-- specify the "importance ranking" of entity and relationship in the schema. Every object is given a rank (currently from 1 to 5).

Schema exploration stage.

During this stage, the user can use the hierarchical subject directory to reach the most relevant part of the schema. From there the schema can be graphically examined at several levels of abstraction and only objects above a specified importance ranking are made visible. The user can also graphically edit the schema so that irrelevant objects can be removed from the screen. With the relevant part of the schema selected and displayed at the desired level of detail the user is now ready to express queries.

Query expression stage.

In the query stage, the user can build up a query in a piecemeal fashion. The database retrieval results of a partial query can be shown if so desired. Examples and explanations on any object on the screen can be requested. Graphically, the user will be traversing a network of objects. The query is a path selected by the user and shown in different colors for each partial query. The user is also encouraged to experiment with different conditions on the schema objects by adding or subtracting conditions and the result of these experimentations are available at any time. Piecemeal formulation of a query is an important facility. It is achieved through the formulation of "local queries." The user can concentrate on several parts of the schema being shown on the screen and can formulate a "local query" on each part so that each local query is completely independent of another. The idea is to allow the user to have a focused vision of small parts of the schema so that local results can be obtained and understood without having to compose a complex query covering a large schema space at the same time. The user can then link local queries to form a complex one. This complex query can then be treated as a local query when the user expresses additional local queries. All the local queries can be linked to form yet another more complex query, etc. This process continues until the final query is formulated. The retrieval results of each local query are always available for display. The result of linking several local queries is also displayable at any time.

Aggregation stage.

During this stage, the user can select graphically the category and summary attributes from entities for aggregation. Facilities are available for the user to examine any entity included in a previous query by displaying the available attributes within that entity, to find out the description on the entity and its attributes, and select aggregate functions for the summary attributes.

Graphics data display stage.

In this stage, the aggregation result can be displayed in various forms such as graphs, bar charts, pie charts, etc., with a set of optional graphics enhancements. The CHART package [8] has been interfaced to GUIDE for graphical displays.

4. An Example Session

The database used in this example is from the Current Population Survey (CPS) which is conducted by the Bureau of the Census for the Bureau of Labor Statistics. It contains comprehensive data on individuals, families and households. Data such as employment, occupation, age, race, sex, educational background are available for each person in the household enumerated. Also, data such as income sources, health insurance, pension plan, food stamps, etc. is provided.

An E-R database for a subset of CPS has been designed with the help of a statistician who uses the CPS frequently. Currently it contains 20 entity and relationship types. This session shows some example interactions between a user and GUIDE when a realistic query is being composed.

Photo 1 shows the configuration of the interaction. The terminal on the left is used as a keyboard. An additional terminal (Ramtek) equipped with a joystick is used as a interactive graphics device. The interaction with GUIDE is through the selection of commands from a menu located on the right side of the screen. The selection is done by manipulating the joystick to move the cursor to the desired item and entering a "hit" through the keyboard.

The first command menu is shown in Photo 2. The second and third commands in the menu are the entry commands to the "explore schema" and "query expression" stages, and will be illustrated later. The first command "use subject directory" can be used to locate a particular object in a schema in a hierarchical, general-to-specific fashion. The top level of the CPS subject directory is shown on the left of Photo 2. A larger segment of the hierarchy is displayed in Figure 1. The non-leafs of the tree are subject categories and the leafs point to specific nodes in the schema. A hit of a non-leaf in the hierarchy will expand to the next level below the hit non-leaf. A hit of a leaf will bring the user to the "explore schema" stage with the node that is pointed to by the leaf of the hierarchy in the center of the screen.

Photo 3 shows the result of the user's selection of the path CPS - demographic characteristics - person. The commands on the right are the "explore schema" stage commands and they are briefly explained below:

examine node -- this command has a submenu (not shown) whose commands can be used to display some sample instances of any user selected entity or relationship type, to describe its meaning and display the attributes using its hierarchical attribute directory (example in figure 2 and more on this later).

set schema detail level L -- the effect of this command is that only nodes with ranking $\leq$ L will appear on the screen and only those links that connect these nodes will appear. In the CPS example, users rank the entity and relationship types according to the users interests and frequency of access to these objects.

focus -- the command allows the user to select a node in the schema as a "focus" and the selected node is moved to the center of the screen.

set radius R -- the effect is to use the current focus as a center point and include those nodes and links that are R links away. Only nodes with ranking $\leq$ L are shown, where L is the current detail level.

hide node or link -- this command allows the user to erase some nodes or links from the screen to unclutter the picture before stating the query.

zoom -- this command allows the user to scale the picture according to a selected scaling factor to achieve the effect of zoom-in and zoom-out. A larger or smaller number of objects may appear on the screen, depending on the zoom scale.

move schema -- this command can be used to move the physical layout of the schema on the screen by moving it in the direction of up, down, left or right.

By using a combination of these commands, the user can display the most relevant part of the schema, and can concentrate on the query expression with only the relevant elements on the screen.

Photo 4 shows the schema with detail level set at 1, radius 5-beyond and entity type PERSON as the focus. This combination presents the focal points of the schema and they constitute the "core" of the database. Note that the selected focus will stay on the screen regardless of its detail level. A dash line indicates the existence of an indirect path between two objects through the connection of some intermediate objects (entities or relationships). A query can now be expressed and the system will augment it with the necessary objects to complete the query.

Photo 5 shows the schema with detail level set at 3 and radius 4. Entity types FAMILY, HOUSEHOLD and relationship types EMPLOYMENT, UNEMPLOYMENT and LAST-YEAR EMPLOYMENT are now displayed. Note that line width is used to represent cardinality of relationships. For example, the relationship IS IN between entities PERSON and FAMILY is a many-to-1 mapping as indicated by the thick line between PERSON and IS IN and a thin line between IS IN and FAMILY. Finally, Photo 6 shows the schema after a series of the commands above. The user is now ready to express the query.

The commands in query stage are displayed on Photo 7. The first command "examine node" is the same command as in the "explore schema" stage.

The command "include node in query" displays a submenu (listed below) after the user points to a node on the screen.

use attribute hierarchy
display attribute value codes
show restriction samples
enter attribute restriction
select attributes for output

Briefly, this submenu deals with the details of the attribute level of entity or relationship types. It allows the user to examine the attributes in a hierarchical, general-to-specific fashion similar to the subject directory; to display the code convention of the attribute value (e.g., age-group 18 to 25 has code value 5); to display sample of attribute restriction conditions; to enter restriction on attribute as part of the query (examples to follow); and to select the output attributes for qualified entity or relationship instances.

Figure 2 shows the attribute hierarchy for the relationship type EMPLOYMENT. The leafs are specific attributes and at this level attribute restrictions can be entered. Note that the user can return from any submenu by hitting the END command. Note also that the code convention of attributes are exactly as contained in the CPS database. However, with the use of menus of English descriptions of code values, the user can express restrictions in a query without having to remember any code convention at all.

After using the command "include node in query" one or more times, the user can use one of the three commands shown below to obtain the result of the database retrieval.

show local result - the purpose of this command is to allow the user to compose "local queries" on several parts of the schema independently from each other. All the "included" nodes along with their restrictions on attributes will constitute a local query when the command is used. The next series of node inclusions will constitute a new independent local query. Multiple local queries can be composed in this way and they can be thought of as building blocks of larger queries. Each local query is painted a different color by GUIDE to communicate the effect and scope of the local query graphically to the user. The intermediate results of these local queries can provide the user with data that either confirms or rejects the intuition or "feeling" of the restriction conditions provided. The confirmation or rejection of the user's intuition is possible because the schema space and the number of restrictions are small in a local query.

show result so far - local queries can be linked together to form a larger query and its retrieval result can be displayed using this command. This larger query will be treated as another local query when a new local query is being composed. The combination of this command and the command "show local result" provide the capability to compose queries in a piecemeal fashion. In the case where there are multiple ways of linking local queries, the system will blink all possible paths and prompts for additional nodes to make a distinct path.

show final result - is the same as "show result so far" except the system expects a entirely new query to be composed after the result is displayed.

Photo 7 shows a local query expressed by the CPS user by first including the entities PERSON and DEPENDENTS and then hitting the "local result" command. The restrictions that the person is female, divorced, with at least 3 children under 18 were also expressed using the "include node in query" command's submenu. Notice that the shading of the nodes within a local query is different from the rest and the purpose is to inform and graphically remind the user of the scope and effect of the local query. Actually, colors are used instead of shading in GUIDE. Notice also that a linear query (expressed in the simple syntax of CABLE [4]) is being built up at the bottom of the screen as the user includes nodes. This serves as a form of documentation for the restrictions entered by the user. The attribute names and their value codes in the restrictions are looked up by using the commands in the "include node in query" submenu also.

The database retrieval results of each local query can be displayed using the following submenu which is available for all three "show result" commands.

    display # of instances
    show result sample
        5
        10
        15
        20
        more
    show all instances on printer
    use aggregation facility
    use graphics data display

These commands are used to control the amount of output from the database retrieval. The command "more" provides the capability of "stepping through" the result.

The aggregation facility is invoked by the command "use aggregation facility". Photo 8 shows the first submenu of this facility. The user can select two sets of attributes, category and summary, and for each summary attribute, a simple aggregate function. The command "category attribute" has a submenu which allows the user to examine any node in the schema and open up the node in order to select the set of attributes as category attributes. Photo 9 is an example of a list of attributes of the entity PERSON which the user has picked. Photo 10 shows the kind of summary functions available for the summary attributes.

The command "use graphics data display" has three graphics submenus corresponding to the following types of options:

Table manipulation -- which includes facilities such as windowing, masking, sequencing, switching, ranking, formating, and grouping of columns;
Graphic form selection -- which offers the user a choice of plotting, bar chart, pie chart, and line;
Display enhancement -- which contains features such as labeling, titling, shading, etc. of the generated graphics.

Photos 11 and 12 are examples of the table report format and the bar chart display format with some limited display enhancement features. They present the result of using the aggregation facility over the first local query. In this example, the user expresses the aggregation "from the collection of persons retrieved from the first local query, tabulate the total number of people, average number of dependents, the eldest child's age, by the age, race, and sex". The expression of aggregation is done first by selecting as category attributes age, race, and sex the entity type PERSON, then selecting as summary attributes are the number of dependents and the age of dependents, both are from the entity type DEPENDENT. The summary functions are the count (of people of each subgroup of the category attributes), average, and maximum. This aggregation represents the result of "total number of people, average number of dependents, maximum age of children by the age, race, and sex of persons".

Photo 13 shows a second local query on INCOME, FAMILY and GEO-LOCATION with the restrictions that the family is poor (below poverty level) and located in the San Francisco area. Notice that the nodes in this local query are shaded in a different pattern. Also a new linear query is composed for this local query on the bottom of the screen.

Photo 14 shows the result of aggregation over the second local query. The aggregation is "from the set of families retrieved, show the sizes of families in descending order by household income, number of household labor force members, and city district". The photo shows a bar chart of descending sizes of families.

Photo 15 shows the effect the command "show result so far". The two local queries are linked together to form a larger query for: "Divorced mothers with at least 3 under 18 children in a poor family located in the San Francisco area." The formal query is expressed again at the bottom of the screen.

Photo 16 shows another local query involving OCCUPATION, INDUSTRY and EMPLOYMENT with the conditions that OCCUPATION be classified as "clerical or unskilled" and EMPLOYMENT as "part-time".

Photo 17 shows the effect of the "show final result" command. The final query is put together as indicated by the shading of the included nodes. Also the whole formal query is displayed along with all the conditions expressed for: "Part-time, clerical or unskilled, divorced mother with at least 3 children under 18 in poor family located in the San Francisco area." Photo 18 shows a pie chart display of such people involving category attributes age, race, and district, from entity types PERSON and GEO-LOCATION.

5. Comparisons with Some Related Systems

The following is a comparison of our system with several graphics-based systems that have similar goals. These include the spatial data management system (SDMS) [9], the CUPID system [10], the E-R user interface [11], and Query by Example (QBE) [12]. In the interest of brevity, it is assumed that the user has some knowledge of these systems.

First, all systems except the E-R user interface are based on the relational model, hence relationships must be specified by the user as discussed. SDMS concentrates on single relations, it is insufficient for our purposes.

Second, when using CUPID, the user must remember relation names, units and formats of attributes, etc.

Third, all systems except SDMS lack the power of presenting information in several levels of detail.

Fourth, the E-R user interface and SDMS emphasize interaction with the user at the instance level (i.e., specific instances of entities and attributes values). Expressing generic queries is not dealt with.

Fifth, all systems lack the facility of allowing the user to compose queries in a piece-by-piece fashion with intermediate feedback at any time.

Sixth, none of the systems encourage the user to traverse and explore meta-data of the database. Cattell's system provides exploration at the instance level, but without the overall picture of the schema available, the user can lose sight of relative position.

Seventh, both CUPID and QBE provide aggregation facility. But the difficulty of expressing aggregate functions in general query languages still exists. The requirement of easy-to-use, simple aggregation and tabulation is not met.

Lastly, the facility of integrated graphics data display is lacking in most systems. This, again, is a serious shortcoming for the data analyst of statistical databases.

6. Conclusions

A preliminary version of GUIDE was reported in [13]. The entire system has been designed and coded in the programming language C and running on the VAX under VMS. This includes the data definition loader, the graphics editor, and the menu builder. The data definition loader is described in Section 3. The graphics editor is built on top of GWCORE, a George Washington University implementation of the SIGGRAPH GSPC '79 standard proposal. The menu builder consists of two parts, a menu loader for menu definition and a menu manipulator for invocation and expansion of menus.

We are currently planning to put together a stable version of GUIDE for real users to try out. Results of feedback from these users will be reported in a forthcoming paper along with a description of an extended version of GUIDE.

Implementation of an interface to SPSS has been started. The idea is to augment GUIDE with powerful, integrated facility of data analysis tools. Implementation of update commands in GUIDE is also being examined.

Longer term plans include the exploration of a more powerful data model (such as an extended E-R model with aggregation and generalization hierarchies) to better model the semantics of databases. More specific constructs useful in modeling statistical databases will also be researched and incorporated into the system. Finally, we are examining a more powerful query language as our underlying query language in order to provide the foundation for more complicated query expressions such as cyclic queries (a cyclic query is a query whose path on the schema network forms a cycle, typically involving range variables in queries such as "employees who make more than their managers") and queries with dynamic link creation for temporary relationship establishment (users can establish dynamic paths among entities or relationships (joins)).

Acknowledgements

We would like to acknowledge the valuable insights of Arie Shoshani, who has contributed some ideas in our system. The contribution from Linda Wong and Ivy Kuo are very much appreciated.

References

[1] Codd, E.F., "A Relational Model for Large Shared Data Banks", CACM 13, 6, 1970.

[2] Johnson, R. "Modelling Summary Data," 1981 SIGMOD Conference.

[3] Chen, P.P.S. "The Entity Relationship Model: Toward a Unifying View of Data," ACM TODS, 1,1.

[4] Shoshani, Arie. "CABLE: A Language Based on the E-R Model," 1979 E-R Conference.

[5] DATATRIEVE Users Guide, DEC.

[6] Chan, P., Arie Shoshani. "Subject: A Directory driven System for Organizing and Accessing Large Statistical Databases," VLDB-81, Cannes, France.

[7] Shoshani, A., "Statistical Databases: Characteristics, Problems, and Some Solutions", this proceedings.

[8] Benson, W.H., B. Kitous. "Interactive Analysis and Display of Tabular Data," *Proceedings of the Fourth Annual SIGGRAPH-ACM Conference on Computer Graphics and Interactive Techniques*, San Jose, California, July, 1977.

[9] Herot, C.F. "Spatial Management of Data," TODS 5,4, 1980.

[10] McDonald, N., M. Stonebraker. "CUPID - The Friendly Query Language," Memo ERL-M487, Berkeley: Univ. of California, October, 1974.

[11] Cattell, R.G.G. "An Entity-based Database User Interface," SIGMOD 1980.

[12] Zloof, M.M. "Query by Example," AFIPS, NCC '75.

[13] Wong, H.K.T., I. Kuo. "GUIDE: Graphical User Interface for Database Exploration", VLDB-82, Mexico City, Sept., 1982.

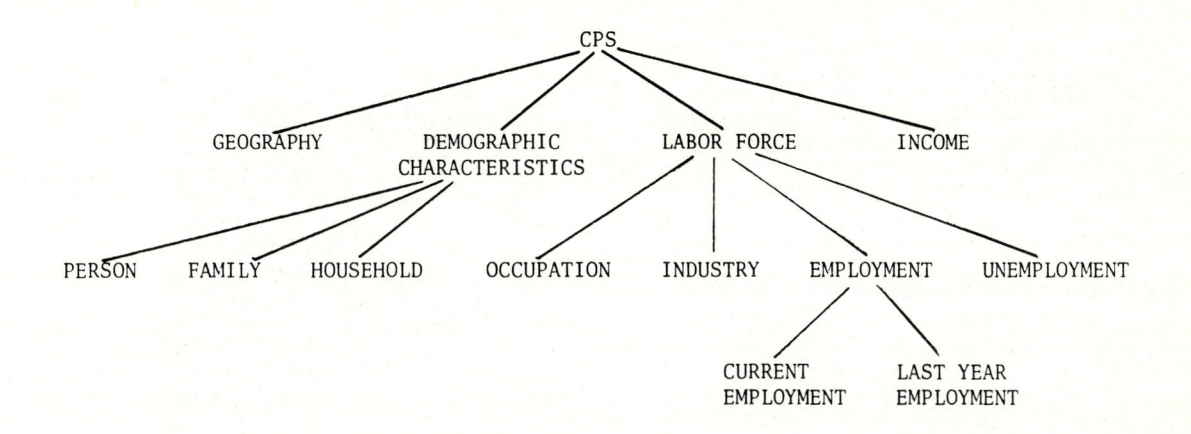

FIGURE 1 - SUBJECT HIERARCHY

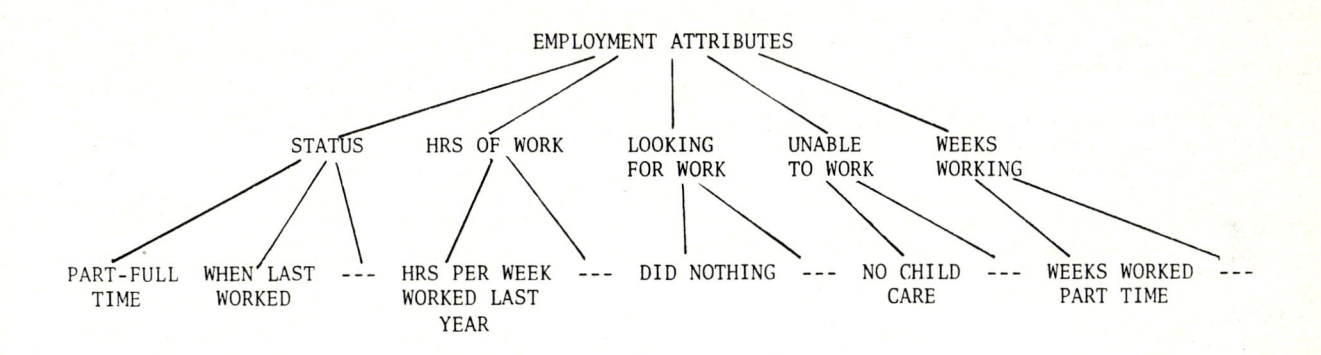

FIGURE 2 - ATTRIBUTE HIERARCHY FOR EMPLOYMENT

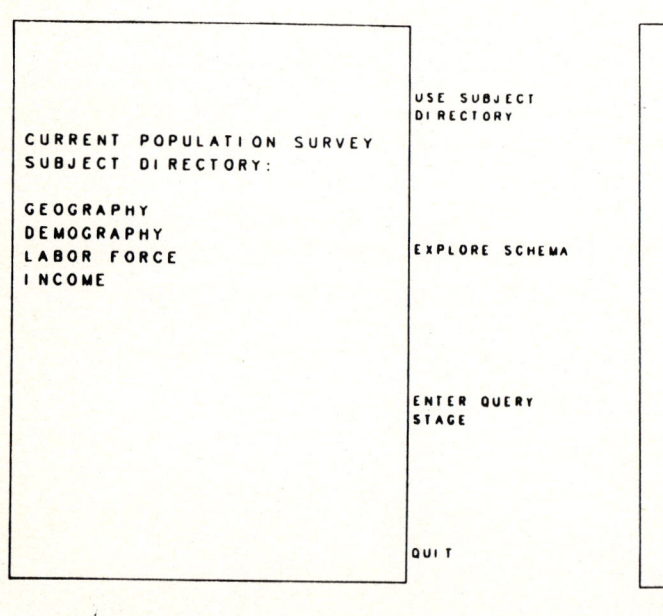

```
CURRENT POPULATION SURVEY          USE SUBJECT
SUBJECT DIRECTORY:                 DIRECTORY

GEOGRAPHY
DEMOGRAPHY
LABOR FORCE                        EXPLORE SCHEMA
INCOME




                                   ENTER QUERY
                                   STAGE




                                   QUIT
```

PHOTO 2

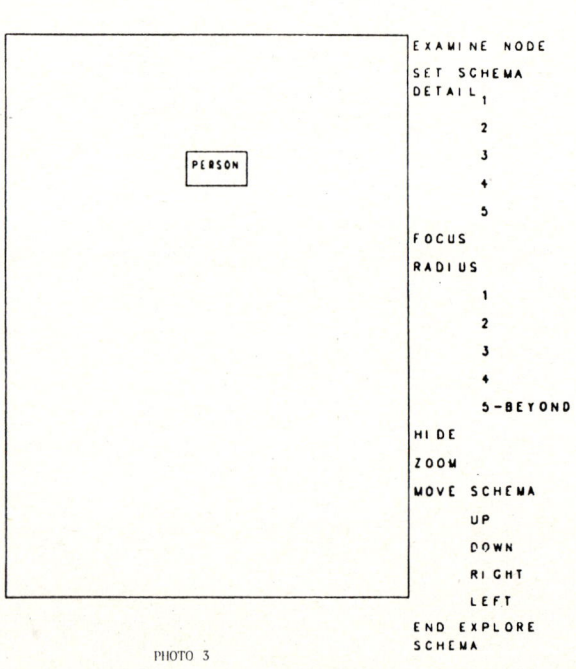

```
                                                    EXAMINE NODE
                                                    SET SCHEMA
                                                    DETAIL
                                                          1
                                                          2
              ┌────────┐                                  3
              │ PERSON │                                  4
              └────────┘                                  5
                                                    FOCUS
                                                    RADIUS
                                                          1
                                                          2
                                                          3
                                                          4
                                                          5-BEYOND
                                                    HIDE
                                                    ZOOM
                                                    MOVE SCHEMA
                                                          UP
                                                          DOWN
                                                          RIGHT
                                                          LEFT
                                                    END EXPLORE
                                                    SCHEMA
```

PHOTO 3

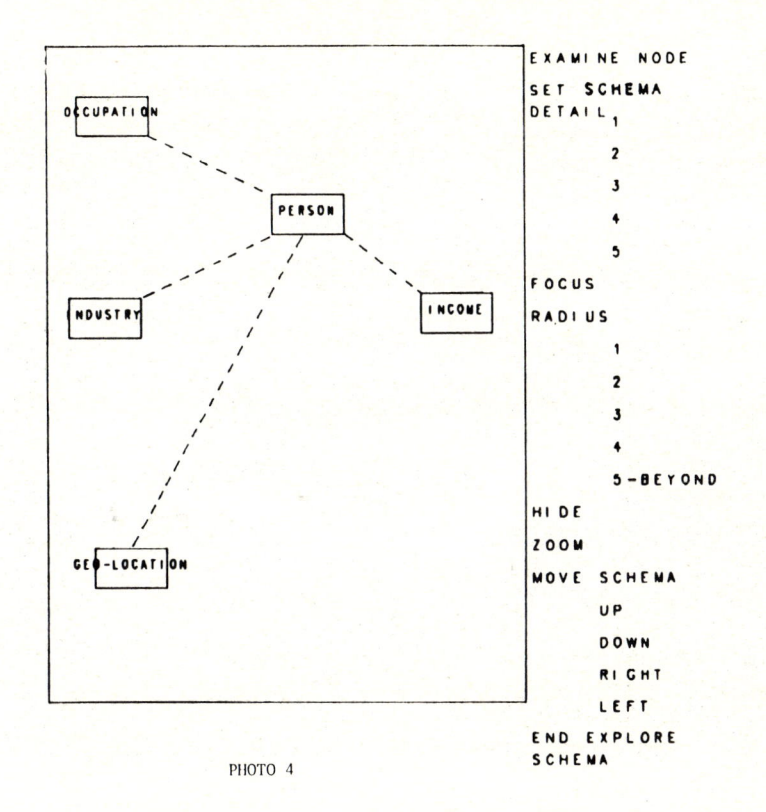

Graphical Query System for Complex Statistical Databases

43

```
EXAMINE NODE
SET SCHEMA
DETAIL 1
       2
       3
       4
       5
FOCUS
RADIUS
       1
       2
       3
       4
       5-BEYOND
HIDE
ZOOM
MOVE SCHEMA
     UP
     DOWN
     RIGHT
     LEFT
END EXPLORE
SCHEMA
```

Diagram nodes: OCCUPATION, PERSON, INDUSTRY, INCOME, GEO-LOCATION (connected by dashed lines to PERSON).

PHOTO 4

```
EXAMINE NODE
SET SCHEMA
DETAIL 1
       2
       3
       4
       5
FOCUS
RADIUS
       1
       2
       3
       4
       5-BEYOND
HIDE
ZOOM
MOVE SCHEMA
     UP
     DOWN
     RIGHT
     LEFT
END EXPLORE
SCHEMA
```

Diagram nodes: OCCUPATION, UNEMPL, EMPL, PERSON, HAS, LAST EMPL, IS IN, INCOME, INDUSTRY, FAMILY, HAS, IS IN, HOUSEHOLD.

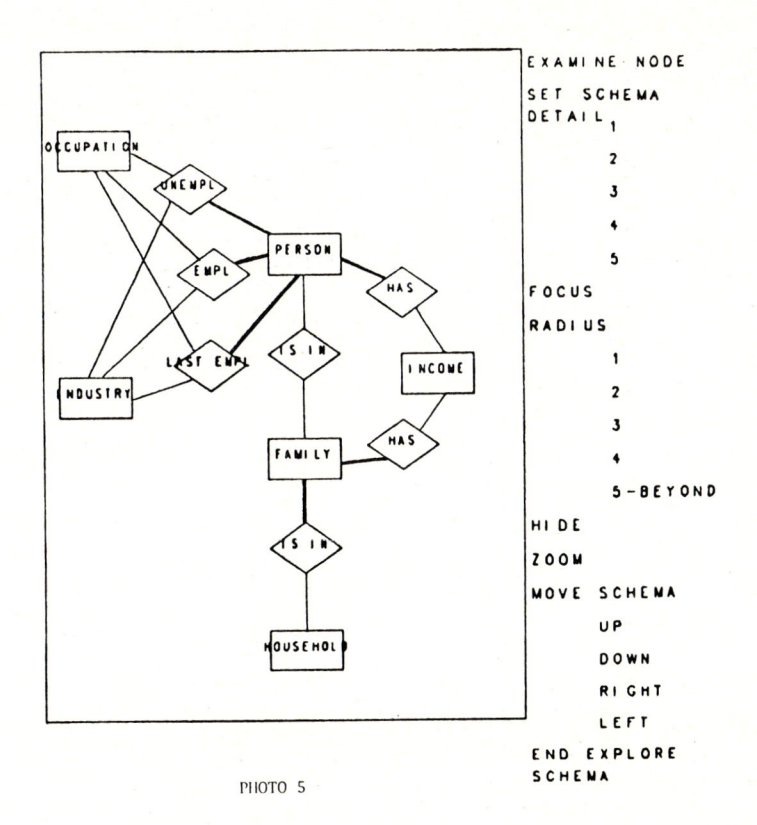

PHOTO 5

PHOTO 6

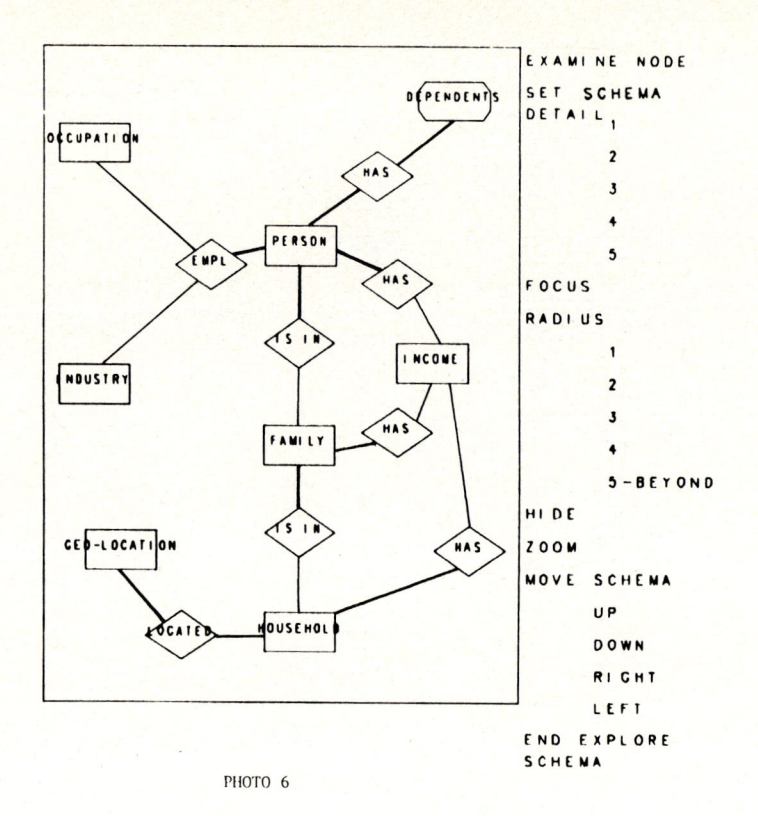

PHOTO 7

PERSON: SEX=2&MARSTAT=6&NODEPS>2
DEPENDENTS: AGE<18

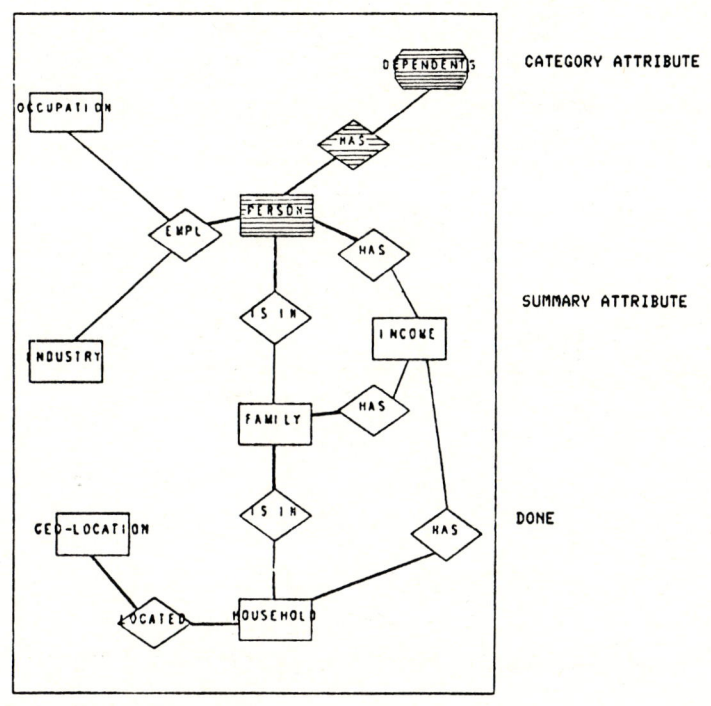

Graphical Query System for Complex Statistical Databases

HIT ANY ITEM ON THE MENU.

PHOTO 8

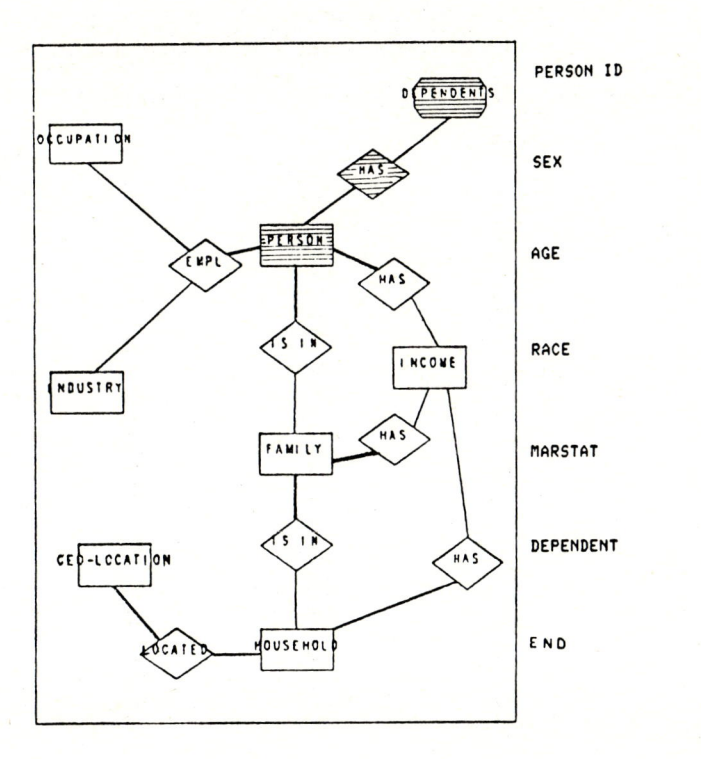

PHOTO 9

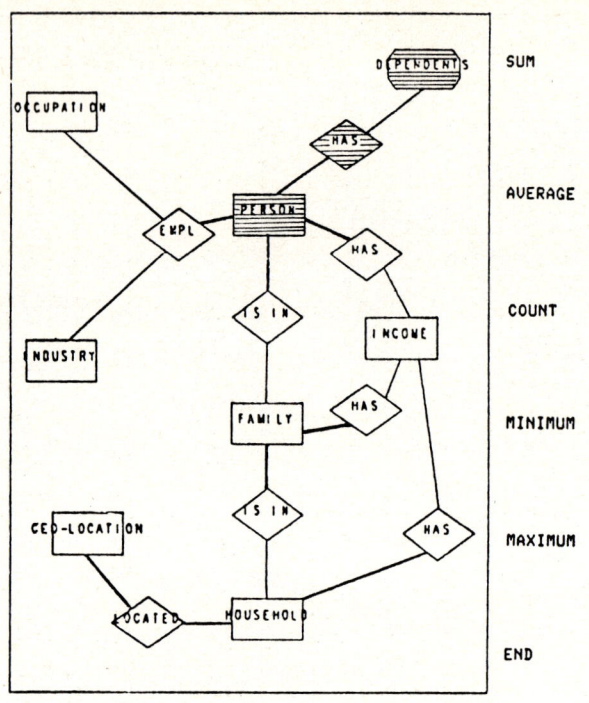

SELECT AGGREGATE FUNCTIONS FOR SUMMARY ATTRIBUTES.

PHOTO 10

THIS IS TABULATION RESULT FOR FIRST LOCAL QUERY.
CATEGORY ATTRIBUTES: PERSON'S AGE, RACE AND DEPENDENT SEX.
SUMMARY ATTRIBUTES : COUNT, AVERAGE OF DEPENDENT NUMBER, DEPENDENT AGE AND MAXIMUM DEPENDENT AGE.

	count	average(dependent)	average(dage)	max(dage)
age1,white,m	11.0	3.5	6.0	15.0
age1,white,f	12.0	3.5	4.0	16.0
age1,black,m	48.0	4.6	3.5	10.0
age1,black,f	31.0	3.2	5.5	17.0
age1,indian,m	5.0	3.0	3.0	8.0
age1,indian,f	.0	.0	.0	.0
age1,hispanic,m	22.0	4.0	7.0	15.0
age1,hispanic,f	5.0	5.0	1.0	12.0
age1,asian,m	10.0	4.0	2.0	7.0
age1,asian,f	41.0	5.0	3.0	10.0
age2,white,m	11.0	3.5	6.0	15.0
age2,white,f	7.0	4.0	14.0	16.0
age2,black,m	11.0	3.0	4.0	8.0
age2,black,f	9.0	3.5	4.0	7.0
age2,indian,m	18.0	5.0	7.5	18.0
age2,indian,f	20.0	4.5	2.8	7.0
age2,hispanic,m	35.0	4.5	6.2	14.0
age2,hispanic,f	52.0	6.0	4.0	8.0
age2,asian,m	19.0	5.0	8.0	15.0
age2,asian,f	9.0	7.0	13.0	18.0
age3,white,m	38.0	4.5	8.0	13.0
age3,white,f	7.0	4.0	14.0	16.0
age3,black,m	11.0	3.0	4.0	8.0
age3,black,f	9.0	3.5	4.0	7.0
age3,indian,m	18.0	5.0	7.5	18.0
age3,indian,f	47.0	4.5	3.5	10.0
age3,hispanic,m	12.0	4.5	5.0	9.0
age3,hispanic,f	26.0	5.0	5.0	9.0
age3,asian,m	1.0	5.0	8.0	15.0
age3,asian,f	6.0	6.0	5.0	8.0
age4,white,m	4.0	7.0	10.0	18.0
age4,white,f	17.0	4.0	10.0	15.0
age4,black,m	18.0	3.0	3.0	5.0
age4,black,f	2.0	3.0	5.0	5.0
age4,indian,m	9.0	6.0	2.0	5.0
age4,indian,f	8.0	4.0	1.0	3.0
age4,hispanic,m	15.0	3.5	2.0	5.0
age4,hispanic,f	12.0	5.0	5.0	8.0
age4,asian,m	5.0	4.0	4.0	7.0
age4,asian,f	12.0	5.0	7.0	10.0

PHOTO 11

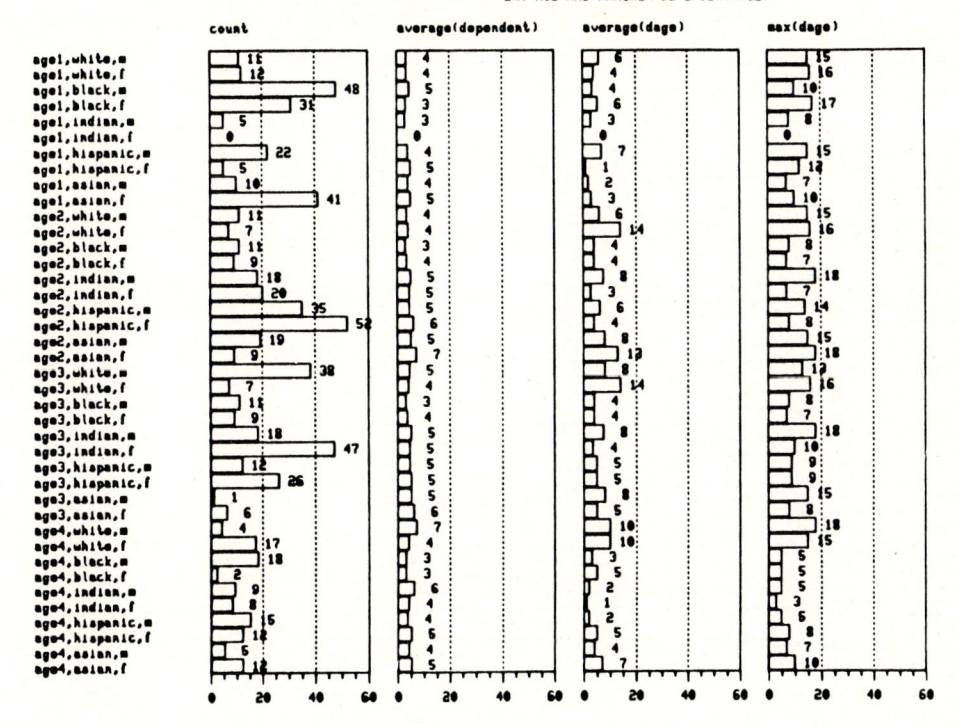

Graphical Query System for Complex Statistical Databases

PHOTO 12

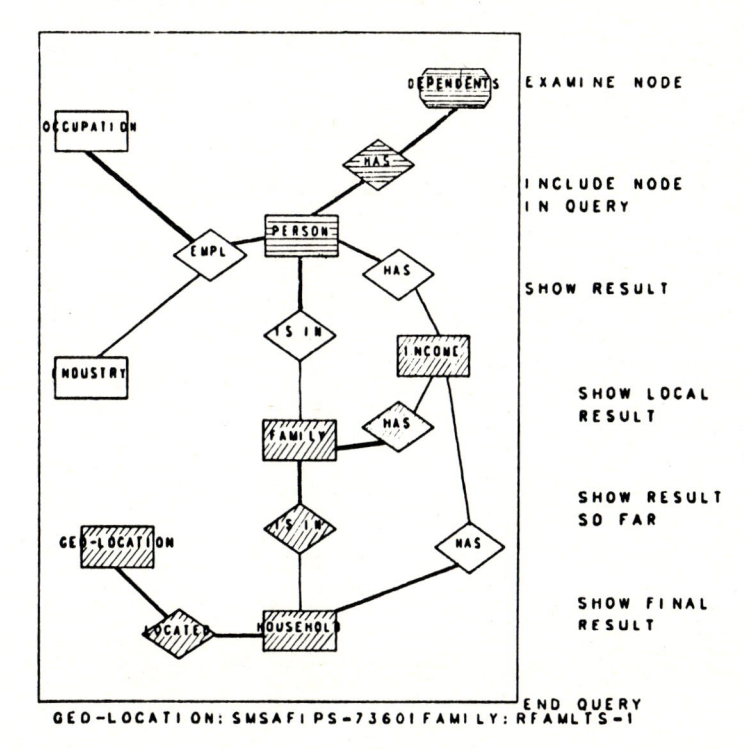

PHOTO 13

PHOTO 14

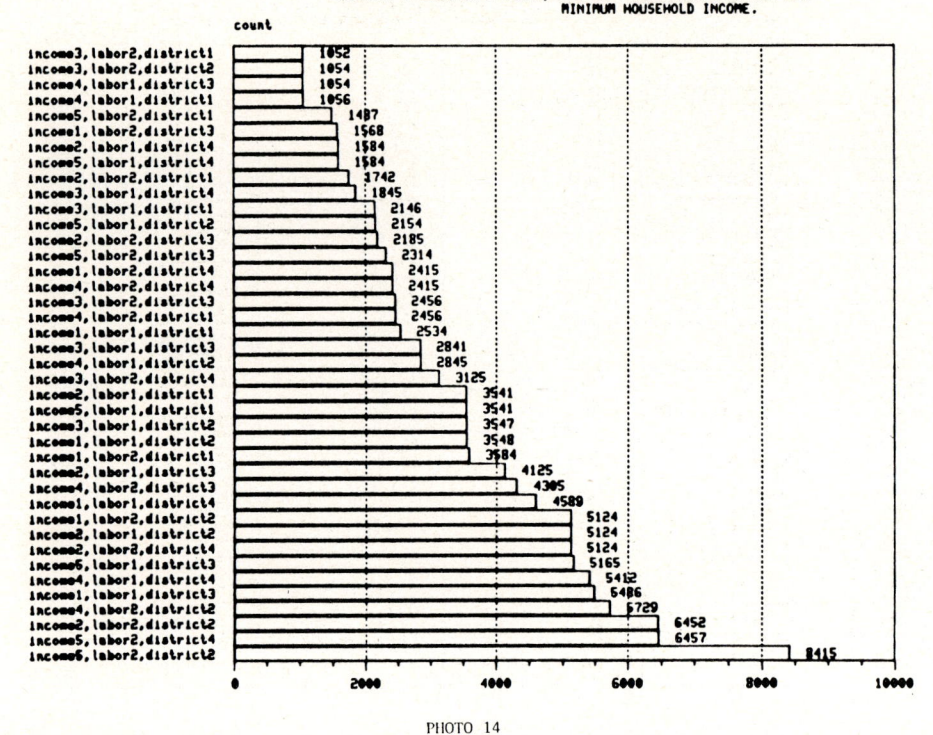

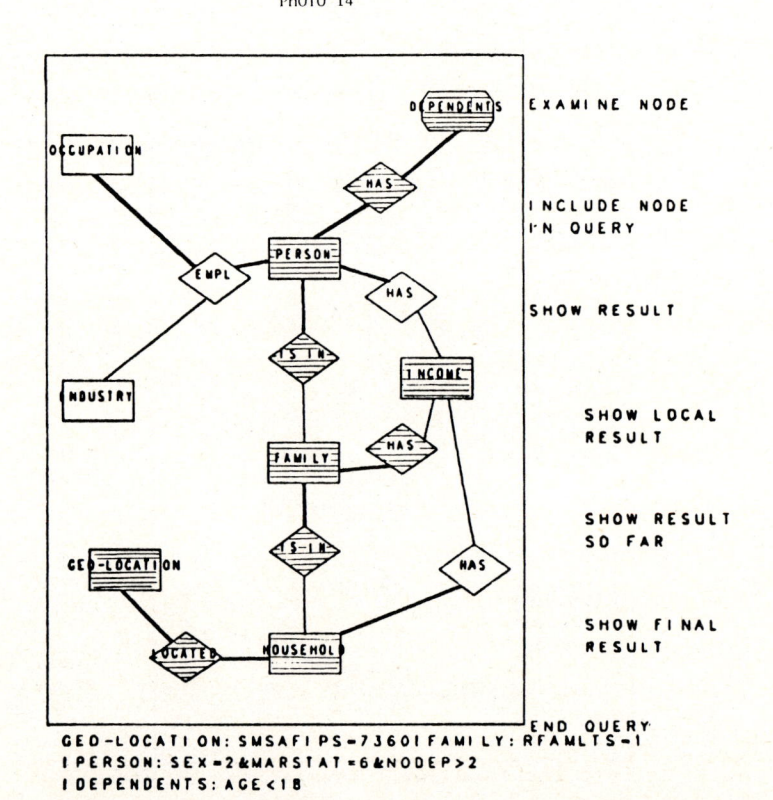

PHOTO 15

Graphical Query System for Complex Statistical Databases

WINDOW FIRST 25 ROWS OF
TABULATION RESULT FOR FINAL RESULT OF 3 LOCAL QUERIES.
AFTER RANKING COLUMN 1 WITH ASCENDING ORDER AND
MASKING COLUMNS 2 TO 4 AND
MASKING THOSE ROWS WITH COUNT=0.
CATEGORY ATTRIBUTES: PERSON'S AGE, RACE, AN AND LIVING DISTRICT.
SUMMARY ATTRIBUTES : COUNT, AVERAGE DEPENDENT NUMBER, MAXIMUM
HOUSEHOLD INCOME AND MINIMUM HOUSEHOLD
INCOME.

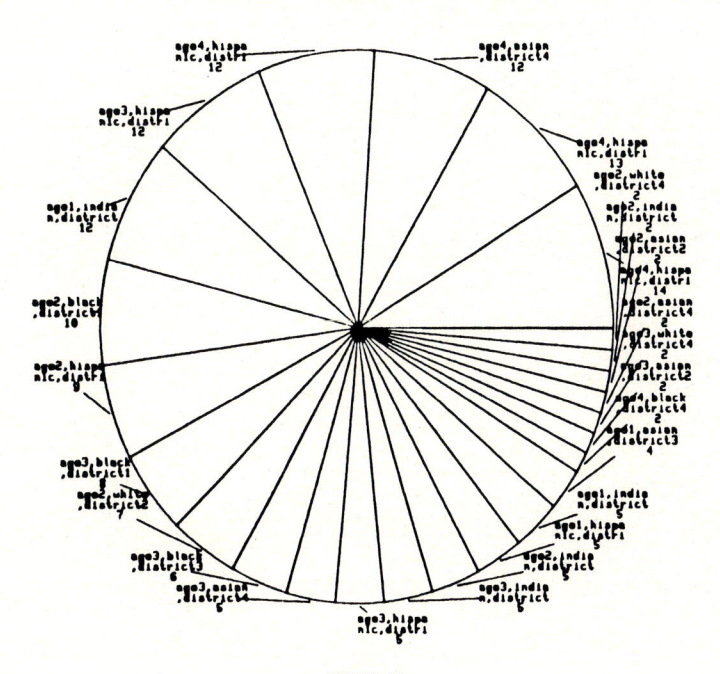

PHOTO 18

Simulation Output Analysis/Graphics

Organizer: P. A. W. Lewis

Invited Presentations:

    SIMTBED: Overview, P. A. W. Lewis, D. G. Linnebur, E. J. Orav, L. Uribe, and H. W. Drueg

    Simulation and Graphics--From Single Purpose Applications to Generic Interfaces, Joseph Polito

SIMTBED: Overview

P. A. W. Lewis, D. G. Linnebur, E. J. Orav, L. Uribe, H. W. Drueg

Naval Postgraduate School
Monterey, California

A graphical test bed in which the results of a simulation experiment can be reported and analyzed is described. The test bed is based on the regression adjusted graphics and estimation methodology developed by Heidelberger and Lewis for regenerative simulation. From the graphics and associated numerics, the experimenter can summarize and see simultaneously relative properties, such as bias, normality and standard deviation, of several estimators of a characteristic of a population for up to 8 sample sizes. The evolution of these properties with sample size is also displayed. The graphics is supported on a line printer to make it and the program portable. The technique is illustrated by examples concerning the effects of changes in data distribution on the behavior of the lag one serial correlation coefficient and the estimation of kurtosis for highly skewed Gamma random variables.

SIMTBED, short for Simulation Test Bed, is a graphical display program that can be used via a simulation on a digital computer to (i) explore the distribution of a statistical estimator for a given sample size, (ii) to compare the properties of that distribution when the estimator is calculated for various sample sizes, and (iii) to contrast these properties under different estimation conditions.

Features (i) and (ii) are demonstrated in Figure 1, where we look at the distribution of the maximum likelihood estimate of the shape parameter of a Gamma(5) population, $\hat{K}(n)$, for samples of size n = 33, 50, 71, 100, 125, 166, 250 and 500. Each boxplot in the graphics portion represents the distribution of 50,000/n estimates of $\hat{K}(n)$, where each estimate of $\hat{K}(n)$ is calculated based on a sample of n simulated Gamma(5) random variables. Numerical summaries below the graphics further describe the distributions at each sample size n. We note that the distributions have positive bias, skewness and kurtosis, all of which fall off toward 0 as the sample size gets large. SIMTBED also performs regressions in order to quantify the changes in the mean and variance of the distribution of $\hat{K}(n)$ as n grows. These regressions estimate α_0, α_1, α_2 and α_3, and β_0, β_1, β_2 and β_3 in the relationships

$$E(\hat{K}(n)) = \alpha_0 + \alpha_1/n + \alpha_2/n^2 + \alpha_3/n^3 + \ldots$$

and

$$Var(\hat{K}(n)) = \beta_0/n + \beta_1/n^{1.5} + \beta_2/n^2 + \beta_3/n^{2.5} + \ldots,$$

respectively. Such expansions hold for many statistical estimators and we can use α_0 to predict the true asymptotically unbiased mean, α_1 to predict the coefficient of the order n^{-1} bias term, and β_0 to compare asymptotic efficiencies. The results of the regressions are reported in Figure 1 below the summary statistics.

Feature (iii) of SIMTBED, the ability of compare different estimation conditions, means, in the current example, that SIMTBED will generate output analogous to Figure 1 for other competing estimators, for instance the method of moments estimator, jackknifed m.l.e. and the jackknifed moment estimator. In general, those varying estimation conditions are left to the imagination of the user, although the most common situations will include competing estimators and changes in the underlying distribution of the simulated data.

To use the program, it is necessary only to define the optional parameters, supply the simulated random variables, and provide the FORTRAN functions which, when passed the data and sample size, transform the data if necessary and compute the desired statistic. SIMTBED will subdivide and feed the data properly into the functions, produce boxplots and summary statistics, and compute regressions for the mean and variance. Up to three estimators can be used with the option to produce equally scaled graphs, or graphs with outliers counted but not plotted in order to emphasize the body of the boxplot, for all the statistics.

SIMTBED uses the same batch of simulated random variables (e.g. Gamma(5)'s) to explore the properties of all the estimators at various sample sizes. This is done for economy of time and could be important on slow computers; the price paid is that the analytical analysis provided by SIMTBED of its graphical output is performed on correlated samples. The graphics is done through FORTRAN print statements on a standard line printer in order to make the program portable. It is this combination of quickness, portability and ease of use which we hope will inspire users to try more diverse and extensive simulation experiments.

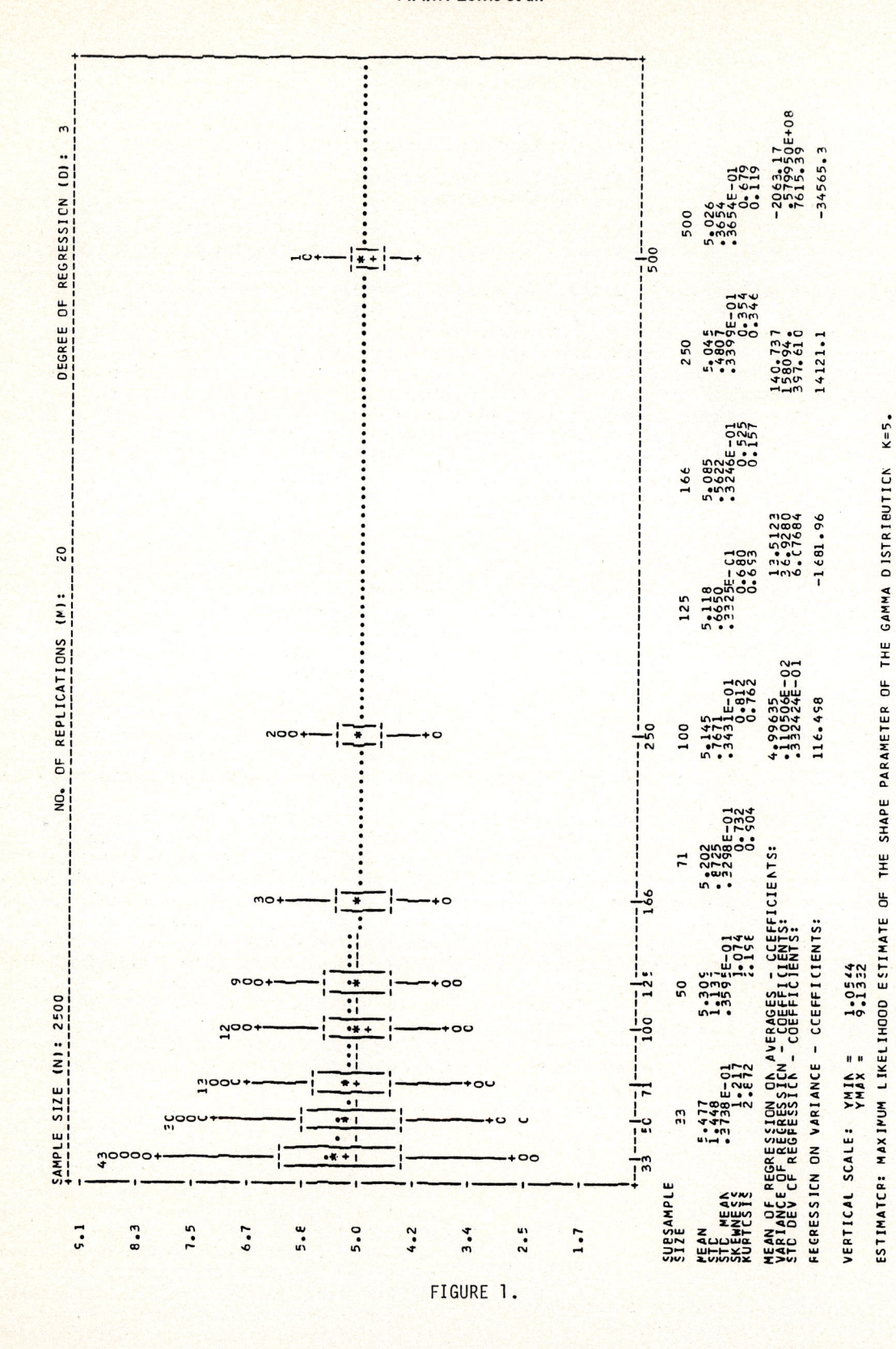

FIGURE 1.

Technical details concerning the SIMTBED software can be found in Linnebur (1982), and application to the analysis of output in a regenerative simulation can be found in Heidelberger and Lewis (1981), and a discussion concerning output interpretation with two extensive applications to the estimation of serial correlation and to the estimation of the shape parameter for a Gamma population can be found in Lewis, et. al. (1983).

References

[1] Heidelberger, P. and Lewis, P.A.W., Regression-adjusted estimates for regenerative simulations, with graphics, Comm. of the A.C.M., 24 (1981) 260-273.

[2] Lewis, P.A.W., Linnebur, D.G., Orav, E.J., Uribe, L. and Drueg, H.W., SIMTBED: A graphical test bed for analyzing and reporting the results of a statistical simulation experiment, Technical Report (1983) Naval Postgraduate School (forthcoming).

[3] Linnebur, D.G., A graphical test bed for analyzing and reporting the results of a simulation experiment, Master's Thesis, Dept. of Ops Research, Naval Postgraduate School (1983).

SIMULATION AND GRAPHICS - FROM SINGLE PURPOSE APPLICATIONS TO GENERIC INTERFACES

Joseph Polito, Ph.D.

Pritsker & Associates, Inc.
P. O. Box 8345
Albuquerque, New Mexico 87198
(505) 255-5597

Early uses of computer graphics with general purpose simulation languages were tailored to particular applications. Recently, however, generalized graphics modules with well defined interfaces have become available for use with general purpose languages. These modules provide a flexible capability to interactively select a variety of graphical displays without re-programming and/or re-running the simulation. This allows output analysis to be "progressive" in that not all of the required displays need have been anticipated prior to performing the run. New graphics/simulation systems are being developed which are fully integrated. In other words, the simulation language, the graphics modules, and a data base management system all communicate through a single user interface.

1. INTRODUCTION

There is an increasing trend to provide "friendly" input formats and output reports for computer users. This tendency has become more pronounced as the availability of microcomputers has increased, and while most of the attention in this area has been directed toward end users of specialized software products, efforts have also been made to develop productivity improving graphical tools for analysts and users of more general purpose simulation languages.(1)

Generally, output graphics aid the analyst in two ways: 1) helping to gain insight, in other words, output analysis, and 2) by speeding preparation of report or presentation quality reports. This paper deals mainly with the first of these, but it is not inappropriate to acknowledge that the second often is very significant in getting the results of simulation used. The capability to meaningfully and quickly display results reduces the turn around cycle time for "what if" analyses and may be the deciding factor in selling the decision maker on the utility of the model.

If graphical simulation output were useful only for presentation materials, however, such applications would be mere curiosities. In fact, a flexible capability to graphically examine simulation output provides a powerful tool to the analyst and decision maker that aids in model verification, validation, and system design. It is the impact of graphics in this area that has lead to the current high level of interest in simulation output graphics.

As is often the case, development of a new capability usually proceeds from the particular to the more general. This has certainly been true with simulation graphics. Section 2 discusses relatively large scale graphical applications of simulation that were tailored to particular problem classes. These efforts contributed to more general capabilities that can be used with general purpose simulation languages. These software packages are discussed in Section 3. Finally, Section 4 presents innovative uses of graphical displays which demand new capabilities in simulation languages to exploit them.

2. SPECIAL PURPOSE GRAPHICS DISPLAYS

One of the most persistent applications of graphics for simulation output analysis is in the preparation of graphical traces. This technique is particularly applicable to modeling of systems which have a spatial dimension. In other words, systems in which objects or people move about. Most often the routing and processing of these entities are the subject of the model and complex rules are applied to a great number of entities. Line printer event traces must be examined in great detail to ensure proper operation of such models before they can be used. Frequently, a graphical plan view of the facility being modeled on which entity movement can be observed will reveal model "bugs" almost immediately. Furthermore, large numbers of runs can be examined much more quickly than even a few runs can be verified without graphics. This capability obviously can lead to a more effective verification and greater confidence in the model.

Examples of such traces are shown in Figure 1 and 2. Figure 1 is a plan view of the Rose Bowl stadium.(2) The graphical display is a special purpose plan view developed directly with a graphics software package. The display is "hard coded" and can not be used for other simulation prupsoes. The model was developed to determine the effect of response times for police, paramedics, fire, ambulances, and bomb squad incidents to various manning and positioning strategies. It allowed the user to interactively set the rules and policies for responses of various types and then observe the outcomes of computer simulations.

The logic of the model was divided into four event types:

(a) generation of incidents
(b) response to incidents
(c) resolution of incidents
(d) change of post assignments for personnel

Data was input to the model with a preprocessor that allowed the user to easily change the three types of data:

(a) incident characteristics
(b) back-up rules
(c) post assignments

The locations of emergency response resources and emergency incidents are shown on the graphical display. Dispatching and movement of the responses are then shown until the incident is resolved.

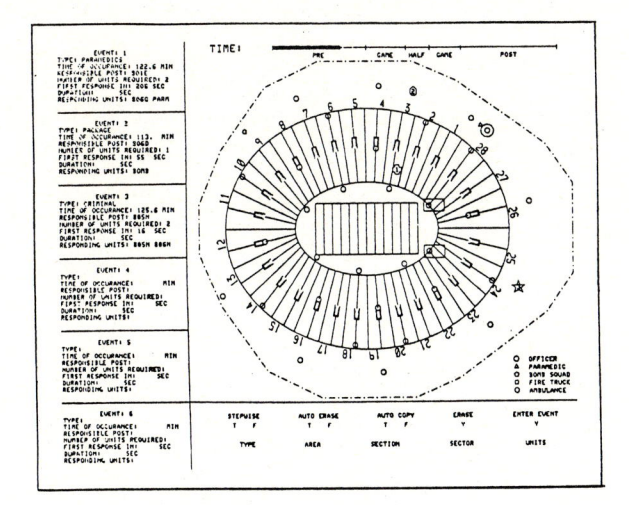

Figure 1: Rose Bowl Trace Graphics.

Figure 2 provides a limited generalization of the Rose Bowl capability. It shows a plan view of a facility generated by SOS.(3,4) SOS is a graphical model building and display tool for the SNAP simulation language.(5) It provides a great deal more capability than output graphics (the interested reader should consult (1) for a brief description), but it has a general trace capability within a limited problem class. SNAP models adversaries attacking a nuclear site to perform sabotage or steal nuclear material. It was developed for the Nuclear Regulatory Commission and the U. S. Navy as a tool to analyze security systems at nuclear sites and on ships. The adversary and guard actions are simulated by SNAP and graphically presented on the site layout by SOS. The interesting feature of SOS is that a digitized representation of the site plan is stored in a data base. It is used during model building and also to draw the graphical trace. Thus, the plan view is not hard coded, but is "data driven" so the same software can be used to produce graphical traces of any site. The data representing the site graphics can be prepared automatically (with a digitizing table) or it can be generated manually with the screen cursor. SOS has the capability of representing facilities with several floors with scenarios involving multiple guard and adversary forces. Force movements and engagements are shown on an event-by-event basis with flexible capabilities for replay and zoom. Guard tactics may be quite complex and the graphical trace feature of SOS allows many runs to be quickly examined to determine the correctness of the model and the effectiveness of tactics.

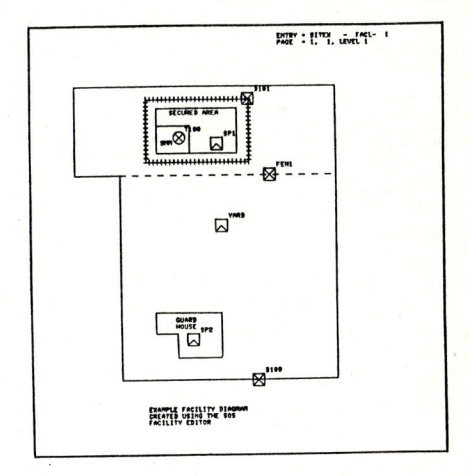

Figure 2: SOS Guard and Adversary Trace.

These examples show a progression from model specific to problem class specific graphics capabilities. The following section presents similar capabilities for general purpose simulation languages.

3. GENERAL PURPOSE GRAPHICS DISPLAYS

Graphics software is available (SIMCHART[TM])(6) which has a direct interface to the SLAM II[TM] simulation language.(7) In other words, SLAM II recognizes input statements that cause it to automatically produce graphic data files for later processing. As an alternative, FORTRAN subprogram calls can be used from other simulation languages to accomplish the same thing. As an example, A Q-GERT^{TM} (8) simulation of an aircraft repair facility at McClellan Air Force Base was developed. A three-way interface between Q-GERT, a simulation data base, SDL^{TM}(9) and the simulation graphics software, SIMCHART, was used. This produced a large variety of graphical displays. Among them was the aircraft routing trace shown in Figure 3.

GRAPHIC PORTRAYAL OF THE A10/192'S SCHEDULE

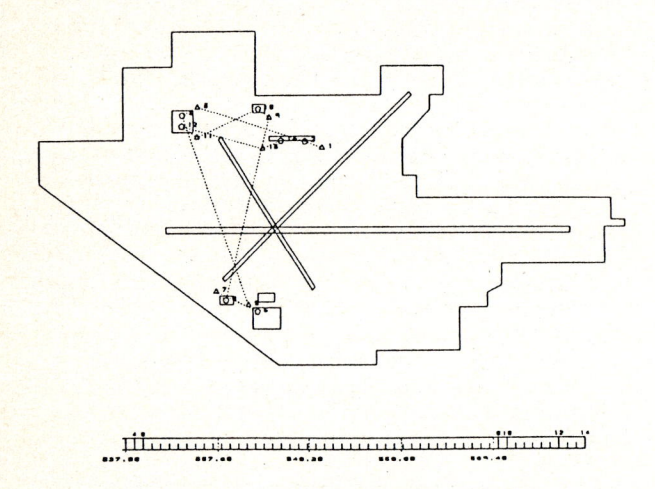

Figure 3: Aircraft Routing Trace.

An important aspect of combining graphics, simulation, and a data base is the ability to write detailed trace information automatically to the data base for later analysis. The analyst can then interactively call up graphical displays after the simulation is completed. Thus, the particular displays and statistics to be calculated need not all have been specified prior to the simulation.

The simulation of the maintenance facility was used to reschedule arrivals to the base. Graphical plots of queue length and server utilization over time revealed that, although average values were low, spikes occurred in queue length that caused serious bottlenecks. Rescheduling smoothed the arrival pattern to the system and improved throughput. The graphical analysis quickly revealed which aircraft (by tail number) would cause the bottleneck to occur.

The software described above is in the form of separate (but compatible) packages. A unified package was, of course the next step.

An integrated simulation language, data base, and graphics capability has been developed called IDEF$_2$(10) which combines many of the capabilities discussed above for SOS, SDL, and SIMCHART. In particular, IDEF$_2$ allows the analyst to interactively develop site layouts and determine the graphical shapes of entities and site facilities which will flow on the graphical trace. It also preserves all of the flexible display options available with SIMCHART and SDL. These capabilities include the ability to display graphs, pie charts, pie graphs, and histograms. Figures 4 and 5 are examples.

The advantage of integrating the three functions is that many of the interface requirements become transparent to the user. Simple input specification suffices to store simulation trace data in the data base for automatic call-up after the simulation is run.

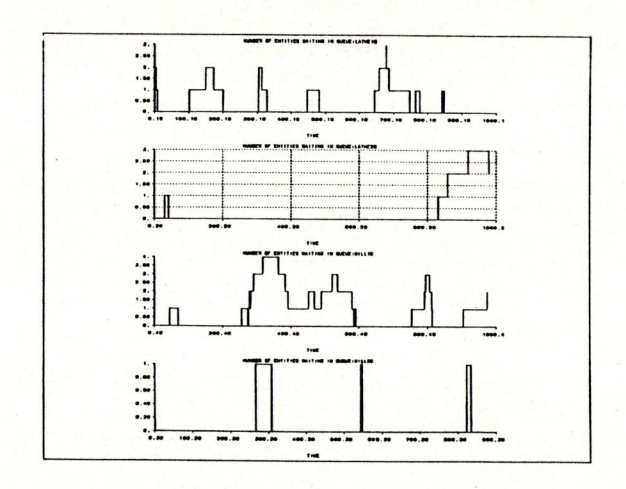

Figure 4: Four IDEF$_2$ Plots on One Screen.

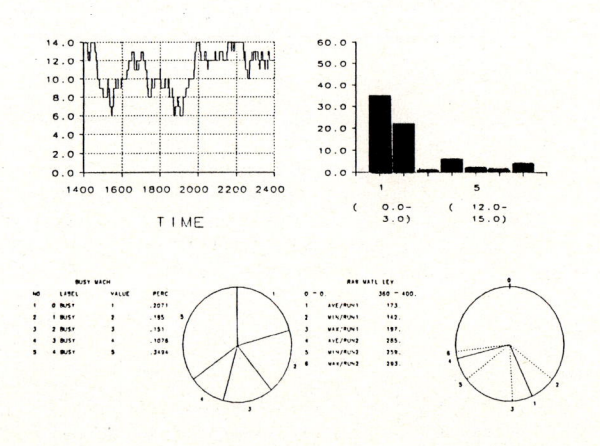

Figure 5: Report Quality Presentations.

It is interesting that the ability to save and plot large quantities of simulation generated data is creating a demand for enhanced capabilities of the simulation languages. Furthermore, this demand arises from an apparent shortcoming of examining simulation traces.

4. INNOVATIVE USES OF GRAPHICAL DISPLAYS

Simulation of stochastic processes is performed to obtain statistical measures of performance. It would seem paradoxical to draw conclusions from the examination of one or several individual runs or realizations of the system. Yet that is one of the main uses of simulation graphics.

In fact, however, a great deal of insight is gained by plotting resource utilizations, queue lengths, schedule deviations, etc. against time for individual runs. If we disregard initial conditions, then simulations of complex systems often exhibit very characteristic behavior even on individual runs. Naturally, the effects of policy changes based on individual runs should be evaluated with aggregate statistics over many runs, but the insights that suggest new policies may be gained by examining an individual trace.

Consider a single simulation run of a job shop. Once the run is complete and the trace data is stored in a data base, then all the processing times for all the parts are known. An interesting possibility is to simulate the system again with the same processing times, but with different routings, part priorities, etc. Of course, any improvements obtained must be tested with a full simulation. Performing quick deterministic runs with the previously generated processing times however, will lead to rapid evaluation of many alternatives.

Current simulation languages do not have a convenient facility for performing these "trace driven" simulations. Work is underway now to examine capabilities needed for the user to easily specify resource allocation rules, routings, queue disciplines, scheduling priorities, etc. and for the simulation languages to access data base information for process times.

One of the most important features of output graphics is that they are meaningful to "clients" who are not modelers. The displays are presented in ways that are easy to understand by the user of the system under study, not by the user of the simulation language. Indeed, interactive output analysis is not beyond the capability of many managers. It is not beyond imagination that models might be prepared by analysts for the use of decision makers. The decision maker could test alternative policies rapidly with "trace driven" simulations. Then the analyst can test the most promising ones with full scale simulations.

5. CONCLUSION

The use of graphics for simulation output analysis is growing. The software has evolved from hard coded, single model applications to flexible, general purpose, fully integrated capabilities. These graphical tools not only reduce verification time and improve understanding of the system under study, but they are also creating a demand for improved simulation capabilities that will allow output graphics to be more fully exploited.

Space has not permitted full description of the graphical application and software described here. The interested reader should consult the cited references.

SIMCHART, SLAM II, Q-GERT, and SDL are trademarks of Pritsker & Associates, Inc.

REFERENCES

[1] Polito, J. and Grant, F. H., Computer Aided Computer Simulation Modeling: Productivity Improvements for Modelers, ORSA/TIMS Conference, Detroit, MI (April 18-21, 1982)

[2] Polito, J. and Starks, D. W., Simulation Application to Base Level and Security Level Planning, Proceedings of the Society for Computer Simulation Multi-Conference, San Diego, CA (January 1983)

[3] Sabuda, J., Polito, J., Walker, J. L., and Grant, F. H., The SNAP Operating System (SOS) User's Guide, Pritsker & Associates, Inc., Report ABQ 008, NUREG/CR-2604, SAND82-7018 (March 1982)

[4] Sabuda, J., Polito, J., Walker, J. L., and Grant, F. H., The SNAP Operating System Reference Manual, Pritsker & Associates, Inc., Report ABQ 009, NUREG/CR-2605, SAND82-7019 (March 1982)

[5] Grant, F. H., Engi, D., and Chapman, L. D., User's Guide for SNAP, NUREG/CR-1245, SAND80-0315 (July 1981)

[6] Duket, S. D., Hixson, A. F., and Rolston, L., The SIMCHART User's Guide, Pritsker & Associates, Inc., West Lafayette, IN (June 1981)

[7] Pritsker, A. A. B. and Pegden, D. C., Introduction to Simulation and SLAM (Halsted Press, New York, 1979)

[8] Pritsker, A. A. B., Modeling and Analysis Using Q-GERT Networks (Halsted Press, New York, 1979)

[9] Standridge, C. R., SDL-Simulation Data Language Reference Manual, Pritsker & Associates, Inc., West Lafayette, IN (June 1981)

[10] Yancey, D. P., Sale, M. W., O'Reilly, J.J., Whitford, J. P., and Lilegdon, W. R., The IDSS Prototype (2.0) - Version 2 User's Reference Manual, Pritsker & Associates, Inc., West Lafayette, IN (January 1983)

SMALLER COMPUTERS: SOFTWARE TRENDS

Organizer: Thomas J. Boardman

Invited Presentations:

    Statistical Packages on Microcomputers, Wayne A. Woodward and Alan C. Elliott

    What More is There to Statistics on a Micro than Precise Estimates of Regression Coefficients? or Mainframe Statistical Software in a Box, MaryAnn Hill and Jerome Toporek

    Machines and Metaphors, Joseph Deken

STATISTICAL PACKAGES ON MICROCOMPUTERS

Wayne A. Woodward and Alan C. Elliott

Southern Methodist University
Dallas, Texas

The results of a survey of statistical packages for microcomputers is presented. Questionnaires were sent to known statistical software producers, and completed questionnaires were obtained on 26 packages. This information is presented in the paper along with partial information on 32 other known packages. A brief listing of packages available on larger microcomputer systems is also presented. We also stress the need for reviews of the microcomputer statistical packages.

1. Introduction

The microcomputer explosion which has taken place during the last few years has brought sophisticated computing capabilities within the reach of many persons to whom it was previously not available. The growth of the popularity of the microcomputer has been accompanied by the development of software designed for use on it. Of course, word processing, spreadsheet, database, and entertainment software are among the most popular forms of microcomputer software. However, programs designed to perform statistical analysis are also abundant, and in this paper, we will discuss the statistical software available on the microcomputer.

2. The Project

Considerable effort has been devoted to the review of statistical packages on the mainframes over the last 10 years, and as a result, the quality of these packages has improved. In the summer of 1981 we decided to begin a review of the microcomputer statistical packages since, to our knowledge, they had not been reviewed by the statistical community, and were to a large extent unknown to statisticians. However, the large number of statistical packages and hardware configurations involved made this a formidable task. Rather than abandon the project entirely, we developed a questionnaire and sent it to all known producers of statistical software. The results of the survey are presented in this paper, and it is hoped that they will provide statisticians with an awareness of "what is available" in much the same way that the Schucany, Minton, and Shannon(1972) survey provided an early awareness of the available statistical packages on the mainframes. Preliminary results were presented at Cincinnati in August 1982 - see Elliott and Woodward(1982), while Neffendorf(1983) has provided a listing of statistical packages available on microcomputers.

3. The Packages

Before discussing the results of our survey, some terms should be defined. Of course, microcomputers are available in a wide variety of prices. We have rather arbitrarily broken this price range into two components:

1) lower priced systems - hardware and statistical software cost less than $8000

2) higher priced systems - hardware and statistical software cost at least $8000

Our study focused on the software for the lower priced systems which include Apple, Radio Shack, IBM Personal Computer and similar systems based on processors such as the Z80, 6502, or 8086. Computers based on the 68000 processor are generally included in the higher priced group.

There are many single purpose programs available for performing regression analysis, etc., but in order for us to classify a program as a statistical package, it should provide a variety of statistical functions. It may be argued that some of the programs listed in our tables should not actually have been classified as packages. Of course, the prices and capabilities listed in the tables are subject to change.

In Tables 1 and 2 we include packages for which questionnaire information is available. If the software producer sent a reference manual instead of completing the questionnaire, we attempted to complete the questionnaire ourselves. Examination of these tables reveals that there is a wide variation in the price of the software (from less than $50 to $2000 or more), a wide variation in the reported capabilities, and that many computers are supported. Box plots, confidence intervals, survival analysis, categorical data analyses other than chi-square contingency table analysis, multiple comparisons, and nonlinear regression are among the features which are rarely supported by these packages.

Most of these programs have data entry modules which make it easy for the user to enter and modify data from the keyboard. However, it has been our observation that many of these programs will not read a rectangular data file created externally to the program. However, on the questionnaire we asked whether data created externally to the program could be read by the program and most responded that it could. It appears that we asked the wrong question. Of course, if the user knew exactly how the program expected to see the data, he could create data externally and store it accordingly. It is still unclear which programs can use data from a rectangular data file which, for example, has been downloaded from a mainframe.

In Table 3 we include "partial" information on another 32 programs of which we are aware but for which we do not have questionnaire information. We have listed all known information on these packages, but ANY cell in this table which is left blank should be interpreted as "unknown". Some of these packages may no longer be on the market.

In Table 4 we list the statistical software which is available on the larger microcomputer systems. It should be noted that several of the producers of statistical packages for the mainframes have, or soon will have, programs available on these larger micro systems.

4. Quality

While we believe that the tables provide useful information concerning REPORTED capabilities of the available packages, the question of product performance is still largely unanswered. Although we abandoned our original review effort, we do believe that such reviews should be performed. Our experience with several of the packages has revealed some glaring errors. One program, for example, has a module called "Normal Distribution - Probability and Frequency". Upon entering 1.96 as the standard normal variate and specifying a sample size of 100, the following output appears:

    Frequency :   .0584410626
    Occurrences :   5.84410627
    Prob. of Occurrence : .474994099

Of course, the last value here is $P[0<Z<1.96]$ although it is not properly described on the output. However, the "Frequency" and "Occurrences" are more disturbing. The frequency given is the height of the normal density at $z=1.96$ with occurrences evidently indicating an expected number of 1.96's out of a sample of size 100. This obviously indicates a severe lack of fundamental understanding by the authors of the program. This same package has similar problems with chi-square contingency table analysis in which output is mislabeled and the chi-square is miscalculated.

A second package, selling for over $200, claims to perform 1-way ANOVA (Kruskal-Wallis), 3-way ANOVA (equal cells) and analysis of covariance among other routines. In the 3-way ANOVA example DISPLAYED IN THE MANUAL, nearly one-half of the sums of squares and F-statistics are NEGATIVE. This apparently caused no concern to the author. When we used the program ourselves, we obtained negative sums-of-squares using the 3-way ANOVA and analysis of covariance. Using the Kruskal-Wallis module, we obtained "divide by zero" errors whenever there were more than two groups. Reference manuals for each of these two programs displayed similar lack of understanding of statistical principles. Neither of these two packages is listed in the tables.

In our opinion, the examples described above are a direct result of the fact that producers of statistical software for micros often do not have the same level of statistical expertise as, for example, the producers of the major statistical packages on the mainframes. In fact, some of these programs appear to have been written by "going through" the examples in an elementary statistics book. Our comments should not be interpreted as a blanket indictment of all of these programs, but we believe that there is a real need for professional review of these packages.

5. Reviews

The reviews suggested in the previous section should prove to be useful in educating statisticians concerning the quality of available statistical packages. Even if a statistician is not using a micro for his own statistical analysis, he is probably asked to recommend such programs. This is one method by which the results of the reviews can be made known to the "end-user" who is not a professional statistician. Also, the reviews should tend to improve the quality of the available software.

From the previous section, it seems obvious that these reviews should begin by determining whether or not the program is "doing what it is supposed to be doing". Beyond this, the review should include comments concerning ease of data entry (as related to our previous discussion), the "user friendliness" of the program, whether or not the program takes advantage of graphics capabilities of the computer, and whether or not algorithms used are appropriate for use on microcomputers with their smaller word size. In addition, it seems that the performance of the programs should be examined on benchmark data sets. It has come to our attention recently that reviews are currently being written by members of the Statisics Special Interest Group of the Washington D.C. area IBM Personal Computer Users Group. It seems that wider distribution of reviews through the American Statistician or the Statistical Software Newsletter of the International Association of Statistical Computing is also needed.

6. Concluding Remarks

In this paper we have provided information concerning the statistical software available on micros. The microcomputer is bringing statistical computing within the reach of many people who previously depended on a statistician for data analysis. While this is a dangerous situation in itself, it is complicated by the fact that most of the statistical software has not been reviewed by the statistical community and several of the packages on the market have been shown to have serious errors which may go undetected by these users. We strongly recommend that statisticians become involved in evaluation of these packages.

REFERENCES

[1] Elliott, Alan C. and Woodward, Wayne A., A Survey of Statistical Packages Available on Microcomputers, Proceedings of the Statistical Computing Section - American Statistical Association, (1982) 252-254.

[2] Neffendorf, Hugh., Statistical Packages for Microcomputers, The American Statistician 37, (1983) 83-86.

[3] Schucany, W.R., Minton, Paul D., and Shannon, Stanley Jr., A Survey of Statistical Packages, Computing Surveys 4, (1972) 65-79.

TABLE 1. STATISTICAL PACKAGES FOR WHICH QUESTIONNAIRE INFORMATION IS AVAILABLE
(Smaller Systems - Hardware and Statistical Software < $8000)

SUPPLIER	NAME OF PROGRAM	TYPE OF COMPUTER	MEMORY REQUIREMENTS	LANGUAGE	PROGRAM DESIGNED FOR		PROGRAM STRUCTURE				PRICE
					EDUCATIONAL USE	STAT. ANALYSIS	CONTAINS USERS MANUAL	MENU DIRECTED	COMMAND DIRECTED	INTEGRATED SYSTEM	SEPARATE PROGRAMS EACH PROCEDURE	
ABT Microcomputer Software	Sample Calc	Apple II, II+	48K	Applesoft BASIC		x		x				$ 50
Action-Research Northwest	AIDA	Apple II	48K	Applesoft BASIC		x	x		x	x		$235
Aerocomp, Inc.	Statistical Analysis	CP/M, ISIS-II	64K	FORTRAN		x	x			x		$175
Brooks/Cole Pblshg.	KEYSTAT	Apple II+, IIe	48K	BASIC	x		x	x				?
Compress	Statistical & Probability Demos	Apple II	?	Applesoft BASIC	x			x		x		$2000-$2500
Conduit	Descriptive Statistics	Apple II, IIe	48K	Applesoft BASIC	x	x	x	x				$ 75
Conduit	Exploratory Data Analysis	Apple II, IIe	48K	Applesoft BASIC		x	x		x			$165
Crunch Software	CRISP	IBM PC	128K	Compiled BASIC		x	x	x			x	$350
Ecosoft, Inc.	Microstat 3.0	Z80 with CP/M IBM PC	48K 64K	BASIC		x	x	x				$325
Great Northern Computer	STATFLOW	CP/M or CP/M 86	56K	Compiled BASIC		x	x	x		x		₤225
Hewlett Packard (developed at Colorado State Univ)	Statistics Library	HP 200 Series Model 16	256K	HP Extended BASIC		x	x	x		x		$1500

TABLE 1. (Continued)

SUPPLIER	NAME OF PROGRAM	TYPE OF COMPUTER	MEMORY REQUIREMENTS	LANGUAGE	PROGRAM DESIGNED FOR			PROGRAM STRUCTURE				PRICE
					EDUCATIONAL USE	STAT. ANALYSIS	CONTAINS USERS MANUAL	MENU DIRECTED	COMMAND DIRECTED	INTEGRATED SYSTEM	SEPARATE PROGRAMS EACH PROCEDURE	
Jerry Hintze	Number Cruncher	TRS80 Model I,II,III CPM 2.2, IBM PC	64K	BASIC		x	x	x			x	$200-$400
Keller Software	Statistics, HAL 3001	IBM PC	64K	BASIC		x	x	x		x		$495
Montana State Univ.	MSUSTAT	CP/M	64K	FORTRAN		x	x	x		x		$750
Northwest Analytical	NWA Statpack	CP/M or MS DOS	64K	Microsoft BASIC		x	x	x			x	$495
Alan Pearman Ltd.	STATPL	IBM PC COLUMBIA PC	320K	APL		x	x	x			x	$650
Questionnaire Serv.	SL-MICRO	CP/M	48K	CB80		x	x	x		x		$250
Radio Shack	Advanced Statistical Analysis	TRS80 I,III II	16K 32K	BASIC		x	x	x			x	$40 $99
Rainbow Computing, Inc.	Statistics with Daisy	Apple II+	48K	Applesoft BASIC		x	x		x			$80
Rosen Grandon Assoc.	A-Stat 79	Apple II,II+	48K	Applesoft BASIC		x	x	x	x	x		$2000
The Software Hill	IFDAS	Cromemco Z80 based Computers	64K	Cromemco 32K structured BASIC		x	x		x	x		$595
STATPAL Assoc.	STATPAL	Apple II+ TRS80 Model I,II IBM PC CP/M	48K 32K 32K 32K	BASIC	x		x	x		x		$35
Statistical Consultants	STAN	8080-Z80 CP/M UCSD - P System	56K	PASCAL		x	x	x		x		$300
Statistical Computing Consultants	SURVTAB	IBM PC	64K	BASIC		x	x	x				$180
Systematica	Micro SURVEY	Any CP/M	64K	FORTRAN					x	x		£1200
The Winchendon Group	ELF	Apple II/III CP/M	48K 56K	BASIC and Assembler		x	x	x			x	$200

2. FURTHER DETAILS CONCERNING THE PACKAGES IN TABLE 1.

PROGRAM	SUPPLIER	DATA ENTRY: STORAGE/RETRIEVAL	TRANSFORMATIONS	EDITING	CAN USE EXTERNALLY CREATED DATA	MISSING VALUES	GRAPHICS: HISTOGRAMS	SCATTER(x-y) PLOTS	BOX PLOTS	DESCRIPTIVE: MEAN	VARIANCE	MIN,MAX	MEDIAN	PERCENTILES	SKEWNESS	KURTOSIS	t-TESTS: ONE SAMPLE	2-SAMPLE(EQUAL VAR)	2-SAMPLE(UNEQUAL VAR)	PAIRED t-TEST	CORR & REG: SIMPLE LINEAR REG	PEARSON CORR	MULTIPLE REGRESSION	NON-LINEAR REG	RESIDUALS CALCULATED	RESIDUALS PLOTTED	CI ABOUT μ	ABOUT $\mu_1-\mu_2$	ABOUT p	ABOUT p_1-p_2	CAT DATA: χ^2 CONTINGENCY TABLE	LOG-LIKELIHOOD χ^2	FISHER'S EXACT TEST	GOODNESS OF FIT	ANOVA: 1-WAY	REPEATED MEASURE	2-WAY	MULTIPLE COMPARISON	COVARIANCE	NONPAR: SIGN TEST	WILCOXON-1-SAMPLE	MANN WHITNEY U	KRUSKAL WALLIS	FRIEDMAN χ^2	CORRELATION	DIST: NORMAL	χ^2	F	BINOMIAL	HYPERGEOMETRIC	POISSON	OTHER: LIFE-TABLE ANALYSIS	TIME SERIES	RANDOM NUMBER GEN	p VALUES WITH TESTS	TEST OF $\sigma_1^2=\sigma_2^2$	LINEAR DISCRIMINANT	DETERMINE n
Sample Calc	ABT Microcomputer Software	X	X							X	X	X						X	X	X	X	X	X								X				X		X			X																X
AIDA	Action Research Northwest	X	X	X	X	X	X	X		X	X	X		X	X	X					X	X			X						X				X		X		X																	
Statistical Analysis	AEROCOMP			X						X	X	X																																											
Keystat	Brooks/Cole	X	X	X		X	X	X		X	X	X	X	X	X	X	X	X	X	X	X	X	X		X						X		X	X	X		X		X				X		X	X	X	X	X	X		X	X				
Stat and Prob Demonstrations	Compress		X	X			X		X	X	X	X		X																	X				X											X	X	X	X	X		X					
Descriptive Stat.	Conduit	X	X	X			X	X		X	X	X	X	X	X	X	X																																			X					
Exploratory Data Analysis	Conduit	X	X	X				X		X	X	X																																													
Crisp	Crunch Software	X	X	X	X	X	X	X			X	X	X	X	X	X	X	X	X	X	X	X	X	X		X	X	X	X	X	X	X		X	X	X	X	X	X	X	X	X	X	X	X	X	X	X	X	X	X		X	X	X		X
Microstat	Ecosoft	X	X	X	X	X	X	X			X	X	X	X	X	X	X	X	X	X	X	X	X	X		X	X	X	X	X	X	X	X	X	X	X	X	X	X	X	X	X	X	X	X	X	X	X	X	X	X		X	X	X		X
Statflow	Great Northern	X	X	X	X	X	X	X			X	X	X	X	X	X	X	X	X	X		X	X	X		X	X	X	X			X		X	X	X	X	X	X	X	X	X	X	X	X	X	X	X	X	X	X		X	X	X		X
Statistics Library	Hewlett Packard/Colorado State	X	X	X	X	X	X	X			X	X	X	X	X	X	X	X	X	X	X	X	X	X		X	X	X	X			X				X	X	X	X	X	X	X	X	X	X	X	X	X	X	X	X		X	X			X
Number Cruncher	Jerry Hintze	X	X	X	X	X	X	X			X	X	X	X	X	X	X	X	X	X	X	X	X	X		X	X	X	X			X		X		X	X	X	X	X	X	X	X	X	X	X	X	X	X	X	X		X	X	X		X
Statistics, HAL 3001	Keller Software	X	X	X	X	X	X	X		X	X	X	X					X	X	X	X	X	X	X		X	X	X	X			X				X		X		X			X		X	X	X	X		X			X			
MSUSTAT	Montana State University	X	X	X	X	X	X	X			X	X	X	X	X	X	X	X	X	X	X	X	X	X		X	X	X	X		X	X				X		X		X					X												
MJA Statpak	Northwest Analytical	X	X	X	X	X	X	X			X	X	X	X	X	X	X	X	X	X	X	X	X	X		X	X					X				X		X		X																	
STATPL	Alan Pearman, Lt.	X	X	X	X	X	X	X			X	X	X									X	X	X		X	X					X						X																			
SL-Micro	Questionnaire Service Co.	X	X	X	X	X	X	X			X	X	X									X	X	X		X	X				X	X		X		X		X		X						X							X				
Advanced Stat. Anal.	Radio Shack	X	X	X	X	X	X	X	X		X	X	X	X	X	X	X	X	X	X		X	X	X		X	X					X				X	X	X	X	X	X		X								X		X				
Statistics w/Daisy	Rainbow Computing	X	X	X	X	X	X	X			X	X	X			X	X	X	X	X	X	X	X	X		X	X					X				X	X	X	X	X			X										X				
A-Stat 79	Rosen Grandom	X	X	X	X	X	X	X			X	X	X	X	X	X						X	X	X		X	X					X				X		X		X																	
IFDMS	Software Hill	X	X	X	X	X	X	X			X	X	X					X	X	X		X	X	X		X	X					X				X		X		X																	
STAN	Statistical Consultants	X	X	X	X	X	X	X			X	X	X					X	X	X		X	X	X		X	X					X				X	X	X	X				X		X	X	X	X		X			X				
STATPAL	STATPAL Associates	X	X	X	X	X	X	X			X	X	X		X	X	X	X	X	X		X	X	X		X	X		X			X				X	X	X	X				X		X	X	X	X		X			X	X			
SURVTAB	Statistical Computing Cons.	X	X	X	X	X	X	X			X	X	X																			X																					X				
Micro Survey	Systematica	X	X	X	X	X	X	X			X	X	X									X	X	X		X	X					X																					X				
ELF	The Winchendon Group	X	X	X	X	X	X	X			X	X	X			X	X	X	X	X		X	X	X		X	X					X				X		X								X											

TABLE 3. OTHER KNOWN STATISTICS PACKAGES FOR SMALLER MICROCOMPUTERS
(Hardware + Software < $8,000)

Supplier	Name of Program	Type of Computer	Data Management	Graphics Output	Descriptive Stats	t-tests	Correlation/Reg.	Confidence Interval	Categ. Data Anal.	ANOVA	Nonpara.	Dist. Functions	Price
Adventure International	Maxi-Stat												
Anderson-Bell	AbStat	CP/M	X	X	X	X			X	X			$400
Basic Business Software	Statistics	CP/M, IBM PC											$75
BHRA	TECPACS	PET, APPLE			X	X		X		X			£80-100 each module
Charles Mann & Assoc.	Statistics Pac	Apple, TRS80	X		X		X					X	$100
Conduit	Laboratory Manual for Probability & Statistical Inf.	TRS80-III, Apple II											Coming fall 83
Creative Discount Software	Statistics Pac	TRS80/Apple	X		X		X						$100
DYNACOMP	ANOVA, Regression I, II	Atari, Apple, TRS80 PET, North Star, CP/M	X	X		X				X			$47 (ANOVA) $70 regression
Ed-Sci Development	ED-SCI Statistics	Apple II, II +	X		X	X	X			X	X		
Hayden Software	Stat Manager	CP/M	X	X	X	X	X						
H&H Associates	Statistical Anal. System	CP/M											£335
Human Systems Dynamics	HSD Stat Programs	IBM PC											$425
Information Analysis Assoc.	MEDLOG	CP/M-86		X	X	X	X			X	X		
International Software	SAM	Apple, CP/M, PET											
Lifeboat	STATPAK	CP/M	X		X	X	X			X	X	X	$495
Lombardy	Statistics Pack.	Apple			X		X			X	X		
Mercator	SNAP	CP/M											£645
Micro Act	Basic Stat. Pkgs.	PET											
Micro Data Collection	MDC Statistics	CP/M											
Minitab	Penn. State Univ.	IBM PC											coming summer 83
Old Bird Software	Statistics Pkg.	CP/M		X	X	X	X			X			$50
On Going Ideas	Statistics Pkg. I	Apple II, TRS80-I											
Personal Computers, Ltd.	Personal Data Anal.	Apple	X	X			X						7 modules £75-£175
Quest Software	MICRO QUEST	Apple											£1000 w/hardware
Patrick Royston	DATA-X	PET	X		X		X				X	X	£350
SPSS, Inc.	IDA	IBM PC											(summer 83)
Statistical Graphics Corp.	STATGRAPHICS.PC	IBMPC	X	X	X	X	X		X	X	X	X	$475
Stat Micro USA	Statistical Micro-programs	Apple	X		X	X	X			X			$150
Total Computer Systems	Basic Stat.	TRS 80 I											
Texas Instruments	Statistics	TI Home Computer	X		X	X	X			X	X	X	$45
Wadsworth Electronic Publishing Company	STATPRO	Apple	X	X	X	X	X			X	X		$1995
Westat Associates	MASS	CP/M	X	X	X	X	X			X	X	X	

TABLE 4. LARGER MICRO SYSTEMS
(Hardware and Software > $8,000)

	Machine	Price	Availability Date
BMDP	Motorolla 6800	$1200 w/hardware	Now
CLYDE - Statistical Analysis System	"small" LSI-11	?	Now
HP	HP 200 Series Models 26, 36	$1500	Now
IMSL			Within next 6 months - on one or more micros
MINITAB	LIS-11 based systems ONYX FORTUNE DEC Professional	$1000/yr.	Now Summer 1983
P-STAT	WICAT 68000	$1500	Mid-Summer 83
SAS			Micro plans will be announced within next year or so- undeveloped at this point
SPSS	DEC Professional Computer	$600	Summer 1983
Tektronix	Tektronix 4050	$1650	Now

WHAT MORE IS THERE TO STATISTICS ON A MICRO THAN
PRECISE ESTIMATES OF REGRESSION COEFFICIENTS?
or
MAINFRAME STATISTICAL SOFTWARE IN A BOX

MaryAnn Hill and Jerome Toporek, BMDP Statistical Software

With the explosion in micro computer sales, what software support for data analysis can users anticipate? Where should they expect micro software to fit in the broad spectrum between mainframe stat packages and programmable calculators? CPU chips for desktop computers are now available for 32 bit computations. Many capabilities for effective data analysis are possible on these machines, including full screen edit access to data and instructions, output managers, forms handlers for data entry and editing, networking, and mainframe communication. BMDP Statistical Software now provides these capabilities along with its mainframe software in the StatCat micro.

Hardware remains the focus of most discussions of microcomputers; the microcomputer hardware explosion promises to alter our society in some fundamental ways, changing the ways we learn, work, plan, and organize our lives. But you have just heard Elliot and Woodward's paper on the proliferation of microcomputer software; without effective software, the microprocessor would remain an engaging toy. Many researchers and data analysts are now purchasing their own systems to tackle messy data that often contain dozens (and dozens) of measurements for hundreds (and hundreds) of subjects. If these systems are to serve them well, they need software that goes beyond an analysis of simple textbook data.

But much of the statistical software for micros is woefully inadequate. For example, a micro software package we saw recently was less capable of performing data analysis than the mainframe software of the 60's -- the multiple regression routine produced only regression coefficients, standard errors, and multiple R^2. Since the 1965 version of BMD, we have been using TOLERANCE statistics, F-TO-ENTER values for stepping in variables, plots of residuals, etc. In our consulting practice, we routinely use Cook's distance measure to identify cases that most influence the regression coefficients, properly standardized residuals, the Mahalanobis distance in the X-space, the correlation matrix of regression coefficients, etc.

The economy and convenience of a micro cannot outweigh the loss of this analytic power. Fortunately, we can have the advantages of both worlds. We have successfully moved the entire 1982 mainframe BMDP package of 40 computer programs (Dixon et al.), plus some 1983 features, to a desktop micro, and BMDP Statistical Software now offers the resulting hardware/software combination as the BMDP StatCat. Users can continue to do state-of-the-art data analysis, while having the convenience and economy of a micro.

We demonstrate that a desktop computer can provide mainframe statistical supports for data analysis, plus the many other features that are necessary to go beyond elementary statistics textbook problems. We discuss the following:

- <u>statistical supports</u> for data analysis

- <u>input managers</u> with immediate full screen edit access to the data and instructions

- <u>output managers</u> for scrolling back and forth through the output and writing to files and the printer

- <u>file management</u> for merging files, updating files, and generating summary records

- <u>forms handler</u> for data entry and <u>BMDP File editing</u>

- <u>mainframe communication</u> for exchanging data or special purpose programs with another computer

- <u>networking</u> for communication among several machines

- <u>other supports</u>, including word processing, graphics boards, and compilers

Obviously, many of the functions described above (such as the forms handler for data entry) can be done effectively on the smaller 8-bit micros, but more power is required to provide the total combination of features on one machine.

The basic hardware for the StatCat includes the widely used Motorola 68000 processor, 768 Kbytes of memory (expandable to 1¼ Mbytes), a 10 Mbyte Winchester disk drive (a 21 Mbyte Winchester is also available), a 600 Kbyte floppy disk drive, and a built-in terminal. It is technically a 16-bit machine, but the registers are 32 bits wide, providing 20 or more bits for addressing (depending on the implementation). Note that the number of addresses available for a program is a function of the size of the address register (8 bits, 16 bits, or more than 20 bits). Since there is no overlay structure, all the program code you want to execute must be in the memory. Thus, memory size affects function as well as performance.

The UNIX™ operating system provides the necessary facilities for multiprocessing and interprocess communication -- these facilities are far more sophisticated than those of other popular systems (such as CP/M) used in other micros. Together, the CPU and the operating system provide a very satisfactory environment for doing data analysis with our mainframe programs.

Before describing the special features of the StatCat, we highlight a few BMDP tools that we use routinely even for relatively simple, everyday data analysis. These same features are available on the StatCat, making it a complete tool for the statistician.

Statistical Supports for Data Analysis

An analysis of data for a real problem is often a complex iterative process that involves locating errors and inconsistencies in the data, assessing the extent and pattern of missing values, and insuring that the data conform to assumptions necessary for subsequent analysis (e.g., the data may require tranformations or separate analysis of heterogeneous subpopulations).

1) ANOVA screening and selection of transformations

As an illustration, consider the use of the analysis of variance program 7D to study the Fisher Iris data. Our null hypothesis is that average petal lengths for Setosa, Versicolor, and Virginia flowers are equal. For this problem, we could use ANOVA software that only prints sums of squares, degrees of freedom, F statistics, and p-values. However, we want to examine the distribution of the data within each cell; for a valid analysis we must satisfy the assumption that the variances for each group are equal. Visually, in the sideways histograms (Figure 1) we see that the spread of the first

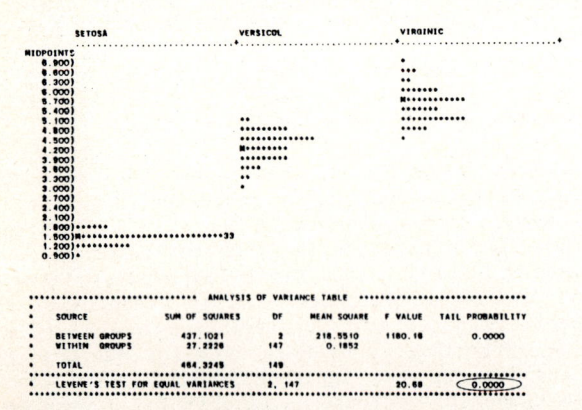

Figure 1. Using 7D histograms and ANOVA to compare the petal length of iris flowers.

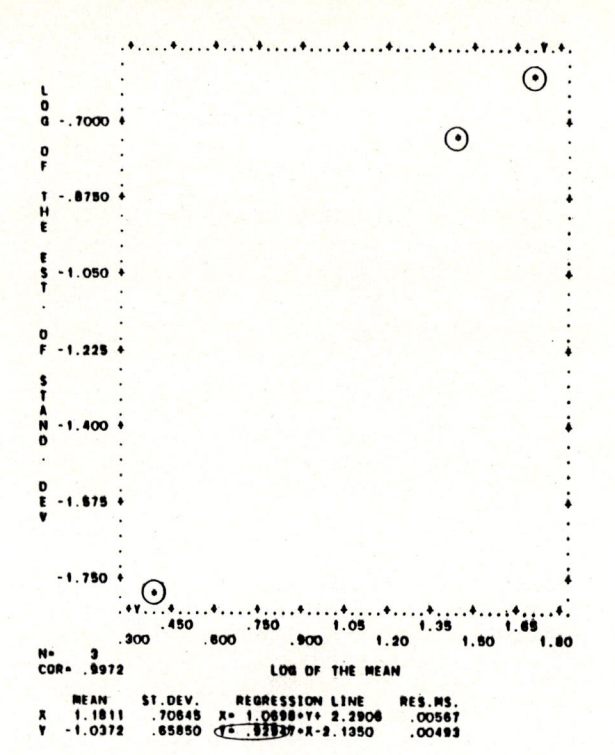

Figure 2. Diagnostic plot for selecting a variance-stabilizing transformation.

group is very small compared with that of the third group. Using the Levene test printed in the output, we reject the equality of variance test (p-value $\leq$ .00005). We then look at the diagnostic plot (Figure 2) produced as part of the same program to identify a power transformation of the form $y^{1-\beta}$, where β is the slope of the equation

$$\log \begin{bmatrix} \text{cell} \\ \text{std.} \\ \text{dev.} \end{bmatrix} = \alpha + \beta \log \begin{bmatrix} \text{cell} \\ \text{mean} \end{bmatrix}$$

The slope is roughly .9, indicating that a log transformation is appropriate. The 7D display in Figure 3 results from using the log of each flower's measurement. The Levene test is no longer significant (p-value = .592), so we have stabilized the cell variances.

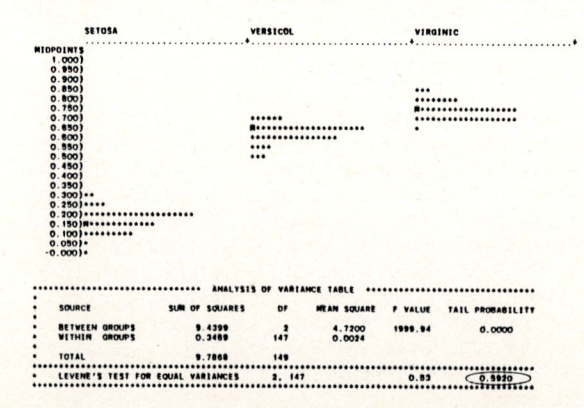

Figure 3. Histogram results after log transforming petal length.

2) Graphical displays to study the influence of subpopulations

Before studying the relationship between petal length and width we should check to see if the presence of subpopulations masks a true relationship or falsely strengthens the apparent relationship. Using the scatterplot program 6D, we use distinct plot symbols to identify three types of iris flowers in one plot of petal length versus width (Figure 4) and in a second plot of sepal length versus width (Figure 5). In each figure we plot all three flower types within one frame, and then the Setosa flowers alone. The lines connecting the Y's on the plot frame mark the least squares regression line. In Figure 4, the correlation is .96 for the complete sample but is only .36 for the Setosa flowers. Thus the presence of subpopulations gives a false impression of the relationship between height and width. The opposite is true for the data in Figure 5. When all flower types are considered as one sample, there appears to be no relation (r = -.018). When only the Setosa flowers are studied, however, the correlation is .74.

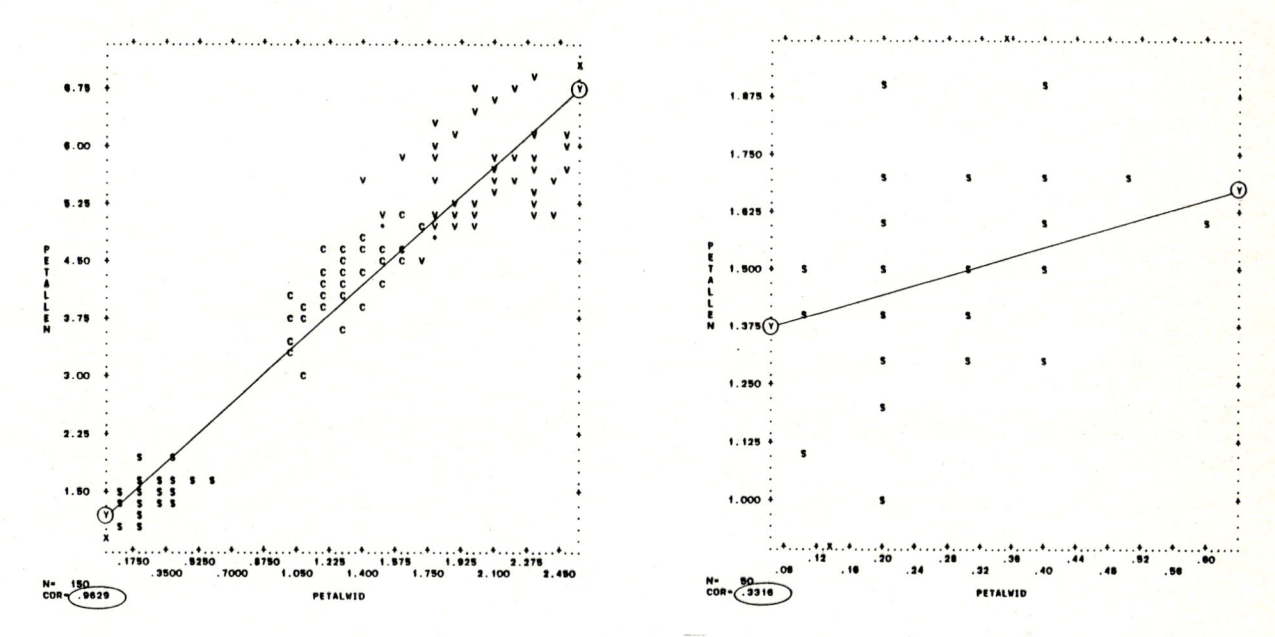

Figure 4. The presence of subpopulations gives a false impression of the relationship between petal length and width.

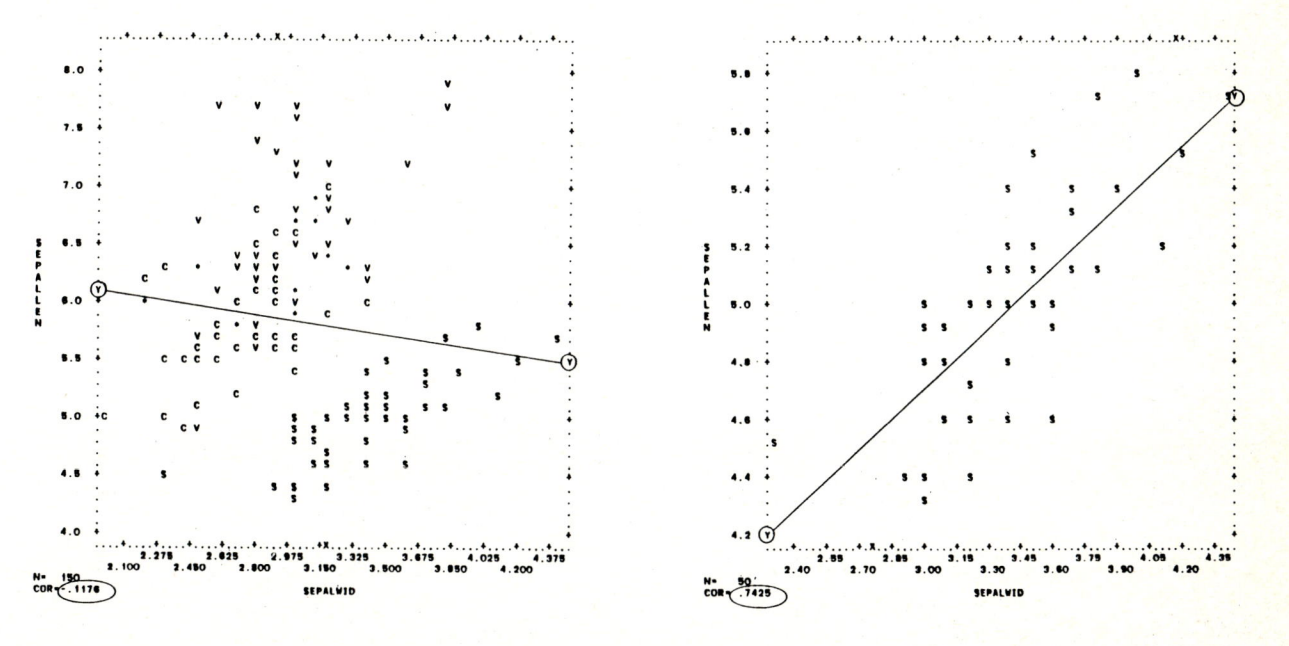

Figure 5. The presence of subpopulations can mask a relationship.

3) Shaded correlation matrices to communicate factor analysis results

In Figure 6, we display a correlation matrix from BMDP Technical Report No. 8, "Annotated Computer Output for Factor Analysis: A Supplement to the Writeup for Computer Program BMDP4M." The density of each overprinted plot symbol indicates the strength of the correlation -- a well-filled character space represents a large correlation, a relatively empty character space, a low correlation. (The results of the factor analysis were used to cluster the variables into their respective factors.) When viewed on a terminal, each correlation is represented by a single digit rather than a set of overprinted characters (an 0 represents very low correlations, a 9, very high). Incidentally, using the smaller version of the StatCat with 3/4 Mbytes, the StatCat can do a factor analysis for 120 variables and 10 factors. For a test run of 100 variables and 200 cases, it took approximately 22 minutes to compute the initial factor loadings for 5 factors, the squared multiple correlation of each variable with all the others, the communalities, eigenvalues, etc. Software floating point arithmetic was used; hardware floating point arithmetic is promised soon, so such analyses will run faster.

4) Regression diagnostics

Just seeing a printout of the regression coefficients, their standard errors and multiple R^2 is often not enough. For each case we like to scan the values of Cook's distance measure, properly standardized residuals, and the Mahalanobis distance to the center of the cases in the x-space. For example, say you are a real estate agent and want a quick formula for estimating a reasonable selling price for homes (you'll adjust the estimate up or down, depending on the current condition of each house) and you've been told to use taxes, living space, number of bathrooms, lot size, etc. as possible predictor variables. You submit data for the last 28 homes sold by your firm to the stepwise multiple regression program 2R, and produce a model with a multiple R^2 of .92. You think, "Ah, that's pretty good." However, you then notice a half-normal probability plot of the modified Cook distance (see Figure 7) and see that cases 28 and 10 appear to be different from the others. The program also prints values of the Andrews-Pregibon statistics, AP1 and AP2 (Draper and John). (These statistics are new in the 1983 release of BMDP.) The first statistic is a function of the decrease in the residual sum of squares produced by deleting the case; the second is one minus the diagonal of the hat matrix (or a measure of the case's distance in the x-space). From this printout you see that case 9 is also unusual. You wonder what difference these cases make to the results, and so you delete the 3 cases and rerun the analysis. For this run, the multiple R^2 is .77, a drop from .92, and the relative importance of variables in the model also changes. You examine plots of the important predictors of price, and identify cases 9, 10, and 28 in the plot of price versus taxes (see Figure 8). Case 28 stands out in all bivariate plots of the other predictors versus taxes, leading you to suspect that the tax information for case 28 is incorrect. Interestingly, for the full model there are 8 cases that have simple residuals larger than that for case 28.

Figure 6. A shaded correlation matrix is used to display the results of a factor analysis.

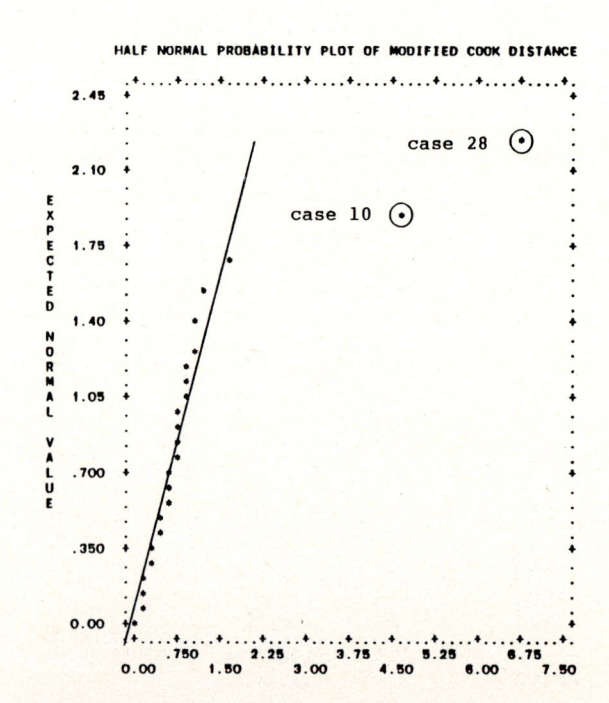

Figure 7. Half normal probability plot of modified Cook distance.

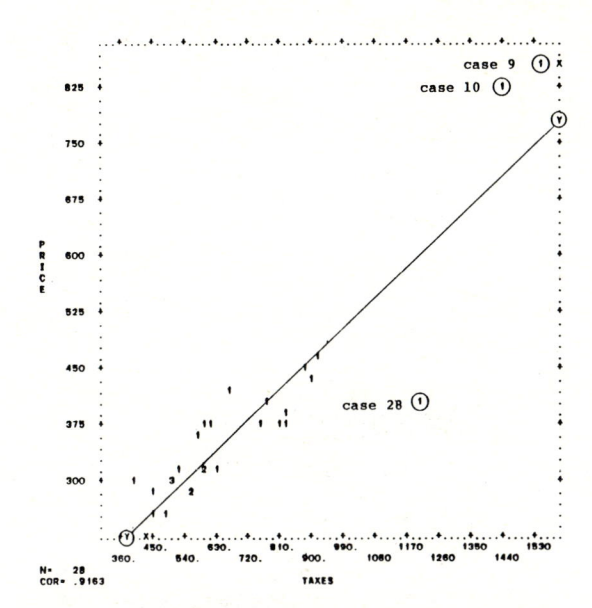

Figure 8. Cases identified as unusual by the regression diagnostics for the complete model are highlighted in a plot of the dependent variable PRICE versus one of the predictors, TAXES.

Also note that cases 9 and 10 are very expensive homes -- their prices are very high compared to the other homes. They act as leverage points in the model, so you must do some hard thinking about the price range within which you want to make estimates.

5) Case and variable selection, transformations, and BMDP Files

Other features of a mainframe statistical system that are required by the data analyst include subsetting variables and cases, convenient transformation capabilities, and flexible file facilities. BMDP has a full complement of these features.

In addition to increasing the speed of data reading, we find BMDP Files useful because they also store variable names, category codes and names, flags for missing values and values out of range, etc. Once specified as part of the file, this information need not be repeated; it is available any time the data are analyzed. Transformed data, residuals, tables for log-linear model building, factor scores, etc., can also be stored in BMDP Files.

All of the features described above are available on the StatCat. They help to decrease the time and effort necessary to produce valid results.

Input Manager

In the StatCat environment, the user has access to all system commands without leaving the program. The input manager takes advantage of the UNIX^{TM} system, allowing immediate access to all input typed by the user. This provides the user with

- a full screen edit capability to access
  - the data file
  - the BMDP instructions

The user can, for example, change a character on the screen without reentering any lines.

- the ability to communicate on a network -- e.g., to bring in a new file from another machine

- automatic recovery from fatal errors

Previously, fatal errors caused BMDP programs to terminate. On the StatCat, the program terminates only when the user indicates that the session is finished. Now, if a "fatal error" is encountered, the program restarts immediately. Thus, the annoying delay associated with the initial loading of the program from the disk is avoided.

Output Manager

On some systems, when output is viewed on the screen it is lost forever when new information appears. Consequently, the user is forced to send output to a file. This means that none of the output can be viewed until after the program terminates. However, with the StatCat output manager, the user can

- view the BMDP output at any point in the program. You can scroll repeatedly up and down through the output and move left and right (for output wider than 80 characters) or return to the top to review the BMDP instructions. (See Figure 9.)

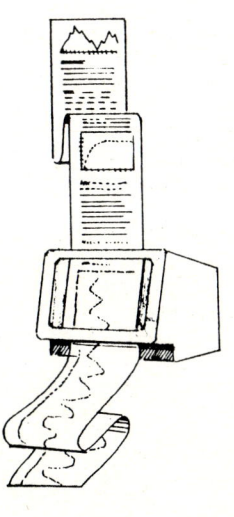

Figure 9. When the user selects 100K for output viewing, he can scroll back and forth over roughly 40 pages of output, marking panels for easy return with a push of a control key.

For example, when the program asks which variable to enter next in a stepwise regression run, you can scan any portion of the previous steps before responding.

- <u>search</u> for specific words or characters. For example, in a long stepwise regression run you may want to jump to the summary table. Just type SUMMARY and hit the FIND function key, and the summary panel will appear on the display screen.

- <u>mark output panels for easy return</u>. For example, before showing the output to a colleague, you can insert marks on stepwise regression output panels for steps of interest. Then, when displaying the output, you can hit a control key to skip ahead to the beginning of the next selected section.

- <u>write to files</u> and <u>send output to the printer</u> by hitting the appropriate control key. (The file or printer name is entered first when there is a change in destination.)

The user can also control
- the dynamic storage area reserved for the data and the output
- how the display screen is split (if at all)

In the smaller 3/4 megabyte configuration, storage is allocated as follows:

- program   400K
- code for output   50K
  manager   } 570K
- input manager   20K
- UNIX™ operating system   100K

Note that these processes run simultaneously -- the operating system is not limited to single tasks. The user can control how the remaining 200K are used for data and output storage. Most BMDP programs do not require that the data be stored in core. When the data are not in core, the minimum amount of space required for data is about 13K for a stepwise regression run with 50 variables collected for 200 subjects. However, for 7D, which <u>does</u> require data in core, the same data require approximately 45 kbytes of storage (50 x 200 x 4 bytes/word = 40K, plus storage of mean and standard deviation vectors).

If the user were to allot only 100K for output, he would be able to scroll back and forth over 40 pages of output (we are counting a "page" as 50 lines) without losing output. With output lines that average 50 characters per line, there are then 2000 lines available for searching, marking, and returning with instant response from the function keys.

Work is underway to allow the user to split the screen to display different types of information -- for example, designate the top 15 lines for output display and the bottom 9 for the associated input.

If the terminal is connected to other terminals and uses the associated networking software, the screen may be split into a maximum of four windows of information from different machines (see Figure 10).

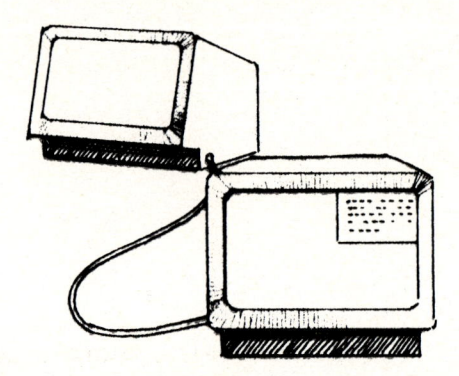

Figure 10. When networking, results from one machine may be displayed on another.

File Management

A major project now under way at BMDP is a file management program. In 1982, outside advisors were consulted about program design, and coding is now in process. This program will be available on the StatCat as well as in the mainframe distribution. Briefly, this program includes procedures for

- side by side merges (Figures 11 and 12)
- adding new records to existing files (Figure 13)
- a batch edit capability to merge in corrections (Figure 14)
- generating summary records (an aggregate procedure) (Figure 15)
- unpacking (Figure 16) -- when multiple values are stored on one record, this procedure generates an individual record for each realization and moves values of other variables (e.g., the ID and category codes) to the new records.

Begin with 3 files

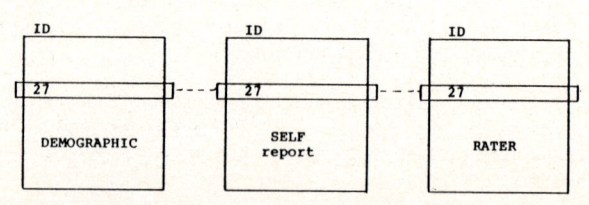

ID	ID	ID
27	27	27
DEMOGRAPHIC	SELF report	RATER

Create new records

| 27 DEMOGRAPHIC variables   SELF total score   RATER score |

Figure 11. Side by side merge.

- case selection   USE = AGE GT 20.
- transformations   SELF_SUM = TOTAL (ITEM_1 TO ITEM_20).
- variable selection   KEEP = 1 TO 20, SELF_SUM, Q_1 TO Q_20.
- padding for missing records   PAD = 'D.R'.
- print ID and match keys for problem records   LIST = '.SR', '..R'.

Figure 12. Examples of options available with the BMDP file merge program.

Begin with 2 files

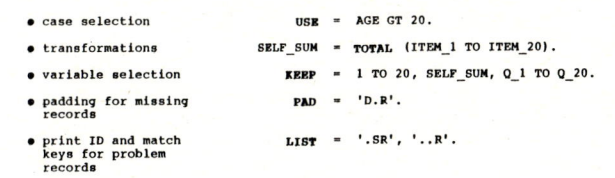

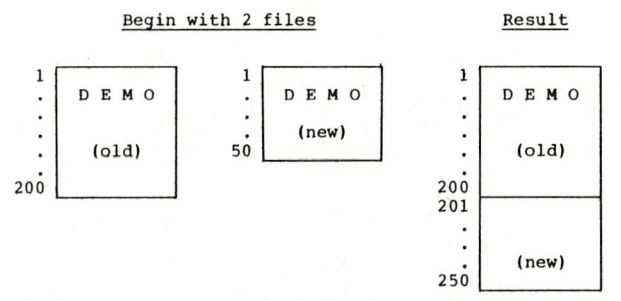

```
1                1                    Result
. DEMO           . DEMO               1
. (old)       50 (new)                . DEMO
.                                     . (old)
200                                   .
                                      200
                                      201
                                      .
                                      (new)
                                      250
```

Figure 13. Adding new records to the data file.

Current file (with errors)

corrections

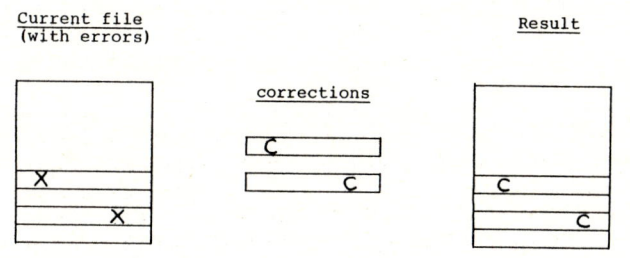

Result

Figure 14. Merging corrections into the data file.

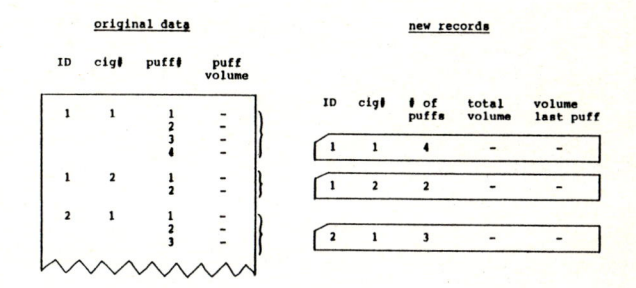

Figure 15. Generating summary records.

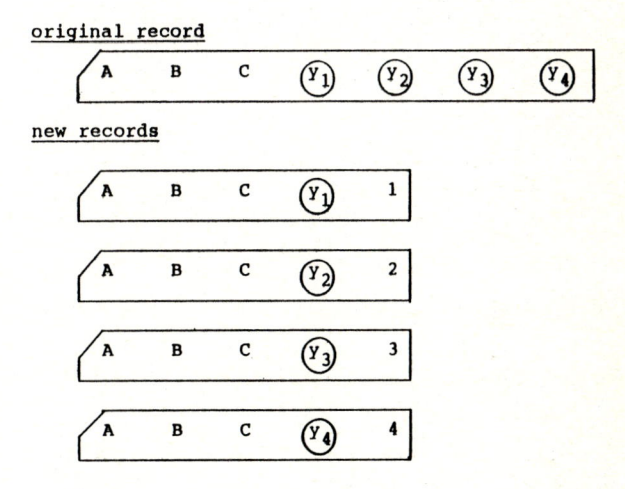

Figure 16. Unpacking multiple values from a single record.

Forms Handler and BMDP File Editor

This StatCat procedure will enable the user to lay out data entry forms on the display screen, and to edit BMDP Files. With the forms display procedure, the user can specify acceptable values for each field; when these specifications are violated, a warning is flashed to the data entry technician. This approach results in fewer errors than when keypunching (with verifying) is used.

BMDP Files not only have the advantage that the data are stored in binary to minimize rereading time, but the files contain variable names, category names, range limits, etc. Once a recording error has been found, it is convenient to edit the BMDP File instead of the raw data file. The data from the file appear on the screen through the forms display, are edited, and returned to the internal BMDP file.

Mainframe Communication

When a user purchases a StatCat, he also receives software that allows communication with any computer that can be "dialed into" from an ASCII terminal. The communications program enables users to transfer data from mainframe computer files or to move already existing special purpose programs from other computers. BMDP moved the source code for our 40 programs from an IBM environment to the StatCat in this manner. The code was compiled and run on the StatCat and the resulting output sent back to the mainframe computer for comparison with our standard output. The StatCat can also be used "plug to plug," that is, it can be plugged into another machine for use as a terminal. It is particularly convenient, of course, to "talk" to other hardware that uses the UNIX™ system.

Networking

When a user wants to have several StatCats talk together (and with speed), Ethernet local area networking support is available. The Ethernet software and hardware is provided by our colleagues at Network Research Corporation. In such an environment, the users can designate specific terminals for special tasks, such as one dedicated to data management.

Any number of StatCats, as well as VAX's, PDP-11's, IBM PC's, and various other machines, may be connected. The software provides the ability to log in to any machine on the net and transfer files at high speed or do remote execution. The user interface is highly developed and the file transfer rate is among the fastest currently obtainable.

Additional Supports

Standard software for the StatCat includes utility programs in the UNIX™ system, such as text formatting, phototypesetting, compilers, and editors. In addition, BMDP also supplies a full screen editor written by Bruce Carneal of Network Research Corporation. Software currently available from other vendors includes a word processing system, spread-sheet software for financial modeling, and a relational data base management system that has been used on larger machines for some time. Users may write special purpose programs in Pascal or FORTRAN. A graphics board may be purchased (currently, however, there is no direct interface with BMDP). The system is compatible with Tektronix 4010 software.

The standard StatCat comes with a single RS-232 serial port, which can be used to attach a printer, terminal, or modem for communication with another machine. Eight ports are available with a serial port expansion board if the user wants to attach more than one peripheral device simultaneously. It is possible to interface directly with instruments whose data are collected in analog form and require conversion to digital.

Full maintenance contracts are available in the U.S. through ITT, or maintenance can be provided through BMDP's hardware affiliate, Network Research Corporation. Most problems can be dealt with by swapping boards via overnight delivery services. The maintenance method chosen depends on how important it is for the user to be operational within 24 hours of a failure, how willing he is to open the box and pull boards (believe us, it's easy), his budget, and other considerations.

Conclusion

While none of the features described above is an earthshaking innovation, we think that collectively "in a box" they are quite something! We have made mainframe data analysis available for researchers to use NOW on a microcomputer. Further, the mainframe data analysis provided by the StatCat incorporates 20 years of development of BMDP statistics, accurate algorithms (one does not simply code textbook formulas in FORTRAN), graphical displays, etc., with extensive debugging across a broad base of user applications.

Being interactive is a desirable attribute for micro software. Because of changes within BMDP and advances in systems, the great difference existing ten years ago between "batch" and interactive use has greatly diminished. Our system advances for the StatCat let the user access system commands without leaving the program. This interface allows

- use of the full screen edit capability to edit instructions and/or data without reloading the program
- windowing back and forth through the output while the program is running
- use of control keys to mark and return to panels of output or send a screen of output to a file or the printer

Such an operating environment is certainly different from the older "batch" approach, where you submitted a deck of cards and two hours later received the output, only to find that you misspelled the instruction PLOT.

Within BMDP, changes also have been made to improve interactive use. For example, the programs with stepping procedures (multiple regression, discriminant analysis, and logistic regression) now allow the user to intervene at any step to pick a variable for entry. Some of the newer programs are written to be interactive. These include the multivariate ANOVA modeling program 4V, the spectral analysis time series program 1T, the Box-Jenkins time series program, 2T, and the new data management program (under development) for merging files and creating summary records.

In addition to the more than 20 years of development and debugging effort that has gone into the BMDP system, substantial effort has gone into input and output managers and other features to make the micro useful for data analysis. For these efforts, our programmers have worked side by side with microcomputer software experts from Network Research Corporation. Their work can be summarized in man years as follows:

- debugging FORTRAN compilers -- 1 year
  On paper, many say they have a FORTRAN compiler. Compiler producers test their products, but not as exhaustively as real use. Some small machines do not implement the full FORTRAN and other machines may have strict FORTRAN'77, which may not be compatible with existing software.

- input manager -- 3 months

- output manager -- 3 months
  The basics might be done in three days, but polishing, testing, and debugging greatly extends the effort. We have actually started over twice and now use the third version of the output manager.

- forms handler -- in progress

As a final summary, in Figure 17 we display rough costs for different systems in order to emphasize that StatCat costs are much closer to the low end of the line (where we list 8-bit machines that execute one task in limited memory) than to the costs of the larger machines. An additional factor that must be incorporated into total dollars for data analysis is "body" time. The cost incurred in having a $20,000- to $35,000-a-year programmer squeeze, adjust, write, or rewrite procedures for completing the data analysis on the smallest machines quickly evens out the cost differences between the StatCat and these machines.

References

Dixon, W.J. et al. (1983). BMDP 1983 Statistical Software Manual, Berkeley, University of California Press.

Draper, N.R. and J.A. John (1981). Influential observations and outliers in regression. Technometrics, 23:1, p. 21.

Frane, J.W. and M.A. Hill (1974). Annotated computer output for factor analysis: A supplement to the writup for computer program BMDP4M. Technical Report No. 8, BMDP Statistical Software, Los Angeles.

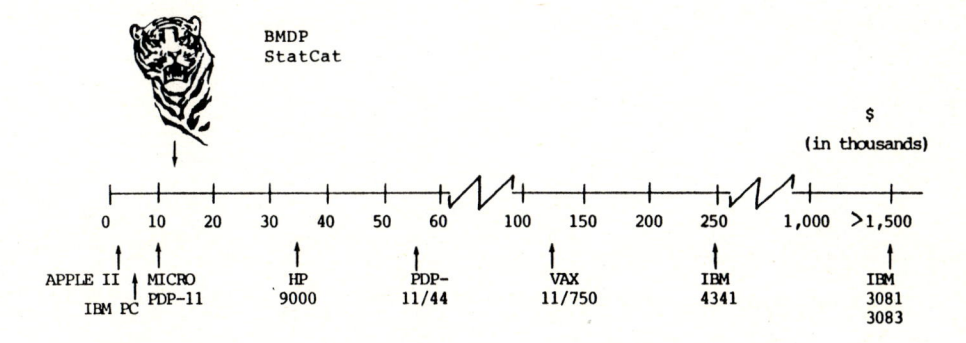

Figure 17. A dollar line (in thousands) for computers.

MACHINES AND METAPHORS

Joseph Deken

Departments of General Business & Computer Sciences
University of Texas
Austin, Texas

Computer hardware developments are causing radical changes in both the power and the availability of personal computers. In the sophisticated computing environments of these new machines, computationally effective new metaphors are emerging. The static and linearly connected models of traditional mathematics are being augmented with the notion of parallel, communicating processes. Object-oriented programming and an application to data analysis are illustrated.

In a session entitled "Software Trends," it is appropriate to have a variety of horizons. On the near horizon, I'm sure that many participants in this session are directly and immediately involved in using today's microcomputers and today's statistical software, to solve problems that need to be done today (or yesterday). For my part, I would like to scan a slightly farther horizon and discuss trends which may be felt widely within the next five to ten years.

The rapid advance of microprocessor technology we are immersed in represents much more, than simply a new era of convenient conventional computing. What we are seeing is a microcomputer revolution, which promises as a byproduct to initiate fundamental changes in both the theory and practice of statistics. As our computing machines change in both power and accessibility, they are making possible the use of new and more appropriate metaphors for data analysis.

In this paper, I would first like to establish a perspective: What is the personal computing phenomenon, and what are the important aspects of the "new machines?" From that technological perspective, I will then propose a new approach to computationally effective metaphors for data analysis. New approaches will inevitably be developed, to adapt to the radically new microcomputer environment. For the benefit of both our profession and our clients, we statistical professionals should be taking an active part.

The Micro Wave

Even if microprocessor technology meant nothing more than old computers in smaller, cheaper boxes, there would still be important repercussions of this technology for the practice of statistics. Those of us who appreciate practical techniques for real-world data analysis, could hope to broaden our definition of what is "feasible" or "interpretable, even if nothing changed in computing, except that every practicing statistician would be given a personal mainframe computer. We would be able to try more alternatives, do more experiments, and deepen our own insight (and perhaps that of our clients) into both traditional and novel methods of data analysis.

The impact of microcomputing reaches though, not just to us as professionals, but to our clients as well. It would be obviously foolish for statisticians to ignore the availability of their own powerful computers to assist them in conducting their analyses of data. It would be equally foolish to ignore the direct use of computers by their final audience. No longer can we realistically think of the statistician as the anointed emissary who takes the data from the client, goes to the computer, and returns with the truth graven on stone tablets. Thanks to microcomputers, the computer is no longer just the professional's tool, but is becoming a familiar part of the client's everyday environment. Like it or not, both non-statistical professionals and non-professional statisticians of all types are going to have computers. And, wisely or not, they _will_ be using them.

A Changing Audience

Effective data analysis rests on the skillful use of both feasible and interpretable techniques. Rather than simply focusing on the effect of our newfound _professional_ computer in determining what is _feasible_, we should also assess the impact of the client's newly discovered _personal_ computer on what is _interpretable._ The importance of both feasible and interpretable procedures is recognized by any successful applied statistician. There is a nice encapsulation in John Tukey's notes on multiple comparisons:

<> Practical power: The product of the mathematical power of a technique and the probability that the technique will be used. (Churchill Eisenhart)

<> Useful power: The product of the mathematical power by the probability that, when a positive result is found ... this result can be understood and clearly interpreted (partial understandings and interpretations duly weighted.) (John Tukey)

Some of the most interesting work in computing today, in the area of graphics and user interfaces, rests on the simple observation that most people are adept at visual processing and unskilled at speaking mathematics. What I am suggesting here is that in the near future, we ought to add the further, simple observation that most of our clients regularly use personal computers. My contention is that the regular use of a personal computer will become as relevant to the way a person _thinks_ as the fact that the person speaks English say, rather than BNF.

New Machines

It is perhaps obvious, but worth emphasizing nonetheless, that the widespread use of personal computers means much more than it would if these computers were of the programmable pocket calculator variety, or even roughly similar to the familiar eight-bit microcomputers which "broke the ice" in personal computing. The personal computer of the near future will at once be much more powerful and simultaneously much more accessible and easy to use. (Machines such as the Apple Lisa, Xerox 8010, SUN workstation, or Symbolics' Lisp Machine are just a few worthy examples to illustrate the _beginnings_ of this trend.)

Besides containing powerful microprocessors, the modern and future personal computer will be more accessible than ever because of other hardware developments. In particular, the quality of computer _displays_ has made a quantum move forward in the last few years. Rather than seeing a screen which is rigidly divided into a small number of row-and-column positions for alphanumerics, the modern computer user can enjoy a flexible _bit mapped_ display. In the bit-mapped display, the screen may easily have a million independent picture elements. (Of course, any few dozen neighboring elements may often be configured to form a letter or number. In fact, effective text presentations, using a variety of character sizes and styles simultaneously on the screen, are one of the most immediate benefits of a bit-mapped display.)

Far more consequential than its ability to change type faces, the bit mapped display has the capability of being divided into _concurrent_, _interacting windows_. These windows add a powerful new dimension to the computer's effectiveness in multi-tasking. Each individual window can display not only text and conventional graphics, but pictorial icons and animated images as well. With such versatile display capability, each window can have a communicative and comprehensible "personality," effectively representing its underlying process to the user.

Communication though, is a two-way street. Hand in hand (or hand-on-screen) with developing graphics, speech synthesis, and other _output_ capabilities, there is a powerful trend emerging toward more flexible _input_ devices. With a modern interface, the human can communicate to the computer by pointing, motioning, touching or even speaking.

When input and output capabilities emerge from the first steps now being taken, into anything like full stride, the personal computer will not be so much a machine as a conversant, powerful array of "companions" able to subtly and powerfully shape people's expectations and thinking styles. An effective computer graphics interface can go far beyond words in making the Euclidean geometric projection metaphor powerful, practical, and directly useful for human data analysts. In the same way, the new style computer, with its multi-faced, responsive and interactive interface, goes far beyond "fossil" imagery in opening up a world of new, <u>actor-like</u> metaphors for understanding data.

For all its beauty and durability, classical mathematics furnishes a restricted source of metaphors for human data analysts. In fact, in the style of mathematics which underlies all of theoretical statistics, there is only one type of "interaction" (function application) and one type of "communication" (function composition). It is a highly serial world, where the symbols are "dead." (The innards of a corpse may be rearranged without apparent effect. The symbols in a mathematical theorem have the same property.) Even the graphics in this mathematical world are "fossils," which can only be manipulated in the same way that a skull could be rotated before pressing into sand. (Hamlet aside, <u>conversation</u> with skulls or fossils is not usually attempted.)

If we could, it would be very effective for us humans to expand our analytic "language" beyond the linear, fossilized realm of classical mathematics, and bring <u>active</u>, <u>parallel</u>, <u>communicating</u>, and <u>conversational</u> metaphors into play. I have tried, in my description of computer technology developments above, to convince you that indeed we can and will develop these new analytic tools. Rather than simply speculating about details of the future though, it would be informative in conclusion to describe just one approach which is being actively pursued, and already produced effective new computing metaphors.

<u>Object-Oriented Data Analysis</u>

There are several different versions of object-oriented programming actively under development, though much of the credit and early work is associated with the Smalltalk group at Xerox Palo Alto Research Center. The details of implementing an object-oriented system can vary considerably, but the underlying concept is constant and most relevant here: Unlike a conventional serial programming language, in which <u>function application</u> is managed by <u>linear</u> control structures, the object-oriented system seeks to establish <u>parallel</u>, <u>communicating agents</u>. These agents are highly modular and independent, and communicate both among themselves and with any human participants. (For example, Smalltalk is organized in terms of actors, methods, and messages. Upon receiving a message, the actor uses one of its methods to take action.)

In an object-oriented environment, humans can use the computer effectively by being good "managers" and good "directors," as well as logicians or mathematicians. The computer is not furnishing functions it seems, as much as a competent and co-operative staff of "assistants." The key is to organize the given agents' capabilities efficiently, to communicate with them effectively.

As a specific example, we can consider the classic case of "regression." (Actually, I have used the word loosely. What is described here could be a framework for doing either familiar types of analyses or entirely new investigations.) I will describe three <u>agents</u>, describe their <u>methods</u>, and indicate their paths of <u>communication</u>. The "staff" structure created here would effectively and interactively assist a human data analyst to understand a quantitative situation.

The <u>Data-Item</u> Agent: There are many of these agents, all with the same methods and intercommunication pattern. Each data-item agent is associated with an observation. The <u>capture</u> methods handle such tasks as data acquisition, verification, missing-value reconstruction, and so on. The <u>compromise</u> methods handle "fitted" approximations to the observation, "voting" with some weight for that observation's "position." The <u>group</u> methods enable an observation to find and communicate with "neighboring" observations, should that be relevant. The communications of a data-item agent are directed to its neighbor data-items, to any active <u>organizers</u> (agents described below), and to any <u>reporters</u> (see below also).

The <u>Organizer</u> Agent: Any attempt to make overall "sense" of the observations is handled by an organizer agent. In the give and take between the organizer and the individual observations, the organizer's methods might be to <u>poll</u> the data-items as to their preferred positions, <u>suggest</u> compromise positions, <u>inform</u> data-items of non-local conditions, and ultimately <u>arbitrate</u> perhaps a final organization.

The <u>Reporter</u> Agent: The reporter might function as the principal agent to manage the human/data interaction. The state of the analysis and the current interactions between the other agents could be <u>presented</u> or <u>recast</u> in a variety of sensible and sensory formats. The reporter might also need methods to <u>interpret</u> questions from the human analyst, or to generate relevant questions for consideration.

Here is a tabular summary of the "staff." It is only a suggested cast of characters. In the dramatic era ahead, we can look forward to many casts, many plays, and much insight:

agent	methods	message
Data Item	capture compromise group	neighbor organizer reporter
Organizer	poll-suggest inform arbitrate	data items reporter
Reporter	present recast interpret -questions	organizer data items

Software for Survival Analysis

Organizer: Richard M. Heiberger

Invited Presentations:

SURVAN A Nonparametric Survival Analysis Package, Barry W. Brown, M. Elizabeth Rozell, and Eula Y. Webster

Data Analysis Considerations and Survival Analysis Programs, Frank E. Harrell, Jr.

Fitting Survival Models Using GLIM, John Whitehead

An Interactive Statistical Analysis System for Cancer Clinical Trials Data, Anne-Therese Leney and Marcello Pagano

Discussion of Papers on "Software for Survival Analysis", Richard M. Heiberger and Milton N. Parnes

SURVAN   A Nonparametric Survival Analysis Package

Barry W. Brown, M. Elizabeth Rozell, Eula Y. Webster

The University of Texas System Cancer Center
M. D. Anderson Hospital and Tumor Institute
Houston, Texas

SURVAN is a computer program for performing commonly used nonparametric analyses on survival data. The capabilities of SURVAN include Kaplan-Meier estimation of survival functions; testing for equality of functions by the log-rank or Gehan/Breslow methods; and proportional hazard estimation. The package is the culmination of two separate development efforts. One is the writing of the survival analysis algorithms; the other is the production of a generalized packaging for statistical algorithms. SURVAN is in the final stages of development.

1. INTRODUCTION

SURVAN represents a culmination of two software development efforts: writing high quality code for survival analyses and creating general and flexible routines for data input, transformation, and output.

The statistical capabilities of SURVAN include all usual nonparametric procedures. Kaplan-Meier estimation of the survival function is available with confidence estimates for failure points and prespecified quantiles of the survival function. Tests for equality of survival of several groups through either the log-rank or Gehan-Breslow procedures is available along with optional adjustment for strata. Proportional hazard estimation for assessing effects of concomitant variables on survival is possible, as is estimation of the underlying hazard function. For convenience, descriptive statistics can be obtained (mean, standard deviation) for continuous variables as can frequency distributions for discrete variables.

The packaging of the computational algorithms allows flexible data input and transformation, the handling of large problems, labeling, and acknowledgment of missing values. The population on which calculations are performed is restricted to observations that do not have missing values in relevant variables as there is no accepted method of adapting the algorithms for missing information.

2. RATIONALE

The statistical capabilities of SURVAN already exist to varying degrees in several packages. BMDP and SAS have all of them. SPSS has Berkson-Gage estimation and a variant of the Gehan-Breslow test. Other packages with these capabilities have been announced. However, some are of very poor quality in some critical respect. Attempts to obtain others, which held high promise at announcement, have failed. SAS is not available to us because we do not have a supported machine.

The primary reason for writing SURVAN is that BMDP and, to a lesser extent, SPSS are batch packages - as is SAS. An appreciable amount of effort must be spent learning the package and performing the set-up operations necessary to use it. Yet many analyses requested are for small data sets listed on a single sheet of paper. The researcher typically wants survival curves drawn and perhaps the survival experience of a few groups compared.

Thus, a very easy to use, interactive package was needed. The package should pose options to the user that let him thread his way through the analysis -- rather than forcing him to read a large manual. It should be possible to interactively enter small amounts of data in a natural fashion. Answers to inquiries about these data should be available immediately.

Advantages to writing a package instead of using an existing one include the fact that quality of code can be assured. It is also possible to add features such as nonparametric hazard estimation.

These are the practical reasons for writing SURVAN. A strong intellectual stimulus to coding the underlying algorithms arose from the realization that all of the methods used in the analysis had very strong common features and could be presented as minor variations on a single data and algorithmic structure.

The first version of SURVAN was written over a year ago. All data must reside in memory in this version, and elaborate input options and transformations are not available. The package is easy to use and provides immediate answers for small data sets. This program was well received and widely used. The many awkwardnesses that were present rarely drew comments. The number of requests to increase the work space from 10,000 storage units was surprising. Users with large problems were apologetic and claimed that their circumstances were unique. It became clear that this assessment was incorrect and that large problems were quite common. Work spaces of sufficient sizes to accommodate these problems were not available, and so algorithms were developed that would perform sequential passes of the data from mass storage.

3. DEVELOPMENT CRITERIA

The reason for writing SURVAN was to meet the somewhat incompatible goals of ease of use, flexibility, and ability to handle large problems. Quality of the package was considered important, and some of the criteria for quality are considered here.

Portability is important. The package should be as widely distributed as possible, which warrants the extra costs of development within a portable framework. One way we achieved portability was to write SURVAN entirely in ANSI FORTRAN 77. An additional advantage of remaining within a standard language is that it assures longevity. It is axiomatic in computer circles that code spontaneously decays with time. Quite often, operating systems evolve, making inoperative code dependent on special features. (That is if one is lucky. If one is unlucky the code remains operative, but produces wrong answers.) Vendors must have ANSI standard compilers to sell their machines, and so are much less likely to damage code written within the standard.

Good numeric methods must be used in order to obtain correct and reasonably inexpensive answers.

There should be a careful separation of levels of software. High-quality routines that embody the basic algorithms should be written without concern for the packaging. This greatly aids modularity and consequently the extendability of the package. It also makes the package more maintainable. Separation of levels assures that the working level routines can be extracted and used in applications totally unrelated to the program for which they were written.

4. STATISTICAL DESIGN DECISIONS

Choices between competing methods must be made in the implementation of a package. Generally, we chose good-enough approximations over computationally expensive or complex, more-exact algorithms.

The methods of Simon and Lee [1] were used for determining confidence limits on the points of the survival function and prespecified quantiles. There are more exact methods, but they tend to involve combinatorial computations that can be lengthy and difficult to program. The user generally wants only a fairly good idea of the worth of the estimate -- precise values are rarely needed.

The method of Breslow [2] was used for handling ties in proportional hazards modelling. The method suggested by Cox [3] can be inordinately expensive if ties are frequent, and a sampling of the combinatorial possibilities as suggested by some authors confuses the user by producing stochastic output.

Breslow's simple suggestion for estimating the underlying hazard function was not followed. The conditional likelihood estimate of Kalbfleish and Prentice [4] is more exact and so was used. Computationally, it involves nothing more than finding zeros of monotone functions on an interval. Algoritms which are fast and are guaranteed to converge to an answer exist in the literature [5].

How we chose to handle missing values was constrained by a lack of good statistical methods so the options were not as open as in the other issues discussed. The initial version of SURVAN ignored missing values -- not a good decision for a package intended for frequent use in clinical research. In the new version, an observation is ignored in a computation if the survival (or event), group, or strata variable is missing (assuming that group and strata variables are relevant). In proportional hazards estimation, an observation is also ignored if any variable in the regression is missing. In addition, an option is provided for the user to specify a set of variables that must be present for an observation to be used. This option is useful when related proportional hazards models are examined.

5. USER VIEW OF SURVAN

SURVAN is primarily menu driven. The exceptions are in the input and transformation packages, for which the user must provide statements in the appropriate language (as discussed later). To give some idea of the options provided in SURVAN, we will describe the input selection and overall action menu.

For input, the user is provided a choice of providing data: (1) by groups; (2) freefield -- blanks separate consecutive variables; (3) by providing data specifications (described later); or (4) by providing a FORTRAN format. Option (1) is intended primarily for interactive entry of small data sets. The user is queried for the number of groups and given the option of entering censored and uncensored data separately or entering both in a combined fashion with a "+" indicating censoring (e.g., 12+ indicates censoring at the twelfth time period). The user signals the end of a set of data by either typing "END" or providing a null line.

The overall action menu allows the user to: (1) review variables and labels and modify labels; (2) transform the data; (3) define a new survival variable; (4) review and change the environment; (5) perform statistical procedures; and (6) quit.

The environment consists of the definition of the survival and event variable, the definition of an optional grouping variable (for comparisons), an optional strata variable, and compulsory variables. Compulsory variables must all have nonmissing values for an observation to be included in a computation. Reviewing or changing the computational environment allows any except the survival variable to be changed.

Changing the survival variable must be done explicitly because it is expensive. When a survival (or event variable) is defined, the entire data file is sorted for reasons explained in the next section. Consequently, all desired work should be done with one survival variable before another is activated.

Transforming data is another potentially expensive procedure. It necessitates reading the entire data file. In addition, transformation annihilates the computational environment, including the survival variable (thus necessitating a sort), so for efficiency, all desired transformations should be performed at one time.

6. THE COMMON ALGORITHMIC STRUCTURE

As mentioned, one of the reasons for writing SURVAN was a realization that the estimation, testing, and proportional hazards likelihood calculation could be expressed as variations on a common algorithmic theme. The algorithm involves traversal of the data file in the reverse order of survival time. Long times are examined before short times. Uncensored observations are examined before the censored observations with which they are tied in length of time. By viewing the data in this order, the program knows the number of subjects at risk at each failure time.

A highly stylized version of the common algorithm is presented below followed by a discussion of the instantiation necessary to make it perform Kaplan-Meier estimation.

```
INITIALIZE-AT-RISK
INITIALIZE-TIED-DEAD
FOR (EACH OBSERVATION)
    IF (DEAD AND NOT TIED WITH PREVIOUS)
        FINISH-TIED-DEAD
        INITIALIZE-TIED-DEAD
    END IF
    UPDATE-AT-RISK
    IF (DEAD)
        ANOTHER-TIED-DEAD
    END IF
END FOR
IF (DEAD AND NOT TIED WITH PREVIOUS)
    FINISH-TIED-DEAD
END IF
WRAP-IT-UP
```

The condition DEAD AND NOT TIED WITH PREVIOUS means that there was a previous observation, the current observation is a death, and the previous observation was not a death at the same time as the current one.

The instantiation of the actions for Kaplan-Meier estimation are as follows. INITIALIZE-AT-RISK sets the number at risk (NATRISK) to zero, INITIALIZE-TIED-DEAD sets the number dead (NDEAD) at the current time to zero. UPDATE-AT-RISK adds one to the number at risk. ANOTHER-TIED-DEAD sets time of death to the current time and adds one to the number dead at this time. FINISH-TIED-DEAD calculates MULT = NDEAD/NATRISK and produces a record containing the failure time and MULT.

WRAP-IT-UP consists of several steps. The file produced by ANOTHER-TIED-DEAD contains survival times in descending order. Consequently, it is reversed and traversed. Each record other than the first has the cumulative survival function replaced by the previous value multiplied by MULT of the current record.

7. THE USE OF TOOLS IN THE ALGORITHMS

To a large extent, the maturity of the field of applied computer science can be measured by the amount of use that is made of tools -- particularly those written by others.

The general unconstrained minimization routine, HUMSOL, written by Gay [6], is used for proportional hazards estimation. Supporting routines of HUMSOL are used in the triangular factorization of the covariance matrix for calculating the statistic in the log-rank and Gehan-Breslow tests (without inverting the covariance matrix). Estimating the underlying hazard function of a proportional hazards model requires the finding of a zero of a monotone function. The fast algorithm R of Bus and Dekker [5], which is guaranteed to converge, is used for this purpose.

The PORT routines of Fox, Hall, and Schryer [7] were modified slightly for FORTRAN 77 and used to provide dynamic memory usage and uniform error handling for SURVAN.

SURVAN has an echo feature that allows the user to copy the entire interactive session or user input (or both) to files for later examination. This echo is invaluable for documentation, teaching, and communication with users who believe the program is malfunctioning.

This feature was added after the development of the program by the writing of appropriate macros in the macro-processor language ML/I created by Brown [8].

Utility routines for sorting a file of records of real numbers by arbitrary keys and reversing a file are used. Efficient code for these purposes was written by Lawrence Newton of our group using the direct access features of FORTRAN 77.

8. TOOLS USED IN PACKAGING

It may seem peculiar to begin the discussion of the packaging with a description of the tools used to produce it, but the method of producing this code is as important as the code itself.

The primary tool used is the LALR(1) parser generator of Donnegan et al, PARSGEN, [9]. (For an excellent discussion of LALR and related grammars, see Aho and Uhlman [10].) Input to PARSGEN is a description of the grammar in Backus-Naur form. The program checks the description for conformance with the LALR constraint, then produces a set of parsing tables for handling sentences in the grammar.

There are several advantages to using a parser generator rather than *ad hoc* methods for handling languages. One, adherence to the non-severe restriction that the grammar must be LALR assures that the language can be parsed. This is not a trivial advantage because the first several efforts at producing grammars for input and transformation were not LALR (and may or may not have been implementable). Two, the parsing tables provide a systematic method for handling the language. Input information is transformed to target form at each reduction step. The set of reduction rules thus forms an automatic decomposition of the problem of implementation. Three, LALR parsers guarantee that all sentences of the language will be accepted and no others will be. With *ad hoc* methods, it is easy to write code that either will not handle some statements in the grammar or will accept and possibly produce strange results from statements not in the grammar.

Possibly the biggest drawback to this methodology is that anyone but the implementor who views the parsing code will find the tables and procedures mysterious. The language is unchangable by any mechanism other than generating new tables with PARSGEN.

In order to use a parser, a lexical analyzer is necessary and was written. The lexical analyzer divides the input stream of characters into tokens such as names, integers, real numbers, delimiters, etc.

9. THE INPUT FACILITY

The novel portion of the input facility is the DATA SPECS form of file description. At its simplist, the specification consists of a list of item descriptors. Each descriptor potentially consists of three parts: (1) a variable name, which must be present; (2) an optional location or length indicator; (3) an optional data type indicator, which can be present only if one of the choices from (2) is present.

Variable names are traditional, one to eight alphanumeric characters.

A location indicator is a beginning column number followed by a colon followed in turn by an ending column number of the field. Both numbers must be present even if the field is only one column long. Length of the field in characters is presented as a single integer. If only the length of the field is specified, the beginning column is taken to follow the default number of spaces from the previously described field or from the beginning of the record if there is no previously described field. The default number of spaces separating items is specified by an integer between angular brackets, "<" and ">". Until specified, it has the value zero.

The type is separated from the previous specification by a hyphen and can consist of a nonnegative number indicating a real type item, the letter "A" indicating an alphabetic item, or a date specification. If a number is specified, the FORTRAN convention is used that this is the number of places to the right of the decimal point to be assumed when the decimal point is not explicitly presented. A date specification consists of the characters "M", "D", "Y", and "I" (for month, day, year, and ignore, respectively). A specification which could read the date 31/01/1983 would be DDIMMIYYYY. Variables whose length is given, but whose type is not are assumed to be of type 0, i.e., to contain integer values unless a decimal point is present in the field.

Variables for whom neither type nor length is specified inherit these characteristics from the next variable for which either is specified. This is handy for specifying groups of contiguous variables of identical size and type. An example is the following specification:

    <0> A01 TO A10 2 - 0,

which specifies ten variables, A01, A02, ..., A10. A01 occupies columns one and two, A02 occupies three and four, and so on. All values represent integers unless a decimal point is in the field. A01 in the above specification could not be replaced by A1 because the number of trailing integers in the beginning and ending variable names must match.

There is also a repetition facility for groups of variables. Suppose the data have a

sequence of fives pairs of entries. The first variable of each pair is an indicator of whether the event referred to by the pair in fact occurs. The second variable is the date, if applicable. The following specification could then be used:
(EVENT1 1-0 DATE1 6-MMDDYY)*5.

Another feature allows input cases to span multiple records.

10. THE TRANSFORMATION FACILITY

The transformation language is patterned after that of SPSS, and only important deviations will be discussed here. Each statement begins with a keyword and ends with a semicolon. Line breaks are equivalent to a single blank character.

Arithmetic expressions are defined in the usual way and with the usual precedence (function evaluation, exponentiation, multiplication and division, and addition and subtraction). Logical expressions are also traditional with two exceptions relating to the handling of missing values. The first feature allows testing for missing values via the two unary trailing logical operators EXISTS and its opposite FAILS. EXISTS yields true when its single argument's value is not the system's missing value, and false otherwise. Other than as arguments of these operators, missing values propagate through arithmetic expressions and force an indeterminate truth value when a missing value is compared to any value. An evident extension of Boolean manipulation is used to evaluate logical expressions. (True or indeterminate is true; false and indeterminate is false; all other combinations are indeterminate.) These features give the user total control over usage of missing values.

The COMPUTE statement can be followed by a block of assignment statements, as can the "IF condition" header. No ELSE IF or ELSE statements are available, a minor inconvenience that allowed us much easier implementation.

A RECODE statement allows values or groups of values to be recoded to other values. In contradistinction to SPSS, variables can be recoded to other variables. There is a CUT facility (identical in action to BMDP's CUTPOINT), which abbreviates the RECODE specification. CUT variable (10,20) produces three values; 1 when the variable has a value less than or equal to 10; 2 when the value is between 10 and 20; and 3 when the value is greater than 20.

The SPREAD function is used to produce zero or one valued variables from categorical data. Its intended use is in regression (in SURVAN, proportional hazards modelling). The function is quite similar to the INDICATOR facility of MINITAB, although the syntax is different.
SPREAD A(2,3,4) A2 A3 A4
defines three zero-one variables A2, A3, and A4. A2 has the value one when A has the value two, etc.

The transformation facility was intended to be used with programs which perform numeric calculations and so its ability to handle alphabetic variables is weak. Even so, the ease and accuracy to be gained from coding sex as "M" and "F" instead of "1" and "2" cannot be ignored. The RECODE ALPHA statement allows alphabetic variables to be recoded into numbers.

11. PLANS

We plan to add a few new algorithms to SURVAN, including logistic regression and nonparametric hazard estimation. Parametric modelling capabilities may be added, but available packages with these abilities must first be surveyed.

Two planned enhancements will provide the ability to do stepwise proportional hazards modelling (under strict user control) and to obtain large numbers of runs with little user effort. Both are needed because of the amount of exploratory activity present in our environment.

12. AVAILABILITY

The described version of SURVAN is under intensive development, and distributable code will not be available for several months. When completed, this code will be sent to any interested parties for a distribution fee, which is intended to cover our costs of producing a distribution package, doing the distribution, and communicating about the package thereafter.

13. ACKNOWLEDGMENTS

This work was supported in part by grants CA16672 and CA11430 from the National Cancer Institute.

REFERENCES

[1] Simon, R. and Lee, Y., Nonparametric confidence limits for survival probabilities and the median, Cancer Treat. Rep. 66 (1982) 37-42.

[2] Breslow, N., Covariance analysis of censored survival data, Biometrics 30 (1974) 89-99.

[3] Cox, D. R., Regression models and life-tables (with discussion), J. R. Stat. Soc. B 34 (1972) 187-220.

[4] Kalbfleisch, J. D., and Prentice, R. L., Marginal likelihoods based on Cox's regression and life model, Biometrika 60 (1973) 267-278.

[5] Bus, J. C. P. and Dekker, T. J., Two efficient algorithms with guaranteed convergence for finding a zero of a function, ACM Trans. Math. Software 1 (1975) 330-345.

[6] Gay, D. M., Subroutines for unconstrained minimization using a model/trust-region approach, Technical Report TR-18, CCREMS, M.I.T. (1980)

[7] Fox, P. A., Hall, A. D., and Schryer, N. L., The PORT mathematical subroutine library, ACM Trans. Math. Software 4 (1978) 104-126.

[8] Brown, P. J., Macro Processors and Techniques for Portable Software (John Wiley and Sons, London, 1976)

[9] Donegan, M., Noonan, R. E., Feyock, S., A code generator language, in Proceedings of the ACM Sigplan symposium on Compiler construction (ACM, New York, 1979)

[10] Aho, A. V., and Ullman, J. D., Principles of Compiler Design (Addison-Wesley, Reading, Mass., 1977)

DATA ANALYSIS CONSIDERATIONS AND SURVIVAL ANALYSIS PROGRAMS

Frank E. Harrell, Jr.

Division of Biometry, Department of Community and Family Medicine
Duke University Medical Center
Durham, North Carolina 27710

Many statistical issues arise in considering how best to implement survival analysis tools in computer program packages. Some of these issues are not well addressed even in more traditional areas such as multiple linear regression. An example is the failure to address multiple comparison problems arising from modeling too many factors. Another important area is the use of programs to assist the data analyst in developing, testing, and describing a model. For this purpose, the ability to output and save many types of estimates for graphing and further analysis is essential. New options in the SAS procedure PHGLM relating to these issues will be discussed.

INTRODUCTION

Since its introduction in 1972, the Cox proportional hazards (PH) regression model[1] has provided the statistician with a new arsenal of tools for handling many problems in survival analysis. Letting Y_i denote the time until failure of the i'th individual under study, the model can be stated in terms of the probability that the i'th individual survives until at least time y:

$$\text{Prob}(Y_i > y) = S_0(y)^{\exp(X_i B)},$$

where X_i denotes a vector of covariates or independent variables for the i'th individual, B denotes a corresponding vector of regression coefficients, and $S_0(y)$ is an arbitrary and unspecified "underlying" survival function. This model provides a robust procedure for estimating the effects of several covariates, as it utilizes only the ranks of the failure times across all individuals. The PH model provides a generalization of other statistical procedures such as the Kaplan-Meier[2] nonparametric survival function estimator and the "log-rank" methods such as the Peto & Peto and Mantel-Haenszel test for survival, including the ability to control for factors one does not wish to model, through stratification[3]. Hence the PH model encompasses and unifies many techniques of survival analysis.

The regression coefficients B may be estimated by the method of maximizing a partial or conditional likelihood, and the underlying survival curve $S_0(y)$ is best estimated by the discrete model method of Kalbfleisch and Prentice[4]. When the sample is homogeneous with respect to the covariates, in other words there is no regression with the covariates, B=0 and the estimate of $S_0(y)$ is exactly the Kaplan-Meier estimate based on Y alone.

When there are essentially categorical factors, say Z, for which one does not want to assume either the linear or the proportional hazards part of the model, the PH model has been adapted[3] by fitting a separate underlying survival function for each group of individuals with like Z. Here, the observations in the sample are "blocked" or "stratified" by Z while the regression coefficients can be constant across blocks. Thus pooled estimates of B can be obtained while allowing Z to enter arbitrarily into the model.

This technique is also useful for checking assumptions of the model. Suppose that one does not know whether potential continuous covariate X_j can be used in the linear part of the model. By grouping X_j into intervals, either by rounding or by computing quintiles, deciles, etc., a separate S_0 can be estimated for each group. Then survival probabilities can be estimated by varying X_j and y. By comparing these probabilities to those estimated by treating X_j as a linear term in the model, one can check whether the model is adequate with respect to X_j. A later example will demonstrate this procedure.

FINDING "SIGNIFICANT" FACTORS ASSOCIATED WITH SURVIVAL TIME

In many areas of biological research, survival analysis techniques are being used to identify which factors from a series of covariates have a significant and independent association with survival time. A convenient method is to use a stepwise variable selection strategy in which variables are added to the model in order of statistical significance, until no more variables have a statistical association with survival time[5].

The inherent problem with this strategy is that when one is examining many factors, chance associations can easily cause factors to be identified that are the result of "noise" in the data that will not replicate with further experimentation. The degree to which this is a problem is a function of the ratio of the number of factors examined to the number of observations in the sample. The problem is present in all regression situations, but even more so with survival data because the effective number of observations is closer to the number of individuals failing than to the total number of individuals. When heavy censoring is present, i.e. when most of the individuals have not failed but their survival times are censored at the last known point of contact before failure, the censored observations do not contribute much information to the sample. Some papers in the medical literature have gone so far as to try to fit a survival model using as many independent variables as there were failures.

If one were to postulate that there is some association between a set of p covariates and time until failure, a straightforward statistical test of $H_0: B_1 = B_2 = ... = B_p = 0$ can be carried out. This global statistic for testing for any regression can be a chi-square statistic with p degrees of freedom, or an F statistic with numerator degrees of freedom equal to p. In the normal linear regression setting, this statistic is calculated from the multiple R^2, and in non-normal situations involving the method of maximum likelihood it can readily be calculated using Rao's efficient score statistic[6], which does not require iterations. To control type I error while asking the question "are any of the independent variables associated with survival time?", this global chi-square statistic can be tested at, say the 0.05 level. If it is not significant, proceeding to identify which factors are associated with survival is risky. This test with p degrees of freedom penalizes the researcher for the number of factors "tossed" into the analysis.

The Statistical Analysis System (SAS) procedure PHGLM[5] computes this global score statistic and optionally will stop with the analysis if it is not significant at a given level. If it is significant, it is warranted to proceed by allowing the single most significant covariate to enter the model. Next, a "residual chi-square" statistic is computed to test the joint significance of all p-1 variables not selected, after adjusting for the variable selected. The residual chi-square is also a score statistic that is calculated without further iteration. If this statistic is significant, one can justify adding another variable to the model. This process continues in like fashion until the residual chi-square is not significant. Although this method does not take into account at a given step what has happened at previous steps, it is a reasonable one that will given greater protection against spurious associations.

The reader should also refer to the Akaike information criterion, aimed at choosing a model with optimal cross-validation accuracy[7]. This criterion specifies an even more conservative stopping rule, i.e., selecting variables until the residual chi-square drops below twice its degrees of freedom. However, cross validation studies performed by the author indicate that this criterion, when used with the residual chi-square, does result in models that validate better than models containing more variables.

Computation of a statistic like the residual chi-square should be a standard feature of every stepwise regression program to help solve multiple comparison problems.

SAS AND MODULAR PROCEDURES

The Statistical Analysis System[8] provides the statistician with many procedures that, although not designed with survival analysis in mind, can be used in conjunction with procedures like PHGLM to accomplish a wide variety of tasks. One can easily take results from one SAS procedure and input those results to another procedure because procedures communicate through the use of SAS datasets, and the user need not visualize these datasets.

Incomplete Principal Component Regression

The use of incomplete principal components regression[9] can provide regression models that will predict more accurately in a new sample because the effective number of parameters estimated is reduced. SAS procedure PRINCOMP can compute principal components and output them in a SAS dataset that can be used as input to any other SAS regression precedure, as follows:

```
PROC PRINCOMP PREFIX=P N=5 OUT=PC;
     VAR X1-X20;
PROC PHGLM DATA=PC;EVENT DEATH;
     MODEL Y=P1-P5;
```

Here 20 variables in the original dataset have been reduced to the 5 most major principal components. One can obtain coefficients in terms of the original variables by asking PHGLM to output the vector of coefficients of P1-P5 and by asking PRINCOMP to output component scores for suitably chosen dummy observations. A small PROC MATRIX program operating on these outputs will give the required result.

Predicted Survival Probabilities

The PHGLM procedure can compute estimates of the probability of surviving until time y for any combination of covariate values, blocking levels, and values of y. To allow the requests for predictions to be of a general nature, PHGLM allows the user to provide an ID variable value for each input observation. When this variable's value is missing, the corresponding observation is used in deriving model estimates but a prediction is not made for that observation. If the ID variable value is 0, that observation is used for deriving estimates and for getting predictions. If the ID variable has a value of 1, that observation is ignored during the estimation phase and is only used to specify that a certain prediction is requested. All predicted values are placed in an output SAS dataset.

The value of the covariates and of the survival time variable in the input dataset specify what predictions are desired. When a covariate value is missing for an observation having a ID variable value of 1, the overall mean value for that covariate is inserted. In this way, values of S_0 can be computed and estimates for varying values of other covariates can be requested with other covariates adjusted to their means. When the time variable is specified for an observation having ID=1, the probability of surviving until that time is computed. When the time variable is missing or unspecified, the entire survival curve (a step function) is computed.

As an example, suppose one wished to obtain survival curves empirically for each sex and race group, but adjusted using the PH model to have an age equal to mean age in the combined dataset. Hence we want to block or stratify on SEX and RACE, and model AGE with a linear PH model. The following SAS code will compute and plot the needed survival curves:

```
*Create dataset to contain prediction
    requests;
DATA EST;I=1;
        DO SEX='F','M'; DO RACE=1 TO 2;
        OUTPUT;
        END; END;
*Note that here T and AGE are not
    specified - PHGLM will output estimates
    for all T and for AGE=mean age;
*Add dummy and original data;
DATA BOTH;SET ORIGINAL EST;
PROC SORT;BY SEX RACE DESCENDING T;
PROC PHGLM BLOCK OUTP=PREDICT;
        ID I;EVENT DEATH;
        BY SEX RACE; MODEL T=AGE;
*Now dataset PREDICT contains the
    predicted survival probabilities as
    a function of T;
PROC PLOT UNIFORM DATA=PREDICT;
        BY SEX RACE;
        PLOT (SURVIVAL HAZARD)*T;
*We also plotted the hazard function;
```

If AGE had been omitted from the MODEL statement, unadjusted Kaplan-Meier survival estimates would have been computed.

The next example shows how one can display the relationship between a continuous covariate and survival probability. Here we plot the probability of surviving 1,3, or 5 years as a function of X (X=0,1,...,50).

```
DATA EST;I=1;
        DO T=1,3,5;DO X=0 TO 50;
        OUTPUT;
        END; END;
DATA BOTH;SET ORIGINAL EST;
PROC SORT;BY DESCENDING T;
PROC PHGLM OUTP=PREDICT;ID I;
        EVENT DEATH;MODEL T=X;
*Plot 3 curves distinguished by
    year=1,3,5 with X on x-axis;
PROC PLOT DATA=PREDICT;
        PLOT SURVIVAL*X=T;
```

Now suppose that one wanted to again display the impact of a continuous variable on survival, but without assuming the model to hold. Here we will block or stratify the variable AGE. In order to obtain sufficient data in each block, we will group the data into deciles of AGE. We will assume that the model holds with respect to SEX (1 or 2) groups, so we will use the model to adjust for sex differences and display predictions only for males (SEX=2) even though females are also used to develop the model. For plotting the empirical relationship between AGE and survival probability, each decile group will be labelled with the mean age in that group.

```
*First compute decile groups AGEDEC (0-9);
PROC RANK DATA=ORIGINAL OUT=DG GROUPS=10;
    VAR AGE; RANKS AGEDEC;
*Compute mean age in each decile group;
PROC SUMMARY;CLASS AGEDEC;VAR AGE;
    OUTPUT OUT=MEANS MEAN=AGE;
*Now dataset MEANS links decile group
    (AGEDEC) with mean age (AGE);
*Add time for prediction and delete
    observation containing grand mean age;
DATA EST;SET MEANS; IF AGEDEC NE .;
T=2; I=1; SEX=2;
DATA BOTH;SET DG EST;
PROC SORT;BY AGEDEC DESCENDING T;
PROC PHGLM BLOCK OUTP=PREDICT;ID I;
    EVENT DEATH;BY AGEDEC;MODEL T=SEX;
PROC PLOT DATA=PREDICT;PLOT SURVIVAL*AGE;
```

Note that dataset EST contains both the decile group indicator (0-9) and the mean age in the decile. This information is carried through to dataset PREDICT, so AGE can be used for plotting the X-axis.

SUMMARY

The Cox proportional hazards regression model unifies many statistical procedures for survival analysis and provides as special cases many other statistical tests. By allowing flexibility in terms of blocking, linear modeling, and outputting estimates of regression coefficients and survival probabilities, a versatile array of tools for analyzing survival data exists.

In the design of statistical software to be part of a system of programs, the designer should examine closely how the new procedure will interact with existing procedures. This will result in a more usable program that can deal with situations (such as predictions by decile groups or obtaining graphs from devices other than the line printer) that are not envisioned when the program is written. A side benefit of such design is that when other programs can do some of the work, a significant amount of programming time can be saved and there is less likelihood for programming errors.

REFERENCES

[1] Cox, D.R. Regression models and life tables (with discussion). *Journal of the Royal Statistical Society,* B, 34, 187-220 (1972).

[2] Kaplan, E.L. and Meier, P. Nonparametric estimation from incomplete observations. *Journal of the American Statistical Association,* 53, 457-81 (1958).

[3] Kalbfleisch, J.D. and Prentice, R.L. *The Statistical Analysis of Failure Time Data,* 87, New York: Wiley (1980).

[4] Kalbfleisch, J.D. and Prentice, R.L., op cit, 85.

[5] Harrell, F.E. The PHGLM procedure, in Reinhardt, P.S. (ed), *SAS Supplemental Library User's Guide, 1983 Edition,* Cary NC: SAS Institute.

[6] Rao, C.R. *Linear Statistical Inference and its Applications,* second edition, 418-9, New York: Wiley (1973).

[7] Atkinson, A.C. A note on the generalized information criterion for choice of a model. *Biometrika,* 67, 413-8 (1980).

[8] Ray, A. A. (ed) *SAS User's Guide: Statistics,* 461-73, Cary NC: SAS Institute (1982).

[9] Marquardt, D.W. and Snee, R.D. Ridge regression in practice. *American Statistician,* 29, 3 (1975).

ACKNOWLEDGEMENTS

This work was supported by grant HS 03834 from the National Center for Health Services Research and the National Center for Health Care Technology, Hyattsville, Md; research grant HL-17670 from the National Heart, Lung, and Blood Institute, Bethesda, Md; training grant LM 07003 and grants LM 03373 and LM 00042 from the National Library of Medicine, Bethesda, Md; and grants from the Prudential Insurance Company of America, Newark, the Kaiser Family Foundation, Palo Alto, and the Andrew W. Mellon Foundation, New York.

FITTING SURVIVAL MODELS USING GLIM

John Whitehead

Department of Applied Statistics, University of Reading, England
Visiting the Fred Hutchinson Cancer Research Center, Seattle, WA

The flexibility and convenience of the GLIM language make it a natural choice for the fitting of linear models, even when these lie outside its range of standard options. In this paper GLIM programs currently available for the analysis of survival models will be reviewed, and new macros for the fitting of piecewise exponential models will be presented. An example illustrating the use of these macros will be given and the results will be compared with those obtained by fitting Cox's model by the program RCOX.

1. INTRODUCTION

During the 1970's rapid progress was made in the development of linear models for the analysis of survival data. Parametric models based on the exponential, Weibull, extreme value and log-logistic distributions were proposed and their theoretical properties explored [1]. The regression model of Cox [2], which is only partially parametric, was particularly extensively investigated. In part, the development of these models was made worthwhile because of the existence of computers sufficiently powerful to fit them to data. The creators of the necessary software had to choose between two strategies: the creation of totally new programs or the adaptation of existing ones. In this paper the adaptations of GLIM that have been made to accommodate survival data will be reviewed.

A program specially written for the analysis of survival data in a language such as FORTRAN can be designed to make efficient use of computer time and storage and to give a specific and appropriate output. Adaptation of GLIM creates an inefficient program with output which may require careful interpretation and perhaps additional calculation before it can be presented in its final form. Despite these considerations, the use of GLIM has many advantages because of its standard features which automatically become a part of the new procedure. Thus flexible data entry is available; calculations can be performed on data and on output; patients can easily be excluded from analyses using the weight facility; the simple model notation, using either factors or covariates, is available; estimates, variance-covariance matrices and residuals can be displayed as required; plots can be produced and combinations of commands can be collected and used as macros. All these features have already been prepared and de-bugged, and are familiar to many potential users of survival procedures. The strategy of adapting existing packages to deal with survival data is efficient in user time and resources rather than in computer time.

2. THE CORRESPONDENCE BETWEEN EXPONENTIAL AND POISSON MODELS

Most procedures for fitting survival models using GLIM make use of an equivalence between linear models for exponential data which include censored observations, and linear models for Poisson data. Suppose that $T_1,\ldots,T_n$ are independent exponentially distributed survival times, some of which are right-censored. The ith observation, T_i, has survivor function $F_i(t;\theta_i) = P(T_i \geq t) = \exp(-\theta_i t)$ where

$$\theta_i = \exp(\eta_i); \quad \eta_i = x_{i1}\beta_1 + \ldots + x_{ip}\beta_p. \quad (1)$$

Here η_i is a linear model involving the known covariate values x_{ij} relating to the ith observation and the unknown parameters β_j. The likelihood of the vector $\underline{\beta}$ of parameters, based on the T_i, is

$$L(\underline{\beta};\underline{T}) = \{\Pi_U \theta_i \exp(-\theta_i T_i)\}\{\Pi_C \exp(-\theta_i T_i)\}$$

$$= \{\Pi_U \theta_i\}\exp(-\sum \theta_i T_i). \quad (2)$$

The products Π_U and Π_C are over the uncensored and censored observations respectively, the sum $\sum$ is over all observations.

Consider now independent Poisson observations $C_1,\ldots,C_n$, where C_i has mean μ_i and

$$\mu_i = \theta_i T_i = \exp(\eta_i + \log T_i) \quad (3)$$

with η_i as defined in (1). Suppose further that the C_i have realised values of 1 if the corresponding T_i is uncensored and 0 otherwise. Then the likelihood of $\underline{\beta}$ based on the C_i is

$$L(\underline{\beta};\underline{C}) = \{\Pi_U \mu_i \exp(-\mu_i)\}\{\Pi_C \exp(-\mu_i)\}$$

$$= \{\Pi_U T_i\} L(\underline{\beta};\underline{T}).$$

Thus the two forms of likelihood are proportional, and it follows that they yield the same maximum likelihood estimates and the same variance-covariance matrix.

The linear Poisson model can be fitted on GLIM using the log link. The censoring indicators, C_i, are declared to be the y-variates and the log T_i are used as offsets.

3. GLIM MACROS FOR PARAMETRIC SURVIVAL MODELS

The approach of the previous Section was described and then extended to the Weibull and extreme value distributions by Aitken and Clayton [3]. In the Weibull model, the ith observation T_i has survivor function $F_i(t;\theta_i) = \exp(-\theta_i t^\alpha)$ where θ_i was defined in (1) and α is a shape parameter common to all observations. If α is known then the simple transformation T_i^α retrieves the exponential model of Section 2. If α is unknown then an iterative procedure is employed: a starting value of α is set and the θ_i estimated, these estimates are then used in updating the value of α. As the final GLIM fit interprets the given value of α as fixed rather than estimated, the displayed variance-covariance matrix is invalid, but other output is correct for the Weibull model.

Aitkin and Clayton discuss the correction of the variance-covariance matrix, the treatment of extreme value models and the display of goodness-of-fit plots. Aitkin and Francis [4] provide listings of the necessary macros.

A relationship similar to that between exponential and Poisson models exists between log-logistic and binomial models. An observation T_i has the log-logistic distribution if its survivor function is $F(t;\theta_i) = (1 + \theta_i t^\alpha)^{-1}$ with θ_i given by (1). The scale parameter α is common to all observations. The relationship allows the log-logistic observations to be censored either on the right or on the left. The utility of the log-logistic model in survival analysis is discussed by Bennett [5], and macros for fitting it on GLIM are provided by Bennett and Whitehead [6].

The two procedures described both require additional iteration, besides that used by GLIM fit statements, in order to accommodate a nuisance parameter α. Both produce invalid variance-covariance matrices. Roger and Peacock [7] describe macros for fitting both models which use the 'own model' option in GLIM to eliminate the extra interation and to produce correct variance-covariance matrices.

A general family of survival models in which the hazard function, or a known function of the hazard function, is expressed as a polynomial is discussed, and fitted to data using GLIM, by Clayton [8].

4. COX'S MODEL AND THE PIECEWISE EXPONENTIAL MODEL

In Cox's model the ith observation T_i has hazard function $h_i(t;\theta_i) = \theta_i \lambda(t)$ where θ_i is as given in (1) and $\lambda(t)$ is an unknown hazard function. An extension of the argument presented in Section 2 allows this model to be fitted, again using the Poisson error structure of GLIM [9]. The procedure requires no extra iteration, and gives correct estimates and variance-covariance matrices. However, it does involve the creation of a large dummy data set in which each individual's record appears once for every risk set in which he or she is included. Although the grouping of individuals can reduce the size of the dummy data set (see also Thompson [10]), direct application of this approach is limited to small samples with few covariates. The next release of GLIM, GLIM4, will be included within the PRISM package and will fit Cox's model as a standard option.

Breslow [11] has pointed out that Cox's model is obtained if one assumes a piecewise exponential model with constant hazards between each failure time. This view allows Cox's model to be seen as an extension of more explicit piecewise exponential models in which the time interval is divided into a small number of subintervals in which hazard is constant. This connection is made by Laird and Oliver [12] who suggest the use of contingency table packages in the analysis of survival data. It is easy to use GLIM to fit piecewise exponential models, and if the number of pieces is small this is a practical procedure even with GLIM3.

5. GLIM MACROS FOR THE FITTING OF PIECEWISE EXPONENTIAL MODELS

In principle the fitting of piecewise exponential models is a simple application of the result given in Section 2. Consider for example a study of the survival of cancer patients from time of diagnosis to death. A 3-piece exponential model will be used with constant hazards over the intervals $(0,\%T]$, $(\%T,\%U]$ and $(\%U,\infty)$. Because of the lack of memory property of the exponential distribution, the time spent by a patient within each of the intervals will be an exponential random variable, right-censored if the patient survives the interval. The macro PRELIM in Listing 1 prepares the data for such a model.

The macro is called after the data have been entered into GLIM. It assumes that the data file consists of a record for each patient containing the survival time T, indicator of

Listing 1: Macros PRELIM and PLOT

```
$MAC PRELIM
$CALC A=%GL(3,%NU/3)
    :   IND1=%IF(%LE(T,%T),1,0)
    :   IND2=%IF(%LE(T,%U),1-IND1,0)
    :   C=%IF(%EQ(A,1),IND1*C,C)
    :   C=%IF(%EQ(A,2),IND2*C,C)
    :   C=%IF(%EQ(A,3),(1-IND1-IND2)*C,C)
    :   T=%IF(%EQ(A,1),IND1*T+(1-IND1)*%T,T)
    :   T=%IF(%EQ(A,2),IND2*T+(1-IND1-IND2)*%U-%T,T)
    :   T=%IF(%EQ(A,3),(1-IND1-IND2)*T-%U,T)
    :   W=%IF(%LE(T,0.5),0,1)
    :   T=%IF(%LE(T,0.5),1,T) : T=%LOG(T)
$FAC A 3
$WEIGHT W
$OFFSET T
$YVAR C
$ERR P $
$END

$MAC PLOT
$CALC ZPL=%FV
$SORT CPL C ZPL: ZPL
$CALC RPL=CPL/(%NU-%GL(%NU,1)+1) : %Y=%CU(CPL)
    :   GPL=CPL*%CU(CPL*RPL) : HPL=CPL*ZPL
$VAR %Y IPL JPL
$CALC IPL(CPL*%CU(CPL))=GPL
    :   JPL(CPL*%CU(CPL))=HPL
    :   IPL=%ANG(%EXP(-IPL))
    :   JPL=%ANG(%EXP(-JPL))
$PLOT IPL JPL $
$END
```

censoring C (=1 for uncensored and 0 for censored) and the covariates. The data file is prepared in triplicate running from the first patient record to the last and then repeating twice. The macro also assumes that the number of units has been set as the number of patients multiplied by three.

Macro PRELIM assigns a variable A to indicate whether a record is the first, second or third copy in the data set. In the first copy T is adjusted to give the time survived by the patient during the first interval $(0, \%T]$ and C is adjusted to take the value 1 if the patient died during the first interval and 0 otherwise. Similar changes are made to the other two copies. When a patient/time interval combination is empty because the patient died or was censored before the beginning of the interval, the corresponding copy of the patient's record is given weight W = 0. The macro also sets A to be a factor with three levels, W to be the weight, log T to be the offset, C to be the y-variate and selects the Poisson error structure. After the use of macro PRELIM further elimination of patients can be achieved by re-calculating W, and covariate transformations can be made before fitting models.

Listing 1 also contains macro PLOT which provides a goodness-of-fit plot. It is based on Altshuler's [13] estimate of the survival curve rather than Kaplan and Meier's, but otherwise it resembles the plots suggested in [3]. If the piecewise exponential assumption is truly valid then the plot produced should be linear with gradient one.

6. EXAMPLE: PEDIATRIC APLASTIC ANEMIA

Listing 2 shows a program written to analyse survival from time of bone marrow transplant for 70 children suffering from aplastic anemia, of whom 14 died during the study period. The program illustrates some of the flexibility of the GLIM language, and with its detailed comments provides an accurate record of coding and exclusions. The data on each patient are PN (patient number), C (censoring indicator), T (survival time in days), AG (grade of acute graft-versus-host disease), TR (transfusion record), PS (patient sex), DS (donor sex), and CR (conditioning regimen). A 3-piece exponential model is used with cut-off points at 100 and 200 days. After the use of macro PRELIM some recoding is performed and some ineligible patients are removed. This part of the program is stored as a subfile, and input into each GLIM session spent analysing these data to set up the system ready for modelling.

Listing 3 shows the results of such an interactive session. The null model in this context is fit by $FIT A and gives the null values of

Listing 2: GLIM program for the aplastic anemia data

```
$UNITS 210
$C
$C      MACRO PRELIM IS INPUT FROM FILE FTN33
$C
$INPUT 33 72
$C
$C      THE DATA ARE INPUT FROM FILE FTN22
$C
$DATA PN C T AG TR PS DS CR
$FORMAT
(I5,X,I1,X,I4,3X,I1,3X,I1,4X,I1,4X,I1,X,I1)
$DINPUT 22 72
$C
$C      MACRO PRELIM IS USED
$C
$CALC %T=100: %U=200
$USE PRELIM
$C
$C      RECALCULATION OF VARIABLES -
$C
$C      PS AND DS=0 FOR MALE
$C              1  "   FEMALE
$C
$C      AG=0 FOR ACUTE GVHD GRADES 0 OR I
$C         1   "    "    "    "    II, III OR IV
$C
$C      CR=0 FOR OTHER CONDITIONING REGIMEN
$C         1  "   BUFFY COAT CELLS
$C         2  "   TBI
$C
$C      SM=0 FOR SEX MATCH
$C         1  "   SEX MIS-MATCH
$C
$CALC PS=PS-1: DS=DS-1
    : AG=%IF(%GE(AG,3),1,0)
    : CR=3-CR
    : SM=(PS-DS)*(PS-DS)
$C
$C      EXCLUSION OF SUBJECTS - WEIGHT ZERO IF
$C
$C      1).   CR=2      (CONDITIONING REGIMEN IS TBI)
$C      2).   PN=1540   (PATIENT RECEIVED CYCLOSPORINE)
$C
$CALC W=%IF(%EQ(CR,2),0,W)
    : W=%IF(%EQ(PN,1540),0,W)
```

the three constant hazards as exp(-7.124) = 0.0008, 0.0002 and 0.0000 respectively. The goodness-of-fit plot deviates from linearity at the left hand end.

Fitting incidence of acute graft-versus-host disease reduces the deviance by 22.15 and is highly significant. The goodness-of-fit plot is more satisfactory in this case. A further fit of patient sex does not reduce the deviance significantly.

The results of fitting Cox's model using the program RCOX [14] are included in parentheses on Listing 3. Once AG is included in the model, so that the goodness-of-fit plot is satisfactory, the two sets of results correspond closely.

ACKNOWLEDGEMENTS

The author wishes to thank Drs. Charles Odoroff and Henry Davis of the Division of Biostatistics, University of Rochester, N.Y. for helpful discussions concerning the use of piecewise exponential models.

This work was supported by NIH grant GM-28314.

Listing 3: Results from the aplastic anemia data

```
$FIT A $

            SCALED
   CYCLE   DEVIANCE      DF
     6      98.57        178

$DIS E $

         ESTIMATE      S.E.      PARAMETER
    1     -7.124      .4472       %GM
    2     -1.543      1.080       A(2)
    3     -2.859      .6322       A(3)
   SCALE PARAMETER TAKEN AS      1.000

$USE PLOT $

    1.50      *                              I
    1.49      *
    1.48      *
    1.46      *                          I
    1.45      *
    1.44      *                        I
    1.43      *                      I
    1.42      *
    1.40      *                 I
    1.39      *                 I
    1.38      *                 I
    1.37      *
    1.36      *         I
    1.34      *        I
    1.33      *
    1.32      *   I
    1.31      *
    1.30      *  I
    1.28      *  I
    1.27      *  I
    1.26      *
    1.25      * I
         .........*.........*.........*.........*.........*.........*
              1.28      1.36      1.44      1.52      1.60      1.68
```

REFERENCES

[1] Kalbfleisch, J.D. and Prentice, R.L., The Statistical Analysis of Failure Time Data (Wiley, New York, 1980).

[2] Cox, D.R., Regression models and life tables, J.R. Statist. Soc. B26 (1972) 187-220.

[3] Aitkin, M. and Clayton, D., The fitting of exponential, Weibull and extreme value distributions to complex censored survival data using GLIM, Appl. Statist. 29 (1980) 156-163.

[4] Aitkin, M. and Francis, B., A GLIM macro for fitting the exponential or Weibull distribution to censored data, GLIM Newsletter (June 1980) 19-25.

[5] Bennett, S., Log-logistic regression models for survival data, Appl. Statist. (In press).

[6] Bennett, S. and Whitehead, J., Fitting logistic and log-logistic regression models to censored data using GLIM, GLIM Newsletter (June 1981) 12-19, and Correction, GLIM Newsletter (December 1981) 3.

[7] Roger, J.H. and Peacock, S.D., Fitting the scale as a GLIM parameter for Weibull, extreme value, logistic and log-logistic regression models with censored data, GLIM Newsletter (In press).

[8] Clayton, D.G., Fitting a general family of failure-time distributions using GLIM, (In preparation).

[9] Whitehead, J., Fitting Cox's regression model to survival data using GLIM, Appl.

Listing 3 (cont.)

```
$FIT A+AG $

          SCALED
  CYCLE   DEVIANCE      DF
    7      76.42        177      χ² for AG = 22.15 (27.83)

$DIS E $

          ESTIMATE      S.E.     PARAMETER
     1    -9.360        1.056    %GM
     2    -1.367        1.093    A(2)
     3    -2.652         .6324   A(3)
     4     3.446        1.038    AG        (3.509, 1.051)
  SCALE PARAMETER TAKEN AS       1.000

$USE PLOT $

     1.52      *
     1.50      *                                              I
     1.47      *                                              I
     1.45      *
     1.42      *                                        I
     1.40      *                                        2
     1.37      *
     1.35      *                                     I
     1.32      *
     1.30      *                              I
     1.27      *
     1.25      *                          I
     1.22      *                        I
     1.20      *              I
     1.17      *
     1.15      *         I
     1.12      *       I
     1.10      *
     1.07      *     I
     1.05      *    I
     1.02      *
     1.00      *
          .........*..........*..........*..........*..........*..........*
               1.02       1.14       1.26       1.38       1.50       1.62

$FIT A+AG+PS $

          SCALED
  CYCLE   DEVIANCE      DF
    7      76.38        176      χ² for PS = 0.04(0.04)

$DIS E $

          ESTIMATE      S.E.     PARAMETER
     1    -9.290        1.112    %GM
     2    -1.369        1.093    A(2)
     3    -2.656         .6328   A(3)
     4     3.422        1.046    AG        (3.486, 1.058)
     5     -.1232        .6305   PS        (-0.1132, 0.6328)
  SCALE PARAMETER TAKEN AS       1.000
```

Statist. 29 (1980) 268-275.

[10] Thompson, R., Letter to the editor, Appl. Statist. 30 (1981) 310.

[11] Breslow, N., Covariance analysis of censored survival data, Biometrics 30 (1974) 89-99.

[12] Laird, N. and Oliver, D., Covariance analysis of censored survival data using log-linear analysis techniques, J. Am. Statist. Soc. 76 (1981) 231-240.

[13] Altshuler, B., Theory for the measurement of competing risks in animal experiments, Math. Biosc. 6 (1970) 1-11.

[14] Peterson, A.V. Jr., Prentice, R.L. and Marek, P.M., Implications and Computational considerations of the Cox partial likelihood analysis, In Proceedings of the Symposium on the Interface: Computer Science and Statistics (In press, 1983).

AN INTERACTIVE STATISTICAL ANALYSIS SYSTEM FOR CANCER CLINICAL TRIALS DATA

Anne-Therese Leney, Marcello Pagano

Dana-Farber Cancer Institute
Harvard University

A statistical analysis system must be efficient, flexible, and complete to serve the needs of its users effectively. If computations are to be feasible, efficient data access methods and high performance algorithms are necessary. Flexibility is also essential, both internally, to take advantage of the inherent structure of the data, and externally, to provide a user-interface for varying levels of expertise. The need for completeness implies that a system have data modification and selection features, be capable of performing standard statistical procedures, and in areas where it may be lacking, be generally compatible with external user programs or special-purpose packages.

Meeting these goals is a complex task, as is exemplified by the lack of such systems today. This paper explores the capabilities of the DASH system, designed to conform to the previous criteria and aimed at performing the calculations necessary to analyze clinical trials data. It examines the internal structure, user-interface, and statistical algorithms currently employed.

1. INTRODUCTION

Traditionally, the analysis of clinical trials data has involved using a standard statistical package along with several stand-alone programs to handle more elaborate survival analyses. Cox regression that handles stratification, competing risks, and time-varying covariates, logistic regression with polychotomous response, nonparametric methods which handle censored observations, etc. are essential to a thorough analysis. It is uncommon to find these methods available in a standard package, nor should they be, given that their application is remote from the general problems which prompted package design. It was this lack of a coherent portable package for the analysis of survival data, which would provide both standard procedures as well as those needed for clinical trials analysis, which spurred the development of the DASH system.

2. SYSTEM OVERVIEW

DASH (Data Analysis System at Harvard) is an interactive data analysis system designed to be efficient, flexible, and complete. Efficiency is essential in any system if computations are to be feasible. The data access methods used, the system storage requirements, and the statistical algorithms employed all affect efficiency. To speed data access, DASH transposes the original data, typically stored as case records, to a working data structure indexed by variables, the units of interest in clinical trials analysis.

Program storage requirements are minimized by a highly modular design, allowing for efficient execution and easy expansion. Though storage constraints will limit the volume of data which may be analyzed, the functionality of the system is unaffected by the hardware environment. As for computational performance, special attention was given to the numerical properties, speed, and stability of all algorithms used.

A flexible user-interface is an important design consideration in a community where users have varying levels of expertise. DASH has incorporated a number of features to accommodate both the naive and sophisticated user. The system is command-driven with each command name selected to reflect its function accurately. There are four basic command types: information commands which guide the user during a session; data commands which handle input and output; utility commands for variable and case selection or transformation; and execution commands for statistical analysis. Control of appropriate command usage is handled internally, so the user may interchange between functions freely. Abbreviations for commands, a prompt/no prompt option, an on-line help system, and extensive error recovery are among the other features provided.

Completeness implies that a system serve the needs of the users for which it was intended. The analysis of clinical trials typically requires data modification capabilities and standard statistical procedures as well as methods particular to survival analysis. DASH

Research supported in part by Grant CA-28066 from the National Cancer Institute, DHHS.

was designed to provide these features and in areas where it is lacking, such as graphics, can easily be interfaced with external user programs or special-purpose packages. The system was written in standard Fortran IV to promote this compatibility as well as machine portability.

3. SYSTEM STRUCTURE

The DASH system can be divided, for purposes of discussion, into its program structure and its file structure. The program is composed of numerous segments forming a tree. The root consists of the general commands, the statistical modules form the nodes, and the commands within each statistical module represent the leaves of the tree. This hierarchical structure, though invisible to the user, facilitates command control and improves performance on machines where overlays are needed.

DASH currently supports a flat file structure. A data file, consisting of fixed-length case records, and a format/labels file must be specified at the start of a session. The data file is transposed to allow random access by variable, the unit of interest in analysis. DASH supports a number of other optional files, specified at the discretion of the user. An output file may be used to redirect the results of an analysis, though prompts and error messages will still appear on the default output device. This file may then be sent to a letter-quality printer or word-processing system for report generation. A copy of the data and format/labels file may be made at any time during an analysis, reflecting their current state. This feature is particularly useful if the data have been transformed and subsetted extensively yet are needed in this new form at a later date. DASH also supports an indirect command file, which allows the user to predefine the sequence of commands and use the system in a batch rather than interactive mode. Furthermore, for each DASH session a command file containing all the executed commands is created by the system. This file may be used indirectly if an analysis needs to be regenerated at a later time, or may simply be kept as a log of a completed analysis.

4. USER-INTERFACE

Because of the diverse levels of computer awareness in a user community, DASH was designed to be highly user-oriented. This implies a need for not only user-friendly options to help the new user but also shortcuts for the more familiar user, a feature often missing in interactive systems.

For the naive user, DASH provides an on-line help system, a detailed users' manual, full-prompting, and information commands which should minimize the need for user-training. In addition, command names reflect the function of the command they enact and may be misspelled, provided the resulting name is not ambiguous. The command argument syntax is also quite flexible, accepting a variety of separators and punctuations.

A number of shortcuts is also provided for the sophisticated user. To minimize keystrokes, abbreviations for commands are accepted, provided they uniquely identify the command in question. The command argument syntax allows for conjunctives and collectives, making it possible to perform multiple operations in a single step. Commands specified during a session are recorded so they may be altered for a modified analysis or saved for a repeated analysis. These command files may then be used indirectly, eliminating the need for re-entering each command at some future date.

Another aspect of the interface which is helpful to all users is extensive error recovery. No error is considered fatal, and the system provides both diagnostics and correction suggestions. Comments for documenting a session, upper/lower case translation, and redirection of output are some other features included for the convenience of the user.

5. ANALYSIS CAPABILITIES

DASH provides a number of powerful utility commands used in preparing the data for analysis. Commands are available for editing, copying, subsetting, creating, censoring, recoding, and displaying data. These commands may be used at any time during a session once the data file has been opened and transposed, thus allowing data transformation and selection within an analysis as well as before. An on-line calculator is also provided within DASH in an attempt to address secondary needs as well as the primary one of analysis software.

Currently there are four statistical modules available in the system: statistical summary, nonparametric methods, logistic regression, and Cox regression. Times does not permit a detailed description of all the analysis modules, so we briefly describe the first three and detail the fourth.

The statistical summary module allows the user to calculate basic statistics and generate contingency tables. The statistics available are mean, median, quartiles, minimum, maximum, standard deviation, correlation, and a summary which gives all the statistics just mentioned. The tables generated may be up to 5-way and optionally give row, column, and total percentages, a chi-square statistic, and expected values, as well as the counts. The commands in the statistics module have a wide range of valid syntactic forms to allow for prompting, case conditioning within the command, collections of variables, and categorizing by variables.

The commands within the nonparametric module invoke two-sample tests for comparing survival distributions based on the permutation distribution for the Wilcoxon, Logrank, Peto (Generalized Wilcoxon), or Permutation scores for censored data. The first three score functions are given by Prentice and Marek [1] while the Permutation scores for censored data are given by Pagano and Tritchler [2]. All four tests assume equal censoring patterns for the two groups.

With censoring the Wilcoxon test is the Gilbert-Gehan test, without censoring it reduces to the Wilcoxon rank-sum test. All tests are approximate unless the user requests exact significance. Optional stratification of the outcome may also be specified.

The logistic regression module is based on an algorithm developed by Pagano and Tritchler for finding the maximum likelihood estimates of the parameters of the logistic regression model for polychotomous response data [3]. This is particularly useful in clinical trials where response to treatment is often analyzed as failure, partial response, and complete response, giving a trichotomous response variable. The methods used in estimation are numerically stable yet have minimal storage requirements. Once the user has specified the response variable and covariates, a number of options are available. To change the covariate list, covariate(s), and/or interaction terms may be explicitly added or dropped, the least significant covariate may be automatically dropped, or covariates may be retained, based on some significance level. Similarly the response variable may be changed either by respecifying it or by collapsing one or more of its categories. Estimation may occur initially and with each alteration of the covariates or response, giving some subset of the following, as determined by the user: the number of observations, number of iterations, maximum loglikelihood when all coefficients except the constant are zero, maximum likelihood estimate (MLE) and associated p-value for each model parameter, test whether all coefficients except the constant equal zero, estimated covariances and correlations of the MLEs within a response category, and the estimated cross-covariances and cross-correlations of the MLEs across response categories.

The commands within the Cox module allow the user to analyze censored survival data, with competing risks, for both fixed and time-varying covariates, using the Cox regression model. The algorithm, based on the Newton method for finding the maximum partial likelihood estimates, is numerically stable and well-suited to large datasets [4].
Specifically, if an individual's covariates are $\underline{z} = (z_1,\ldots,z_p)^T$, then Cox [5] models the hazard function

$$h(t, \underline{z}) = h_o(t) \exp(\underline{z}^T \underline{\beta})$$

where $\underline{\beta}$ is a p x 1 vector of unknown parameters and $h_o(t)$ is an arbitrary and unknown hazard function. He proposes that inference about the $\underline{\beta}$ be based on the "partial likelihood" function

$$L(\underline{\beta}) = \prod_{i=1}^{k} (e_i / \sum_{j \in R(t_i)} e_j)$$

where $t_1 < t_2 < \ldots < t_k$ are the distinct failure times, $R(t)$ indexes the set of individuals who have not failed by time $t-0$, and $e_j = \exp(\underline{z}_j^T \underline{\beta})$, $j = 1,\ldots,n$, n the number of observations. To relax the assumption that failure times are distinct, suppose that m_i failures occur at t_i, $i = 1,\ldots,n$. There is some disagreement in the literature as to how to handle ties. We choose to maximize the approximant to the loglikelihood

$$\log L(\underline{\beta}) = \prod_{i=1}^{k} \ell_i$$

with

$$\ell_i = \sum_{j \in M_i} \underline{z}_j^T \underline{\beta} - m_i \log \sum_{j \in R(t_i)} e_j,$$

where the individuals indexed by M_i,

$$M_i = \{j : t_j = t_i \text{ and } t_i \text{ is a failure time}\}.$$

One desirable property of the $\underline{\beta}$ that maximize this approximation to the partial likelihood is that they are clone invariant. That is, if for a fixed positive integer m every observation is replaced by m copies of itself, then the log likelihood is multiplied m-fold and hence the maximizing solution is unchanged. This invariance is shared by the common location and regression estimators, but does not hold true for other partial likelihood approximants.

To maximize the log likelihood with respect to the $\underline{\beta}$, consider the first two derivatives

(1) $\partial \ell_i / \partial \beta_j = \sum_{\ell \in M_i} z_{\ell j} - m_i (\sum_{\ell \in R(t_i)} z_{\ell j} e_\ell) \div \sum_{\ell \in R(t_i)} e_\ell$

and

(2) $-\partial^2 \ell_i / \partial \beta_j \partial \beta_k = m_i \{(\sum_{\ell \in R(t_i)} z_{\ell j} z_{\ell k} e_\ell) \div (\sum_{\ell \in R(t_i)} e_\ell) - (\sum_{\ell \in R(t_i)} z_{\ell j} e_\ell)(\sum_{\ell \in R(t_i)} z_{\ell k} e_\ell)\} \div \sum_{\ell \in R(t_i)} e_\ell.$

One, of course, would not use these formulae to compute the quantities involved, since the second derivative may involve the difference of two very large almost equal numbers and the round-off errors can be substantial. The subsequent effect is to have a matrix which is not positive definitive determine the Newton step in the optimization.

Alternative formulae exist for the first two derivatives:

(3) $\underline{C}_i = \sum_{j \in M_i} (\underline{z}_j - \underline{\bar{z}}(i))$

and

$$(4) \quad V_i = M_i \sum_{j \in R(t_i)} (\underline{z}_j - \underline{\bar{z}}(i))(\underline{z}_j - \underline{\bar{z}}(i))^T W_j(i)$$

where

$$\underline{\bar{z}}(i) = \sum_{j \in R(t_i)} \underline{z}_j W_j(i),$$

and

$$W_\ell(i) = e_\ell \div \sum_{j \in R(t_i)} e_j, \quad \text{for } \ell \in R(t_i).$$

Analogously to the calculation of sample covariances, these formulae have better numerical properties than (1) and (2). For one, V_i calculated according to (4) is at least positive semidefinite, whereas if it is calculated according to (2) no such guarantee can be made. Unfortunately, if (3) and (4) are implemented directly, they would require either considerable storage or k sweeps through the data for each iteration of the Newton method. To overcome this difficulty and still retain the desirable property of (4), one can employ the easily verified updating formulae:

$$s_{i-1} = s_i + \sum_{j \in R(t_{i-1}) - R(t_i)} e_j$$

$$W_j(i-1) = e_j \div s_{i-1} \quad \text{for } j \in R(t_{i-1}) - R(t_i)$$

$$\underline{\bar{z}}(i-1) = s_i \underline{\bar{z}}(i) \div s_{i-1} + \sum_{j \in R(t_{i-1}) - R(t_i)} \underline{z}_j W_j(i-1)$$

$$V_{i-1} = M_i \{ s_i V_i \div s_{i-1} +$$
$$(\underline{\bar{z}}(i) - \underline{\bar{z}}(i-1))(\underline{\bar{z}}(i) - \underline{\bar{z}}(i-1))^T s_i \div s_{i-1}$$
$$+ \sum_{j \in R(t_{i-1}) - R(t_i)} (\underline{z}_j - \underline{\bar{z}}(i-1))(\underline{z}_j - \underline{\bar{z}}(i-1))^T W_j(i-1) \}.$$

Note that the index starts at the last risk set and then proceeds backwards. The Newton algorithm then requires the solution of the linear equations

$$V \underline{\Delta} = \underline{C}$$

where

$$V = \sum_{i=1}^{k} V_i$$

and

$$\underline{C} = \sum_{i=1}^{k} \underline{C}_i .$$

Since V is positive semidefinite, the Cholesky method can be used to solve for $\underline{\Delta}$, and the obtained values of $\underline{\Delta}$ may then be checked for convergence. The Cholesky factors of V at the last iteration may also be saved to find the inverse of V, which in turn can be used as an estimator of the asymptotic covariance matrix of the estimated $\underline{\beta}$.

This algorithm can easily be extended to handle competing risks, stratification, and time varying covariates. Furthermore, covariates may be added or dropped from the model as in the logistic regression. The statistics available following estimation are also comparable to those discussed previously, though the program will additionally, at the users request, estimate the survival function within a stratum, standardize the survival curves, or perform a goodness-of-fit analysis.

6. DISCUSSION

In many respects DASH has met its goal of providing a flexible and efficient package for the analysis of clinical trials data, though additional work is needed before it can be considered complete. Since possible enhancements are infinite with any system, specific directions must be chosen. Plans are currently underway to improve the software offerings and to increase the diversity of hardware implementations.

In the software realm, support of multiple data structures is one area where the system is lacking. Flat files are not designed to take advantage of the hierarchical relationships which frequently exist in clinical trials data. An alternative hierarchical structure is currently being developed which would provide more efficient data access methods when the data are inherently in this form and give more flexibility to the user in data management.

DASH was designed to be highly modular and portable in order to minimize hardware concerns. The system is running currently on a DEC-2060 at the Dana-Farber Cancer Institute in Boston, and is being installed at the Peter Bent Brigham Hospital, also in Boston, on a VAX 11/780. Dash is also being implemented on a microprocessor.

REFERENCES

1. Prentice, R. L. and P. Marek, "A Qualitative Discrepancy Between Censored Data Rank Tests", Biometrics, 35, 1979.
2. Pagano, M. and D. Tritchler, "On Obtaining Permutation Distributions in Polynomial Time", Technical Report (1980), Department of Biostatistics, Sidney Farber Cancer Institute, Boston, MA.
3. Pagano, M. and D. Tritchler, "An Algorithm for Logistic Regression with Polychotomous Response", Technical Report (1980), Department of Biostatistics, Sidney Farber Cancer Institute, Boston, MA.
4. Chang, A. I. and M. Pagano, "An Algorithm for Maximizing Cox's Likelihood Function for Censored Survival Data", Technical Report (1980), Department of Biostatistics, Sidney Farber Cancer Institute, Boston, MA.
5. Cox, D.R.: Regression models and life tables. J. Roy. Statis. Soc. B, 34, 187-220, 1972.

Software for Survival Analysis

Richard M. Heiberger and Milton N. Parnes

Temple University
Department of Statistics
Philadelphia, Pa. 19122

Survival analysis software is a special case of software. The software design considerations and solutions are the same as those for other types of software. The speakers have discussed these issues and shown a variety of solutions, all of which depend on various tools for software development. In addition, the speakers have advanced the specifics of algorithms for survival analysis itself.

Packaging

The speakers in this session are representatives of four of the preeminent institutions interested in survival analysis. These institutions have the resources to develop software for their needs and are willing and able to make it available to others.

The speakers addressed the issue as to what software design decisions they made in forming a survival analysis package. Two have worked within the framework of large existing systems and two report on systems developed ab initio. In the first category, John Whitehead has used the macro feature of GLIM and written macros for survival analysis constructed entirely of GLIM statements. Frank E. Harrell has written a new PROC for proportional hazard capabilities for the SAS package. SAS procedures are written in Fortran or PL/1 and use subroutine calls provided by the SAS system to link with the rest of the package. In the second category, Barry W. Brown, M. Elizabeth Rozel, and Eula Y. Webster report on their development of SURVAN, and Anne-Therese Leney and Marcello Pagano report on their development of DASH.

Existing Packages

John Whitehead's perceived user is a GLIM user or would be GLIM user. He noted that in order to gain the use of GLIM's general features there is a loss of computing efficiency. He says this loss does not recommend it for use in a group whose primary computation need is for survival analysis. There is, of course, a consequent gain in overall efficiency. The survival analysis user can go directly to survival analysis and not worry at all about the details of the computation. The paper further tells us that the MACRO feature also allows the immediate use of alternative parametric methods, mentioning in particular the use of the Weibull distribution and a recent piecewise exponential method of Laird and Oliver. We believe these are valuable features and that they show how flexible a well designed system with macros, such as GLIM, can be.

Whitehead also reports that a new version of GLIM is soon to be released with more efficient survival capabilities. They will make the GLIM system even more attractive.

Frank Harrell's paper shows his PHGLM is as flexible for Proportional Hazard models as is SAS's GLM for regression models. In fact, there is much to be learned about the use of regression in general from the reading of his paper. Harrell has succeeded in maintaining the features that lead to SAS's popularity. One major strength and drawback of SAS is that the SAS system is presently tied to a fairly expensive bit of hardware (the IBM 370 or equivalent) which many institutions (including our own Temple University) do not have. The strength is the efficiency that can be gained by designing for a specific computing environment. The weakness is the lack of portability. The intended users are SAS users who need an additional data analysis capability.

Custom-designed Packages

SURVAN and DASH are being custom designed to suit the particular needs of their institutions. It has taken them many man-months of work to bring these packages to their present stage and they are not yet complete. The developers have designed interactive data handling packages with some general features and, of course, the specific survival analysis algorithms. They have freely adopted procedures, algorithms, and programming styles from the literature and from their own experience. Both groups are the pioneers in survival analysis and have both the needs and resources to justify such an undertaking. We ought to remember that these two groups have been doing survival analysis since long before there were any survival subroutines on the large packages.

The DASH group designed a command language for ease of usage (e.g., mnemonic command names, complete HELP files, tolerance of misspellings, easy interaction with the computer system's file structure and other programs (e.g., for graphical output)). They also developed faster and more accurate algorithms, particularly emphasizing an improved algorithm for maximizing Cox's Likelihood function.

The SURVAN group chose a menu driven command structure for the same goal of ease of use. They were partially stimulated by the realization that all the methods in the non-parametric survival package could be presented as "minor variations on a single data and algorithmic structure".

The Medical End User

Brown noted that the medical people are also users of the system and have an effect on its development. It is possible to display the variance of the estimates at the various percentiles of the estimated distributions. The medical people were surprised at how large the variance is in the tails and they did not want to use the variance display in their publications.

It is our own experience that the Kaplan-Meier product-moment estimator is often used as a lead display in articles and almost as an EDA tool to medical people (a sometimes forgotten end-user of the software). We note that the separation in the tails of hte display looks just about as striking as separation in the beginning. This is very deceptive to the uninitiated and something should be done to clarify the false impression. One method that can be helpful is to occasionally write down the number of patients at risk along the curve. Another practice that is sometimes used, and that can give a misleading impression, is to stop a curve at the point of last death. The short curve is often not perceived as the good sign it probably is. To stretch this point to its mathematical limit the curve for a great success with no deaths consists of a single point above the origin (or perhaps no point at all).

Software Tools

Brown gives a nice discussion of the need for and the use of software development tools, such as automatic parser generators. We would like to note that all of the programs have effectively made use of such tools. Both SURVAN and DASH use the tools explicitly. PHGLM uses them implicitly by embedding itself within the SAS system and using the many subroutine calls to these tools that SAS has made available to developers of new SAS procedures. Both types of usage give the programmer of the basic statistical algorithm identical access to the results of the parsed input. The difference is that in the first case the program developer may interact with the parser developer (perhaps both are the same person), while in the second case the programmer is working with a fixed parsing scheme.

One major advantage to the use of parser generator technology is the ease with which the definition of the user language can be changed. Errors can be corrected, and additional capabilities can be added by minor changes to the grammar definition in the Backus-Naur form. All the hard work of setting up tables is done by the computer. Heiberger and Laurance (1982) found this to be of major importance in the construction and development of user level languages. Thus we disagree with Brown's assessment that automatic generation of parser code and tables leads to mysterious tables. Anyone wanting to find out what the code does would go directly to the BNF specification used as input to the parser generator. No one would look at the generated Fortran listing, any more than they would look at the assembly listing of generated object code.

The construction of macros, such as Whitehead's in GLIM, uses a different set of sophisticated programming tools. These consist of both the MACRO facility which was designed into GLIM, and the power of the individual GLIM statements which have been used in the MACRO. The more fundamental software tools are not directly accessed at all. In fact, the macro writer need not even be aware of their existence.

All four speakers used other tools, primarily numerical optimization routines. The three programmers used them directly, the macro writer indirectly through the routines in GLIM.

Conclusion

In the keynote address of this meeting, Richard Hamming decried the over production and duplication in the literature. It is obvious that much of the preparation of survival programs is and will continue to result in duplication of effort and end results. Is this really wasteful? We think not. Survival analysis will be around for long time. We already see that there is borrowing of good features from one development effort to the next. We believe this cross-fertilization will lead to better programs.

References

Laurance, David and Heiberger, Richard M. (1982), Software Design for Conversational Statistical Analysis, Proceedings of the American Statistical Association, Statistical Computing Section.

Computing for Time Series

Organizer: H. Joseph Newton

Invited Presentations:

    Computing for Autoregressions, *H. Joseph Newton and Marcello Pagano*

    Time Series ARMA Model Identification by Estimating Information, *Emanuel Parzen*

    What Should Your Time Series Analysis Program Do?, *H. Joseph Newton*

Computing for Autoregressions

H. Joseph Newton
Texas A & M University

Marcello Pagano*
Dana-Farber Cancer Institute
and
Harvard University

When computing the parameter estimates for an autoregressive model, it is important that the algorithm be numerically stable, recursive, so that one can fit successively higher orders, and lead to stationary solutions. We present a number of common estimators and a unifying orthogonal autoregression algorithm that has these desirable properties.

1. INTRODUCTION

Let $Y(1),\ldots,Y(n)$ be a sample from the zero mean, covariance stationary autoregressive process Y of order p,

$$\sum_{j=0}^{p} \alpha_p(j) Y(t-j) = \varepsilon(t), \quad t \in Z,$$

with $\alpha_p(0) = 1$ and Z the set of all integers. The white noise series, ε, has variance σ_p^2 and the complex polynomial

$$g_p(z) = \sum_{j=0}^{p} \alpha_p(j) z^j$$

has zeroes outside the unit circle. We write $Y \sim AR(p, \alpha_p, \sigma_p^2)$. The aim of this paper is to comment on some methods for estimating the parameters of this scheme.

To this end, define the autocovariance function $R(v) = \text{cov}[Y(t), Y(t+v)]$, $v \in Z$, and the spectral density

$$f(w) = \sum_{v=-\infty}^{\infty} R(v) e^{-ivw} \div 2\Pi, \quad w \in [-\Pi, \Pi].$$

Then the relation between these various parametrizations is given by

$$f(w) = \sigma_p^2 |g(e^{iw})|^{-2} \div 2\Pi$$

and the Yule-Walker equations

$$\sum_{j=0}^{p} \alpha_p(j) R(j-v) = \delta_v \sigma_p^2, \quad v \geq 0,$$

with δ_v the Kronecker delta. The first $(p+1)$ Yule-Walker equations allow one to solve for the α_p and σ_p^2 from the R.

Autoregressions are interesting models in their own right, but they also serve as a general method for estimating a spectral density. In this latter case, p also becomes a parameter that must be fit. The most common method for fitting these models is:

a) Choose $M \geq p$.
b) Estimate $R(0),\ldots,R(M)$ by

$$R(v) = \sum_{t=1}^{n-v} Y(t) Y(t+v) \div n, \quad v = 0,\ldots,M.$$

(The Y are assumed zero mean. If they are not, a mean correction is required).

c) Find estimators $\hat{\alpha}_k$ and $\hat{\sigma}_k^2$ from the sample analogues of the Yule-Walker equations for $k = 1,\ldots,M$.

d) Use $\hat{\sigma}_1^2,\ldots,\hat{\sigma}_M^2$ to determine p.

e) Estimate R and f by

$$\hat{R}_p(v) = \begin{cases} \hat{R}(v), & v = 0,\ldots,p \\ -\sum_{j=1}^{p} \hat{\alpha}_p(j) \hat{R}_p(v) & v = p+1,\ldots \end{cases}$$

$$\hat{f}_p(w) = \hat{\sigma}_p^2 |\hat{g}_p(e^{iw})|^{-2} \div 2\Pi, \quad w \in [-\Pi, \Pi].$$

Step (d) is easily carried out because of Levinson's (1947) algorithm:

$$\alpha_1(1) = -R(1)/R(0), \quad \sigma_1^2 = R(0)[1-\alpha_1^2(1)]$$

For $k = 2,\ldots,M$:

$$\alpha_k(k) = -[R(k) + \sum_{j=1}^{k-1} \alpha_{k-1}(j) R(k-j)] \div \sigma_{k-1}^2$$

$$\alpha_k(j) = \alpha_{k-1}(j) + \alpha_k(k) \alpha_{k-1}(k-j), \quad j = 1,\ldots,k-1$$

$$\sigma_k^2 = \sigma_{k-1}^2 [1-\alpha_k^2(k)].$$

These equations also illustrate that if one has $R(0)$ and the partial autocorrelations $\alpha_1(1),\ldots,\alpha_M(M)$ then one can determine $\alpha_k(j)$ and σ_k^2 for $1 \leq j \leq k \leq M$. Furthermore, $R(0)$ and the partial autocorrelations are the only parameter estimates we need to determine the order to be fit.

---

*The work of this author was supported in part by Grant #CA-28066 awarded by the National Cancer Institute, DHHS.

Unfortunately, using this Yule-Walker approach can often lead to very poor estimators of the parameters. This is exemplified in Table 1 where the Yule-Walker estimates for five samples of length n = 100 are given for a gaussian AR (4, -2.7607, 3.8106, -2.6535, .9238, 1.0) process.

TABLE 1   Yule-Walker Estimates for Five Realizations of Length 100 for the AR(4) Process

Sample	$\alpha(1)$	$\alpha(2)$	$\alpha(3)$	$\alpha(4)$
1	-1.58	1.16	-.14	-.01
2	-1.76	1.62	-.54	.10
3	-1.36	.91	.03	-.04
4	-1.54	1.19	-.18	-.01
5	-1.25	.83	.11	-.03
True Values	-2.7607	3.8106	-2.6535	.9238

The reason for this poor performance, in general, is well known, the Yule-Walker equations for order p are the normal equations for the ordinary least squares estimators in the model $\underline{Y}_p = -X_p \underline{\alpha}_p + \underline{\varepsilon}$, with

$$\underline{Y}_p^T = (Y(1),\ldots,Y(n), 0,\ldots,0)$$

$$X_p = (L\,\underline{Y}_p, L^2\,\underline{Y}_p,\ldots,L^p\,\underline{Y}_p)$$

where if $\underline{a} = L\,\underline{b}$, then $a(1) = 0$ and $a(j) = b(j-1)$ for $j = 2,\ldots,n$. The vector $\underline{Y}_p$ is of length (n+p). The "end effects", i.e. the effect of placing p zeroes at the beginning or end of $Y(1),\ldots,Y(n)$, can be devastating on the estimation if the zeroes are poor proxies for $Y(0),\ldots,Y(1-p)$ and $Y(n+1),\ldots,Y(n+p)$. This is particularly important if R(0) is large and the zeroes of g(z) are close to the unit circle. For the AR(4) model above, we have R(0) = 761.3, zeroes of g(z) have moduli 1.01998, 1.01998, 1.019798, and 1.019798 and the ratio of the maximum of f to the minimum of f is 6×10^6. Thus, this is a very ill-conditioned model.

For well-conditioned models, the Yule-Walker equations perform adequately as exemplified in Table 2 where the estimators for five realizations of length 100 are given for a gaussian AR (2, -.4, -.45, 1) model. For this model, R(0) = 2.66, the zeroes have moduli 1.11 and 2.00, and the spectral ratio is 111.5.

TABLE 2
Yule-Walker Estimates for Five Realizations of Length 100 for the AR(2) Process

Sample	$\alpha(1)$	$\alpha(2)$
1	-.323	-.478
2	-.482	-.322
3	-.403	-.454
4	-.387	-.411
5	-.390	-.380
True Values	-.40	-.45

Several procedures have been proposed to handle the end effects problem including:

1) **Least Squares Estimates** (LSE); i.e. drop the first and last p rows of $\underline{Y}_p$ and X_p.

2) **Exact Maximum Likelihood Estimation** (MLE); i.e. determine $\hat{\underline{\alpha}}_k$ and $\hat{\sigma}_k^2$ to maximize the gaussian likelihood of an AR(k). Note that this procedure does not requires one to drop any information.

3) **The Burg (1968) Algorithm** (BG). We describe this below.

Unfortunately, MLE and LSE are not recursive and are not guaranteed to lead to estimators whose characteristic polynomials have zeroes outside the unit circle. Also, the numerical procedure used to maximize the likelihood function can fail to converge properly in precisely the situations where the Yule-Walker method is poor.

The Burg method is recursive, does lead to stationary parameters, and seems to work quite well in general.

In the next section, we describe a method for calculating the Yule-Walker estimators that is based on the modified Gram-Schmidt decomposition of X_M and that:

1) Is numerically superior to Levinson's algorithm since it is a method based on X_M instead of $X_M^T X_M$.

2) Recursively generates $\hat{\alpha}_k(k)$ and $\hat{\sigma}_k^2$.

3) Can easily provide evidence of ill-conditioning.

4) Can be easily modified to produce more robust estimators of partial autocorrelations.

5) Allows the Burg algorithm to be succinctly explained.

2. An Orthogonal Autoregression Algorithm

Let $\underline{Y} = X \underline{\beta} + \underline{\varepsilon}$ be an ordinary full rank (n x p) regression model. Let $X = QR$ be the Gram-Schmidt decomposition of X, i.e. R is a (p x p) unit upper triangular matrix while Q is an (n x p) matrix with $Q^T Q = D = \text{Diag}(\underline{q}_1^T \underline{q}_1, \ldots, \underline{q}_p^T \underline{q}_p)$ where $\underline{q}_1, \ldots, \underline{q}_p$ are the columns of Q. The modified Gram-Schmidt algorithm (MGSD) is a good procedure for performing this decomposition (Bjorck (1967)).

Reparametrize $\underline{Y} = X \underline{\beta} + \underline{\varepsilon}$ as $\underline{Y} = Q \underline{\Theta} + \underline{\varepsilon}$ where $\underline{\Theta} = \underline{\beta}$. Then the elements of $\underline{\Theta}$ are the orthogonal parameters of the regression problem, so $\hat{\underline{\Theta}} = (Q^T Q)^{-1} Q^T \underline{Y} = D^{-1} Q^T \underline{Y}$ and $\text{Var}(\hat{\underline{\Theta}}) = \sigma^2 (Q^T Q)^{-1} = \sigma^2 D^{-1}$, i.e. the elements of $\hat{\underline{\Theta}}$ are uncorrelated.

In the autoregressive model, we have:

LEMMA (Pagano (1971))

For the AR regression model $\underline{Y}_M = -X_M \underline{\alpha}_M + \underline{\varepsilon}$, which is orthogonalized as above, then $\hat{\underline{\Theta}}_M = (\hat{\alpha}_1(1), \ldots, \hat{\alpha}_M(M))^T$.

Thus, the orthogonal parametrization is important, especially when making inference about the order to be fit, in an AR model. But the consequences of the special structure of an autoregression does not stop there.

The ordinary modified Gram-Schmidt decomposition has p steps. At the kth step, one obtains the kth column of Q, makes all subsequent columns of Q orthogonal to the kth column, and obtains that part of $\underline{Y}$ which is orthogonal to the first k columns of Q. Thus, there are $(p - k + 2)$ inner products at the kth step. Furthermore, the storage required is of the order of $n(p + 1) + p^2/2$. Because of the structure of an autoregression the amount of work and storage is greatly reduced. Consider the following algorithm:

Orthogonal Autoregression Algorithm:

1° Set $\underline{Y}_{(1)} = \underline{Y}_M$, $\underline{x}_{(1)} = L \underline{Y}_M$

2° For $k = 1, \ldots, M$
$$a(k) = -\underline{Y}_{(k)}^T \underline{x}_{(k)} \div \underline{x}_{(k)}^T \underline{x}_{(k)} \quad (1)$$
$$\underline{Y}_{(k+1)} = \underline{Y}_{(k)} + a(k) \underline{x}_{(k)}$$
$$\underline{x}_{(k+1)} = L(\underline{x}_{(k)} + a(k) \underline{Y}_{(k)}).$$

We then have:

THEOREM (Davis, Newton, and Pagano (1982))

a) $a(k) = \hat{\alpha}_k(k)$, $k = 1, \ldots, M$

b) $\underline{x}_{(k)}$ is the kth column of Q in the MGSD of X_M and $\underline{Y}_{(k)}$ is that part of $\underline{Y}_M$ orthogonal to the first (k - 1) columns of Q.

c) $\underline{1}^T \underline{x}_{(k)} = \underline{1}^T \underline{Y}_{(k)} = 0$

d) $\underline{x}_{(k)}^T \underline{x}_{(k)} = \underline{Y}_{(k)}^T \underline{Y}_{(k)} = (n) \hat{\sigma}_{k-1}^2$ ($\hat{\sigma}_0^2 \equiv \hat{R}(0)$)

e) From (b) $\underline{x}_{(k)}$ and $\underline{Y}_{(k)}$ are the order (k-1) forward and backward forecast errors assuming the required k unobserved observations on the ends of the data are zero.

f) From (e), (c), and (d), $\hat{\alpha}_k(k)$ can be written as either the negative of the ordinary correlation coefficient of $\underline{x}_{(k)}$ and $\underline{Y}_{(k)}$ or the negative of the regression of $\underline{Y}_{(k)}$ on $\underline{x}_{(k)}$ or $\underline{x}_{(k)}$ on $\underline{Y}_{(k)}$. Further,

$$\hat{\alpha}_k(k) = \frac{-2 \underline{Y}_{(k)}^T \underline{x}_{(k)}}{\underline{x}_{(k)}^T \underline{x}_{(k)} + \underline{Y}_{(k)}^T \underline{Y}_{(k)}} . \quad (2)$$

g) The hat matrix H_M for the regression $\underline{Y}_M = -X_M \underline{\alpha}_M + \underline{\varepsilon}$ is $H_M = X_M(X_M^T X_M)^{-1} X_M^T = Q D^{-1} Q^T$ and thus, the vector of hat diagonals can also be accumulated easily in the orthogonal autoregression algorithm since the columns of Q are $\underline{x}_{(1)}, \ldots, \underline{x}_{(M)}$.

We thus get all the information available from the Gram-Schmidt decomposition with much less work (only 2p inner products) and much less storage (only need to operate on two vectors).

The effect of putting zeroes on the ends of the data is dramatically illustrated by considering Table 3 and 4 where we list the first and last four hat diagonals for the AR(4) model (Table 3) and the first and last two hat diagonals for the AR(2) model (Table 4). The last column in each Table is the proportion of the sum of the hat diagonals made up by the displayed diagonals. Thus, in Table 3 seven (h(1) is necessarily zero) of the one hundred and three equations account for twenty to forty one percent of the sum. Whereas, in the better behaved Table 4, the three artificial equations account for only about two percent of the sum.

TABLE 3 First and Last Four Hat Diagonals for the Five AR(4) Series in Table 1

Sample	h(1)	h(2)	h(3)	h(4)	h(101)	h(102)	h(103)	h(104)	Prop
1	0.0	.71	.72	.06	.01	.06	.07	.01	.41
2	0.0	.18	.23	.07	.02	.11	.16	.05	.20
3	0.0	.01	.02	.01	.04	.19	.61	.50	.35
4	0.0	.51	.49	.07	.01	.14	.28	.10	.40
5	0.0	.01	.01	.01	.03	.18	.69	.60	.38

TABLE 4 First and Last Two Hat Diagonals for the Five AR(2) Series in Table 2

Sample	h(1)	h(2)	h(101)	h(102)	Prop
1	0.0	.021	.006	.009	.018
2	0.0	.016	.021	.000	.019
3	0.0	.008	.008	.014	.015
4	0.0	.018	.018	.006	.021
5	0.0	.004	.004	.002	.005

The Burg Algorithm:

Equation (2) provides the framework for understanding the Burg algorithm. The Burg algorithm is exactly the same as the orthogonal autoregression algorithm except that the inner products in (1) are calculated using (2) and not for the whole length of $\underline{Y}_{(k)}$ and $\underline{x}_{(k)}$ but for their $(k+1)^{st},...,(n-k)^{th}$ elements, i.e. the elements for which observations exist in the forward and backward regressions.

This is sensible, not only because of the end effects, but also because this is the answer that one would obtain if one were not using an ordinary regression, but rather an errors-in-variable regression; the same observations appear in the left- and right-hand side of the regression equations.

We illustrate, in Table 5, the improvement given by the Burg algorithm using the same five data sets as in Table 1.

TABLE 5 The Burg Estimates for The Same Five AR(4) Data Sets as in Table 1

Sample	α(1)	α(2)	α(3)	α(4)
1	−2.68	3.55	−2.36	.78
2	−2.76	3.82	−2.68	.94
3	−2.62	3.45	−2.29	.76
4	−2.82	3.86	−2.67	.89
5	−2.75	3.77	−2.60	.89
True Values	−2.7607	3.8106	−2.6535	.9238

In Table 6, we give the Burg estimates for the five AR(2) series.

TABLE 6 The Burg Estimates for the Five AR(2) Data Sets as in Table 2

Sample	$\alpha(1)$	$\alpha(2)$
1	-.324	-.482
2	-.485	-.320
3	-.388	-.478
4	-.380	-.425
5	-.391	-.379
True Values	-.40	-.45

Thus, when the end effects are important in the Yule-Walker equations, the Burg estimates are different, when they have little effect, the Burg estimates are similar.

Alternate Autoregression

The logical framework provided by the orthogonal autoregression algorithm, which is both recursive and isolates the condition for stationarity ($|\alpha_k(k)| < 1$), suggests using the same algorithm but with different regression or correlation estimators for $\hat{\alpha}_k(k)$. To illustrate this, we added to a randomly selected number of the observations in each of the five data sets $N(0, 4R(0))$ random variables and then used Boscovich's (5) algorithm to find $\hat{\alpha}_k(k)$ as the least absolute value (LAV) regression coefficient of $\underline{Y}_{(k)}$ on $\underline{x}_{(k)}$. If any $\hat{\alpha}_k(k)$ was greater than one in absolute value, it was replaced by sgn$(\hat{\alpha}_k(k)) \times .95$ and the process continued. The results are shown in Table 7.

Thus, for the AR(2) case, both Burg and LAV perform adequately for both contaminated and uncontaminated data. However, the AR(4) case illustrates the following progression:

1) For uncontaminated poorly conditioned models, the Yule-Walker estimates are bad while the Burg and LAV estimates are both good.
2) For contaminated poorly conditioned models, LAV, while not very good, is significantly better than Burg.
3) If data from an ill-conditioned model is contaminated enough, then none of the presented methods work well (see data set 4 for the AR(4) model).

TABLE 7 Burg and LAV Estimates for Contaminated and Uncontaminated Data Sets

AR(4)		$\alpha(1)$	$\alpha(2)$	$\alpha(3)$	$\alpha(4)$
Burg (Uncontaminated)					
1		-2.683	3.553	-2.362	.782
2		-2.755	3.823	-2.677	.943
3		-2.622	3.449	-2.295	.764
4		-2.815	3.865	-2.673	.898
5		-2.751	3.773	-2.604	.891
LAV (Uncontaminated)					
1		-2.652	3.514	-2.344	.785
2		-2.685	3.714	-2.622	.949
3		-2.844	3.823	-2.613	.839
4		-2.243	2.733	-1.691	.585
5		-2.820	3.826	-2.610	.846
Burg (Contaminated)					
1	(1)	-1.779	1.438	-.311	.004
2	(2)	-2.121	2.361	-1.231	.348
3	(2)	-1.667	1.328	-.244	-.031
4	(3)	-1.101	.433	.461	-.132
5	(2)	-1.696	1.396	-.268	-.054
LAV (Contaminated)					
1		-2.274	2.712	-1.596	.526
2		-2.129	2.772	-1.778	.722
3		-2.411	2.953	-1.844	.607
4		-1.609	1.249	-.192	-.051
5		-1.922	2.049	-1.007	.308
True Value		-2.7607	3.8106	-2.6535	.9238

Table 7 (Continued)

AR(2)	α(1)	α(2)
Burg (Uncontaminated)		
1	−.324	−.482
2	−.485	−.320
3	−.388	−.478
4	−.380	−.425
5	−.391	−.379
LAV (Uncontaminated)		
1	−.353	−.497
2	−.434	−.341
3	−.416	−.402
4	−.322	−.536
5	−.387	−.331
Burg (Contaminated)		
1 (2)	−.282	−.489
2 (3)	−.499	−.297
3 (6)	−.333	−.388
4 (1)	−.386	−.377
5 (2)	−.274	−.293
LAV (Contaminated)		
1	−.306	−.550
2	−.459	−.354
3	−.381	−.414
4	−.322	−.534
5	−.368	−.305
True Value	−.40	−.45

REFERENCES

[1] Bjorck, A. (1967), "Solving Linear Least Squares Problems by Gram-Schmidt Orthogonalization," BIT, 7, 1-21.

[2] Burg, J.P. (1968), "A New Analysis Technique for Time Series Data," presented at NATO Advanced Study Conference on Signal Processing, Enschede, Netherlands.

[3] Clayton, D.G. (1971), "Gram-Schmidt Orthogonalization of Vectors," Applied Statistics, 20, 335-338.

[4] Davis, H.T, Newton, H.J., and Pagano, M. (1982), "A Toeplitz Gram-Schmidt algorithm for autoregressive modeling", Technical Report, Dana-Farber Cancer Institute.

[5] Eisenhart, C. (1961), "Boscovich and the Combination of Observations," in Roger Joseph Boscovich (L.L. Whyte, Ed.), Forham Univ. Press, N.Y.

[6] Levinson, N. (1947), "The Wiener RMS Error Criterion in Filter Design and Prediction," Journal of Mathematical Physics, 25, 261-278.

[7] Pagano, M. (1972), "An Algorithm for Fitting Autoregressive Schemes," Applied Statistics, 21, 274-281.

TIME SERIES ARMA MODEL IDENTIFICATION BY ESTIMATING INFORMATION

Emanuel Parzen

Institute of Statistics
Texas A&M University

Statisticians, economists, and system engineers are becoming aware that to identify models for time series and dynamic systems, information theoretic ideas can play a valuable (and unifying) role. Models for time series $Y(t)$ can be formulated as hypotheses concerning the information about $Y(t)$ given various bases involving past, current, and future values of $Y(\cdot)$ and related time series $X(\cdot)$. To determine sets of variables that are sufficient to forecast $Y(t)$, and especially to determine an ARMA model for $Y(t)$, an approach is presented which estimates and compares various information increments. We discuss how to non-parametrically estimate the $MA(\infty)$ representation, and use it to form estimators of the many information numbers that might compare to identify an ARMA model for a univariate time series.

1. Information Measures

The information approach to model identification formulates a model (or hypothesis about the probability law of random variables or time series) as a hypothesis that an information number is zero. Information measures for random variables are defined in terms of information measures for probability densities. The latter can be regarded as defining "distances" between probability measures.

Let $f(y)$ and $g(y)$ be two probability densities on a real line, $-\infty < y < \infty$. The information divergence of index α of a (model) g from (a true density) f is defined for $\alpha = 1$ (index 1) by

$$I_1(f;g) = \int_{-\infty}^{\infty} \{-\log \frac{g(y)}{f(y)}\} f(y) \, dy$$

and for $\alpha > 0$ (but $\alpha \neq 1$) by

$$I_\alpha(f;g) = \frac{-1}{1-\alpha} \log \int_{-\infty}^{\infty} \{\frac{g(x)}{f(x)}\}^{1-\alpha} f(x) \, dx \quad .$$

Information divergence of index 1 has a preferred role because it has an important decomposition

$$I_1(f;g) = H(f;g) - H(f)$$

defining

$$H(f;g) = \int_{-\infty}^{\infty} \{-\log g(y)\} f(y) \, dy,$$

$$H(f) = H(f;f) = \int_{-\infty}^{\infty} \{-\log f(y)\} f(y) \, dy \quad .$$

We call $H(f;g)$ the cross-entropy of f and g, and $H(f)$ the entropy of f. Information divergence of index 1 is usually referred to just as information divergence $I(f;g)$.

The information $I(Y|X)$ about a continuous random variable Y in a continuous random variable X is defined by

$$I(Y|X) = I(f_{Y|X}; f_Y) = E_X I(f_{Y|X=x}; f_Y) \quad .$$

The entropy of Y and conditional entropy of Y given X are defined by

$$H(Y) = H(f_Y)$$

$$H(Y|X) = H(f_{Y|X}) = E_X H(f_{Y|X=x}) \quad .$$

One can establish a fundamental decomposition:

$$I(Y|X) = H(Y) - H(Y|X).$$

The most fundamental concept used in identifying models by estimating information is $I(Y|X_1; X_1, X_2)$, the information about Y in X_2 conditional on X_1; it is defined

(I) $I(Y|X_1; X_1, X_2) = H(f_{Y|X_1}) - H(f_{Y|X_1,X_2})$
$= H(Y|X_1) - H(Y|X_1,X_2) \quad .$

A fundamental formula to evaluate $I(Y|X_1; X_1, X_2)$ is

(II) $I(Y|X_1; X_1, X_2) = I(Y|X_1, X_2) - I(Y|X_1) \quad .$

When X and Y are jointly normal random variables $f_{Y|X=x}(y)$ is a normal distribution whose variance (which does not depend on x) is denoted $\Sigma(Y|X)$. The variance of Y is denoted $\Sigma(Y)$. The entropy and conditional entropy of Y are

$$H(Y) = \frac{1}{2} \log \Sigma(Y) + \frac{1}{2}(1 + \log 2\pi)$$

$$H(Y|X) = \frac{1}{2} \log \Sigma(Y|X) + \frac{1}{2}(1 + \log 2\pi) \quad .$$

The information about Y in X when X and Y are bivariate normal, with correlation coefficient ρ, can be expressed

(III) $I(Y|X) = -\frac{1}{2} \log \Sigma^{-1}(Y)\Sigma(Y|X) = -\frac{1}{2}\log(1-\rho^2)$.

When Y and X are jointly multivariate normal random vector, let Σ denote a covariance matrix. One can show that

(IV) $I(Y|X) = (-\frac{1}{2}) \log \det \Sigma^{-1}(Y)\Sigma(Y|X)$
$= (-\frac{1}{2}) \text{ sum log eigenvalues } \Sigma^{-1}(Y)\Sigma(Y|X).$

Research supported by Office of Naval Research under contract no. N00014-82-MP-20001.

To illustrate the information approach to model identification (or determining relations between random variables) consider the general problem of testing the hypothesis H_0: X and Y are independent. One could express H_0 in any one of the following equivalent ways:

H_0: $f_{X,Y}(x,y) = f_X(x)f_Y(y)$ for all x and y;

H_0: $f_{Y|X=x}(y) = f_Y(y)$ for all x and y;

H_0: $I(f_{X,Y}; f_X f_Y) = 0$;

H_0: $I(Y|X) = 0$.

The information approach to testing H_0 is to form an estimator $\hat{I}(Y|X)$ of $I(Y|X)$, and test whether it is significantly different from zero. One can distinguish several types of estimators of $I(Y|X)$: (a) fully parametric, (b) fully non-parametric; (c) functionally parametric which uses functional statistical inference smoothing techniques to estimate $I(Y|X)$ [see Woodfield (1982)].

An example of fully parametric estimators arises when one assumes X and Y are bivariate normal with correlation coefficient ρ. Given a random sample $(X_1,Y_1),\ldots,(X_n,Y_n)$ a fully parametric estimator of $I(Y|X)$ is the maximum likelihood estimator

$$\hat{I}(Y|X) = -\frac{1}{2} \log (1-\hat{\rho}^2)$$

where $\hat{\rho}$ is the sample correlation coefficient.

2. Information and Memory Approach to Time Series Model Identification

The approach to time series analysis developed by Parzen distinguishes four general types of time series models:

1. No memory or white noise
2. Short memory or stationary
3. Long memory (or non-stationary)
3a. Long memory: transform to short memory
3b. Long memory: long memory plus short memory.

Memory type can be defined in terms of the information numbers I_m

$$I_m = I(Y|Y_{-1},\ldots,Y_{-m})$$
$$= I(Y(t)|Y(t-1),\ldots,Y(t-m)) ;$$

in words, I_m is the information about a time series $Y(t)$ at time t in the m most recent values $Y(t-1),\ldots,Y(t-m)$. Let Y^- denote the infinite past $Y(t-1), Y(t-2),\ldots$. As m tends to ∞, I_m tends to

$$I_\infty = I(Y|Y^-) = I(Y(t)|Y(t-1),\ldots) .$$

We define a time series $Y(t)$, $t=0,\pm 1,\ldots$ to be:

no memory if $I_\infty = 0$
short memory if $0 < I_\infty < \infty$
long memory if $I_\infty = \infty$.

The models we build for a time series depend on its memory type. A model corresponds to a transformation of the time series to a no memory (white noise) series. Therefore a no memory (white noise) time series requires no further modeling, although one may be interested in determining such statistical characteristics as the mean, variance, and probability distribution.

A short memory time series $Y(t)$ is modeled by an invertible filter which transforms it to white noise:

$$Y(t) \longrightarrow \boxed{\text{innovations filter } g_\infty} \longrightarrow \varepsilon(t) = Y^\nu(t)$$

where $Y^\nu(\cdot)$ is the <u>innovation</u> series, or series of infinite memory one-step ahead prediction errors, defined by [using ν to connote "what's new"]

$$Y^\nu(t) = Y(t) - Y^\mu(t) .$$

The predictor $Y^\mu(t)$ is denoted

$$Y^\mu(t) = E[Y(t)|Y(t-1), Y(t-2), \ldots]$$
$$= (Y|Y_{-1},\ldots,Y_{-n},\ldots)(t) .$$

We use μ as the superscript for a predictor to indicate that it is an averaging operator.

The infinite memory mean square prediction error is defined as the normalized variance

$$\sigma_\infty^2 = E[|Y^\nu(t)|^2] \div E[|Y(t)|^2] .$$

The appropriateness of normalizing is justified by the formula for information:

$$I_\infty = -\frac{1}{2} \log \sigma_\infty^2$$

if a time series $Y(t)$, $t=0,\pm 1,\ldots$ is a zero mean Gaussian stationary time series. Its probability law can be described by the covariance function

$$R(v) = E[Y(t)Y(t+v)] ,$$

and correlation function

$$\rho(v) = \frac{R(v)}{R(0)} = \text{Corr}[Y(t),Y(t+v)] .$$

Alternatively the probability law of $Y(\cdot)$ can be described by the spectral density function f which is defined by

$$f(\omega) = \sum_{v=-\infty}^{\infty} e^{-2\pi i v \omega} \rho(v), \quad 0 \leq \omega \leq 1$$

when $\sum_{v=-\infty}^{\infty} |\rho(v)| < \infty$. The frequency variable ω is usually assumed to vary in the interval $-0.5 < \omega < 0.5$. But only the interval $0 \leq \omega \leq 0.5$ has physical significance. We prefer the interval $0 \leq \omega \leq 1$ for mathematical reasons.

Perhaps the most insightful way to model a short memory time series is by representing it, or approximating it, by an ARMA(p,q) scheme:

$$Y(t) + \alpha_p(1)Y(t-1)+\ldots+\alpha_p(p)Y(t-p)$$
$$= \varepsilon(t) + \beta_q(1)\varepsilon(t-1)+\ldots+\beta_q(q)\varepsilon(t-q)$$

where the polynomials

$$g_p(z) = 1+\alpha_p(1)z+\ldots+\alpha_p(p)z^p$$
$$h_q(z) = 1+\beta_q(1)z+\ldots+\beta_q(q)z^q$$

are chosen so that all their roots in the complex z-plane are in the region $\{z:|z|>1\}$ outside the unit circle. Then $g_p(z)$ and $h_q(z)$ are the transfer functions of invertible filters. $\varepsilon(t)$ is assumed to be a white noise time series which we identify with the innovations $\varepsilon(t) = Y^\nu(t)$;

$$\sigma_{p,q}^2 = E[\varepsilon^2(t)] \div E[Y^2(t)]$$

is an estimator of σ_∞^2. The spectral density of an ARMA (p,q) scheme is

$$f_{p,q}(\omega) = \sigma_{p,q}^2 \frac{|h_q(e^{2\pi i\omega})|^2}{|g_p(e^{2\pi i\omega})|^2}$$

The process of identifying ARMA(p,q) schemes which are adequate (and parsimonious) approximating models for a time series can be studied by determining information characterizations of when the exact (or true) model is an AR(p) or ARMA(p,q).

Let $\Sigma(Y|Y_{-1},\ldots,Y_{-p}, Y_{-1}^\nu,\ldots,Y_{-q}^\nu)$ denote the mean square prediction error of $Y(t)$ when predicted by $Y(t-1),\ldots,Y(t-p), Y^\nu(t-1),\ldots,Y^\nu(t-q)$, or equivalently the conditional variance of $Y(t)$ given $Y(t-1),\ldots,Y(t-p), Y^\nu(t-1),\ldots,Y^\nu(t-q)$. Normalize it to form

$$\sigma_{p,q}^2 = \Sigma^{-1}(Y)\Sigma(Y|Y_{-1},\ldots,Y_{-p}, Y_{-1}^\nu,\ldots,Y_{-q}^\nu)$$
$$I_{p,q} = -\frac{1}{2}\log\sigma_{p,q}^2 .$$

The information difference between $Y_{-1},\ldots,Y_{-p}$, $Y_{-1}^\nu,\ldots,Y_{-q}^\nu$ and Y^- for prediction of $Y(t)$ satisfies

$$I(Y|Y_{-1},\ldots,Y_{-p},Y_{-1}^\nu,\ldots,Y_{-q}^\nu;Y^-) = I_\infty - I_{p,q} .$$

The following two hypotheses are equivalent:
H_0: $Y(\cdot)$ is ARMA(p,q)
H_0: $I_\infty - I_{p,q} = 0$.

3. Information Calculation for ARMA Schemes

Given a sample $\{Y(t), t=1,2,\ldots T\}$, we would like to estimate, for many values of p,q, the information differences (assuming normality)

$$I_\infty - I_{p,q} = -\frac{1}{2}\log\sigma_\infty^2 - \{-\frac{1}{2}\log\sigma_{p,q}^2\} .$$

We need to estimate σ_∞^2 and $\sigma_{p,q}^2$. To understand the method we would like to propose, let us first discuss how to compute the true value of $\sigma_{p,q}^2$. The MA(∞), or infinite order moving average, representation of $Y(t)$ will play a central role:

$$Y(t) = Y^\nu(t) + \beta_1 Y^\nu(t-1) + \beta_2 Y^\nu(t-2) + \ldots$$

Note that $E[|Y(t)|^2] = E[|Y^\nu(t)|^2]\{1+\beta_1^2+\ldots\}$ so that

$$1 = \sigma_\infty^2\{1+\beta_1^2+\beta_2^2+\ldots\} .$$

The correlations $\rho(v)$ can be computed by
$$\rho(v) = \sigma_\infty^2\{\beta_v + \beta_1\beta_{v+1} +\ldots\} .$$

By using matrix sweep operations on the joint covariance matrix of $Y, Y_{-1},\ldots,Y_{-p}, Y_{-1}^\nu,\ldots,Y_{-q}^\nu$ one can determine (in a stepwise manner) the conditional variance $\Sigma(Y|Y_{-1},\ldots,Y_{-p}, Y_{-1}^\nu,\ldots,Y_{-q}^\nu)$ required to compute the information $I_{p,q}$.

We illustrate the approach being proposed in the case p=1, q=1. The covariance matrix of Y, Y_{-1}, Y_{-1}^ν is

$$\Sigma = \begin{bmatrix} 1 & \rho(1) & \sigma_\infty^2\beta_1 \\ \rho(1) & 1 & \sigma_\infty^2 \\ \sigma_\infty^2\beta_1 & \sigma_\infty^2 & \sigma_\infty^2 \end{bmatrix}$$

Sweep Σ on Y_{-1} to obtain

$$\Sigma_1 = \begin{bmatrix} 1-\rho^2(1) & \rho(1) & \sigma_\infty^2\beta_1-\rho(1)) \\ -\rho(1) & 1 & -\sigma_\infty^2 \\ \sigma_\infty^2(\beta_1-\rho(1)). & \sigma_\infty^2 & \sigma_\infty^2(1-\sigma_\infty^2) \end{bmatrix}$$

Sweep Σ on Y_{-1}^ν to obtain

$$\Sigma_2 = \begin{bmatrix} 1-\sigma_\infty^2\beta_1^2 & \rho(1)-\sigma_\infty^2\beta_1 & \beta_1 \\ \rho(1)-\sigma_\infty^2\beta_1 & 1-\sigma_\infty^2 & 1 \\ -\beta_1 & -1 & (\sigma_\infty^2)^{-1} \end{bmatrix}$$

Sweep Σ_1 on Y_{-1}^ν or sweep Σ_2 on Y_{-1} to obtain a matrix which we write in the following form:

$$\begin{bmatrix} (1-\rho^2(1))-\dfrac{(\beta_1-\rho(1))^2\sigma_\infty^2}{1-\sigma_\infty^2} & \dfrac{\rho(1)-\sigma_\infty^2\beta_1}{1-\sigma_\infty^2} & \dfrac{\beta_1-\rho(1)}{1-\sigma_\infty^2} \\ -\dfrac{\rho(1)-\sigma_\infty^2\beta_1}{1-\sigma_\infty^2} & \dfrac{1}{1-\sigma_\infty^2} & \dfrac{-1}{1-\sigma_\infty^2} \\ -\dfrac{\beta_1-\rho(1)}{1-\sigma_\infty^2} & \dfrac{-1}{1-\sigma_\infty^2} & \dfrac{1}{\sigma_\infty^2(1-\sigma_\infty^2)} \end{bmatrix}$$

We conclude that
$$\Sigma(Y|Y_{-1}) = 1-\rho^2(1), \quad (Y|Y_{-1})(t) = \rho(1)Y_{-1}(t)$$
$$\Sigma(Y|Y_{-1}^\nu) = 1 - \sigma_\infty^2\beta_1^2 \quad (Y|Y_{-1}^\nu)(t) = \beta_1 Y_{-1}^\nu(t)$$
$$\Sigma(Y|Y_{-1},Y_{-1}^\nu) = (1-\rho^2(1)) - \frac{(\beta_1-\rho(1))^2\sigma_\infty^2}{1-\sigma_\infty^2}$$

$$(Y|Y_{-1}, Y^\nu_{-1})(t) = \frac{\rho(1) - \sigma^2_\infty \beta_1}{1 - \sigma^2_\infty} Y_{-1}(t)$$
$$+ \frac{\beta_1 - \rho(1)}{1 - \sigma^2_\infty} Y^\nu_{-1}(t) \ .$$

These coefficients of $Y_{-1}(t)$ and $Y^\nu_{-1}(t)$ can be used as initial (or perhaps even final) values for an efficient parameter estimation algorithm for an ARMA(1,1).

As a check on these formulas, note that for an MA(1), $Y(t) = \varepsilon(t) + b\varepsilon(t-1)$, $\beta_1 = b$, $\sigma^2_\infty = (1+b^2)^{-1}$, $\rho(1) = b/(1+b^2)$. The coefficients of $Y_{-1}(t)$ and $Y^\nu_{-1}(t)$ in the predictor are respectively 0 and b.

For a numerical illustration of these formulas, consider the ARMA(1,1) model $Y(t) - aY(t-1) = Y^\nu(t) + bY^\nu(t-1)$. Then $\beta_1 = a+b$. $1 = \sigma^2_\infty(1 + (\beta_1^2/(1-a^2)))$, $\rho(1) = [\beta_1 + (\beta_1^2 a/(1-a^2))]\sigma^2_\infty$. For $a = b = 0.5$, $\beta_1 = 1$, $\sigma^2_\infty = 3/7$, $\rho(1) = 5/7$. The general formulas yield the values assumed in the model.

To test whether a time series $Y(\cdot)$ obeys an ARMA(1,1), form

$$I(Y|Y_{-1}, Y^\nu_{-1}; Y^-) = \frac{1}{2} \log\{\frac{1-\rho^2(1)}{\sigma^2_\infty} - \frac{\{\beta_1 - \rho(1)\}^2}{1 - \sigma^2_\infty}\}$$

This information number equals 0 if the time series obeys any one of the schemes AR(1), MA(1), or ARMA(1,1). The information numbers for an AR(1) and MA(1) are respectively

$$I(Y|Y_{-1}; Y^-) = \frac{1}{2} \log \{\frac{1-\rho^2(1)}{\sigma^2_\infty}\} \ ;$$

$$I(Y|Y^\nu_{-1}; Y^-) = \frac{1}{2} \log \{\frac{1}{\sigma^2_\infty} - \beta_1^2\} \ .$$

One accepts H_0: $Y(\cdot)$ is ARMA(1,1) if the last two information numbers are different from zero, but $I(Y|Y_{-1}, Y^\nu_{-1}; Y^-) = 0$.

For the ARMA(1,1) model $Y(t) - 0.5 Y(t-1) = Y^\nu(t) + 0.5 Y^\nu(t-1)$,

$$I(Y|Y_{-1}; Y^-) = \frac{1}{2} \log \frac{8}{7} = .067$$

$$I(Y|Y^\nu_{-1}; Y^-) = \frac{1}{2} \log \frac{4}{3} = .143 \ .$$

When information $I_{p,q}$ is estimated from a sample of size T, a penalty term $(1+p+q)/T$ is subtracted from the estimated information $I_{p,q}$ in the Akaike information approach. If .067 were an estimated value of $I(Y|Y_{-1}; Y^-)$ it would be regarded as significantly different from zero if $.067 - (2/T) \geq 0$, which is true for $T \geq 30$.

To identify the best orders p,q of approximating ARMA(p,q) one could use subset regression techniques to steer the calculation of $I_{p,q}$.

Alternatively one could compute the information numbers of AR(p), MA(q), ARMA(p,q) for p,q = 1,...,M (a specified upper limit). Subtract from estimated information number a penalty $(1+p+q)/T$. Then sort the array of penalized estimated information numbers $I_{p,q}$ to determine the orders (p,q) of schemes with the largest amount of information (and which therefore minimize $I_\infty - I_{p,q}$ and correspond to best approximating ARMA schemes by this measure of divergence between probability distributions).

4. Nonparametric Estimation of MA(∞) Representation

An information approach to computing $I_{p,q}$ and thus identifying best fitting schemes has been described which is based on estimating the coefficients of the MA(∞) representation. Two possible methods for non-parametric MA(∞) estimation are described in this section: (1) approximating long autoregressive schemes; (2) cepstral correlations. The two methods may be used simultaneously for greater confidence in the results obtained. Both methods require further theoretical investigation [compare Bhansali (1982)].

Denote the MA(∞) representation of $Y(t)$ by

$$Y(t) = b(0) Y^\nu(t) + b(1) Y^\nu(t-1) + \ldots$$

where $b(0) = 1$. Denote the AR(∞) representation by

$$a(0) Y(t) + a(1) Y(t-1) + \ldots = Y^\nu(t)$$

where $a(0) = 1$.

The approximating long autoregressive scheme estimates the AR(∞) representation of a time series $Y(\cdot)$ by a finite order AR(p) scheme

$$Y(t) + a_p(1)Y(t-1) + \ldots + a_p(p)Y(t-p) = \varepsilon(t)$$

whose order p is determined by an order determining scheme [such as AIC, due to Akaike, or CAT, due to Parzen]. The generating functions

$$h_\infty(z) = 1 + b(1)z + b(2)z^2 + \ldots$$
$$g_\infty(z) = 1 \ a(1)z + a(2)z^2 + \ldots$$
$$g_p(z) = 1 + a_p(1) z + \ldots + a_p(p) z^p$$

satisfy

$$g_\infty(z) \ h_\infty(z) = 1.$$

One can solve recursively for b(j) using the recursion

$$a(0) b(k) + a(1) b(k-1) + \ldots + a(k) b(0) = 0.$$

When $g_\infty(z)$ is approximated by $g_p(z)$, one replaces a(k) by $a_p(k)$; note that $a_p(k) = 0$ for k>p. The approximating autoregressive method of estimating the MA(∞) representation often yields reasonable results in practice. However it is difficult to study its properties theoretically.

The cepstral correlation method is available for

short memory time series; then $\log f(\omega)$ is integrable, and can be used to compute I_∞ using the fundamental formula (due to Kolmogorov and Szego)

$$\log \sigma_\infty^2 = \int_0^1 \log f(\omega) \, d\omega \quad .$$

The <u>cepstral correlations</u> are defined by, for $v=0, \pm 1, \ldots,$

$$\psi(v) = \int_0^1 e^{2\pi i \omega v} \log f(\omega) \, d\omega \quad .$$

The name "cepstral correlations" is intended to connote that $\psi(v)$ is the Fourier transform of $\log f(\omega)$. However the sequence $\{\psi(v)\}$ does not share an essential property of the sequence $\{\rho(v)\}$ of correlations; the cepstral-correlations are not non-negative definite since $\log f(\omega)$ is not non-negative. Define

$$\Psi(z) = \sum_{k=1}^\infty \psi(k) z^k, \quad \Psi^*(z) = \sum_{k=1}^\infty \psi(-k) z^{-k} \quad .$$

Then
$$f(\omega) = \sigma_\infty^2 \, |h(e^{2\pi i\omega})|^2$$
and
$$\log f(\omega) = \Psi_0 + \Psi(e^{2\pi i\omega}) + \Psi^*(e^{2\pi i\omega}) \quad .$$

A very important relation [which goes back to the dawn of modern time series analysis, due to Kolmogorov (1939)] is

$$h_\infty(z) = \exp \Psi(z).$$

One can obtain an explicit formula for $b(k)$ in terms of $\psi(k)$; thus Janacek (1982) writes

$b(1) = \psi(1),$
$b(2) = \psi(2) + \psi^2(1)/2!,$
$b(3) = \psi(3) + \psi(1)\psi(2) + \psi^3(1)/3!$.

A more useful representation of the formula for $b(k)$ in terms of $\psi(k)$ has been given by Pourahmadi (1982):

$$b(n+1) = \sum_{j=0}^n \left(1 - \frac{j}{n+1}\right) \psi(n+1-j) \, b(j).$$

We outline Pourahmadi's proof; differentiate with respect to z the relation $h_\infty = \exp \Psi$. Obtain $h_\infty' = h_\infty \Psi'$; explicitly

$$\sum_{n=1}^\infty n b(n) z^{n-1} = \{\sum_{n=0}^\infty b(n) z^n\} \{\sum_{n=1}^\infty n\psi(n) z^{n-1}\}$$

or

$$\sum_{n=0}^\infty (n+1) b(n+1) z^n = \{\sum_{n=0}^\infty b(n) z^n\}$$
$$\{\sum_{n=0}^\infty (n+1)\psi(n+1) z^n\} \quad .$$

Therefore
$$(n+1) b(n+1) = \sum_{k=0}^n (k+1)\psi(k+1) b(n-k)$$
$$= \sum_{j=0}^n b(j) (n+1-j) \psi(n+1-j) \quad .$$

Divide by $n+1$ to obtain the desired conclusion.

Pourahmadi (1982) also states a recursive formula for computation of the AR(∞) coefficients $a(k)$ from $\psi(k)$:

$$a(n+1) = -\sum_{j=0}^n \left(1 - \frac{j}{n+1}\right) \psi(n+1-j) a(j) \quad .$$

The properties of cepstral correlations can be understood by examining their values in the case of an AR(1); then

$$f(\omega) = \sigma_\infty^2 \, |1 - \rho e^{2\pi i\omega}|^{-2}$$

where $|\rho| < 1$. Then, for $k \geq 1$,

$$\psi(k) = \int_0^1 -\log \{1 - \rho e^{-2\pi i\omega}\} e^{2\pi i k\omega}$$
$$= \frac{1}{k} \rho^k \quad .$$

The rate of decay of $k\psi(k)$, $k=1,2,\ldots$, is a measure of the memory of the time series.

To estimate $\psi(k)$ from a sample $Y(t)$, $t=1,\ldots,T$ one could take the logarithm of the sample spectral density (computed for $\omega = k/Q$, where one should choose $Q \geq 2T$)

$$\tilde{f}(\omega) = |\sum_{t=1}^T Y(t) \exp(2\pi i\omega t)|^2 \div \sum_{t=1}^T |Y(t)|^2$$

or a smoothed estimator $\hat{f}(\omega)$ of $f(\omega)$. Then

$$\hat{\psi}(v) = \frac{1}{Q} \sum_{k=0}^{Q-1} \log \tilde{f}\left(\frac{k}{Q}\right) \exp(2\pi i v k/Q)$$

A convenient formula for $\hat{f}(\omega)$ is the windowed periodogram of bandwidth $1/T$ defined by

$$\hat{f}(\omega) = \sum_{|v|<T} k\left(\frac{v}{T}\right) \hat{\rho}(v) \exp(2\pi i v\omega)$$

where $\hat{\rho}(v)$ is the sample correlation function computed by

$$\hat{\rho}(v) = \frac{1}{Q} \sum_{k=0}^{Q-1} \tilde{f}\left(\frac{k}{Q}\right) \exp(2\pi i v k/Q)$$

and $k(t)$ is a suitable kernel (providing non-negative estimators) such as the Parzen window

$k(t) = 1 - 6t^2 + 6t^3$, $|t| \leq 0.5$,
$\quad = 2(1 - |t|)^3$, $0.5 \leq |t| \leq 1$,
$\quad = 0$, $1 \leq |t|$.

A kernel with superior properties (but not necessarily non-negative estimates) is the spline-equivalent window [Parzen (1958), Cogburn and Davis (1974), Wahba (1980)]

$$k(t) = \frac{1}{1+t^{2r}}$$

where r is usually chosen to equal 2 or 4.

An obvious moral of the foregoing formulas is that modern time series model identification requires the scientist to integrate time domain and frequency domain techniques. The cepstral correlations approach to ARMA model identification also may provide a unification of ARMA models and the exponential spectral models introduced by Bloomfield (1973).

5. Conclusion

Given a sample of time series, one should estimate its correlations $\rho(v)$ and cepstral correlation $\psi(v)$ through Fast Fourier transformation from the sample spectral density $f(\omega)$ and its logarithm $\log f(\omega)$.

Using the estimated correlations, the Yule-Walker equations are solved to estimate innovation variances σ_m^2, m=1,2,... . Order determining criteria, such as AIC and CAT, are applied to this sequence to determine orders $\hat{m}$ of approximating AR schemes, to determine the memory type of the time series [Parzen (1982)], and to form autoregressive estimators of $f(\omega)$, $\log f(\omega)$, and $\psi(v)$.

When a time series is classified as short memory the estimated cepstral correlations are used to form the MA(∞) coefficients b(k). They are used to form information numbers (via sweep or subset regression procedures) for determining best fitting ARMA schemes, and the corresponding ARMA spectral density estimator.

We do not believe that spectral estimation is a non-parametric procedure to be conducted independently of model identification. The final form of spectral estimator should be based on an identification of the type (AR, MA, or ARMA) of the whitening filter of a short memory time series.

Statistical computing has a vital role in time series analysis in two important ways: (1) to rapidly make available to the broader scientific community new algorithms for time series analysis; (2) to make old theoretical ideas of time series analysis practically useful and to stimulate the integration of old and new techniques of time series analysis.

For other aspects of the role of entropy and information measures in model identification, see Akaike (1977) and IFAC (1982). For modeling of multiple time series, see Parzen and Newton (1980), Newton (1983), and Cooper and Wood (1982). A review (and power study) of some standard statistical procedures for determining the orders p and q of an ARMA scheme is given by Clarke and Godolphin (1982).

REFERENCES

Akaike, H. (1977). On entropy maximization principle, Application of Statistics, P. R. Krishnaiah, ed., North-Holland: Amsterdam, 27-41.

Bhansali, R. J. (1982). A comparison of the Wiener-Kolmogorov and the Yule-Walker methods of fitting autoregressions for prediction. Time Series Analysis: Theory and Practice I, ed. O. D. Anderson, North Holland: Amsterdam, 31-52.

Bloomfield, P. (1973). An exponential model for the spectrum of a scalar time series. Biometrika 60, 217-226.

Clarke, B. R. and Godolphin, E. J. (1982). Comparative Power Studies for Goodness of Fit Tests of Time Series Models, Journal of Time Series Analysis, 3. 141-151.

Cogburn, R. and Davis, H. T. (1974). Periodic splines and spectral estimation. Annals of Statistics, 2, 1108-1126.

Cooper, D. M. and Wood, E. F. (1982). Identifying multivariate time series models. Journal of Time Series Analysis, 3, 153-164.

IFAC (1982). Symposium on Identification and System Parameter Identification, Arlington, Virginia, June 7-11, 1982. Session on "New Ideas in System Identification" emphasizing information theory and entropy function approaches.

Janacek, G. (1982). Determining the degree of differencing for time series via the log spectrum. Journal of Time Series Analysis, 3, 177-183.

Kolmogorov, A. (1939). Interpolation and extrapolation of stationary random sequences (in French). Comptes Rendus Ac. Sci. (Paris), 208, 2043-2045.

Newton, H. Joseph (1983). An Introduction to the Methods of Time Series Analysis in the Time and Frequency Domains: A User's Guide to the TIMESBOARD Computing Library. Manuscript.

Parzen, E. (1958). "On asymptotically efficient consistent estimates of the spectral density function of a stationary time series" J. Roy. Statist. Soc., B., 20, 303-322.

Parzen, E. (1982). Time Series Model Identification and Prediction Variance Horizon, Applied Time Series Analysis II, ed. D. Findley, Academic Press: New York, 415-447.

Parzen, E. and Newton, H. J. (1980). Multiple Time Series Modeling, II Multivariate Analysis - V, edited by P. Krishnaiah, North Holland: Amsterdam, 181-197.

Pourahamdi, Mohsen (1982). Exact factorization of the spectral density and its application to forecasting and time series analysis. Manuscript (in press).

Wahba, Grace (1980). Automatic smoothing of the log periodogram, Journal of the American Statistical Assn., 75, 122-132.

Woodfield, Terry J. (1982). Statistical Modeling of Bivariate Data. Ph.D. Thesis. Institute of Statistics, Texas A&M University.

WHAT SHOULD YOUR TIME SERIES ANALYSIS PROGRAM DO?

H. Joseph Newton

Institute of Statistics
Texas A&M University

Abstract

This paper seeks to lay a foundation for a discussion of the ideal time series analysis program. Some of the basic scientific problems where time series analysis is used are described, some of the basic approaches to the solution of those problems are discussed, currently available time series analysis software is reviewed, and an ideal time series analysis program is proposed.

Key Words

Time series analysis software, ARMA models, Expert software, Interactive computing.

1. INTRODUCTION

The interface between Computer Science and Statistics has always been very important in time series analysis. Some of the most important contributions of statistical computing to science at large have been made in this area. The most obvious example of this is the Fast Fourier transform algorithm whose popularization was accomplished by Cooley and Tukey (1965).

In studying the contents of the book <u>Directions in Time Series</u>: the Proceedings of IMS Special Topics Meeting on Time Series Analysis held at Ames Iowa in May, 1978, one is struck by how computer intensive many of the areas of time series were perceived by the authors. The time series literature of the following five years does nothing to disturb this perception. Thus, for example, a large number of articles on time series graphics, methods for maximizing likelihoods, and algorithms for robustly doing time series have appeared.

Time series analysis is also somewhat unusual in Statistics in that important theoretical and computational results are developed in a wide variety of scientific fields.

The purpose of this paper is to 1) describe the basic scientific problems that time series attacks, 2) survey some of the basic approaches to solution of these problems, 3) describe currently available statistical time series software, and 4) suggest an ideal time series analysis program.

2. BASIC TIME SERIES PROBLEMS

In this section we discuss some of the scientific problems that can be considered using time series analysis. Obviously many of these problems overlap and this list is assuredly not exhaustive.

Data Description

Many scientific areas have large amounts of data observed over time that need to be described in succinct numerical and graphical ways. Correlation analysis and spectral density estimation are basic tools used in this description.

Time Series Regression Analysis

Given several time series what relationships can be determined among them?

Prediction and Interpolation

Given a sample realization $\underset{\sim}{Y}(1),\ldots,\underset{\sim}{Y}(T)$ from a time series $\underset{\sim}{Y}$, how can one predict a future value $\underset{\sim}{Y}(T+h)$? Alternatively, given that there is a gap in the series $\underset{\sim}{Y}(1),\ldots,\underset{\sim}{Y}(T)$, how can it best be filled in? One might want to do this only using the information in $\underset{\sim}{Y}$ itself or by using its relationship with some other series $\underset{\sim}{X}$.

Data Reduction

Many areas have huge masses of data to contend with. Time series can be used to reduce the amount of data required for an analysis.

Seasonal Adjustment

Given a series $\underset{\sim}{Y}$ can it be decomposed into a "trend", a "seasonal", and an "irregular"?

Time Series Deconvolution

Given a time series Y that is a filtered version of white noise X how can one design a filter which recovers X from the observed Y?

Control Theory

How can one control some output variable of a process as close to some desired value when the input variables and the process itself are subject to variability?

Change Detection

Does the nature of a time series change at some time too? (This problem is often termed intervention analysis (Box and Tiao (1975)).

Spatial Time Series

How can data that is observed over a regular or irregular spatial grid be analyzed using time series?

These problems go by a wide variety of names in the various scientific fields in which they are used. Thus, for example, a seasonal adjustment can be thought of as a special case of signal processing, *i.e.* a decomposition of an observed time series (signal) into frequency component parts.

3. BASIC TIME SERIES METHODS

Time series analysis is basically concerned with finding a mathematic framework in which one can describe the behavior of the set of all possible realizations $\{Y(t), t \in T\}$ of a time series Y given only one part of one possible realization. The time series Y can be univariate (one observation per index t) or multivariate (d observation per index t). The index set T can be continuous (*i.e.* a subset of Euclidean space) or discrete (*i.e.* T has countably many elements). One might seek to describe Y in terms of other time series or only in terms of its own values.

The traditional methods of time series can be classified as linear and Gaussian. Thus one usually assumes that the finite dimensional joint distributions of Y can be adaquately described by its first two moments (as a Gaussian process can be described), *i.e.* the mean value function

$$m(t) = E(Y(t)), t \in T$$

and covariance Kernel

$$K(s,t) = Cov(Y(s), Y(t)), s, t \in T$$

and that these distributions are not effected by a common time shift (strict stationarity) or at least $m(\cdot)$ is constant and $K(s,t)$ is only a function $R(\cdot)$ of $t-s$ (covariance stationary).

A function that is mathematically equivalent to R is its Fourier Transform f, *i.e.* the spectral density function. The spectral density is an important tool in further modeling R, is an important descriptive device, is an important theoretical tool, and is important in itself in many applications.

To parsimoniously model R or f, the autoregressive moving average model (henceforth we consider explicitly only discrete parameter series)

$$Y(t) + A(1)Y(t-1) + \ldots + A(p)Y(t-p) = \varepsilon(t) + B(1)\varepsilon(t-1) + \ldots + B(q)\varepsilon(t-q)$$

for ε a white noise time series, has become very popular and a great deal of attention has been paid in a wide variety of scientific disciplines to the estimation of the parameters $A(1),\ldots,A(p),B(1),\ldots,B(q)$.

An example of the use of ARMA models is afforded by time series regression. Let Y and X be d_y and d_x dimensional time series and suppose one wants to determine a filter $\{B(j), j \in Z\}$ to approximate Y by

$$Y(t) = \sum_{j=-\infty}^{\infty} B(j)X(t-j)$$

so that

$$E\{[Y(t) - \sum_{j=-\infty}^{\infty} B(j)X(t-j)][Y(t) - \sum_{j=-\infty}^{\infty} B(j)X(t-j)]^T\}$$

is minimized.

Then

$$B(j) = \frac{1}{2\pi} \int_{-\pi}^{\pi} f_{YX}(\omega) f_{XX}^{-1}(\omega) e^{ij\omega} d\omega,$$

$$\ddagger = Var[Y(t) - \Sigma B(j)X(t-j)] = \int_{-\pi}^{\pi} [f_{YY}(\omega) - f_{YX}(\omega) f_{XX}^{-1}(\omega) f_{XY}(\omega)] d\omega$$

where f_{XX}, f_{YY}, and f_{XY} are the autospectra of X and Y and cross spectra of X and Y respectively. Then one can estimate B and $\ddagger$ by using estimators of f_{XX}, f_{YY}, and f_{XY}.

4. CURRENT TIME SERIES SOFTWARE

The traditional time series analysis procedure can be thought of as having 5 steps:

1) Preprocessing of data including handling of missing data, obvious nonstationarity, etc.

2) Initial, non-model oriented descriptive statistics usually correlogram and/or periodogram and smoothed periodogram analysis.

3) Fitting of a tentative model or models.

4) Verification of the model or models by examination of residuals.

5) Solution to the scientific problem, *i.e.* forecasting, regression, cross-spectral analysis, control, etc. based on verified model.

This type of analysis is well illustrated by the so-called Box-Jenkins Method (1970). In the Box-Jenkins Method the five steps are defined by:

1) Preprocessing: power transformations and differencing operators.

2) Descriptive statistics: sample correlogram and partial autocorrelations.

3) Tentative Models: seasonal ARMA models suggested by visual inspection of correlogram and partial autocorrelations.

4) Verification of Model: Portmanteau test.

5) Scientific Problems Considered: forecasting, regression of a univariate time series on a univariate time series without cross spectral analysis, intervention analysis.

Thus the Box-Jenkins procedure is easily implemented on a computer (although most of the difficult modeling decisions are left to the user) and each of the major statistical packages (SPSS, SAS, BMDP, and IMSL) have a Box-Jenkins procedure (SPSS: BOX-JENKINS, SAS: PROC ARIMA, BMDP: P2T, IMSL: Subroutines FTARPS, FTAUTO, FTCAST, FTCMP, FTMXL, etc.). Also several specialized software producers have Box-Jenkins software (see Pack (1980) for a survey).

In the rest of this section we describe the other capabilities of SPSS, SAS, BMDP, and IMSL as well as two other packages, TIMSAC-78 and TIMESBOARD, which were not commercially produced.

BMD Procedure P1T

Procedure P1T is entitled "Univariate and Bivariate Spectral Analysis" and has the following features as described in Dixon (1981):

1) data screening, missing value replacement, trend removal, and seasonal adjustment.

2) filtering by bandpass filters, user specified filters, or autoregressive filters (with user specified order).

3) univariate and bivariate spectral estimation using smoothed periodogram (using the Parzen, Hanning, or Hamming windows) or autoregressive spectral estimator.

SAS

As described in Ray (1982), SAS has 5 procedures in addition to the ARIMA procedure:

FORECAST: uses exponential smoothing or a simple autoregressive model to forecast a univariate series.

AUTOREG: regression with autoregressive errors.

STATESPACE: models a vector time series by state space methods.

SPECTRA: cross spectral analysis using pure averages of periodogram.

X11: seasonal adjustment.

SPSS

According to Hull and Nie (1981) there are no other time series analysis programs in SPSS other than BOX-JENKINS.

IMSL

The IMSL library has the following subroutines that are useful in analyzing time series:

FTCROS: Correlations and cross correlations for multiple series

FTFPS: Cross spectral analysis using Parzen window

FTFREQ: Time series regression of a univariate series on another univariate series.

FTKALM: General Kalman filter algorithm

FTWENM: Wiener forecast for multiple series

TIMSAC-78

As described by Kitagawa and Akaike (1981), TIMSAC-78 (which is available through the Division of Mathematical Sciences of the University of Tulsa) is a package of Fortran subroutines and mainline programs that implement the work of Akaike. The following main programs are available:

UNIMAR   Univariate case of minimum AIC method of AR model fitting

UNIBAR   Univariate case of Bayesian method of AR model fitting

BSUBST   Bayesian type all subset analysis of time series by a model linear in parameters

MULMAR   Multivariate case of minimum AIC method of AR model fitting

MULBAR   Multivariate case of Bayesian method of AR model fitting

PERARS   Periodic autoregression for a scalar time series

MLOCAR   Minimum AIC method of locally stationary AR model fitting

BLOCAR   Bayesian method of locally stationary AR model fitting

MLOMAR   Minimum AIC method of locally stationary multivariate AR model fitting

BLOMAR   Bayesian method of locally stationary

multivariate AR model fitting

NADCON Noise adaptive controller

EXSAR Exact maximum likelihood method of scalar AR model fitting

XSARMA Exact maximum likelihood method of scalar AR-MA model fitting

TSSBST Test of subroutine SUBSET

TSWIND Test of subroutine WINDOW

TSCHCK Test of subroutine CHECK

TSROOT Test of subroutine SHROOT

TSTIMS Test of subroutine EDTIMS

TSCANC Test of subroutine CANCOR

<u>TIMESBOARD</u>

As described by Newton (1983), TIMESBOARD (which is now available by contacting the Institute of Statistics, Texas A&M University) is also a package of Fortran subroutines and main programs. TIMESBOARD implements the time series modeling techniques suggested by Parzen, Pagano, and Newton.

There are approximately 250 subroutines and ten main programs including:

ARSPID: Univariate time series modeling, forecasting, and spectral estimation using <u>Auto</u>regressive <u>Spectral Identification</u> techniques (CAT criterion, best and second best AR spectra, etc.). ARSPID attempts to use the same modeling philosophy as the Box-Jenkins method but with a host of semi-automatic criteria to guide the various stages of the analysis.

MULTSP: Spectral analysis of multiple time series using the Parzen window, autoregressive spectral estimation, and periodic autoregressive spectral estimation (Newton (1982)).

DTFORE: Univariate Forecasting using ARARMA models (Parzen (1982a)).

MODIDM: Identify linear relationships between multiple time series $\underline{Y}$ and $\underline{X}$ using information numbers. (Parzen (1982b)). Thus the dependence of $\underline{Y}$ and $\underline{X}$ on their own past, on the past of the other series, on the past of both series, on their own past and the past and present of the other series, and on the entire record of the other series only are measured using information numbers. These numbers give a succinct summary that can be used to consider questions of statistical causality and feedback between $\underline{Y}$ and $\underline{X}$.

<u>Summary of Existing Software</u>

There is certainly other time series analysis software available (for example RATS (Doan and Litterman (1982)), TIMESERIES (Shahabuddin (1979)), and TSERIES (Meeker (1978)), but the ones listed above are the ones we are most familiar with.

All are batch oriented and except for TIMESBOARD do not contain high resolution graphics. The major packages are working on developing interactive versions and at least SPSS and IMSL have plans for major expansion of their time series capabilities.

5. AN "IDEAL" TIME SERIES PROGRAM

In the light of sections 1-4 then, what properties should an ideal time series analysis program have?

1) It should be <u>interactive</u>. Time series analysis is by its very nature an interactive science/art. This is perhaps the most important lesson of the Box-Jenkins philosophy; an iterative procedure of model guessing, diagnostic checking, and analysis (prediction, regression, spectral analysis, etc.) seems to be an excellent framework in which to do time series analysis.

2) It should be <u>expert</u> (Chambers (1981)), <u>i.e.</u> the analysis should be guided by the program using generally accepted attacks on various problem types. Thus the first menu in the program should be a list of possible ultimate goals of the current run of the program.

3) It should incorporate a wide variety of methods for each task. The major statistical results for time series analysis are asymptotic. Also, many of the methods require the user to specify certain fine-tuning parameters (<u>e.g.</u> bandwidth of spectral windows, Autoregressive orders, etc.). Thus for each possible task there is a bewildering array of possible methods and parameters. The program should allow the user to try out several methods and see if there is any consistency in the results.

4) The program should be capable of operating at several levels of user sohpistication. Thus for an inexperienced user the program could carry on almost a tutorial dialog with the user, suggesting possible courses of action and the implications of these actions. Also, warnings about the possible abuse of statistical inferences made (<u>e.g.</u> p-values and confidence intervals) could be given. On the other hand an experienced user should be able to proceed through the program quickly.

5) The program should be designed for a moderate size minicomputer. One of the major difficulties in current software design is not knowing what computing will be like by the

time any products were completed. However with the development of large high speed hard disks for microcomputers it would appear that the personal computer of the future will have the capabilities of the current minicomputer.

6) Great attention must be paid to graphical compatibility. The major difficulty in transferring a program from one computer to another is not in the compatibility of the language used for code but rather in the graphics. The graphics in the ideal program is crucial as most of the results are displayed as graphs and most decisions are based on inspection of graphs.

7) The program should be expandable. Because the program will be interactive it could be designed to guide a sophisticated user through an expansion of itself.

8) The program must devote a lot of attention to the problems of missing values and irregularly spaced observations including the case where one has sampled a continuous record and the properties of the continuous time series are of primary importance in the analysis. (See Parzen (1984) for a summary of the problems involved.)

9) The recent developments in the robust analysis of time series should be incorporated in the program (see Martin (1980) for example).

10) The wide variety of computational methods available for carrying out certain tasks should be incorporated. For example there are a large number of numerical methods for estimation of ARMA parameters for missing data, including the Kalman filter algorithm, the EM algorithm, Householder transformations, Cholesley decomposition, etc. Until the numerical and statistical properties of competing procedures are better understood one should have the ability to try them all.

How Should the Ideal Program be Created?

Development of the program described above is a massive task. This paper is an attempt to create a framework for discussion of a possible program. I have tried to emphasize two major points:

1) There is little consensus among researchers as to the best possible means of doing various time series analyses.

2) There are a wide variety of users and developers of time series methods, many of whom are not aware of each other or the existence of competing methodology.

Thus it would appear that a required first step would be for some organization or organizations (e.g. ASA, IEEE, SIAM, etc.) to arrange a symposium on the development of expert time series software (the first and second Applied Time Series Symposiums in 1976 and 1980 organized by Findley (1978, 1981) illustrate the variety of researchers that would be required). The hope would be that such a symposium would result in a fairly detailed outline of a possible program. How the program actually got created is then another problem.

We conclude by noting that anytime there is a discussion of statistical software development there is a discussion to the effect that the availability of software leads inevitably to the abuse by some users of the statistical method embodied in the software. We would hope that the interactive nature of the ideal time series analysis program could be used to minimize this abuse.

Acknowledgement

This research was supported by grant number N00014-82-MP-2001 from the Office of Naval Research.

6. REFERENCES

(1) Box, G.E.P, and Jenkins, G.M. (1970). Time Series Analysis, Forecasting and Control. Holden-Day, San Francisco.

(2) Box, G.E.P. and Tiao, G.C. (1975). "Intervention Analysis with Applications to Economic and Environmental Problems," JASA, 70, 70-79.

(3) Brillinger, D.R. and Tiao, G.C. (1980). Directions in Time Series. Institute of Mathematical Statistics.

(4) Chambers, J.M. (1981). "Some Thoughts on Expert Software", 13th Interface Symposium, ed. by W.F. Eddy. Springer-Verlag, New York.

(5) Cooley, J.W. and Tukey, J.W. (1965). "An Algorithm for the Machine Calculation of Complex Fourier Series", Math. Comput., 19, 297-301.

(6) Dixon, W.J. (1981). BMDP Statistical Software 1981, University of California Press, Berkeley.

(7) Doan, T.A. and Litterman, R.B. (1982). "RATS-Regression Analysis of Time Series", American Statistician, 36, 63.

(8) Findley, D.F. (1978). Applied Time Series Analysis, Academic Press, New York.

(9) Findley, D.F. (1981). Applied Time Series Analysis II, Academic Press, New York.

(10) Hull, C.H. and Nie, N.H. (1981). SPSS Update 7.9, McGraw-Hill, New York.

(11) International Mathematical and Statistical Libraries, Inc., Houston, Texas.

(12) Kitagawa, G. and Akaike, H. (1981). "On TIMSAC-78", in *Applied Time Series Analysis II*, ed. by D.F. Findley, Academic Press, New York, 499-548.

(13) Martin, R.D. (1980). "Robust Estimation of Autoregressive Models", in *Directions in Time Series*, ed. by D.R. Brillinger and G.C. Tiao, IMS, 228-254.

(14) Meeker, W.Q. (1978). "TSERIES-A User-Oriented Computer Program for Time Series", *American Statistician*, *32*, 111-112.

(15) Newton, H.J. (1982). "On Using Periodic Autoregressions for Multiple Spectral Estimation", *Technometrics*, *24*, 109-116.

(16) Newton, H.J. (1983). *An Introduction to the Methods of Time Series Analysis in the Time and Frequency Domains*, Institute of Statistics, Texas A&M Univ.

(17) Pack, D.J. (1980). "What Will Your Time Series Analysis Computer Package Do?", in *Directions In Time Series*, ed. by D.R. Brillinger and G.C. Tiao, IMS, 60-79.

(18) Parzen, E. (1982a). "ARARMA Models for Time Series Analysis and Forecasting", *J. of Forecasting*, *1*, 67-82.

(19) Parzen, E. (1982b). Comment on "The Measurement of Linear Dependence and Feedback Between Multiple Time Series" by J. Geweke, *JASA*, *77*, 320-322.

(20) Parzen, E. (1984). *Time Series with Missing and Irregularly Spaced Data*, Springer-Verlag Lecture Notes in Statistics, New York.

(21) Ray, A.A. (1982). *SAS/ETS Users Guide*, SAS Institute, Inc. Cary, North Carolina.

(22) Shahabuddin, S. (1979). "TIMESERIES-An Interactive Computer Program for Time Series Analysis", *American Statistician*, *33*, 223.

NUMERICAL ALGORITHMS

Organizer: W. J. Kennedy, Jr.

Invited Presentations:

A Generalization of the Proposed IEEE Standard for Floating-Point Arithmetic, W. J. Cody

A Package for Solving Large Sparse Linear Least Squares Problems, Alan George and Esmond Ng

Exact Computation with Order-N Farey Fractions, R. T. Gregory

Recent Advances in Generating Observations from Discrete Random Variables, Bruce Schmeiser

A Generalization of the Proposed IEEE Standard for Floating-Point Arithmetic

W. J. Cody[1]

Mathematics and Computer Science Division
Argonne National Laboratory
Argonne, Illinois 60439

Abstract

Several years ago the Microprocessor Standards Committee of the IEEE Computer Society established a Floating-Point Working Group to draft a standard for binary floating-point arithmetic on 32-bit microprocessors. As that task neared completion, a second working group was established to generalize the proposed binary standard for other radices and wordlengths. We discuss the emerging generalization, its influence on final deliberations on the proposed binary standard, and the implications for numerical computation.

"The practical numerical analyst with high standards is...inextricably involved with the arithmetic behavior of his "digital computer hardware and accompanying operating systems."

G. E. Forsythe [12]

1. INTRODUCTION

The Microprocessor Standards Committee of the IEEE Computer Society currently sponsors two efforts to produce standards for floating-point arithmetic. The first, Task P754, began deliberation in late 1977 on a proposed standard for binary floating-point arithmetic. It finally reported Draft 10.0 [29] out of committee in December, 1982. The second, Task P854, first convened in January 1981, charged with producing a draft standard that is independent of the computer wordlength and floating-point radix and that is upward compatible with the draft binary standard. That group hopes to have a draft proposal for public review by mid-1983.

Each of these projects has benefited from the work of the other. P754 settled most of the controversial issues, such as handling of infinities and underflow, in hard-fought committee meetings before P854 began work; in return, the abstract parameterized description of arithmetic vital to the success of P854 has recently contributed to the clarity of exposition in P754. Mindful of the requirement to be upward compatible, P854 has also struck out into areas P754 neglected to consider, such as the details of handling exceptional strings encountered during I/O conversion.

Our purpose here is to discuss the current status of this work. We start by presenting the important features of the draft binary standard

in the next section, including significant changes introduced since the appearance of Draft 8.0 [28]. Section 3 discusses and motivates some of the generalizations for P854 [5]. Section 4 suggests the numerical benefits of these proposals by looking at computational examples, and Section 5 discusses important issues not directly addressed by the committees, such as language support. Finally, the list of references at the end contains as many relevant publications as we could locate, whether cited in the body of this paper or not.

We emphasize that the ideas presented here are the result of deliberations by many talented people. In particular, the elegant argument presented in Section 3 justifying the magic constant 7 was first put forth by Fred Ris. We cannot hope to credit everyone who has contributed, but point to the extensive bibliography at the end of this article and to the credits in Draft 10.0 as indicating the broad participation in these standardization efforts.

2. THE DRAFT BINARY STANDARD

The drafters of the binary standard have specified a dramatically new programming environment to foster the development of high-quality numerical software and to simplify things for Forsythe's "practical numerical analyst with high standards." By fiat the committee limited their attention to binary arithmetic on 32-bit microprocessors. They addressed such problems as data types, arithmetic and exception handling without concern for conventions of existing hardware. The resulting draft, in the words of the chairman, D. Stevenson [28], places "the interest of the user

---

[1] This work was supported by the Applied Mathematical Sciences Research Program (KC-04-02) of the Office of Energy Research of the U.S. Department of Energy under Contract W-31-109-ENG-38.

community...above the goal of industrial continuity." There is space here only to describe briefly the highlights of that achievement. Interested readers must consult the documents cited in the bibliography for more details.

The draft rigidly defines a total arithmetic system, i.e., data representation and arithmetic engine, with few implementation options. It specifies four data formats in two groups, basic and extended, each having two widths, single and double. A standard-conforming system must implement single format, and should implement the extended format corresponding to the widest basic format supported.

Each basic format uses a compact signed magnitude representation. In single precision, for example, an algebraic sign, an 8-bit biased exponent and a 24-bit significand are squeezed into the 32-bit format by making the leading bit of the normalized significand implicit instead of explicit. A normalized significand is thus represented as $\pm 1.f$, where the 1 is implicit and $0 \leq f < 1$ is explicit. Further, the exponent bias is chosen so that $\lambda\sigma \geq 1$, where λ is the largest finite floating-point number and σ is the smallest normalized floating-point number. This provides some overflow protection for n/x where n is a small integer.

Two exponent fields are reserved to represent special quantities. Data with the largest biased exponent represent either ∞ or NaN (not-a-number) while data with the smallest biased exponent represent tiny denormalized numbers introduced by graceful underflow. We return to these special quantities later.

There is a similar prescription for double precision data with an algebraic sign, an 11-bit biased exponent and a 53-bit significand fitted into a 64-bit format. The extra exponent and significand width are chosen to offer reasonable overflow/underflow and accuracy protection for such computations as double-length accumulation of single-length vector inner products.

For similar reasons, the extended formats must have at least 3 or 4 more bits in their exponent field than the corresponding basic format, and must have a significand at least as wide as the total width of the corresponding basic format. However, the leading significand bit is not required to be implicit in extended formats.

In addition to the usual four basic arithmetic operations, the draft standard requires operations to extract the square root, find the remainder, round to a floating-point integer, convert between floating-point numbers and integers, convert between floating-point representations and decimal strings, and compare. All these operations must support the basic formats, and many of them must also support extended formats. Conversion between decimal strings and an extended format is an example of an operation that is not specified. Another omission is the copy operation.

The draft standard prescribes four different rounding modes. These are round to nearest and three directed roundings: round toward $+\infty$, round toward $-\infty$ and round toward 0. In round to nearest, the representable value nearest to the infinitely precise result is delivered with ties broken by rounding to the nearest representable value with least significant bit 0. Infinitely precise results exceeding the largest representable finite number by at least a half unit in the least significant bit round to ∞ in this mode. The directed rounding modes are self-explanatory. Their primary usage is for special purposes, such as to support interval arithmetic.

These rounding modes apply fully to each of the operations described earlier except comparison, remainder and binary$\leftrightarrow$decimal string conversion. In the last case, conversion results are to be correctly rounded for data within a certain specified range. For other data, the rounding error in the round to nearest case cannot exceed by more than 0.47 units in the destination's least significant digit the error prescribed for usual round to nearest.

The range for correct rounding is defined as follows. Let p_c denote the significand length of the precision in which the conversion is to be carried out (normally this corresponds to an extended precision). Then define N such that $5^N < 2^{p_c}$, i.e., such that 10^N can be represented exactly (because scaling of 5^N by powers of 2 is exact) in the conversion precision. Similarly, let D be the largest integer such that $M \leq 10^D-1$ can be represented exactly in the conversion precision. Then numbers of the form $\pm M \times 10^{\pm N}$ can be represented in the conversion precision with at most one rounding error. Coonen [8] gives the details of this analysis as well as the genesis of the magic constant 0.47.

The special quantities mentioned earlier provide closure of sorts to the arithmetic system and permit sensible responses to exceptional conditions. For example, because ∞ is the mathematical limit of a sequence of arbitrarily large reals, machine ∞ provides a default result for overflow and division by zero; arithmetic on it preserves much of its mathematical behavior. For mathematical reasons, Draft 8.0 provided both affine ($+\infty \neq -\infty$) and projective ($+\infty = -\infty$) modes for infinity arithmetic. In a major simplification, the projective mode has now been eliminated.

One logical consequence of signed ∞ is that the sign should be preserved when $-\infty$ is reciprocated twice. This implies that 0 must also carry an algebraic sign, i.e., that $1/(-\infty) = -0$. Signed 0 makes sense for other reasons too. If

0 is construed as the limiting case for underflow, the algebraic sign permits distinction between approaching that limit from the left and from the right, a concept important for interval arithmetic. Of course, signed zero occurs naturally in a sign magnitude representation, so there is no penalty for including it among the representable quantities.

The treatment of underflow leads to tiny denormalized numbers. In traditional arithmetic systems, underflow is detected when the exponent of a number becomes too small. The default action is to flush the significand of the number abruptly to zero. In mathematical terms, this catastrophic loss in significance introduces a singular point into the error analysis. Graceful underflow and denormalized numbers mark a radical departure from this traditional approach. The draft standard denormalizes numbers below the underflow threshold, gracefully shifting low order bits off the registers one at a time instead of abruptly shifting all off at once. This approach elevates loss of significance from underflow to a level comparable to that of rounding, thus removing the singularity from error analysis.

There are side effects, however. Denormalized numbers may be promoted back into the range of normalized numbers through multiplication, for example. Because the number of significant bits in the normalized result may be limited by the number of significant bits in the denormalized operand, the significance of the result may be much less than normally expected. Draft 8.0 provided two options for the user here. The normalizing mode simply normalized the result and continued computation after signalling an exception. The warning mode was a complicated alternative in which, depending on the details of the operation and operands, the result might remain denormalized in an extended arithmetic register, or an exception might be signalled and a NaN be provided as a default result. In another major simplification, the warning mode has now been deleted from the draft standard.

NaNs are special entities introduced to handle otherwise intractable situations. They come in two guises, signalling and quiet. The major difference between them is that signalling NaNs signal an exception when encountered as operands and quiet NaNs normally do not. Quiet NaNs are used as default results for operations when numerical results would be nonsensical. For example, the draft standard specifies that the mathematically meaningless computations $0 \times \infty$ and $0/0$ signal exceptions and deliver quiet NaNs as default results. In contrast, signalling NaNs must be deliberately introduced by the user and are therefore reserved for initializing special actions through an exception handler.

That brings us to the last major component of the draft that we will discuss, the exception handling. There are five types of exceptions that are signalled: invalid operation, division by zero, overflow, underflow and inexact. Examples of invalid operations are 0/0 and any operation involving a signalling NaN. The inexact exception is raised whenever the rounded result is not exact. Other possible exceptions are self-explanatory.

Signalling an exception entails setting a status flag, taking a trap, or possibly doing both. Interrogation, setting, resetting, saving and restoring of status flags are under user control. Traps may be user-supplied, opening the door for implementation of special arithmetics, for example, accessible through signalling NaNs. The draft specifies default results to be delivered to active trap handlers; when traps are not taken, computation continues with (possibly different) default values.

This, then, is the essence of the draft standard. We believe that the committee has acted responsibly by requiring only those things that are useful and achievable at reasonable cost, and by publishing information on how to implement them [6,8,20]. Where early drafts were unnecessarily complicated or unclear, corrective action has been taken (the latest draft includes clarified discussions of underflow, comparisons and exception handling in addition to improvements mentioned earlier). The importance and success of this effort is emphasized each time a new chip appears implementing all or part of the draft standard.

3. THE DRAFT GENERALIZED STANDARD

The committee working on the generalized standard has tried to retain all the features of the binary draft as they have been described here, while eliminating all dependence on radix and wordlength from the description of those features. Of necessity, the generalized draft cannot discuss how a floating-point number is to be represented in storage; it can only discuss the values of floating-point numbers and constraints on parameters defining permissible values.

The draft thus discusses data precisions instead of data formats. Four parameters specify each of the four precisions discussed. These are

b - the radix,
p - the number of base-b digits in the significand,
E_{max} - the maximum exponent, and
E_{min} - the minimum exponent.

(Following this lead, the draft binary standard now describes and discusses its formats using the parameters p, E_{min} and E_{max}.) Because the committee could find no valid technical reason for b to be anything except 2 or 10, it is limited to those two values. This is a departure from the apparent charter for the group, but there have been no complaints yet from the parent organization. The committee believes this deviation will be acceptable to everyone provided it can show that alternatives were seriously considered, and were rejected on technical grounds.

There are restrictions imposed on the other parameters defining a precision to guarantee that the defined precision is useful. Specifically, the parameters for basic precisions must satisfy the conditions

$$b^{p-1} \geq 10^5, \text{ and}$$

$(E_{max}-E_{min})/p$ shall exceed 5 and should exceed 10.

These guarantee enough significance to support serious computation, and enough exponent range relative to significance to minimize underflow/overflow problems in mid-range computations. It is also recommended that

$$b^{E_{max}+E_{min}+1} \geq 4 ,$$

thus balancing the exponent range and guaranteeing $\lambda\sigma \geq 1$. These parameter restrictions are based on the collective experience of the committee, but are otherwise arbitrary.

Relations between parameters for single and double precisions are

$$b^{p_d} > 8b^{2p_s} ,$$
$$E_{max_d} \geq 8 E_{max_s} + 7 ,$$
$$E_{min_d} \leq 8 E_{min_s} .$$

Relations between the parameters defining single/double extended precision and those defining single/double precision are similar, except that

$$p_e \geq 1.2p$$

replaces the above condition on p, where here and in the following p_e denotes an extended precision supporting the basic precision p. In addition,

$$p_e \geq p+7$$

is required for proper conversion to and from decimal strings when b = 2, and

$$p_e > 1 + p + \frac{\ln\{3\ln(b)[E_{max}+1]\}}{\ln(b)}$$

is recommended for controlling error in y^x.

The derivation of the first of these additional conditions typifies the technical work involved in drafting these standards. For b = 2, the condition $10^D < 2^{p_e}$ defines D such that D digit decimal integers can be represented exactly in the extended precision. In addition, because there are p-10 distinct floating-point numbers and D-4 distinct decimal numbers between 1000 and 1001, the mapping from floating point to decimal will be unique only if

$$10^{D-4} \geq 2^{p-10} ,$$

or

$$D \geq \lceil p \log_{10}(2)+1 \rceil .$$

When rounding to nearest, conversion from floating point to decimal string and back to floating point will be the identity if both conditions on D are satisfied, i.e., if

$$\lfloor p_e \log_{10}(2) \rfloor \geq D \geq \lceil p \log_{10}(2)+1 \rceil ,$$

or

$$\lfloor p_e \log_{10}(2) \rfloor \geq \lfloor p \log_{10}(2)+2 \rfloor .$$

This reduces to the condition

$$p_e \geq p + 6.64...$$

or, because p_e and p are integers,

$$p_e \geq p + 7 .$$

Because formats cannot be discussed in the generalized draft, there are no constraints on the uniqueness of representation. All that is required is that the same results be obtained when operand precisions and values are the same. Similarly, the draft requires that there be special quantities (±∞, at least one signalling NaN and at least one quiet NaN) without specifying how they are to be represented.

Because the concept of normalization does not exist in this context, denormalized and unnormalized numbers occur naturally. Graceful underflow is, therefore, much less controversial from this abstract viewpoint than from the rigid viewpoint of the draft binary standard.

In addition to the technical matters just discussed, the committee has wrestled with other problems. For one, what does "upward compatible" mean? If P754 has failed to consider some topic, how is P854 governed in its treatment of that topic? The binary draft states that overflow/underflow and NaNs and infinities encountered during binary to decimal conversion should be indicated to the user by appropriate strings, and says nothing about dealing with NaNs encoded in decimal strings. In the generalized standard, we define the "appropriate strings" and specify how NaNs encoded in decimal strings are to be handled. Must such new stipulations be only recommendations, or can they be requirements? The parent committee has opined that these can be requirements, a viewpoint that contradicts my interpretation of upward compatibility. If this opinion should be reversed, we may have to reword some sections of the draft.

Like all committees, P854 must justify technical decisions. For example, based on what we perceive to be the intended use of the standard, namely, serious scientific computation, a lower limit of 6 decimal digits of arithmetic significance has been inserted in the draft. But is there a technical reason for requiring at least 6 digits and precluding 5? Might there be non-frivolous situations in which 5 digits is acceptable? If so, then how about 15 bits? Are there overriding engineering arguments for or against some of these alternatives? As they are settled, these and similar questions are addressed in technical notes in an appendix to the draft. We hope these notes will help us to avoid repetitious arguments over the inclusion of some feature, and will provide ammunition to answer critics intelligently when the time comes.

4. COMPUTATIONAL IMPLICATIONS

The floating-point systems just described are intended to provide the characteristics and flexibility in arithmetic systems long sought by numerical software professionals, and to afford additional protection to the novice using "obvious algorithms." They are a significant advance over previous systems, but they do have limitations.

How advantageous the new systems are depends on the system configurations and the algorithms to be implemented. In general, graceful underflow will benefit algorithms that might malfunction because of destructive underflow; extended precision will benefit algorithms that might malfunction because of underflow, overflow and certain subtraction errors. Algorithms that are fundamentally unstable will remain unstable and dangerous to use with IEEE arithmetic, although they may successfully solve more problems using IEEE arithmetic than they could using traditional arithmetic.

Consider, for example, complex division, $z = w/t$, where

$$z = x + iy,$$
$$w = u + iv,$$

and

$$t = r + is.$$

Ideally, an algorithm for this computation should return a quotient that is correct to within rounding error in each component whenever those results are representable in the host arithmetic system. That ideal is not achievable with traditional arithmetic systems at acceptable cost. Instead, a quotient correct to within a few rounding errors in $|z|$ is frequently returned, provided that $|z|$ is not "too large" and there has been no destructive underflow or overflow.

The obvious algorithm for this computation provides

$$x = \frac{ur + vs}{r^2 + s^2},$$

where there is a possibility of overflow or underflow in any of the intermediate products using traditional arithmetic.

R. Smith [27] devised a better algorithm:

$$x = \frac{u + v(s/r)}{r + s(s/r)},$$
$$y = \frac{v - u(s/r)}{r + s(s/r)},$$

when $|r| \geq |s|$, and

$$x = \frac{v + u(r/s)}{s + r(r/s)},$$
$$y = \frac{-u + v(r/s)}{s + r(r/s)},$$

otherwise. This scheme provides additional protection against intermediate overflow and underflow, and usually guarantees errors within a few roundings in $|z|$, but it does not guarantee that individual components are correct to within a few units in their last digits. It is still possible, for example, for (s/r) to underflow and thus to destroy an otherwise usable product vs/r in the first equation for x. In this case, the computation could be saved by instead computing $(vs)/r$ or $(v/r)s$, as appropriate.

Modifications to Smith's algorithm to handle this situation were proposed several years ago at Argonne, but never polished. Recently, Stewart [30] suggested a particularly elegant modification that complicates things only slightly. These modified Smith algorithms return quotients correct to within a few rounding errors in each component except in extreme situations. Overflow is still possible when both components of the numerator or denominator are close to λ, and partial underflow with accompanying loss of significance is possible when both components are close to σ.

The effect of IEEE arithmetic on these computations varies. The Smith algorithm has been analyzed by Hough [15] for IEEE single precision unsupported by extended. He shows that despite roundoff, the computed result is always, instead of usually, correct to within rounding error in $|z|$ unless $|u|+|v|$ or $|r|+|s|$ would overflow. While this still does not guarantee that each component of z is correct to within rounding error, it is a stronger and simpler statement than can be made assuming a traditional arithmetic system.

On the other hand, if IEEE single is supported by extended, all the algorithms discussed, including the obvious simple algorithm, return z correct to within rounding error in each component, provided only that computation is done in extended and the components are within range. That is a powerful result.

Inherently unstable algorithms are still not completely safe on IEEE machines. The solution of a quadratic equation using the well-known quadratic formula has been shown to be numerically dangerous on traditional machines [13], leading to possible destructive underflow or overflow or to loss of all significance in the smaller (in magnitude) root. The algorithm remains numerically dangerous on IEEE machines [16], although its performance is improved. The underflow/overflow problem is eliminated by extended precision, but it is still possible to lose 8 bits of significance out of 24 in the smaller root using single precision supported by extended.

Those working on the binary standard have studied many different algorithms [9,16,19,21] for use on IEEE machines in addition to those required by the draft standard [8,20]. Algorithms designed to work on minimal implementations (machines without an extended precision) are often similar to their counterparts on traditional machines, except that they are less devious, while those designed to use extended precision are usually simpler than their counterparts on traditional machines. The big advantages are that the new algorithms can solve more problems, exception handling is more reasonable and error analysis is simpler.

5. SUPPORT PROBLEMS

Of course, the specification of a standard and the appearance of standard-conforming arithmetic systems are not enough; new features must also be supported by operating systems, language compilers and other software if they are to be properly exploited. That presents problems. Just as the IEEE committees have rightfully regarded the specification of the arithmetic system as their responsibility, they must respect others' responsibility for specifying languages and operating systems. While the IEEE committees have ideas on how to support extended precision and exception handling, they cannot force their views on others; they can only advise.

Recognizing this, the P854 committee plans to extend its work to include specific recommendations for language support, suggesting reserved names for the special operands and for the various status flags and mode settings that are to be accessible to the user, suggesting intrinsic functions and perhaps even discussing semantics. Of course, these recommendations must avoid unnecessary conflict with existing language standards. Basic reserves the string INF to denote the largest finite floating-point number in a system, so we must find another string for the special quantity ∞.

Recommendations on intrinsic functions already exist in appendices to both of the draft arithmetic standards. Where additional recommendations will appear has not been decided; they may appear in a separate document or in a second appendix to the generalized standard. Wherever they appear, the suggestions must be made tactfully.

At the moment relations between the various standardization committees are cordial. Everyone recognizes that language extensions will be necessary, but the language people are not willing to do anything hastily. That is proper. Until the arithmetic people can decide what they want from languages, all that they can ask is that new language standards not include things that would prohibit necessary extensions later. If it is too early for language standards to lead the way in supporting the new arithmetic, and it probably is, at least they must not block the way.

6. REFERENCES

[1] ACM, Special Issue on the Proposed IEEE Floating-Point Standard, ACM SIGNUM Newsletter (October 1979).

[2] W. J. Cody, Impact of the proposed IEEE floating point standard on numerical software, ACM SIGNUM Newsletter, Special Issue (October 1979), 29-30.

[3] W. J. Cody, Towards sensible floating-point arithmetic, in Proceedings COMPCON Spring 80 (1980), 488-490.

[4] W. J. Cody, Analysis of proposals for the floating-point standard, Computer 14(3) (March 1981), 63-68.

[5] W. J. Cody, Chairman, A proposed radix-independent standard for floating-point arithmetic, Draft 0.5e, IEEE Radix-Free Floating Point Subcommittee Working Document P854/82-6.3 (1982).

[6] J. T. Coonen, An implementation guide to a proposed standard for floating-point arithmetic, Computer 13(1) (January 1980), 68-79. (Errata in Computer 14(3) (March 1981), 62.)

[7] J. T. Coonen, Underflow and the denormalized numbers, Computer 14(3) (March 1981), 75-86.

[8] J. T. Coonen, Accurate, economical binary ↔ decimal conversion, submitted for publication.

[9] J. Demmel, Effects of underflow on solving linear systems, submitted for publication.

[10] J. Demmel, Underflow and the reliability of numerical software, IEEE Floating Point Subcommittee Working Document P754/82-6.14 (1982).

[11] S. I. Feldman, The impact of the proposed standard for floating point arithmetic on languages and systems, ACM SIGNUM Newsletter, Special Issue (October 1979), 31-32.

[12] G. E. Forsythe, Algorithms for scientific computation, Communications of the ACM 9 (1966), 255-256.

[13] G. E. Forsythe, What is a satisfactory quadratic equation solver?, in Dejon, B. and Henrici, P. (eds.), Constructive Aspects of the Fundamental Theorem of Algebra (Wiley-Interscience, New York, 1969, 53-61).

[14] R. Fraley, S. Walther, Proposal to eliminate denormalized numbers, ACM SIGNUM Newsletter, Special Issue (October 1979), 22-23.

[15] D. Hough, Errors and error bounds, IEEE Floating-Point Subcommittee Working Document P754/80-3.2 (1980).

[16] D. Hough, Applications of the IEEE-754 standard for floating-point arithmetic, Computer 14(3) (March 1981), 70-74.

[17] Intel, The 8086 Family User's Manual, Numerics Supplement, Intel Corp., Santa Clara, Cal. (1980).

[18] W. Kahan and J. Palmer, On a proposed floating-point standard, ACM SIGNUM Newsletter, Special Issue (October 1979), 13-21.

[19] W. Kahan, Interval arithmetic in the proposed IEEE floating point arithmetic standard, in Nickel, Karl E. (ed.), Interval Mathematics 1980 (Academic Press, New York 1980, 99-128).

[20] W. Kahan, Software square root for the proposed IEEE floating-point standard, submitted for publication.

[21] W. Kahan, Why do we need a floating-point standard?, submitted for publication.

[22] W. Kahan and J. Coonen, The near orthogonality of syntax, semantics and diagnostics in numerical programming environments, in Reid, J. K. (ed.), The Relationship Between Numerical Computing and Programming Languages (North Holland, Amsterdam 1982, 103-113).

[23] W. Kahan, J. Demmel and J. Coonen, Proposed floating point environmental inquiries in Fortran, IEEE Floating Point Subcommittee Working Document P754/82-2.17 (1982).

[24] J. Palmer, The INTEL standard for floating-point arithmetic, in Proc. COMPSAC 77 (1977),107-112.

[25] M. H. Payne, Floating point standardization, in Proceedings COMPCON Fall 1979 (1979), 166-169.

[26] F. N. Ris, A unified decimal floating-point architecture for the support of high-level languages, SIGNUM Newsletter 11(3) (October 1976), 18-23.

[27] R. L. Smith, Algorithm 116: complex division, Communications of the ACM 5 (1962), 435.

[28] D. Stevenson, Chairman, A proposed standard for binary floating-point arithmetic, Draft 8.0 of IEEE Task P754, Computer 14(3) (March 1981), 51-62.

[29] D. Stevenson, Chairman, A proposed standard for binary floating-point arithmetic, Draft 10.0, IEEE Floating Point Subcommittee Working Document P754/82-8.6, (1982).

[30] G. W. Stewart, A note on complex division, Computer Science Technical Report TR-1206, University of Maryland, College Park, Maryland (August, 1982).

A Package for Solving Large Sparse Linear Least Squares Problems

Alan George

Esmond Ng

Department of Computer Science
University of Waterloo
Waterloo, Ontario, CANADA

ABSTRACT

There is a large amount of statistical software available for dealing with linear least squares problems. Apparently, few (if any) of these packages exploit sparsity in the design matrix X. In this paper we describe a package for solving these problems which takes advantage of such sparsity, and we provide some numerical experiments comparing its performance to that of several standard packages (BMDP, SAS, SPSS). A discussion of some of the design considerations and implementation techniques is also included.

1. Introduction

Our objective in this paper is to describe a numerical method and a package for solving the linear least squares problem

$$\min_{\beta} \| X\beta - y \|_2 \; , \tag{1.1}$$

where β and y are vectors of length p and n respectively, and X is an n by p real matrix with $n \geq p$. There are a number of important applications which yield an X which is very large and sparse. Two examples are surface fitting [1] and geodetic adjustment [8]. In order to solve such problems efficiently (or at all), it is crucial that the sparsity in the design matrix be exploited.

Assuming X has full column rank, a standard approach is to solve the normal equations

$$X^T X \beta = X^T y \; , \tag{1.2}$$

where well-developed techniques for exploiting sparsity for positive definite systems of equations can be used. However, it is well-known that the condition number of $X^T X$ is the square of the condition number of X, and for some problems this may make it impossible to obtain a solution to (1.2). The method employed in the package uses a more desirable method (from a numerical point of view) based on the use of rotations to reduce X to upper trapezoidal form [7]. That is, we find an n by n orthogonal matrix Q such that

$$QX = Q \begin{pmatrix} R \\ O \end{pmatrix} \; , \tag{1.3}$$

where R is p by p and upper triangular. If c denotes the first p components of Qy, the least squares solution β is then obtained by solving the triangular system

$$R\beta = c \; . \tag{1.4}$$

An outline of the paper is as follows. In Section 2 we review an efficient method for computing the orthogonal decomposition, and discuss some of the associated problems and generalizations. Some design considerations and implementation techniques in the software package are described briefly in Section 3. The interface of the package and a set of subroutines that can be used to assess the sensitivity of the solution to perturbations and to obtain other statistical quantities are described in Sections 4 and 5 respectively. In Section 6 the performance of the new package is demonstrated by providing some numerical experiments on some sparse problems. Finally some concluding remarks are provided in Section 7.

The package is intended to be both a research tool and useful production software. It is still under development and testing.

2. Sparse orthogonal decomposition and its applications

2.1. An efficient algorithm for computing orthogonal decomposition

An efficient algorithm for decomposing X into $Q\begin{pmatrix}R\\O\end{pmatrix}$ is described in [2]. The basic approach is as follows. Let $R^0 = O$ be the p by p zero matrix. A sequence of p by p upper triangular matrices $R^1, R^2, \cdots, R^n = R$ is then computed, where R^k is obtained by eliminating the k-th row of X using the rows of R^{k-1} and Givens transformations. These transformations are applied to the vector y simultaneously and therefore need not be saved. Furthermore, note that the rows of X are processed sequentially. Thus it is not necessary to store X in main memory since the rows can be stored in secondary storage and read in one at a time when they are needed. This approach is attractive because the observation matrix for many least squares problems is generated row by row.

The scheme described above is attractive whether X is sparse or dense. The novel aspect of the package is the way in which sparse matrix techniques for solving symmetric positive definite systems of equations are exploited in the computation of R. To explain this, first note that

$$X^T X = \begin{pmatrix} R^T & O \end{pmatrix} Q^T Q \begin{pmatrix} R \\ O \end{pmatrix}$$
$$= R^T R, \qquad (2.1)$$

which shows that R is the Cholesky factor of the symmetric positive definite matrix $X^T X$ (apart from possible sign differences in some rows). Second, for n by n and p by p permutation matrices P_r and P_c, we have

$$\begin{pmatrix} P_r X P_c \end{pmatrix}^T \begin{pmatrix} P_r X P_c \end{pmatrix} = \begin{pmatrix} \bar{R}^T & O \end{pmatrix} \begin{pmatrix} \bar{R} \\ O \end{pmatrix}$$
$$= \bar{R}^T \bar{R}. \qquad (2.2)$$

This shows that $\bar{R}$ depends on X and the *column* permutation P_c, but not on the *row* permutation P_r.

The package makes use of reliable methods for finding "good" orderings for sparse positive definite systems of equations. That is, given the structure of the symmetric positive definite matrix $X^T X$, there are reliable schemes for finding P_c such that $P_c^T X^T X P_c$ will yield a sparse $\bar{R}$, assuming such an ordering can be found. Moreover, given P_c and the structure of $X^T X$, there are very efficient algorithms for determining the structure of $\bar{R}$, and setting up an efficient data structure to store $\bar{R}$. See [4] for details.

The existence of these highly efficient algorithms has led to the following five-step procedure which is employed by the package.

1. Determine the structure (not the numerical values) of $X^T X$.
2. Find an ordering for $X^T X$ (column ordering for X) which has a sparse Cholesky factor R.
3. "Symbolically" factorize the reordered $X^T X$, generating an efficient row-oriented data structure for R.
4. Compute R and c by processing the rows of $\{X \mid y\}$ one by one using Givens rotations.
5. Solve $R\beta = c$, and then permute the components of β back into the original column ordering of X.

A few remarks are in order. First, the first three steps in the procedure above are identical to those for solving sparse normal equations. Second, even though R is the Cholesky factor of $X^T X$, we never form $X^T X$ explicitly; R is computed from X using rotations. This avoids cancellation and roundoff in computing $X^T X$ and the conditioning of R should be better than the Cholesky factor computed from $X^T X$ directly. Third, it should be clear that the storage requirement is the same as that used in solving the normal equations. The data structure is fixed after step 3 has been executed. Note that the data structure is row-oriented because of the way in which the elements of R are accessed. Fourth, secondary storage can be used easily and in a natural manner.

2.2. Effect of row ordering

Even though the choice of row permutation P_r has no effect on the sparsity of R, it does affect the cost of (or the amount of arithmetic in) computing R [2,6]. The problem of finding a good P_r to reduce the cost in general is not well-understood. George and Ng have shown that certain column permutations P_c automatically induce a good P_r [6, 10].

2.3. Effect of dense rows

Recall from Section 2.1 that the structure of R is determined from the structure of $X^T X$ (and the permutation P_c). Implicitly we have assumed that the matrix $X^T X$ is sparse if X is sparse. This is often true, but there are instances in which X is sparse but $X^T X$, and consequently R, is dense (at least structurally). This is usually caused by the existence of a *relatively small* number of *relatively dense* rows of X. One way to handle this problem is to *withhold* these dense rows in the orthogonal decomposition. We partition X into $\begin{pmatrix} A \\ B \end{pmatrix}$, where A and B contain respectively the sparse and dense rows of X. Let y be partitioned in a similar manner; that is, $y = \begin{pmatrix} e \\ f \end{pmatrix}$. Instead of solving directly the least squares problem

$$\min_\beta \left\| \begin{pmatrix} A \\ B \end{pmatrix} \beta - \begin{pmatrix} e \\ f \end{pmatrix} \right\|_2 , \qquad (2.3)$$

we first solve

$$\min_{\bar\beta} \| A\bar\beta - e \|_2 \qquad (2.4)$$

using an orthogonal decomposition of the sparse matrix A. (We assume that the upper triangular matrix obtained in the orthogonal decomposition of A is sparse.) Then the solution to (2.3) is obtained by modifying the solution to (2.4) using f and the withheld rows B. Only small dense systems have to be solved in the updating process. Algorithms for performing the update can be found in [2, 9]. This capability has been included in the package, and experience has shown that the savings in both space and time can be large.

The techniques used to handle dense rows can be modified to handle addition and deletion of rows in linear least squares problems. Our intention is to include these in the package in the future.

2.4. Rank-deficient problems

In some applications, the design matrix X may be rank-deficient in which case the least squares problem does not have a unique solution. Note that the upper triangular matrix R is no longer nonsingular; some of the rows will be null. Suppose X has rank r and assume $(p-r)$ is small. It is possible to rearrange the rows and columns of X so that

$$R = \begin{pmatrix} R_{11} & R_{12} \\ O & O \end{pmatrix} , \qquad (2.5)$$

where R_{11} is r by r upper triangular and R_{12} is r by $(p-r)$. In [9], Heath has shown that the minimal norm least squares solution to (1.1) is given by the minimum residual vector for the small dense least squares problem

$$\min_z \left\| \begin{pmatrix} R_{11}^{-1} R_{12} \\ -I \end{pmatrix} z - \begin{pmatrix} R_{11}^{-1} c \\ 0 \end{pmatrix} \right\|_2 , \qquad (2.6)$$

where c contains the first r components of Qy. We have included this algorithm in the package and we have also generalized it to handle the case when X contains some dense rows.

However, since finite precision arithmetic is used in computing R, it is unlikely to have the form given in (2.5). One is more likely to obtain

$$R = \begin{pmatrix} \bar R_{11} & \bar R_{12} \\ O & \bar R_{22} \end{pmatrix} , \qquad (2.7)$$

where $\bar R_{11}$, $\bar R_{12}$ and $\bar R_{22}$ are respectively r by r, r by $(p-r)$ and $(p-r)$ by $(p-r)$ matrices, and the elements of $\bar R_{22}$ are considered to be small enough to be neglected. Unfortunately, there is no reliable procedure available for determining how small the elements of $\bar R_{22}$ should be before they can safely be neglected. At the moment, the package uses a simple comparison of the elements against a user-supplied tolerance. Although our experience with some test problems has shown that the rank can be determined quite reliably, a more sophisticated mechanism seems desirable.

2.5. Constrained problems

The package also handles constrained least squares problems of the form

$$\min_\beta \| X\beta - y \|_2$$
$$\text{subject to } D\beta = f , \qquad (2.8)$$

where D is m by p and f is an m-vector. Heath proposed an algorithm to solve (2.8) using an orthogonal decomposition of X [9]. The basic idea is to solve first the unconstrained least squares problem. Then the effect of the constraints is included by updating the solution to the unconstrained problem using the constraints. The sparsity of the constraints is ignored and the constraints are assumed to be linearly independent. Furthermore, the design matrix X is assumed to have full column rank. It is hoped that these restrictions will be removed in the future.

2.6. Square systems of linear equations

Sparse square nonsingular systems can be solved in exactly the same way as least squares problems. Suppose X is p by p and has an orthogonal decomposition

$$X = QR , \qquad (2.9)$$

where Q is p by p orthogonal and R is p by p upper triangular. Then the solution to the system

$$X\beta = y \qquad (2.10)$$

is simply given by the solution to the triangular system

$$R\beta = Q^T y . \qquad (2.11)$$

Dense rows can be handled as in the case of least squares problems [9].

2.7. Underdetermined systems of linear equations

The technique used to solve sparse linear least squares problems can be used to solve sparse underdetermined systems of linear equations. Suppose X is n by p, with $n<p$. We can compute a decomposition of X^T using the method described in Section 2.1.

$$X^T = Q \begin{pmatrix} R \\ O \end{pmatrix} . \qquad (2.12)$$

Here Q is p by p orthogonal and R is n by n upper triangular. The underdetermined system

$$X\beta = y \qquad (2.13)$$

has an infinite number of solutions, and it can be shown that the solution having the minimal Euclidean norm is given by

$$\beta = X^T (R^T R)^{-1} y . \qquad (2.14)$$

Dense rows in X^T (i.e., dense columns in X) can be withheld from the orthogonal decomposition in order to reduce the storage requirement and the cost of computing the upper triangular matrix. George and Ng have proposed algorithms for handling these dense rows in the solution of underdetermined systems [5]. These algorithms have been incorporated in the package.

2.8. Truncated problems

Let $X^{(1)}$ denote some of the columns of X. In many applications, apart from solving the original problem (1.1), it is often necessary to solve a *truncated* linear least squares problem

$$\min_{\beta^{(1)}} \| X^{(1)} \beta^{(1)} - y \|_2 . \qquad (2.15)$$

One way to handle this situation is as follows. Partition X into $\{X^{(1)} \; X^{(2)}\}$, and assume $X^{(1)}$ and $X^{(2)}$ are n by k and n by $(p-k)$ respectively. Suppose X is decomposed into

$$X = Q \begin{pmatrix} R_{11} & R_{12} \\ O & R_{22} \\ O & O \end{pmatrix} , \qquad (2.16)$$

where R_{11} is k by k upper triangular, R_{12} is k by $(p-k)$, R_{22} is $(p-k)$ by $(p-k)$ upper triangular, and Q is n by n orthogonal. Partition Q in the form $\{Q^{(1)} \; Q^{(2)}\}$, where $Q^{(1)}$ and $Q^{(2)}$ are respectively n by k and n by $(n-k)$. Then $X^{(1)} = Q^{(1)} R_{11}$ and it can be shown that the solution to (2.15) is given by

$$R_{11} \beta^{(1)} = (Q^{(1)})^T y . \qquad (2.17)$$

The implementation for sparse problems is more complicated than that for dense problems, and efficient techniques are still under development. Thus, this capability has not yet been included in the package.

In some situations, it may be desirable to force certain variables (or columns of X) to appear first. The technique described above can also be used to handle these situations. We simply put those columns in $X^{(1)}$. The columns of $X^{(2)}$ are then reordered so that the upper triangular matrix R will be sparse.

3. Design considerations and implementation techniques

The software package SPARSPAK-B is a collection of FORTRAN subroutines for solving large sparse least squares problems. The design and implementation are similar to those of SPARSPAK (*sparse* matrix *pac*kage), which is a package for solving large sparse symmetric positive definite systems [3]. The subroutines of SPARSPAK-B can roughly be divided into two classes: the interface subroutines and the internal subroutines.

The internal subroutines form the core of the package. These subroutines are responsible for performing the various tasks, such as finding the column and row permutations, determining the structure of R and setting up an efficient storage scheme, and computing the QR-decomposition and the least squares solution. In order to exploit sparsity, sophisticated data structures and storage management are involved. Consequently the subroutines have long and complicated calling sequences.

The interface subroutines are designed to be easy to use and have very simple calling sequences. They form an interface between the user and the internal subroutines and will be described in detail in the next section. Each subroutine initiates a major task and it is responsible for invoking the various internal subroutines to carry out that task. Thus these subroutines effectively insulate the user from the complications of the internal subroutines.

A problem that arises immediately is the communication among the internal subroutines and the communication between the interface and internal subroutines. Our solution involves the use of a one-dimensional floating-point array which we denote by T, labelled common blocks, and secondary storage. The observation matrix X, together with the vector y (and other information that describes the problem, such as weights) are written row by row onto a sequential file. This file is accessible by any subroutine of the package. The logical unit number of the sequential file is supplied by the user through a call to an interface subroutine FILEB1. The data structure of R and the numerical values are stored in the floating-point array T. Other (integer and floating-pointing) arrays required by the internal subroutines are also allocated from T. Thus it is important that the user not modify the array T between calls to the interface subroutines. In order to keep track of where the individual arrays are, pointers to the beginning of the arrays are needed. These pointers are stored in labelled common blocks. They are accessible by all the interface subroutines and are used when the internal subroutines are invoked.

Certain control information is also stored in the labelled common blocks. This includes timing information, storage requirements, sequence control and some system parameters. Sequence control is necessary to make sure that the interface subroutines are invoked in the proper sequence. System parameters include logical unit numbers for output files and the ratio of the number of bits in a floating-point number to that in an integer number. The latter parameter is needed in the allocation of integer arrays from the floating-point array T.

It should be noted that the user has the option of using the internal subroutines directly. This allows the user to modify or replace any modules if the user develops new algorithms. Such flexibility also allows the user to tailor the package to her/his applications.

4. The Interface

In this section a brief description of the interface subroutines is provided. The solution process can conveniently be broken into several steps. In the following discussion, T denotes the floating-point array from which storage is allocated. The length of T is stored in a variable in one of the common blocks and it has to be initialized by the user at the beginning of the program.

Step 1. Initialization

Initialization is done by invoking two interface subroutines, SPKB and FILEB1. The FORTRAN statements are as follow.

    CALL SPKB
    CALL FILEB1 (*FILE*)

The first subroutine SPKB needs to be called only once in each program. It is responsible for initializing the timing routine and setting the logical unit numbers for output files. The second subroutine is called for each problem to be solved. The integer variable *FILE* contains the logical unit number for a sequential file which will be used by the package.

Step 2. Problem Input

After the initialization subroutines have been called, the user then invokes the interface subroutine INPXYW to input each observation in the least squares problem. The calling sequence has the form shown below.

    CALL INPXYW (*ROWNUM, TYPE, NSUBS,*
        SUBS, VALUES, RHS, WEIGHT, T)

The integer variable *ROWNUM* contains a positive integer that is less than $n+m+1$ (n is the number of observations and m is the number of constraints). It is the row index of the current observation. If the row belongs to the design matrix and if it is sparse, the integer variable *TYPE* has the value zero. If the row is dense, *TYPE* should be given the value two. If *TYPE* has the value one, the equation is a constraint in a constrained least squares problem. The integer variable *NSUBS* contains the number of nonzeros in the current observation (excluding the entry in y). The column indices and numerical values of the nonzeros are stored respectively in the integer array *SUBS* and floating-point array *VALUES*. The corresponding entry in y is stored in the floating-point variable *RHS*. If the problem is a weighted least squares problem, the floating-point variable *WEIGHT* contains the weight; otherwise, it should be set to 1.0.

Thus the problem can be provided to the package by calling INPXYW repeatedly. Note that the observations can be supplied to the package in any order.

Step 3. Column ordering

This step is initiated by invoking the subroutine ORDCOL. The FORTRAN statement to be used is shown below.

CALL ORDCOL (T)

This subroutine serves several purposes. First, it indicates that Step 2 has finished; that is, the user has supplied the entire problem to the package. Second, the package reorders the columns of X. The minimum degree algorithm which is an effective algorithm in reducing fill-in in $\bar{R}$ is used. See [4] for details. Third, the package determines the nonzero structure of $\bar{R}$ from that of $X^T X$ and the column permutation matrix P_c, and sets up an efficient storage scheme for $\bar{R}$.

Step 4. Row ordering

The subroutine to be invoked is ORDROW. The calling sequence is shown as follows.

CALL ORDROW (OPTION, T)

The integer variable OPTION indicates the order in which the rows should be processed. By setting OPTION=0, the user specifies that the rows should be processed in the order they were input. Other possibilities (OPTION≠0) include

(1) arranging the rows in the order of the parameter ROWNUM,
(2) arranging the rows in the order of increasing number of nonzeros,
(3) arranging the rows in the order of increasing weight (this is useful when the weights vary widely in value).
(4) reordering the rows in order to attempt to reduce the cost of computing R [6, 10].

Step 5. Solution

The numerical computation of the QR-decomposition and the final triangular solution are initiated by calling the subroutine LSQSLV. The FORTRAN statement is

CALL LSQSLV (TOL, TYPTOL, T, RESID)

Here TOL is a tolerance used to determine which diagonal elements of $\bar{R}$ should be regarded as numerically zero. This is useful in detecting rank-deficiency in X. The test can be done in absolute or relative terms. If TYPTOL=0, $\bar{R}_{ii}$ is regarded as numerically zero whenever $|\bar{R}_{ii}| < TOL$. If TYPTOL=1, $\bar{R}_{ii}$ is regarded as numerically zero only when $|\bar{R}_{ii}| < TOL * \max_{1 \leq j \leq p} |\bar{R}_{jj}|$.

After LSQSLV is called, the first p elements of T contain the least squares solution and the floating-point variable RESID contains the residual.

5. Additional interface subroutines

The interface contains a few other subroutines which we have not discussed in the previous section.

5.1. Save and restart

Sometimes it may be necessary to save the intermediate information provided by SPARSPAK-B so that the user can perform something else before returning to the solution of the least squares problem. For example, each major task in the solution process requires a different amount of space from the storage array T. There is a possibility that the computation cannot proceed at some stage because there is not enough space in T. In this case, it is desirable to be able to save what has been computed successfully so far before terminating the program. Then the user can change the size of the array T, recompile the program and resume the computation by restoring the information that has been saved. This avoids the need to execute those tasks that were successfully carried out before.

In order to make sure that everything is saved so that execution can be resumed at a later point, a subroutine SAVEB has been provided. The FORTRAN statement to be used is

CALL SAVEB (K, T)

where K is a logical unit number of a sequential file. The contents of the floating-point array T and the values of the variables in the labelled common blocks are written onto the sequential file specified.

When the user is ready to resume execution, the following statement should be used to restore everything saved by SAVEB.

CALL RSTRTB (K, T)

Note that after SAVEB is called, the user can use the floating-point array T in other computations.

5.2. Condition number estimator

Since finite precision arithmetic is used in the computation, the computed solution may not necessarily be close to the exact solution. One way of estimating the accuracy of the computed solution is to compute a quantity known as the condition number of X, defined by $cond(X) = \|X\|_2 \|X^+\|_2$, where X^+ is a pseudo-inverse of X. This quantity measures the sensitivity of the computed solution to perturbations in the original problem. If $cond(X)$ is large, it means that the computed solution may differ substantially from the exact solution even for a slight change in X and/or y. However, it may be expensive to compute $cond(X)$ since one has to compute $\|X^+\|_2$. An estimate of $cond(X)$ is often sufficient. We plan to include an algorithm for computing such an estimate in the future.

5.3. Estimate of the covariance matrix

The covariance matrix is a multiple of the matrix $(X^TX)^{-1}=(R^TR)^{-1}$. (We assume at the moment that X has full column rank and does not have dense rows.) For large sparse problems, it is not economical to compute the entire matrix $(X^TX)^{-1}$. It is often true that only some of the elements of $(X^TX)^{-1}$ are needed. Note that the (i,j)-element of $(X^TX)^{-1}$ can be obtained as

$$e_i^T(R^TR)^{-1}e_j \ , \tag{5.1}$$

where e_i and e_j are respectively the i-th and j-th columns of the identity matrix. We plan to include software to carry out this computation in the future.

6. Performance

In this section we use SPARSPAK-B to solve a few large sparse least squares problems and compare the performance with some commercial statistical packages. Some are real problems arising in survey applications, and some are artificial. Their characteristics are shown in Table 6.1.

Problem number	Number of rows	Number of columns	Number of nonzeros	Remarks
1	210	85	438	Holland survey data
2	331	104	662	Scotland survey data
3	608	188	1216	England survey data
4	313	176	1557	Sudan survey data
5	144	49	576	7 by 7 grid problem
6	196	64	784	8 by 8 grid problem

Table 6.1 Characteristics of test problems.

Table 6.2 contains the times required by the various packages for solving the problems described in Table 6.1. The programs were run on an IBM 4341. The results show that the execution time can be reduced substantially by exploiting the sparsity in the design matrices. The commercial statistical

problem	BMDP time	SAS time	SPSS time	SPARSPAK-B time
1	21.55	25.30	19.87	1.24
2	39.47	48.85	38.51	2.39
3	173.16	234.56	220.27	4.60
4	92.67	126.59	143.08	2.81
5	7.38	7.82	5.07	0.89
6	13.00	14.38	10.53	1.28

Table 6.2 Execution times (in seconds) for various packages.

packages do not take advantage of the sparsity structures and the corresponding execution times are very much larger than those for SPARSPAK-B.

Comparison of space requirements is a little complicated. The space requirements reported by some commercial packages include the space for the programs, while others do not. Since most of these packages are available only in compiled form, we could not modify them so as to get a uniform reporting format. However, results indicate that the space required by SPARSPAK-B to solve a sparse problem (excluding the space for the programs) is *much* less than that required by the commercial packages.

7. Concluding remarks

In this paper we have described a software package for solving large sparse least squares problems. This package, which is written using a portable subset of ANSI Standard FORTRAN, contains a set of interface subroutines that are easy to use and insulate the user from the complicated data structures and storage management involved in the actual solution subroutines.

There are several reasons for designing the package as a "library" of interface subroutines for different major tasks in the solution process, rather than as a "stand-alone" program. First, it allows the package to be embedded in a "super-package" where different tasks may be carried out at different instances. Second, it provides flexibility that is not available in stand-alone programs, particularly in terms of storage management. Note that each major task may require a different amount of work space from the storage array T. In a stand-alone program, if the computation cannot proceed at some stage because of the lack of space in T, the user can only terminate the program, change the size of the work space and run the program again. There is often no effective way to save the output from those tasks that were executed successfully. This is not desirable in the solution of large problems. In the package, the design of the interface allows the user to insert checkpoints in the main program. Output from tasks that are successfully executed can be saved before the program is terminated and the program can be restarted by restoring the information that is saved. This avoids the need to "re-do" those successful tasks. Finally, the design provides an efficient way to experiment with new ideas. For example, if we want to compare two different algorithms for handling dense rows, we do not have to carry out the column ordering (and row ordering) twice. All we need to do is to save the ordering information when the first method is tested. Then we simply restore the ordering information before we use the second method. Such flexibility is much less easy to provide if the package is designed as a stand-alone program.

Our numerical experiments show that it is important to exploit sparsity in large sparse linear least squares problems, and it appears to be worthwhile and important to develop software specifically for handling such problems.

8. Acknowledgements

This research was supported in part by the Canadian Natural Sciences and Engineering Research Council under Grant A8111 and also supported in part by the U.S. Dept. of Energy under contract W-7405-eng-26 with the Union Carbide Corporation. We are grateful to Dr. M.T. Heath for numerous discussions and advice in the design of the package.

9. References

[1] V. ASHKENAZI, "Geodetic normal equations", in *Large sparse sets of linear equations*, ed. J.K. Reid, Academic Press, London (1971), pp. 57-74.

[2] J.A. GEORGE AND M.T. HEATH, "Solution of sparse linear least squares problems using Givens rotations", *Linear Algebra and its Appl.*, **34** (1980), pp. 69-83.

[3] J.A. GEORGE AND J.W.H. LIU, "The design of a user interface for a sparse matrix package", *ACM Trans. on Math. Software*, **5** (1979), pp. 134-162.

[4] J.A. GEORGE AND J.W.H. LIU, *Computer solution of large sparse positive definite systems*, Prentice-Hall Inc., Englewood Cliffs, N.J. (1981).

[5] J.A. GEORGE AND E.G.Y. NG, "Solution of sparse underdetermined systems of linear equations", Research Report CS-82-39, Dept. of Computer Science, University of Waterloo (1982).

[6] J.A. GEORGE AND E.G.Y. NG, "On row and column orderings for sparse least squares problems", *SIAM J. Numer. Anal.*, **20** (1983), pp. 326-344.

[7] G.H. GOLUB, "Numerical methods for solving linear least squares problems", *Numer. Math.*, **7** (1965), pp. 206-216.

[8] G.H. GOLUB AND R.J. PLEMMONS, "Large-scale geodetic least squares adjustment by dissection and orthogonal decomposition", *Linear Algebra and its Appl.*, **34** (1980), pp. 3-27.

[9] M.T. HEATH, "Some extensions of an algorithm for sparse linear least squares problems", *SIAM J. Sci. Stat. Comput.*, **3** (1982), pp. 223-237.

[10] E.G.Y. NG, "Row elimination in sparse matrices using rotations", Research Report CS-83-01, Dept. of Computer Science, University of Waterloo (1983).

EXACT COMPUTATION WITH ORDER-N FAREY FRACTIONS

R. T. Gregory

The University of Tennessee, Knoxville

A bijective mapping is established between the set of order-N Farey fractions
$$F_N = \{a/b : \gcd(a,b) = 1, 0 \leq |a| \leq N, 0 < b \leq N\}$$
and a subset of the set of integers $I_p = \{0, 1, \ldots, p-1\}$, where p is a prime satisfying $p \geq 2N^2 + 1$. Operands from F_N are mapped into $\hat{I}_p$, the set of images of F_N in I_p; arithmetic operations are carried out in the finite field $(I_p, +, \cdot)$; and the results (which will be in $\hat{I}_p$ unless we have "overflow") are mapped back into F_N. Computation in the finite field is free of rounding errors because only integer arithmetic is involved.

1. INTRODUCTION

It is well known that the standard floating-point arithmetic used on essentially every electronic digital computer is subject to rounding errors, and this fact makes it extremely difficult to obtain accurate numerical approximations to the solutions of many ill-conditioned problems. For example, the solution to the set of linear algebraic equations

(1.1) $Ax = b$,

if the coefficient matrix A is a finite segment of the (infinite) Hilbert matrix, is essentially impossible when the number of equations is $n > 8$, and standard floating-point arithmetic is used.

This is due to the fact that the spectral condition number for this problem is approximately $e^{3.5n}$ and if n = 8,

(1.2) $e^{28} > 10^{12}$.

On the other hand, if exact arithmetic is used, these ill-conditioned problems can be solved. (See Smyre [1983], where (1.1) is solved exactly with a Hilbert matrix of order n = 20.) For this reason many researchers have begun to examine error-free computation as an approach to solving ill-conditioned problems.

For several years both single-modulus and multiple-modulus residue arithmetic, and quite recently, finite-segment p-adic arithmetic, have been investigated as methods for carrying out scientific computation free of rounding errors. See Gregory [1978], [1980], [1981], Howell and Gregory [1969], [1970], Krishnamurthy et al. [1975a], [1975b], Rao et al. [1976], Szabó and Tanaka [1967], and Young and Gregory [1973], for example. However, since finite-segment p-adic arithmetic is mathematically equivalent to single-modulus residue arithmetic, where the modulus has the form $m = p^r$, with p a prime and r a positive integer, we shall restrict this discussion to residue arithmetic.

We present a generalization of residue arithmetic which does not require that the operands be integers only, but allows us to use a restricted set of rational numbers as well. The main thrust of this discussion deals with single-modulus residue arithmetic (for rational operands) but we also include an introduction to the use of multiple-modulus residue arithmetic.

2. SINGLE-MODULUS RESIDUE ARITHMETIC

This is the arithmetic of residue classes. Let I

represent the set of integers and let p be an odd prime. We denote the set of least non-negative residues modulo p by

(2.1) $I_p = \{0, 1, 2, \ldots, p-1\}$

and the set of symmetric residues modulo p by

(2.2) $S_p = \{-(p-1)/2, \ldots, -2, -1, 0, 1, 2, \ldots, (p-1)/2\}$.

2.3 <u>Definition</u> We define the mapping $|\cdot|_p : I \to I_p$ by writing $|a|_p = r$ if and only if $r \in I_p$ and $a \equiv r \pmod{p}$.

2.4 <u>Definition</u> We define the mapping $/\cdot/_p : I \to S_p$ by writing $/a/_p = s$ if and only if $s \in S_p$ and $a \equiv s \pmod{p}$.

For example, $|28|_5 = 3$ and $/28/_5 = -2$.

We define the one-to-one mappings between I_p and S_p as follows:

(2.5) $|a|_p = \begin{cases} /a/_p & \text{if } 0 \le /a/_p \le p/2 \\ /a/_p + p, & \text{otherwise,} \end{cases}$

and

(2.6) $/a/_p = \begin{cases} |a|_p & \text{if } 0 \le |a|_p \le p/2 \\ |a|_p - p, & \text{otherwise,} \end{cases}$

The finite fields $(I_p, +, \cdot)$ and $(S_p, +, \cdot)$, where $+$ and $\cdot$ represent addition and multiplication modulo p, respectively, are isomorphic and they are both isomorphic to the Galois field $GF(p)$. Arithmetic in these two finite fields is integer arithmetic and this can be performed free of rounding errors. In order to avoid the monitoring of algebraic signs, it is common to

(i) map the positive and/or negative integers (which describe the problem) from S_p into I_p using (2.5),
(ii) carry out the computation in the finite field $(I_p, +, \cdot)$, and
(iii) map the results back into S_p using (2.6).

<u>Rational numbers as operands</u>

Up to this point we have assumed that the data which describe a problem are positive and/or negative integers, and in many applications this is the case. If the data are rational numbers, it is often possible to use scaling in order to produce data which are integers. For example, see Howell and Gregory [1969], [1970] and Young and Gregory [1973], chapter 13, where residue arithmetic is used to solve systems of linear algebraic equations whose coefficients are integers.

On the other hand, it is possible to solve problems in which the data consist of positive and/or negative rational numbers without bothering to scale, if we restrict the operands to the set of order-N Farey fractions

(2.7) $F_N = \{a/b : \gcd(a,b)=1, 0 \le |a| \le N, \text{ and } 0 < |b| \le N\}$,

where N is the largest integer satisfying

(2.8) $N \le \{(p-1)/2\}^{\frac{1}{2}}$.

What we do here is establish a bijective mapping between the set F_N and a subset of the elements in I_p. To do this we define $|a/b|_p$ to be the integer

(2.9) $|a/b|_p = |a \cdot b^{-1}|_p$,

where $b^{-1} \equiv b^{-1}(p)$ denotes the multiplicative inverse of b modulo p. This gives us the bijection

(2.10) $|\cdot|_p : F_N \to \hat{I}_p$,

where $\hat{I}_p$ is the set of images of the order-N Farey fractions, and $\hat{I}_p \subset I_p$.

2.11 <u>Example</u> Let $p = 19$ which implies $N = 3$. Then the bijective mapping between F_3 and $\hat{I}_{19}$ is exhibited in the following table:

0	0		
1	1	-1	18
2	2	-2	17
3	3	-3	16
-1/3	6	1/3	13
-2/3	7	-2/3	12
-3/2	8	3/2	11
-1/2	9	1/2	10

Notice that $\hat{I}_{19} \subset I_{19}$ because 4, 5, 14, and 15 are not images of any elements in F_3 under the mapping (2.10).

If we use a very large prime, so that N is large enough to insure that both the data which describe the problem and the solution to the problem lie in F_N, then we can
  (i) map the data from F_N into I_p using (2.10),
  (ii) carry out the computations (free of rounding errors) in the finite field $(I_p, +, \cdot)$, and
  (iii) map the solution back into F_N.

2.12 <u>Example</u> Consider the computation
$$x = 2/3 - 1$$
$$= 2/3 + (-1)$$
using single-modulus residue arithmetic in $(I_{19}, +, \cdot)$. Since $p = 19$ and $N = 3$, we can use the mapping exhibited in the table in Example 2.11. Hence,
$$|x|_{19} = |2/3 + (-1)|_{19}$$
$$= |7 + 18|_{19}$$
$$= 6.$$
When we use the inverse mapping in the same table, we obtain
$$x = -1/3,$$
and this is the correct answer.

3. THE BIJECTION BETWEEN F_N AND $\hat{I}_p$.

Obviously, if N and p are large, it is not computationally feasible to construct a table, such as the table exhibited in Example 2.11, in order to perform the mappings between F_N and $\hat{I}_p$. In practice, we use the algorithms described below.

The forward mapping

It is possible to describe the mapping $F_N \to \hat{I}_p$ in terms of an extension of the Euclidean algorithm. To see that this is true, suppose we generate the sequence of integer pairs $\{(a_i, b_i)\}$ by using the recursion

(3.1) $\begin{cases} a_i = a_{i-2} - q_i a_{i-1} \\ b_i = b_{i-2} - q_i b_{i-1} \end{cases}$ $i = 1, 2, \ldots, n+1$,

with the seed matrix (or initial matrix)

(3.2) $\begin{bmatrix} a_{-1} & b_{-1} \\ a_0 & b_0 \end{bmatrix} = \begin{bmatrix} p & 0 \\ d & c \end{bmatrix}$,

and q_i, for $i = 1, \ldots, n+1$ defined by

(3.3) $q_i = \left\lfloor \dfrac{a_{i-2}}{a_{i-1}} \right\rfloor$.

The computation is exhibited in the following table. (For example, see Kornerup and Gregory [1983].)

3.4 <u>Table</u>

	p	0
	d	c
q_1	a_1	b_1
q_2	a_2	b_2
.	.	.
.	.	.
q_n	a_n	b_n
q_{n+1}	0	b_{n+1}

The elements $a_1, a_2, \ldots, a_n$ are merely the nonzero remainders in the Euclidean algorithm for finding gcd(p,d) and, since a_n is the final nonzero remainder,

(3.5) $a_n = \gcd(p, d)$.

In addition,

(3.6) $|b_n|_p = |c/d|_p = |c \cdot d^{-1}|_p$.

In other words, if c/d is an order-N Farey fraction, then $|b_n|_p$ is its image in $\hat{I}_p$.

3.7 <u>Remark</u> As a special case, if c = 1 in Table 3.4,
$$|b_n|_p = d^{-1},$$
the multiplicative inverse of d modulo p.

3.8 <u>Example</u> Let $p = 19$ which implies $N = 3$. The order-3 Farey fraction $-2/3$ can be mapped into $\hat{I}_{19}$ as follows:

	19	0
	3	-2
6	1	12
3	0	-38

Since $a_1 = 1$ is the last nonzero remainder in the column headed by 19,
$$\gcd(19, 3) = 1.$$

Finally, since $b_1 = 12$,
$$|-2/3|_{19} = |-2 \cdot 3^{-1}|_{19}$$
$$= |12|_{19}$$
$$= 12 ,$$
and this result agrees with the result in the table in Example 2.11.

3.9 Remark There is only one order-3 Farey fraction mapped onto the element $12 \in \hat{I}_{19}$, but there are infinitely many rational numbers not in F_3 which are mapped onto 12 by this forward mapping. For example, $-1/11$, 31, 50, -7, $3/5$, and $5/2$ are not order-3 Farey fractions and yet each is mapped onto 12. We call this infinite set the generalized residue class Q_{12} since it contains the ordinary residue class of integers (which are congruent to 12 modulo 19) as a proper subset. It can be shown (see Gregory and Krishnamurthy [1983], for example) that
$$|a/b|_p = |c/d|_p$$
if and only if $ad \equiv bc \pmod{p}$, and it is this condition which characterizes a generalized residue class modulo p.

The inverse mapping

From Remark 3.9 it is clear that for every integer t in I_p we have an infinite set of rational numbers Q_t (a generalized residue class), each element of which is mapped onto t. It is shown in Gregory and Krishnamurthy [1983] that Q_t contains at most one order-N Farey fraction, and that the number of elements in F_N is less than p. This explains why the set of images of the elements in F_N is $\hat{I}_p$ and $\hat{I}_p \subset I_p$. Thus, the mapping between F_N and $\hat{I}_p$ is a bijection.

To implement the inverse mapping $\hat{I}_p \to F_N$, we also use an algorithm based on the Euclidean algorithm. This algorithm appears in Kornerup and Gregory [1983], and is based on a theorem which states that the elements of the sequence of rational numbers $\{a_i/b_i\}$ (which are generated by the sequence of integer pairs $\{(a_i,b_i)\}$ in Table 3.4) all belong to the same generalized residue class, the generalized residue class containing d/c. As a consequence, if we choose (d,c) to be (t,1), where $t \in \hat{I}_p$, then the integer pairs (a_i,b_i) in Table 3.4 represent elements in the generalized residue class Q_t. From a theorem of Kornerup, the unique order-N Farey fraction corresponding to t always appears in Table 3.4, if the seed matrix is
$$\begin{bmatrix} p & 0 \\ t & 1 \end{bmatrix} .$$

3.10 Example In Example 3.8 we showed that
$$|-2/3|_{19} = 12 .$$
We now demonstrate that the integer $12 \in \hat{I}_{19}$ is mapped back onto $-2/3 \in F_3$. The computation is exhibited in the following table:

	19	0
	12	1
1	7	-1
1	5	2
1	2	-3
2	1	8
2	0	-19

Observe that 12, -7, $5/2$, $-2/3$, and $1/8$ all belong to the generalized residue class Q_{12}. However, $-2/3$ is the only element of this generalized residue class which lies in F_3 and it is easy to select this element from the finite set generated by the table.

3.11 Summary We have discussed both the forward mapping $F_N \to \hat{I}_p$ and the inverse mapping $\hat{I}_p \to F_N$ and presented simple algorithms for each. Hence, if p and N are large enough so that both the data describing a problem, and the solution to the problem, are order-N Farey fractions, then the solution can be obtained using (error-free) integer arithmetic in the finite field $(I_p, +, \cdot)$.

4. MULTIPLE-MODULUS RESIDUE ARITHMETIC

It is well known that multiple-modulus residue

arithmetic using several moduli is equivalent to single-modulus residue arithmetic using the least common multiple of the moduli as the single modulus. See Knuth [1969], for example.

In order to discuss this, consider the ordered n-tuple
$$(4.1) \quad \beta = [p_1, p_2, \ldots, p_n]$$
whose components are the (distinct) primes $p_1, p_2, \ldots, p_n$. In this case we call β the <u>base vector</u> for the multiple-modulus residue number system using moduli $p_1, p_2, \ldots, p_n$. Since the n moduli are primes, their least common multiple is simply their product
$$(4.2) \quad P = p_1 p_2 \ldots p_n \, .$$
We can repeat the entire discussion of Sections 2 and 3 using the single modulus P, as long as we take into consider the fact that P is not a prime (by definition) and $(I_p, +, \cdot)$ is, therefore, not a finite field, but merely a finite commutative ring with identity. This means that the multiplicative inverse $a^{-1}(P)$ exists if and only if
$$(4.3) \quad \gcd(a, p_i) = 1 \, ,$$
for $i = 1, 2, \ldots, n$.

It turns out (see Gregory and Krishnamurthy [1983], for example) that if we let N be the largest integer satisfying
$$(4.4) \quad N \leq \{(P-1)/2\}^{\frac{1}{2}} \, ,$$
then it is always possible to choose two primes with $p_1 > N$ and $p_2 > N$ such that (4.3) is satisfied for any a satisfying $0 < |a| \leq N$. This means that in a two-modulus system with $P = p_1 p_2$, a bijective mapping between F_N and $\hat{I}_p$ exists (similar to the one we described for the single-modulus system in Sections 2 and 3).

Unfortunately, it can be shown that if more than two moduli, $p_1 < p_2 < \ldots < p_n$, are used, then we always have
$$(4.5) \quad p_1 \leq N$$
for N defined in (4.4). This means that we cannot guarantee that (4.3) is satisfied, which implies some uncertainty about the existence of multiplicative inverses modulo P, in which case there is uncertainty about the mapping $F_N \to \hat{I}_p$.

Obviously, if we use more than two prime moduli, we must make certain that we can always satisfy (4.3) for the problem we are attempting to solve. This is one of the drawbacks in using multiple-modulus residue arithmetic with more than two moduli.

4.5 <u>Example</u> Suppose we take another look at the computation in Example 2.12, and this time let us use the two moduli $p_1 = 5$ and $p_2 = 7$. (Observe that $P = 35$, and this value is greater than the value of the single modulus $p = 19$ used in Example 2.12.) For $P = 35$, $N = 4$, and both p_1 and p_2 are greater than N.

If we write
$$x = 2/3 + (-1) \, ,$$
then
$$|x|_5 = \left| |2/3|_5 + |-1|_5 \right|_5$$
$$= |4 + 4|_5$$
$$= 3 \, .$$
Likewise
$$|x|_7 = \left| |2/3|_7 + |-1|_7 \right|_7$$
$$= |3 + 6|_7$$
$$= 2 \, .$$
Thus, since $\beta = [5, 7]$, the residue representation of x in the two-modulus residue system is
$$|x|_\beta = [|x|_5, |x|_7]$$
$$= [3, 2] \, ,$$
and, if we use the Chinese Remainder Theorem, we obtain
$$|x|_{35} = 23 \, .$$
The integer $23 \in \hat{I}_{35}$ is mapped into F_4 as follows:

	35	0
	23	1
1	12	-1
1	11	2
1	1	-3
11	0	35

Thus,
$$x = -1/3,$$
and this is the correct solution.

4.6 <u>Remark</u> The motivation for using multiple-modulus residue arithmetic is that, when $P = p_1 p_2 \ldots p_n$, the computation can be carried out (in parallel) using the smaller moduli $p_1, p_2, \ldots, p_n$. Only when the final results are to be determined do we use the Chinese Remainder Theorem to map the residue representations into $\hat{I}_p$ and then into F_N.

REFERENCES:

Gregory, R. T. [1978], The use of finite-segment p-adic arithmetic for exact computation, BIT 18, 282-300.

Gregory, R. T. [1980], Error-free Computation, (Robert E. Krieger Pub. Co. Melbourne, Florida).

Gregory, R. T. [1981], Error-free computation with rational numbers, BIT 21, 194-202.

Gregory, R. T. and Krishnamurthy, E. V. [1983], Methods and Applications of Error-Free Computation, (Springer-Verlag, New York, to appear).

Howell, J. A. and Gregory, R. T. [1969], An algorithm for solving linear algebraic equations using residue arithmetic, Parts I and II, BIT 9, 200-224, and 324-337.

Howell, J. A. and Gregory, R. T. [1970], Solving linear equations using residue arithmetic-Algorithm II, BIT 10, 23-37.

Knuth, D. E. [1969], The Art of Computer Programming: vol II Semi-numerical Algorithms (Addison-Wesley, Reading, MA).

Kornerup, P. and Gregory, R. T. [1983], Mapping integers and Hensel codes onto Farey fractions, BIT 23, (to appear).

Krishnamurthy, E. V., Rao, T. M., and Subramanian, K. [1975a], Finite-segment p-adic number systems with applications to exact computation, Proc. Indian Acad. Sci. 81A, 58-79.

Krishnamurthy, E. V., Rao, T. M., and Subramanian, K. [1975b], P-adic arithmetic procedures for exact matrix computations, Proc. Indian Acad. Sci. 82A, 165-175.

Rao, T. M., Subramanian, K. and Krishnamurthy, E. V. [1976], Residue arithmetic algorithms for exact computation of g-inverses of matrices, SIAM J. Numer. Anal. 13, 155-171.

Smyre, J. L. [1983], Exact computation using extended-precision single-modulus residue arithmetic, M.S. thesis, Department of Computer Science, The University of Tennessee, Knoxville.

Szabó, N. S. and Tanaka, R. I. [1967], Residue Arithmetic and its Applications to Computer Technology (McGraw-Hill, New York).

Young, D. M. and Gregory, R. T. [1973], A Survey of Numerical Mathematics: vol II (Addison-Wesley, Reading, MA).

RECENT ADVANCES IN GENERATING OBSERVATIONS FROM DISCRETE RANDOM VARIABLES[1]

Bruce Schmeiser

School of Industrial Engineering
Purdue University
West Lafayette, Indiana 47907

The state of the art of random variate generation has changed dramatically in the last ten years. We discuss two new approaches: Acceptance/complement and (2) aliasing. We compare algorithms based on these ideas, both of which are based on the fundamental idea of probability mixing, with algorithms based on acceptance/rejection. The Poisson and binomial distributions are considered. Issues of exactness, speed, memory and ease of use are emphasized.

1. INTRODUCTION

In considering the problem of generating observations from discrete distributions (random variate generation), we assume here the availability of a source of random numbers independently and uniformly distributed over (0,1), denoted by U(0,1). This assumption is an idealization, in that typically only deterministic sequences of pseudo-random numbers that are approximately independent and uniform are available, as discussed, e.g., in Fishman (9) or Kennedy and Gentle (20).

We further assume that the probability mass function $p(x)$, satisfying $p(x) \geq 0$ and $\sum_x p(x) = 1$, for the distribution of interest is known. This assumption is not general, but suffices for discussion of most of the literature. Under this assumption, the problem is to generate independent realizations x of the random variable X. In general, X may be a multivariate time series

$$x = \begin{bmatrix} x_{11} & x_{12} & x_{13} & \cdots & x_{1i} & \cdots \\ x_{21} & x_{22} & x_{23} & \cdots & x_{2i} & \cdots \\ \vdots & \vdots & \vdots & & \vdots & \\ x_{k1} & x_{k2} & x_{k3} & \cdots & x_{ki} & \cdots \end{bmatrix}$$

Most of the literature, and most of this survey, considers the generation of independent scalars $x = x_{11}$. Exceptions include Kemp and Loukas (18,19), who consider bivariate vectors; Jacobs and Lewis (12,13,14,15), who consider discrete time series; Hoffman (11), who considers binary time series; and Boyett (5), who considers random RxC tables.

[1] This research was supported by the Office of Naval Research under Contract N00014-79-C-0832.

There are several good references that are more general than this paper. Kennedy and Gentle (20, chapter 6) provide a detailed discussion and extensive references for all of random variate generation. Fishman (9) and Law and Kelton (23) discuss simulation in general. Johnson (16) considers discrete random variate generation in detail. Topics in this paper without references can be pursued by checking Schmeiser (26), which has more than 350 references and considers all of random variate generation.

2. CRITERIA

In comparing random variate generation algorithms, many criteria are relevant. In a subjective ranking of descending importance, the following list is suggested for comparing algorithms. Of course the importance of any particular criterion is dependent upon the particular application and context.

a. Mathematical Validity -- Obtained if the algorithm would be exact if the U(0,1) random numbers were ideal and the computer arithmetic were perfect.

b. Numerical Stability -- Obtained if the algorithm performs well in the actual computing environment.

c. Execution time -- There are two ways in which execution time can be considered:
   - initialization: The time to calculate constants as functions of the given parameters.
   - marginal: The time to generate one additional observation.

d. Memory -- Total memory required for both logic and tabled constants.

e. Ease of Implementation -- The effort of implementing an otherwise preferred algorithm may not be worthwhile.

f. Portability -- The ability of the algorithm to perform well on a wide variety of computers. (Algorithms working at the bit level, e.g., tend to be faster but less portable than algorithms working at the word level.) Sometimes portability is taken to mean the ability to generate identical, to within the ability of the computer arithmetic, results on different computers. This interpretation usually precludes the use of conditional branching based on floating point variables.

g. Relationship to Variance Reduction -- Some variance reduction techniques require either negative or positive correlation to be induced between generated observations. Such correlation induction is very difficult unless the inverse transformation is used, as discussed below.

h. Generality -- The range of distributions within a family of distributions for which the algorithm is valid.

3. FUNDAMENTAL CONCEPTS

There are four fundamental concepts of random variate generation: (1) inverse transformation, (2) deterministic transformation of intermediate observations, (3) composition, and (4) acceptance/rejection. We discuss each in turn, including common variations and applications.

3.1 Inverse Transformation

For a scalar discrete random variable X, let $F(x_i) = P(X \leq x_i) = \Sigma_{x < x_i} p(x_i)$. The inverse transformation is $x = x_i$, where $F(x_{i-1}) < u \leq F(x_i)$, $F(x_0) = 0$, u is a $U(0,1)$ observation, and the number of distinct values x_i may be infinite.

For many common distributions, such as the Poisson, binomial and hypergeometric, simple recursive formulas can be used to make $F(x_i)$ implicit. Iteratively comparing u to $F(x_i)$ typically works well when the expected number of comparisons is small. When the expected number is large, numerical instability and inefficiency tend to arise.

Several suggestions have been made to improve the inverse transformation when the expected number of comparisons is large. Generally, these are (1) where to begin the search for x, (2) how to search, and (3) closed-form solutions for x. The two most common examples of closed-form solutions are $x = \lceil (\ln(1-u)) / (\ln(1-p)) \rceil$ for the geometric distribution (the number of trials until the first successful Bernoulli trial occurs, when the probability of success is p) and $x = \lceil un \rceil$ for the discrete uniform distribution over $\{1, 2, ..., n\}$.

When closed-form solutions are not available, efficient search methods are helpful. Index tables, as suggested by Chen and Asau (6) and studied by Fishman and Moore (10), aid in beginning the search close to x. The method chooses a discrete uniform variate from $\{1, 2, ..., k\}$ using $y = \lceil uk \rceil$ and begins the search at $T(y)$, where $T(y)$ is the smallest value of x_i such that $F(x_i) \geq (y-1)/k$. The method of search, within the context of index search, is not particularly important, if k is approximately as large as the effective range of X. The author has found index search with large k to nearly dominate other methods applicable to general discrete distributions. (The alias method, discussed below, requires a bit less marginal execution time.) When index search is not used, the method of search becomes important. Binary search is the most often suggested alternative.

However implemented, the inverse transformation has advantages in both variance reduction and in generating order statistics. To induce the maximum correlation between two random variables X and Y, the same random numbers $\{u_i\}$ are input to the inverse transforms of X and Y to obtain $\{x_i\}$ and $\{y_i\}$. Such positive correlation is called common random numbers, and is useful when comparing two alternatives based on X and Y. Antithetic variates is another variance reduction technique that induces correlation. In the simplist form, $\{x_i\}$ is generated using $\{u_i\}$ and $\{y_i\}$ is generated using $\{1-u_i\}$. Averaging $\{x_i\}$ and $\{y_i\}$ when X and Y are identically distributed yields an estimate of $E(X) = E(Y)$ with less variance than independent sampling. Correlation induction outside the context of the inverse transformation is difficult.

The use of the inverse transformation is also useful when order statistics $x_{(1)} \leq x_{(2)} \leq \cdots \leq x_{(n)}$ are required. Since the inverse transformation is monotonic, using $U(0,1)$ order statistics $u_{(1)} \leq u_{(2)} \leq \cdots \leq u_{(n)}$ in the inverse transformation yields the desired order statistics directly. $U(0,1)$ order statistics may be obtained directly using $u_{(n)} = v_1^{1/n}$, $u_{(n-1)} = u_{(n)} v_2^{1/(n-1)}$, $u_{(n-2)} = u_{(n-1)} v_3^{1/(n-2)} \cdots$, where $v_1, v_2, ..., v_n$ are independent $U(0,1)$.

3.2 Deterministic Transformation of Intermediate Observations

A common method is to first generate some intermediate observations $y_1, y_2, ..., y_m$, where m may be random. These values are then deterministically transformed using $x = g(y_1, y_2, ..., y_m)$. Methods of this type are typically a simulation of the process giving rise to the distribution and are usually dominated by other methods. Examples are

a. Poisson with fractional mean $\gamma + f$

$y_1 \sim$ Poisson (γ) where $\gamma \in \{0, 1, 2, ...\}$

$y_2 \sim$ Poisson (f)

$x = y_1 + y_2$

b. Poisson with mean λ

$y_i \sim$ exponential with mean 1 and Y_i independent of Y_j if $i \neq j$

$x = \min_m \{m \mid \sum_{i=1}^{m} y_i > \lambda\}$

c. Binomial with m trials and probability of success p

$y_i \sim$ Bernoulli trial with probability of success p for $i = 1, 2, \ldots, m$.

$x = \sum_{i=1}^{m} y_i$

d. Negative binomial with k successes and probability of success p

$y_i \sim$ Bernoulli trial with probability of success p for $i = 1, 2, \ldots$,

$x = \min_m \{m \mid \sum_{i=1}^{m} y_i = k\}$

Shanthikumar (28) presents an interesting idea for a discrete random variable as the first passage time of a discrete stochastic process associated with a point process generated by independent identically distributed geometric random variables.

3.3 Composition

Composition is the idea of probabilistically mixing multiple (possible infinite) discrete distributions to obtained the desired distribution. There are both discrete and continuous composition.

Discrete: $P(X=x) = \sum_{i=1}^{m} p_i P(X=x|i)$ for all x

Continuous: $P(X=x) = \int_{-\infty}^{\infty} f_Y(y) P(X=x|y)$ for all x

In the discrete case the method is to sample i from the distribution $p_1, p_2, \ldots, p_m$ and then to generate x from the distribution $P(X=x|i)$. Similarly in the continuous case, y is generated from the density function $f_Y(y)$ and x is generated from $P(X=x|y)$. Examples are

a. Poisson with mean μ $(\mu \leq \lambda)$

$i \sim$ Poisson with mean λ

$x \sim$ binomial with i trials and probability of success μ/λ

(This method is very efficient for changing values of μ and a small value of λ. Typically $\lambda = 1$.)

b. negative binomial with k successes and probability of success p

$y \sim$ gamma with shape parameter k and mean $k(1-p)/p$

$x \sim$ Poisson with mean y

In addition to these examples, which make use of special relationships between some common distributions, general methods applicable to any discrete distribution are available. These general methods are (1) crude tables, (2) Marsaglia tables, (3) aliasing, and (4) acceptance/complement type methods. An interesting recent discussion of general mixture methods is Peterson and Kronmal (25).

3.3.1 Crude Tables

The fastest method for generating any discrete distribution, with the exception of the discrete uniform distribution, is to use crude tables. Let $k = \text{l.c.d}\{p(x_i); i = 1, 2, \ldots\}$. The initialization step fills a table, or vector, of length k with $kp(x_i)$ entries of x_i, for $i = 1, 2, \ldots$. For example, consider

$p(0) = 1/9$
$p(1) = 3/9$
$p(2) = 2/9$
$p(3) = 2/9$
$p(5) = 1/9$

The crude table has k=9 entries as follows:

x_i	0	1	1	1	2	2	3	3	5
i	1	2	3	4	5	6	7	8	9

The generation algorithm is simply $i = \lceil uk \rceil$ and $x = x_i$.

Of course, the problem is that even a slight change in the distribution can cause the value of k to become unacceptably large.

Note that crude tables are a form of discrete composition in that we are mixing k degenerate equally likely distributions. The use of equally likely distributions as part of a larger algorithm will reappear in discussions below.

3.3.2 Marsaglia's Tables

Marsaglia (24) suggested an ingenious method to reduce the table size of the crude tables. We present the method in a simplified form and assuming a binary representation of the distribution $p(x_i)$. The basic idea is to mix distributions, one for each power of two in the binary representation of $p(x_i)$. The distributions are each composed of equally likely outcomes.

Consider the example

p(X=0) = 7/32 =	.00111 =		1/8	+ 1/16	+ 1/32
p(X=1) =10/32 =	.01010 =	1/4		+ 1/16	
p(X=2) =10/32 =	.01010 =	1/4		+ 1/16	
p(X=3) = 3/32 =	.00011 =			1/16	+ 1/32
p(X=5) = 2/32 =	.00010 =			1/16	
32/32	1.00000	2/4	1/8	5/16	2/32

The method is to use four distributions, each with probability 2/4, 1/8, 5/16, and 2/32, respectively. The first distribution has two equally likely values, 1 and 2. The second distribution is degenerate at 0. The third distribution has five equally likely values, 0, 1, 2, 3, and 5. The four distribution has two equally likely values, 0 and 3.

3.3.3 The Alias Method

The alias method was proposed by Walker (29, 30) and further developed by Kronmal and Peterson (21). The underlying result is

Result: Any discrete distribution with n (finite) outcomes can be expresses as a mixture of n equiprobable two-point distributions.

Applying the result, once the initialization procedure has calculated the representation indicated in the result, requires only to select one of n equally likely distributions and then to select one of two possible outcomes associated with that distribution. Figure 1 may help to understand the method. Beginning at the root of the tree, select one of the n equally likely branches. Having chosen branch x, return the alias leaf value a_x with probability f_x; otherwise, return the other leaf value x. More precisely, the algorithm is

1. u = un   where u ∿ U(0,1)
2. x = ⌈u⌉ (x = 1, 2, ..., or n)
3. u = x−u now again, u ∿ U(0,1)
4. if u < f_x, then x=a_x
5. return x

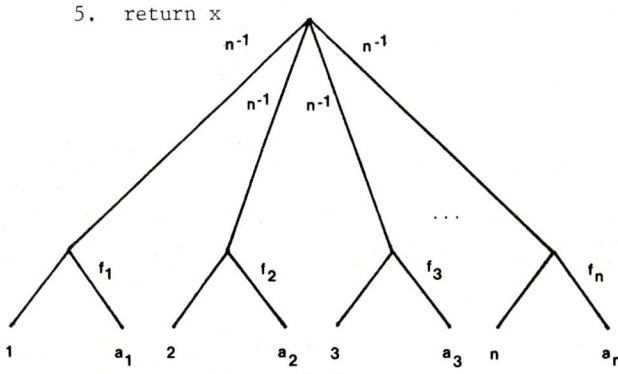

Figure 1: The alias method

Ignoring initialization, the alias method is the fastest general method for generating discrete variates, with the exception of crude tables. The method has the advantage of requiring exactly one U(0,1) random number for each observation of X generated. Initialization time is proportional to n and memory requirements are proportional to 3n. The result can be made a bit tighter, in that actually only n-1 equally likely distributions are necessary. Extensions of the result to equally likely distributions each having k outcomes have been made, but would seem to be dominated by the original result.

3.3.4 Acceptance/Complement

The acceptance/complement method, or very similar ideas, have been proposed independently by three groups of researchers: Ahrens and Dieter (2), Deák (7), and Kronmal and Peterson (22). The method's strength is that the probability of choosing mixture distribution i is not evaluated explicitly in the initialization step. Instead, the mixing probabilities are noted to be equal to probabilities of events arising naturally in the acceptance/rejection method. The resulting algorithms tend to be most competitive in terms of execution time when the distribution parameters change with each value generated. Details may be found in the referenced papers.

3.4 Acceptance/Rejection

The acceptance/rejection method, which is both widely known and widely applied, is based on accepting or rejecting candidate values of x generated from a tractable substitute distribution. Let r(x) be any mass function and c a constant satisfying $c\,r(x) \geq p(x)$ for all x. The algorithm is to (1) generate x ∿ r(x) and y ∿ U(0,1), (2) if y ≥ p(x) /(c r(x)), then reject x by going to step 1; otherwise accept x as the generated observation. Here X may be scalar or multivariate. The acceptance/rejection method is the structure of several state-of-the-art algorithms, as discussed in the next section.

4. Common Discrete Distributions

Until recently, no mathematically valid algorithm existed for Poisson, binomial or hypergeometric random variables that had the property of being "uniformly fast." This property is the ability to generate one variate for any parameter value in finite time. Beginning in 1979, however, a spate of papers appeared, all describing uniformly fast Poisson algorithms. These were Ahrens and Deiter (1,2), Atkinson (3,4), Devroye (8), and Schmeiser and Kachitvichyanukul (27). All but the second Ahrens and Dieter paper are acceptance/rejection; it is acceptance/complement based on the normal distribution. In terms of execution time, the Schmeiser and Kachitvichyanukul paper's algorithm PTPE dominates the others, except when very fast normal variates are available, in which case the Ahrens and Dieter acceptance/complement algorithm is as fast, or a bit faster.

Because of PTPE's speed and our own obvious bias, we briefly describe this algorithm. PTPE; which is an acronym for Poisson, Triangle, Parallelogram, Exponential; is illustrated in Figure 2. The density function from which we generate candidate values is composed of a large triangle, two parallelograms, and two exponential tails. PTPE is fundamentally an acceptance/rejection algorithm. Composition is used in step 1 to generate the candidate values. Roughly 80% of the time the candidate value is simply the sum of two $U(0,1)$ random numbers rescaled. More importantly, since the triangle is entirely under $p(x)$, the candidate value is always returned, with no comparison required. For the parallelogram and exponential regions, some time is saved by using simple one-sided approximations to $p(x)$.

A similar algorithm, BTPE, for binomial random variables, is described in Kachitvichyanukul (17). BTPE (Binomial, Triangle, Parallelogram, Exponential) is the fastest method for generating binomial random variates, expect when the mean is small, in which case methods based on the inverse transformation work well.

Kachitvichyanukul (17) also describes the only uniformly fast algorithm for the hypergeometric distribution, H2PE (Hypergeometric, 2 Point, Exponential). H2PE fits a rectangular region around the body of the distribution and again uses exponential tails in an acceptance/rejection framework.

5. CONCLUSIONS

There have been great changes in the state of the art of generating random variates in the last ten years, both for discrete distributions as discussed here, and for continuous distributions. In particular, the alias method and acceptance/complement have been developed. However, most state-of-the-art algorithms today remain based on acceptance/rejection frameworks.

Good algorithms exist for most common discrete distributions (as well as for continuous distributions). Substantial research questions remain for multivariate distributions and time series.

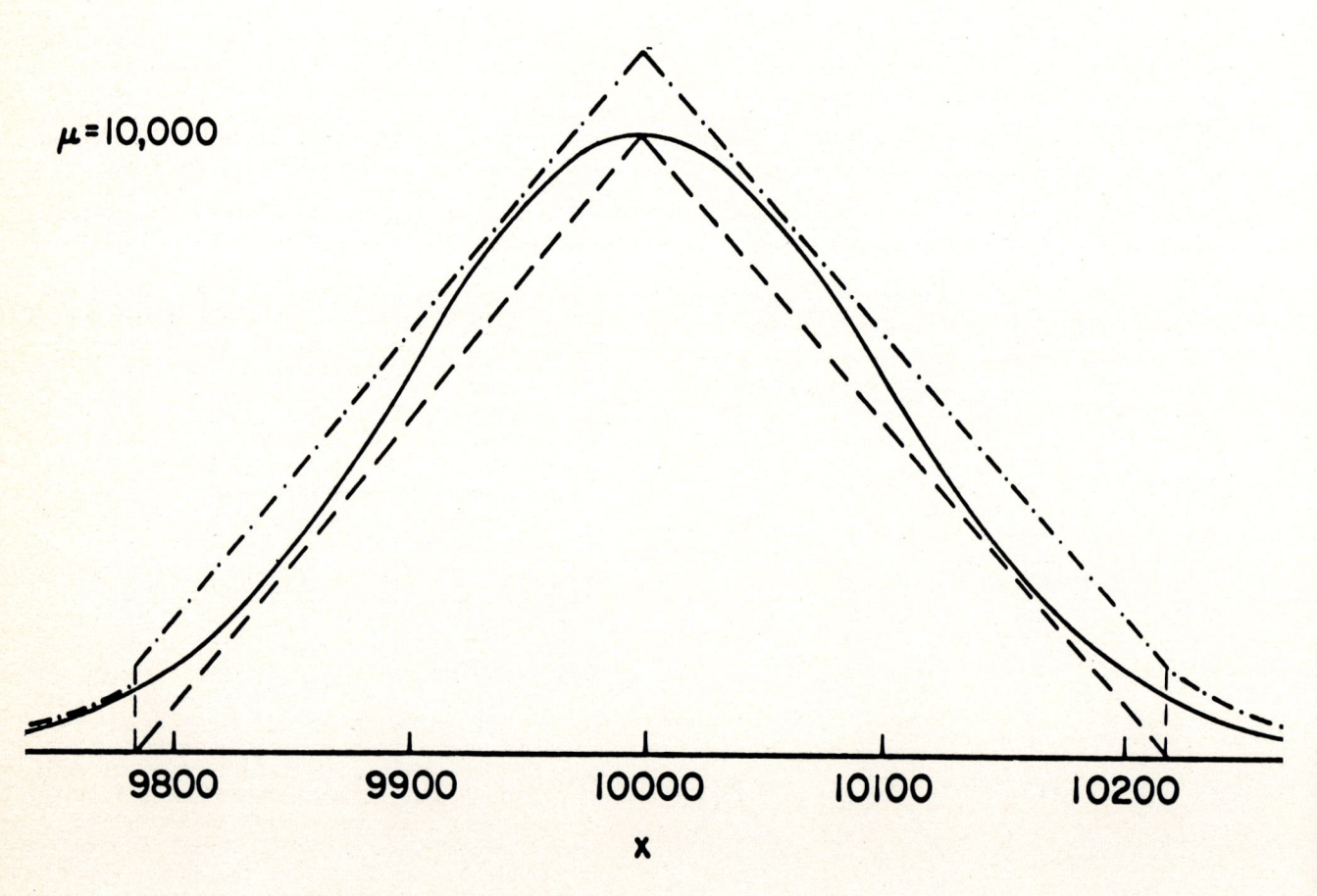

Figure 2. Algorithm PTPE for mean equal to 10,000.

REFERENCES

[1] Ahrens, J.H. and Dieter, U., Sampling from binomial and Poisson distributions: a method with bounded computation times, Computing, 25, 3 (1980) 193-208.

[2] Ahrens, J.H. and Dieter, U., Computer generation of Poisson deviates from modified normal distributions, ACM Transactions on Mathematical Software, 8, 2 (June 1982) 163-170.

[3] Atkinson, A.C., The computer generation of Poisson random variables, Applied Statistics, 28, 1 (1979) 29-35.

[4] Atkinson, A.C., Recent developments in the computer generation of Poisson random variables, Applied Statistics, 28, 3 (1979) 260-263.

[5] Boyett, James M., Random R x C tables with given row and column totals, Applied Statistics, 28, 3 (1979) 329-332.

[6] Chen, H.C. and Asau, Y., On generating random variates from an empirical distribution, AIIE Transactions, 6 (1974) 163-166.

[7] Deák, I., An economical method for random number generation and a normal generator, Computing, 27 (1981) 113-121.

[8] Devroye, Luc, The computer generation of Poisson random variables, McGill University (1980).

[9] Fishman, G.S., Principles of Discrete Event Simulation (Wiley Interscience, NY, 1978).

[10] Fishman, G.S. and Moore, III, L.R., Sampling from a discrete distribution while preserving monotonicity, Technical Report No. UNC/ORSA/TR-81/7, Curriculum in Operations Research and Systems Analysis, University of North Carolina at Chapel Hill (March 1981).

[11] Hoffman, R.G., The simulation and analysis of correlated binary data, Proceedings of the Statistical Computing Section, American Statistical Association (1979) 340-343.

[12] Jacobs, P.A. and Lewis, P.A.W., Discrete time series generated by mixtures I: correlational and runs properties, Journal of the Royal Statistical Society B, 40 (1978) 94-105.

[13] Jacobs, P.A. and Lewis, P.A.W., Discrete time series generated by mixtures II: asymptotic properties, Journal of the Royal Statistical Society, 40 (1978) 222-228.

[14] Jacobs, P.A. and Lewis, P.A.W., Discrete time series generated by mixtures III: autoregressive processes (DAR(p)), Technical Report NPS55-78-022, Naval Postgraduate School (1978).

[15] Jacobs, P.A. and Lewis, P.A.W., Stationary discrete autoregressive-moving average time series generated by mixtures, Technical Report NPS55-82-003, Naval Postgraduate School (January 1982). Journal of Time Series Analysis, forthcoming (1983).

[16] Johnson, M.E., A Unifying Theory for Discrete Random Variable Generation, Master's Thesis, Department of Industrial and Management Engineering, The University of Iowa (1974).

[17] Kachitvichyanukul, Voratas, Computer Generation of Poisson, Binomial, and Hypergeometric Random Variates, Ph.D. Thesis, School of Industrial Engineering, Purdue University (1982).

[18] Kemp, C.D. and Loukas, S., Computer generation of bivariate discrete random variables using ordered probabilities, Proceedings of Statistical Computing Section, American Statistical Association (1978) 115-116.

[19] Kemp, C.D. and Loukas, S., The computer generation of bivariate discrete random variables, J.R.R.S. A, 141 (1978) 513-519.

[20] Kennedy, W.J. and Gentle, J.E., Statistical Computing (Marcel Dekker, Inc., NY, 1980).

[21] Kronmal, R.A. and Peterson, Jr., A.V., On the alias method for generating random variables from a discrete distribution, American Statistician, 33, 4 (1979) 214-218.

[22] Kronmal, R.A. and Peterson, Jr., A.V., A variant of the acceptance-rejection method for computer generation of random variables, JASA, 76, 374 (June 1981) 446-451.

[23] Law, A.M. and Kelton, W.D., Simulation Modeling and Analysis (McGraw-Hill, NY, 1982).

[24] Marsaglia, G., Generating discrete random variables in a computer, CACM, 6 (1963) 37-38.

[25] Peterson, Jr., A.V. and Kronmal, R.A., On mixture methods for the computer generation of random variables, The American Statistician, 36, 3 (August 1982) part 1, 184-191.

[26] Schmeiser, B.W., Random variate generation: a survey, in Oren, T.I., Shub, C.M. and Roth, P.F. (eds.), Simulation with Discrete Models: A State-of-the-Art View (1980) 79-104. Papers presented at the 1980 Winter Simulation Conference.

[27] Schmeiser, B.W. and Kachitvichyanukul, Voratas, Poisson random variate generation, Research Memorandum 81-4, School of Industrial Engineering, Purdue University (1981).

[28] Shanthikumar, J.G., Discrete random variate generation using uniformization, Working Paper No. 83-002, Department of Systems and Industrial Engineering, The University of Arizona (January 1983).

[29] Walker, A.J., New fast method for generating discrete random numbers with arbitrary frequency distribution, Electronics Letters, 10 (1974) 127-128.

[30] Walker, A.J., An efficient method for generating discrete random variables with general distributions, ACM Transactions on Mathematical Software, 3, 3 (September 1977) 253-256.

SMALLER COMPUTERS: WORKSTATIONS AND THE HUMAN INTERFACE

Organizer: *Thomas J. Boardman*

Invited Presentations:

    Hardware for Kinematic Statistical Graphics, *Jerome H. Friedman and Werner Stuetzle*

HARDWARE FOR KINEMATIC STATISTICAL GRAPHICS*

Jerome H. Friedman and Werner Stuetzle

Stanford Linear Accelerator Center and Department of Statistics,
Stanford University, Stanford, California 94305

The hardware requirements for a computer graphics system capable of supporting kinematic statistical graphics are specified. The various options are discussed, and the Orion-1 workstation currently in use at the Stanford Linear Accelerator Center is described.

1. INTRODUCTION

When the pioneering PRIM-9 system for kinematic display of high-dimensional statistical data (Fisherkeller, Friedman and Tukey, 1974) was conceived and implemented at the Stanford Linear Accelerator Center (SLAC), it required specialized and expensive communications and graphics hardware and a non-negligible part of the computing power of an IBM 360/91. This was probably the reason why the system, despite receiving widespread attention, saw little practical use and did not become a generally accepted tool for data analysis. Since then, the price of computers and graphics equipment has dropped drastically, so that a hardware configuration supporting kinematic graphics can now be built for about $25,000. In view of these developments, the ORION-1 workstation for kinematic graphics was designed and built at SLAC. It is based on new and inexpensive microprocessor and raster graphics technology.

In this report, we specify the requirements on the hardware of a computer system capable of supporting kinematic graphics of the type used in PRIM-9. We describe the ORION-1 workstation now in use at SLAC, and we motivate our design decisions.

2. PRIM-9: A BRIEF DESCRIPTION

We will give a brief description of PRIM-9, mainly for the sake of completeness and because the original publication is not easily accessible. No written account can do justice to the system; to appreciate its qualities one has to see the film produced at SLAC and available upon request (write to Computation Research Group, Stanford Linear Accelerator Center, Stanford, CA 94305).

PRIM-9 was a system for visual inspection of up to nine dimensional data, mainly intended for the detection of clusters and hypersurfaces. The computer system on which it was implemented allowed the modification of pictures in real time and thus made it possible to generate movie-like effects. PRIM is an acronym for the basic operations of the system:

Projection: The observations could be projected on a subspace spanned by any pair of the coordinates. The resulting scatterplot was shown on a display screen.

Rotation: A subspace spanned by any pair of coordinates could be rotated. If the projection subspace and the rotation subspace shared a common coordinate, the rotational motion caused the user to perceive a spatial picture of the data as projected onto the three-dimensional subspace defined by the coordinates involved. When the user terminated rotation in a particular plane, the old coordinates in that plane were replaced by the current (updated) coordinates. This make it possible to look at completely arbitrary projections of the data, not necessarily tied to the original coordinates.

Masking: Subregions of the observation space could be specified, and only points inside the subregion were displayed. Under rotation, points would enter and leave the masked region.

Isolation: Points that were masked out (i.e., not visible) could be removed, thus splitting the data into two subsets.

We will call any system for kinematic display of high dimensional data using these or similar basic operations a PRIM system.

3. HARDWARE REQUIREMENTS

Hardware supporting PRIM must have the ability to compute and draw new images at a rate of at least five times per second. Such a low update rate is aesthetically not very pleasing, but it can be sufficient if the controls, the means by which the user interacts with the program, are suitable designed. Operations that have to be performed for every update of the image are called real time operations.

---

*This work was supported by the Department of Energy under contracts DE-AC03-76SF00515 and DE-AT03-81-ER10843 and by the Office of Naval Research under contract ONR N00014-81-K-0340.

The basic real time computation for PRIM is rotation of a set of points. Let $x_1...x_n$ denote a sample of points in R^3 (generally obtained by projecting higher dimensional observations on a 3-dimensional subspace), and let R denote a 3*3 rotation matrix. The user controls the rotation matrix, and the rotated points $Rx_1...Rx_n$ have to be computed for each new image displayed on the screen. In the simplest case, only two coordinates of the rotated points (corresponding to the horizontal and vertical screen axis) are needed. It is sometimes desirable to use the 3rd coordinate for depth cueing, or to work with 4*4 rotation matrices in homogeneous coordinates to draw perspective images of the point cloud, but we will ignore this here. Computing two coordinates of $Rx_1...Rx_n$ requires 6n multiplications, 4n additions plus operations to access and store the matrix elements and the coordinates. One would probably write the rotation routine in assembly code, and the total number of operations will vary significantly from processor to processor, depending on the number of registers, available address modes, etc. In many cases, for example if the rotations are performed on a microprocessor, the time critical operation is the multiplication, and rotating 1000 points 10 items a second require a multiplication time of substantially less than 10 microseconds when the overhead and the time necessary to draw the images is taken into account.

A PRIM system for which rotation is the only real time computation can certainly be useful, but it is highly desirable to have additional capabilities, for example, real time scatterplot smoothing (Cleveland, 1979; Friedman and Stuetzle, 1982), computation of near-neighbor density estimates (Tukey and Tukey, 1982), computation of classification rules, etc.

The basic requirement for the display part of a PRIM system is the ability to draw scatterplots in real time. Assume that we want to draw a scatterplot of 1000 points 10 times a second. If the coordinates of each point on the screen are specified by two 16 bit integers, this requires a data transmission rate of more than 300 K bits per second (300 Kbaud).

An important choice is between color display and monochrome display. The availability of color opens up a whole new range of possibilities. It is much more efficient for distinguishing groups than blinking, highlighting, or use of special symbols, and it allows one to represent a fourth dimension on the screen. We have found color to be highly desirable. Also highly desirable is the ability to draw vectors in real time, for example, to draw the minimal spanning tree of the observations or to draw a smooth through a scatterplot. s a minimal spanning tree has n-1 edges, drawing it for 1000 points 10 times per second requires the ability to put up 10,000 vectors per second.

Ability to draw not only scatterplots and vectors, but also color maps, in real time would be very useful. Color mas could, for example, be used to represent two-dimensional marginal densities. For technical reasons to be discussed in Section 5, only very little energy has been spent so far on techniques requiring real time drawing of color maps.

The purpose of the control part of the PRIM hardware is the translation of a motion of the hand into a motion on the screen. Almost any input device can be used for control if it is suitably used by the program and if its effect is adapted to the available update rate. We will further discuss this point in Section 6.

4. COMPUTER HARDWARE

A basic choice that has to be made is between implementing PRIM on a multi-user mainframe connected to a graphics terminal (terminal concept) or implementing it on a workstation (workstation concept). A workstation by definition performs the real time computations for PRIM on its own dedicated processor, whereas a terminal has very little local computing capability and thus relies heavily on the mainframe it is connected to. Up till now all PRIM systems were implemented on mainframes. (The Evans and Sutherland Picture System 2, which was used in two of the systems, has a built-in matrix multiplier and thus can rotate points, but it is not a general purpose processor.) The adherence to the terminal concept was dictated by technical necessity; there were no inexpensive processors available on which a workstation could be based. Unfortunately, as noted above., the graphics terminal has to be connected to the mainframe with a line allowing for a transmission rate of several hundred Kbaud, and ports of that speed are highly non-standard and are generally not supported by mainframe manufacturers. The terminal concept poses some additional problems.

The update rate will depend on the current load on the mainframe, and the remedy of giving the graphics job a high priority tends to aggravate other users. If there is a charge for CPU time, PRIM will be expensive to run because kinematic graphics, as noted above, is very CPU intensive. The conclusion thus seems to be that PRIM should be implemented on a graphics workstation. The workstation can have various degrees of self sufficiency. At the one extreme, its computer part can consist of a single CPU board, relying on a mainframe for editing, storage and compilation of programs and for preparation of data, and being connected to the mainframe by a standard serial communication line used to download programs and data in binary form. At the other extreme, it can be a self sufficient personal computer, with its own operating system, disks, editors, compilers, etc. We will further discuss the pro's and con's of the two approaches below.

The next question is on what kind of processor the workstation should be based. The eight bit microprocessors that have been commercially available for some time are the first possibility that comes to mind. However, they generally do not even have hardware integer multiplication, and their small address range and low overall speed make them unattractive for development work starting now, especially in view of the availability of 16 bit microprocessors like the Zilog Z-8000 and the Motorola MC-68000, which are about 10 times as powerful as, for example, the Z-80. The C-68000 running at an 8 Mhz clock rate can perform a 16 bit integer multiplication in about 8 microseconds. It thus has enough computing power for the basic real time computation of PRIM, rotation of a set of points

In contrast to CPU boards, personal computers based on the MC-68000 are just now becoming available. Therefore, there is very little experience with them. Implementing PRIM on a personal computer based on the MC-68000 seems very attractive. There are, however, two possible pitfalls:

- The operating system and utilities of these personal computers are not written with real-time applications in mind. If, due to certain idiosyncrasies in the system these applications run too slowly, this might not be easy to repair because the manufacturers generally are very reluctant to distribute circuit diagrams and source code of operating systems and utilities.

- If the display is not integrated, it has to be interfaced, which is more difficult for a personal computer than for a single board processor, in particular when documentation, as usual, is rudimentary or non-existent.

Real time computation of smooths, density estimates, classification rules, etc., requires a processor with fast real arithmetic. Floating-point chips like the AMD 9511 are much too slow. Commercially available array processors are unsuitable for the following reasons:

- They are very expensive ($50,000 +).
- They usually are not easy to program, and the utilities that come with them are geared toward application in digital signal processing.
- If they are programmable at all, the program memory is insufficient.

At the moment, there seems to be no commercially available processor that could serve as basis for a workstation, and would be fast enough to allow advanced real time computations.

5. DISPLAY HARDWARE

There are two kinds of displays suitable for real time applications, vector displays and raster displays.

A vector display basically consists of a graphics processor, beam deflection electronics and a CRT tube. The program for the graphics processor is called the display file; it is usually stored in the display's private memory in binary form. The graphics processor understands instructions like "move beam to screen location (x,y)", "draw vector from location (x_1,y_1) to (x_2,y_2)", "jump to location i in the display file," and it drives the beam deflection electronics accordingly. The display file is executed periodically to refresh the image on the CRT.

A raster display is based on the same principle as a television set; the three electron beams (resp the single beam in a monochrome display) scan the entire screen a fixed number of times per second. The screen is divided into little dots, called pixels, 640 * 480 of them in standard TV format. Associated with each pixel are one or more bits of display memory containing the information on the color and intensity of the pixel. As the electron beams scan the screen, electronic circuits read the display memory and modulate the beams accordingly.

Vector displays have the advantage of high precision: they have up to 4096*4096 addressable screen locations. They can draw very complicated images quickly; the Evans and Sutherland Picture System 2, for example, permits updating of an image with 10,000 vectors or points 20 times per second.

On the other hand, vector displays cannot draw areas, only points and vectors. Most of them are monochrome and, if they have color, it is of poor quality. Furthermore, they are expensive.

Raster displays usually allow for high quality color. They can draw areas, not just vectors and points. Raster displays are less expensive than vector displays. A large fraction of the costs of a raster display goes into the display memory. (A system with 640*480 pixels and 8 bits per pixel for color and intensity information requires about 300 K bytes of display memory.) As memory prices continue to drop, the price of raster displays can be expected to drop accordingly. Personal computers with integrated raster display will then start to appear on the market.

On the other hand, raster displays generally have lower resolution than vector displays, with 512*512 or 640*480 being common and 1024*1024 being the maximum. It is not possible to update vectors very quickly. Note that if a vector is given by its beginning and end coordinates, it has to be determined which pixels the vector crosses, and the corresponding display memory locations have to be changed. Although most raster displays have built in hardware to do that, drawing a vector can take anywhere from 0.2 to 80 microseconds per pixel. If we assume two microseconds per pixel for the moment and an average vector length of, say, 50 pixels, drawing 1000 vectors takes 1/10th of a second, and thus an image with 1000 vectors can be updated, at most, ten times per second. (This estimate is highly optimistic, totally neglecting overhead and the time needed to erase the previous picture.)

We can conclude that raster displays can satisfy the basic requirements of PRIM, and that they are the preferable output device. There is a general tendency towards use of raster displays, with vector displays becoming more and more restricted to very specialized applications, for example, computer aided design. The only tempting advantage of vector displays is the ability to draw pictures with many vectors; the high resolution is not crucial.

For the commonly used raster displays, it takes several seconds to rewrite the entire display memory. Such displays, thus, do not allow real time drawing of color maps. Suitable systems are available, but their costs are still in the million dollar range.

6. CHARACTERISTICS OF A RASTER DISPLAY

We ill not list the most important characteristics that have to be considered inthe choice of a raster display:

<u>Resolution</u>: The commonly available resolutions are 512*512, 640*480 and 1024*1024. For direct viewing, the lower resolutions are sufficient. High resolution results in crisper characters, less staircasing in vectors, and might be advantageous if the screen is filmed with a movie camera. The larger display memory offers the possibility to, for example, store four subpictures, each still with 512*512 resolution, and to view them together or separately, using Pan and Zoom (see below). On the other hand, high resolution systems are much more expensive. They require four times as much display memory, and the monitor is about four times a expensive as a low resolution monitor. The drawing of vectors takes longer because a vector of the same length crosses twice as many pixels (see below). Even drawing points takes longer because, in order to be clearly visible, points have to consist of several pixels, and more pixels are needed in a high resolution system than in a low resolution system. Not, however, that a high resolution system with Pan and Zoom can always be used like a low resolution system.

<u>Refresh Rate</u>: Refresh can either be interlaced or non-interlaced. Interlaced refresh means that the entire screen is refreshed every 130th of a second; in the first half of that time, the even lines are refreshed; in the second half, the odd lines. In non-interlaced refresh, the entire screen is writen every 1/60th of a second. Non-interlaced refresh is preferable. It allows the use of a monitor with short persistence phosphor without incurring flicker, thus reducing streaking on the screen, and it is particularly advantageous if one plans to take movies off the screen with a regular movie camera. On the other hand, non-interlaced refresh requires an expensive high bandwidth monitor, and there is at the time no inexpensive system that offers 1024*1024 resolution with non-interlaced refresh.

<u>Number of Color Planes</u>: The number o color planes (bits of display memory per pixel) determines how many colors can be shown on the screen simultaneously. Four planes, allowing for 16 colors, seem to be the minimum; eight planes are desirable. If eight planes are available, they can be divided into two sets of four planes each and the image can be double buffered: while the image in one set of planes is currently displayed, the new image is written into the other set of planes. This is advantageous because it makes it possible to erase planes instead of selectively overwriting the previous points and vectors with the background color. Plane erase saves time but if it is not used in conjunction with double buffering, it results in disturbing interference effects between update and refresh.

<u>Lookup Tables</u>: A distinction has to be made between pseudo color displays and full color displays. In a pseudo color display, the memory planes are directly assigned to the colors. In a three plane system, for example, plane 1 could control the red gun, plane 2 the green gun, and plane 3 the blue gun. This system would allow showing 8 different colors, but which colors these are would be determined once and for all, depending on the intensity settings of the guns. In a full color s system, the bit pattern in the display memory location corresponding to a given pixel is interpreted as an address defining a location in each of three lookup tables, one for red, one for blue, and one for green. The value stored at that location in the red lookup table determines the intensity of the red gun, the value at the same location of the blue lookup table determines the intensity of the blue gun, etc. If each word in the lookup tables is six bits long, the system can display 2^{18} colors. How many of those can be shown simultaneously depends on the number of memory planes. If there are four planes, each lookup table has to be 16 words long, and for fixed setting of the lookup tables one can choose between 16 out of the total of 2^{18} colors. Displays also vary in the word length of the lookup table and, accordingly, in the precision of the digital-to-analog converters that control the gun intensities. Commonly used word lengths are six bits and eight bits. A word length of six bits seems to be enough for all practical purposes.

Full color is a necessity if color is to be used for displaying a fourth dimension on the screen, or if it is used as a depth cue. Pseudo color is a severe restriction.

Hardware Blink: It allows for random pixels blink, depending on the value written into the corresponding display memory locations. In some systems, a memory plane is permanently assigned to blink control and is not used in addressing the lookup tables, whereas in others, this assignment can be done in software, thus avoiding the loss of a memory plane in sitations where blink is not actually used. Hardware blink is not crucial in a color display and it is certainly not important enough to waste a memory plane for it.

Pixel Update Speed: Pixel update speed is the speed at which information can be written into the display memory. One has to keep in mind, however, that an update speed of two microseconds per pixel does not mean that a scatterplot of 1000 points can actually be drawn in 2 milliseconds. The time needed will, to a large extent, depend on the driver software, the speed of the interface, etc.

Vector drawing speed: Drawing vectors on a raster display is not entirely trivial; it is necessary to calculate from the coordinates of the endpoints of the vector which pixels it crosses. The speed at which a vector can be written into display memory is usually given in microseconds per pixel. The simplest displays rely on their host computer for this task. They are unsuitable if one intends to draw a sizable number of vectors in real time. The same applies to displays that use a standard eight bit microprocessor like the Z-80 for vector drawing. They usually take between 50 and 100 microseconds per pixel. Displays that can be used to draw pictures with a moderate number (500) vectors in real time either contain a custom designed microprocessor or hard-wired logic for vector drawing. Their speeds range from 0.1 to 4 microseconds per pixel.

Local intelligence: Some of the more recent raster displays contain high speed processors allowing local operations other than vector drawing. Examples of additional capabilities are drawing of multiple pixel points and multiple width vectors, area fils, and Newman and Sproull's raster op (see Newman and Sproull, 1979, Chapter 18). These extra capabilities are not crucial, but desirable.

Connection to the host: For use with PRIM, the possibility of connecting the display to the host via a parallel port is a necessity.

Hardware cursor: Displays with hardware cursor put up a cross or other symbol on the screen at a position controlled by an input device like a trackerball or joystick, without the help of the host. The host can read the cursor position but it does not have to bother with drawing the cursor. A hardware cursor is highly desirable.

Pan and zoom: Pan and Zoom allows choosing part of the image and magnifying it so that it fills up the entire screen. This does not bring a gain in resolution; a linear magnification of 2 is achieved by displaying the information for each pixel in the selected region in a square of 4 pixels on the screen, a magnification of 3 by blowing up each pixel into a square of 9 pixels, etc. For a low resolution display, Pan and Zoom do not seem to be worthwhile. If the display has high resolution, Pan and Zoom can be useful because they allow, for example, the simultaneous or separate viewing of four subpictures, each still with 512*512 resolution.

External synchronization: It allows the supply of the vetical synch signal from an external source instead of an internal generator. We have not yet found an example where this can be useful.

RGB to NTSC converter: The video output generated by most raster displays is not the same as the video signal used in color television; the signals for the red, blue and green gun are brought out separately, whereas in color television they are encoded onto a single carrier. An RGB to NTSC converter takes as its input the red, blue and green signals and produces an output that can be shown on a standard television set or recorded on a video recorder. Such a converter can only be used for 640*480 resolution with interlaced refresh. Conversion results in a substantial loss of quality. (The bandwidth o of a commercial TV set or video recorder is about 2.5 Mhz, whereas a usual RGB monitor has amplifiers with a bandwidth of 20 Mhz.) It is not clear whether the combination of converter and video recorder can provide a recording of useful quality.

7. CONTROLS

The design of the controls is a very important part of the overall design of PRIM; it determines to a large extent how easy or difficult the system is to learn and to use. Input devices commonly associated with interactive graphics are graphics tablet, joystick, trackerball and buttons.

A graphics tablet is a rectangular pad, usually about the size of a legal pad with a pen connected to it. The user moves the pen over the tablet. The display or host computer, to which the terminal is attached, can find out at any given time whether the pen is pressed down on the tablet, touches the tablet without being pressed down, or is away from the tablet and, in the first two cases, it can read the coordinates of the contact point. It is probably not necessary to describe a joystick; joysticks are used as input device in many of the recently popular video games. As the name indicates, a trackerball is a ball of about 2 to 3 inches in diameter that is 2/3 embedded in a casing, with the remaining 1/3 protruding on top. The ball can be spun in its casing by sliding the hand over it, and the position of the ball is translated into x and y coordinates. (The ball cannot be spun around the vertical axis of the casing.) Buttons or switches do not output a position; they can only be open or closed.

In PRIM, the output devices are mainly used for two tasks: controlling the rotation and positioning the cursor (for example, to identify a point or to pick an item off a menu). These tasks can be performed with any one of the input devices, given suitable programming. We are thus discussing here only, which ones are the most convenient ones to use.

For controlling rotation, the trackerball clearly is best suited, especially if the program is written in such a way that rotation of the ball translates into rotation of the point cloud in the natural manner. For cursor positioning and menu selection, it has to be combined with buttons or switches - the user positions the cursor and then presses a button (enter button) to tell the system that he is done. The main disadvantage of a trackerball is its price, generally around $1000.

A tablet is as nice for cursor positioning as a trackerball. An enter button is not needed because the pen can be moved to the correct position and then pressed down; the microswitch in the pen acts as the enter button. A tablet is less natural than a trackerball for controlling rotation. However, it has other advantages. It can be used easily to input curves and, combined with a character recognizer, it can even be used to enter numbers.

A joystick is fundamentally different from a trackerball and a tablet in that it is only suitable to control speed and direction of motion, not position. It is good for controlling rotation but positioning of a cursor can be awkward.

Buttons or switches alone (not in combination with another input device) can be used for control. Indeed, the original PRIM-9 system was entirely controlled by buttons. No choice was being made at that time; only buttons were available. Today a combination of trackerball and buttons seems to be the preferable choice, with joystick and buttons second. There is not enough experience with tablet control to warrant a judgment.

What was said above is only valid for moderate to high display update rates (more than, say, five updates per second). If the update rate is very low, the feedback between motion of the hand and motion on the screen gets lost. To prevent overshooting, the speed of motion, for example the angle by which the point cloud is rotated between one update and the next, has to be controlled by the program, leaving the user with the choice of direction. It is clear that in this situaton the trackerball looses its advantage for rotation control.

8. THE ORION-1 WORKSTATION AT SLAC

The ORION-1 workstation used at SLAC was designed in the summer of 1980 and was assembled between October 1980 and March 1981. This is important to keep in mind because new hardware is coming on the market at a rapid rate and a system designed today would certainly look different.

The workstation consists of a SUN68000 microprocessor board, a 168E floating point processor, a Lexidata 3400 raster display, a trackerball and 6 switches for control, 16 bit parallel interfaces between the SUN board and the 168E and between the SUN board and the 3400, and a high speed serial link to the SLAC mainframe, and IBM 3081 (see diagram 1).

The SUN board is a MULTIBUS board developed by the Stanford Computer Science Department for the Stanford University Network and now manufactured by several companies. (MULTIBUS is a trademark of INTEL Corporation.) The board contains an 8 Mhz Motorola MC-68000 microprocessor, 256 K or RAM, two serial and one parallel port, a memory mapping and protection mechanism and a small monitor in ROM. The MC-68000 is the master processor for PRIM, handling overall system control. One of the serial ports is connected to a terminal, the other one via a regular terminal line to the IBM 3081.

The 168E is basically an IBM 370 CPU without channels and interrupt capabilities. It was developed by SLAC engineers for the processing of particle physics data and has about half the speed of the genuine IBM 370 or about twice the speed of a VAX 11/780. As it does not have any input/output facilities, it is used strictly as a slave processor. The MC-68000 loads program and data into the 168E memory via the MULTIBUS and then starts the 168E. The 168E executes its program and signals when it is finished. The MC68000 then reads the results of the computation out of the 168E memory, enters new input data, starts the 168E, etc.

The Lexidata 3400 has a resolution of 640*480 pixels and interlaced refresh. It is a full color display with 8 bit digital-to-analog converters. The pixel update speed is 2 microseconds per pixel, and a custom designed microprocessor allows vectors to be drawn at a speed of 3 microseconds per pixel. It has hardware blink, a hardware cursor and it comes with a 16 bit parallel port.

The high speed (100 Kbaud) serial interface between the MULTIBUS and the IBM 3081 emulates an IBM 3270 terminal. It was developed by the SLAC Computation Research Group.

Programs are edited and stored on the IBM 3081 mainframe. The MC-68000 is programmed mostly in PASCAL, with time critical routines written in assembly language. The programs are translated into object code on the IBM 3081 using a cross compiler respectively a cross assembler, and are downloaded to the MC-68000 via the high speed serial link. The 168E is programmed in FORTRAN. The programs (which must not contain input/output statements) are translated by the IBM FORTRAN compiler, linked into an object module, translated into 168E microcode, and loaded into the 168E via the serial link.

The data to be analyzed are also stored on the IBM 3081, converted to binary representation and loaded through the serial link.

It is important to note that the high speed link is only used for downloading program and data into the 168E and the MC-68000. This could in principle be (and was until recently) accomplished through the regular RS-232 terminal connection between the IBM 3081 and the SUN board. The only difference is that this takes substantially longer. At execution time there is no interaction between the IBM 3081 and the ORION-1 workstation; the workstation runs strictly stand-alone.

9. DISCUSSION

As indicated in Section 8, the ORION-1 workstation at SLAC is not based on a personal computer. At the time of its design, inexpensive personal computers based on 16 bit microprocessors were not yet available.

The choice of a particular graphics system (Lexidata 3400) was based on considerations of the moment (price, rapid delivery, company record, etc.). There are many companies making comparable equipment today. The decision for a low resolution display with interlaced refresh was due to financial constraints. Non-interlaced refresh certainly would have been preferable.

As far as the 168E is concerned, there was no choice to be made. The 168E was then, and still is, the only fast and inexpensive ($6000) numbercruncher in existence. Unfortunately, it is not commercially available. It can be argued that it is not a good idea to base a tool for data analysis on components that are inaccessible to other users. Our consideration was that, first of all, the ORION-1 workstation at SLAC is mainly a tool for research in statistical methodology, and it seemed unwise to limit the range of exploration because a particular piece of hardware is not yet generally available. This does not imply that an alternative system with more limited capabilities, based entirely on standard components, would not be useful. It could help to make PRIM available to the community at large, and the reactions of others would provide valuable insight. Secondly, about 75% of the code run on the ORION-1 workstation resides on the MC-68000 with a clear separation of tasks between the 168E and the MC-68000 being enforced. The 168E could be replaced by a different floating point processor as soon as a suitable one becomes available.

REFERENCES

Cleveland, W.S. (1979). "Robust locally weighted regression and smoothing scatterplots," J. Amer. Statist. Assoc., 74, 829-836.

Fisherkeller, M.A., Friedman, J.H., tukey, J.W. (1974). "PRIM-9: An interactive multidimensional data display and analysis system." SLAC-PUB-1408.

Friedman, J.H. and Stuetzle, W. (1981). "Projection pursuit regression," J. Amer. Statist. Assoc., 76, 817-823.

Newman, W.M. and Sproull, R.F. (1979). "Principles of interactive computer graphics," McGraw-Hill, Inc.

Tukey, P.A. and Tukey, J.W. (1982). "Graphic display of data sets in 3 and more dimensions." To appear in Interpreting Multivariate Data, V. Barnett, Editor, Wiley, London.

Pattern Recognition and Density Estimation

Organizers: *Michael Tarter and Sidney Yakowitz*

Invited Presentations:

    Probability Density Estimation in Higher Dimensions, *David W. Scott and James R. Thompson*

    Pattern Recognition Applications to River Flow Analysis, *S. J. Yakowitz, T. E. Unny, and A. Wong*

    Methods for Bandwidth Choice in Nonparametric Kernel Regression, *John Rice*

    On Curve Estimation and Pattern Recognition in the Presence of Incomplete Data, *Michael Tarter and William Freeman*

    NASA's Fundamental Research Program in MPRIA, *R. P. Heydorn*

    A Mathematical Experiment, *Ulf Grenander*

PROBABILITY DENSITY ESTIMATION IN HIGHER DIMENSIONS

David W. Scott and James R. Thompson
Rice University, Houston, Texas

For the estimation of probability densities in dimensions past two, representational difficulties predominate. Experience indicates that we should investigate the locations of the modes and proceed to describe the unknown density using these as local origins. The scaling system to be employed should also be data determined. Using such a philosophy, density estimation has been successfully carried out in the three dimensional case. Color and motion can be used as enhancement devices so that estimation in dimensions past three becomes feasible.

1. INTRODUCTION

Much of the work in nonparametric probability density estimation is oriented to pointwise estimation. At any point in R_n we can do a good job of estimating the density function and its derivatives if only the sample size is sufficiently large. Unfortunately, almost all of the supporting numerical investigations have been done in R_1 where even a rusty nail works reasonably well. As we increase the dimension of the domain of the density past 2, we find that asymptotic pointwise rates of convergence are at best matters of secondary importance. Issues of L_1, L_2, or L_∞ convergence in density estimation are overwhelmed in importance by representational considerations. It is hard to perceive that this is so when the domain of the density is R_1 or R_2, for these lend themselves to two and three dimensional representations, respectively (recall that the density function itself takes up a dimension), and for these dimensions the _empty space phenomenon_ is usually not a problem.

But as we proceed to densities with domains in spaces of dimension higher than 2, we observe that if we impose a Cartesian mesh on the domain which is of width in each dimension roughly appropriate for the one dimensional marginal, then most of the multivariate parallelipipeds contain no sample points. Moreover, we lose, in great measure, our geometrical intuitions, which are largely three dimensional. Therefore, we cannot simply rely on computing a density function estimate on a grid chosen by dead reckoning, trusting in our visual perceptions somehow to get us out of the soup. We are forced, somehow, to face the task of estimating the density where it is relatively large. To some extent, then, the problem of density estimation in higher dimensions involves first of all finding where the action is.

Let us attempt to illustrate this point by an example. In the case of a one dimensional density, we are given the choice of two information packets:

A: a random sample of size 100

or

B: exact knowledge of the density on a equispaced mesh of size 100 between the .5 and 99.5 percentiles.

Most of us, most of the time, will choose B. Indeed the main thrust in nonparametric density estimation has been to use information of the A type to obtain a degraded version of B.

Now, if the density in question has domain R_4, then chopping each axis into a mesh with three nodes will exhaust 81 of the 100 mesh points available in B. Information packet B is now rather unattractive. A random sample of size 100, however, is much more useful. Since the data points themselves tend to come from those parts of R_4 where the density is high, they can be used as a means of focusing our attention for regions where the density should be estimated.

It is tempting, then, to go with the data set itself, to abandon the density estimation problem altogether, and stick with the scattergram. There are several reasons why we have chosen not to take this approach. First of all, we take some issue with those who would like to treat the scattergram sui generis. The scattergram will obviously change as sample size changes. The density function is the fixed entity to which the scattergram merely points. If it is argued that most data generating mechanisms are, in fact, somewhat nonstationary even over the time frame where the data is being collected, we agree. But this nonstationarity is seldom so great as to negate the practical value of looking for a density function model.

The second reason for going beyond the scattergram is that an unprocessed collection of data points does not fully exploit the natural continuities generally present in a data generating process. It is not advisable, in our view, to insist on software which assumes that a human observer is always to be available to smooth by eye a screen output, as in the case of a scattergram display. Naturally, in the formative stages, we will rely a good deal on using our visual perceptions to assist us in software creation. But our goal should not be cyborg data analysis (CDA). We hope to reach a point where our software does much more than interactively provide graphical displays to a human observer. The density function appears to have much greater potential for full automation than does the scattergram and its variants.

2. THE AVERAGED SHIFTED HISTOGRAM

The number of techniques for nonparametric density estimation is large. They include series estimators [1], kernel estimators [2,3,4], maximum penalized likelihood estimators [5,6,7], isotonic estimators [8]-- to name only a few. Some are much better than others. Some are bound to the one dimensional case. However, in this paper, we shall focus on a modified version [9] of the oldest density estimator--the histogram [10].

The usual histogram in R_1 is constructed by dividing the interval of significant density mass $[a,b]$ into M intervals of length $h=(b-a)/M$. We then count, of the total of the N data points, the number which fall into the jth interval--say N_{1j}. Then the histogram estimator at each point x in the jth interval is

$$\hat{f}_{1H}(x) = N_{1j}/(Nh).$$

To form the averaged shifted histogram we shift the left hand end point to $a+h/m$, where m is the number of shiftings we wish to perform. We then carry out the usual histogram procedure on the shifted support $[a + h/m, b + h/m]$ again using intervals of length h. This leads us to the second histogram estimator $\hat{f}_{2H}$. We continue in this manner until we have m histogram estimators $\hat{f}_{1H}, \hat{f}_{2H}, \ldots, \hat{f}_{mH}$. The averaged shifted histogram estimator at each point x in $[a + h,b]$ is given by

$$\hat{f}_{ASH}(x) = \frac{1}{m} \sum_{i=1}^{m} \hat{f}_{iH}(x).$$

Experience indicates that a m of 5 produces density estimates which are visually nearly indistinguishable from the best of the usual nonparametric density estimates. In Figure 1, we show a sequence of average shifted histograms as m doubles for a Chondrite data set [11] with $n = 22$ and $h = 2$. The improvement in statistical efficiency is apparent. The limiting estimator is, in fact, a kernel estimator.

Extensions of the shifted histogram to higher dimensions are obvious. The main advantage of the technique in several dimensions is its speed--typically 50 times faster than that of kernel techniques. The speed makes possible real time displays for handling density estimation in five dimensions when using a minicomputer such as the VAX 780.

To indicate the graphical representation of a density in higher dimensions, we note first that the two dimensional case is straightforward. We simply use the density for the vertical axis and the variables for the horizontal and depth dimensions.

Let us go then to the first representationally difficult case--that of three dimensional data. A trivariate data set may be viewed directly by a rotating scattergram or clever use of color to permit visualization of the third dimension on a two dimensional graphics terminal. The underlying smooth density function may be viewed as a thunderstorm where the density $f(x,y,z)$ at a spatial point (x,y,z) gives the relative intensity of the storm (or the density of moisture, etc.). In other words, the thunderstorm gives the 3-D representation of the 4-D density surface $f(x,y,z)$. Any particular contour level $f(x,y,z) = c$ gives a surface or disjoint surfaces in R_3. These contour surfaces are displayed by intersecting them with a series of equally spaced planes perpendicular to the x-axis (and occasionally the y-axis). No hidden line algorithms are used in these plots.

For quadravariate data, we extend the thunderstorm analogy by arbitrarily assigning one variable to represent time. Hence the contours of the five-dimensional density surface $f(x,y,z,t)$ may be observed by time-lapse photography of the thunderstorm's progression from calm beginning through the height of the storm to its inevitable conclusion. Clearly, in a reactive environment, each of the four variables may be sequentially assigned to time.

3. Trivariate Example

The data in this section were obtained by processing remote sensing satellite data measured over North Dakota during the summer growing season of 1977 and were furnished by Dr. Richard Heydorn of NASA/Houston. The sample contains approximately 24,000 points, each representing a 1.1 acre pixel, over a 5 by 6 nautical mile region. On each Landsat satellite overpass, the 4 spectral reflectance intensities on an individual pixel were combined into a "greenness" variable. The greenness of a pixel was then plotted as a function of time and Badhwar's [12] growth model was fitted to these points. From this model three variables were extracted: the time of maximum greenness (x); the ripening period of the crop (y); and the value of the greenness at time x (z). In the left column of Figure 2, we display each of the three variables

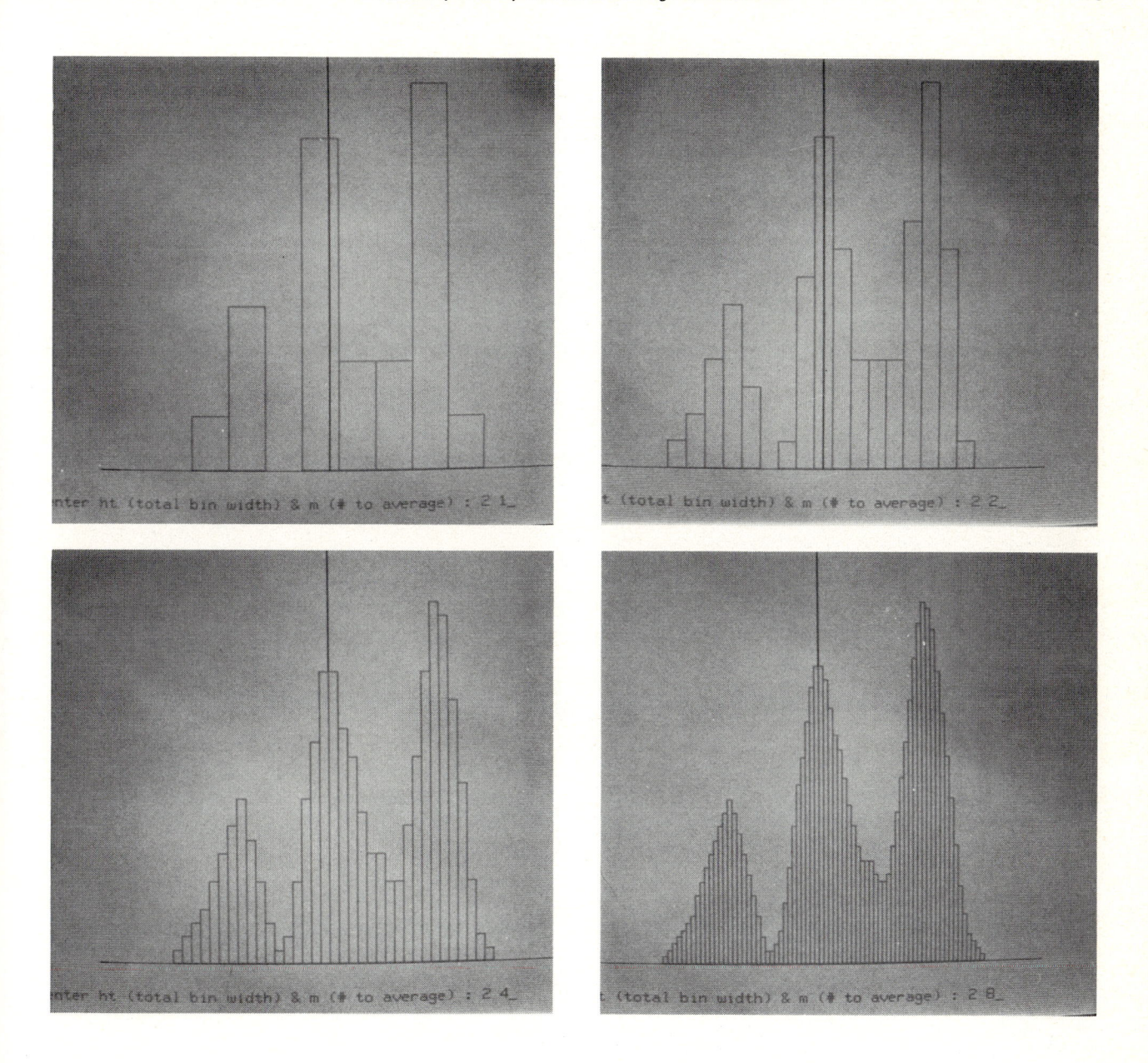

Figure 1. Averaged shifted histograms of Chondrite data
using one, two, four and eight spacings, respectively

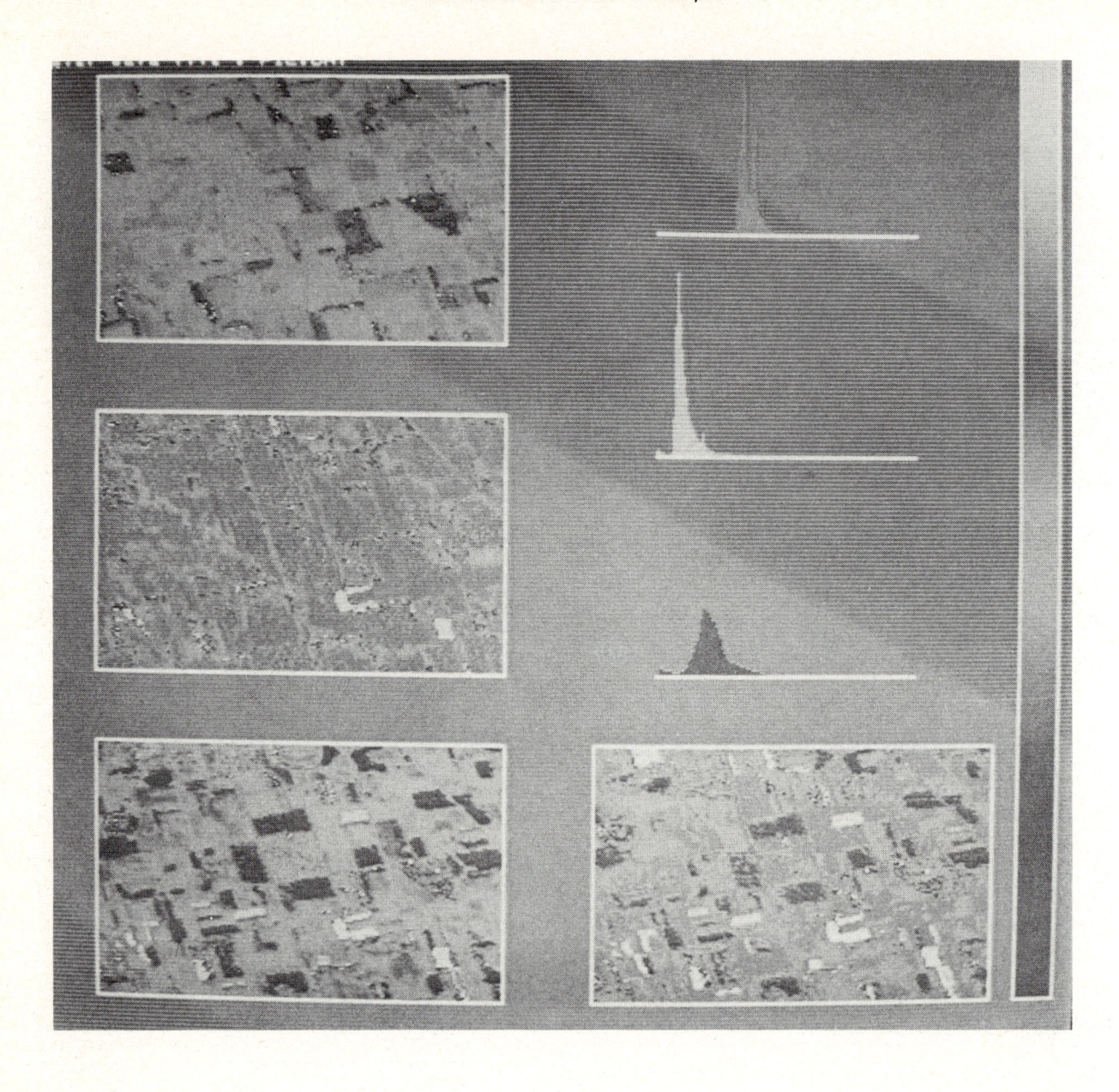

Figure 2. Remotely sensed agricultural data

positioned over the 5 by 6 nautical mile region. Dark pixels represent relatively large values of the pixels which were actually drawn in color. Field structure is easily seen in the color versions. In the second column are histograms of the three variables and an arbitrary composite of the variables.

It is difficult to present a satisfactory picture of a three-dimensional scatter diagram of these data shown in Figure 3. Color was used to represent the value of the component orthogonal to the viewing plane. Notice that the data have been projected; the original axes are shown projected and at the origin. Most of the data are hidden in the picture, only 3000 points are displayed on this 512 by 512 screen.

In Figure 4 we display the two disjoint contours of the averaged shifted histogram with m = 2 at density level 1% of the mode. Comparing Figures 3 and 4 which are drawn on the same scale we see how many of the points observable in the scatter diagram are outside the 1% contours. In Figure 3 our eyes are drawn away from the large majority of the data. The small cylindrical 1% contour represents the line of points behind the large cloud and largely represents fields of sugar beets (which are harvested by plowing rather than cutting). In Figure 5 we zoom in on the large 1% contour shape and plot the 20% contour which is bimodal. Also drawn within the front 20% contour is the single 50% contour. Many modes are apparent in these data, perhaps corresponding to different types of vegetation. On a color graphics terminal we may simultaneously view these and other contours by using different colors to draw each contour level. The density estimate was computed over a 30 by 30 by 30 mesh as the data were sequentially read from disk in less than a minute of real time. In comparison a straightforward kernel estimate over this mesh was projected to required several hours of CPU time.

4. Quadravariate Example

The 500 4-variable physics particle data have been analyzed by Friedman and Tukey using projection pursuit methods [13] and more recently by Tukey and Tukey using other multivariate graphical techniques [14]. Examination of a series of representations of scatter diagrams of three of the four variables suggests that the data lie predominately on two line segments in the shape of a "V" in four-space. For our thunderstorm analysis we first re-expressed the data variables x, y, z, and t in order to minimize skewness of the variables. The transformations were log(x), sqrt(log(1+y)), sqrt(log(1-z)), and sqrt(log(1-t)). After transformation the first variable was bimodal. For the thunderstorm analysis we let the fourth variable represent time and formed an averaged shifted histogram with m = 5 over a 30^4 mesh. In Figure 6 we display the 1%, 15%, 50%, and 80% contours (if any) at selected time points 9, 14, 19, 24, and 29 from the time lapse sequence. The axes are fixed in the same orientation as in Figures 3-5. The "V" shape is easily seen although the "left" segment is both longer and contains relatively more density mass. There appears to be relatively little data where the two segments join, at least in this re-expression. If we choose the bimodal first variable to represent time, we actually can transverse the segments individually. Other interesting shapes appear along these segments. Notice that many data points fall in the region between the segments. One feature not easily seen in this orientation without watching the figure being drawn is that the 1% contours are extremely flat like a pancake.

5. Higher Dimensions

We believe we can extend the thunderstorm representation at least one more dimension. For data sets with large numbers of variables we would first project the data onto a three or four dimensional space by a convenient method such as principal components or projection pursuit. The power of the averaged shifted histogram approach is that it removes many of the limitations of scatter diagrams which are useful only for data sets with sample sizes within a relatively narrow range. We have also nearly achieved the statistical efficiency of kernel methods without resorting to numerical approximation techniques. This speed makes density estimation in several dimensions feasible within the framework of a reactive computer environment of powerful individual workstations with extensive color capabilities. Data analysis has a powerful new tool.

REFERENCES

[1] Kronmal, R.A. and M.E. Tarter (1968), "The estimation of probability densities and cumulatives by Fourier series methods," J. American Statistical Assoc. 1:223-235.

[2] Parzen, E. (1962), "On estimation of a probability density function and mode," Annals of Math. Statistics 33:1065-1076.

[3] Rosenblatt, M. (1956), "Remarks on some nonparametric estimates of a density function," Annals of Math. Statistics 27:832-835.

[4] Scott, D.W., R.A. Tapia, and J.R. Thompson (1977), "Kernel density estimation revisited," J. Nonlinear Analysis 1:339-372.

[5] Good, I.J. and R.A. Gaskins (1972), "Global nonparametric estimation of probability densities," Virginia J. Science 23:171-193.

[6] Scott, D.W., R.A. Tapia, and J.R. Thompson (1980), "Nonparametric probability density estimation by discrete maximum penalized-likelihood criteria," Annals of Statistics 8:820-832.

[7] Scott, D.W. (1976), NDMPLE, IMSL, Inc. Houston, Texas.

[8] Wegman, E.J. (1970), "Maximum likelihood estimation of a unimodal density function," Annals Math. Statistics 41:457-471.

[9] Scott, D.W. (1983), "Averaged Shifted Histograms for Multidimensional Density Estimation," Technical Report No. 83-101, Rice University, Houston, Texas.

[10] Graunt, J. (1662), Natural and Political Observations on the Bills of Mortality.

[11] Good, I.J. and R.A. Gaskins (1980), "Density estimation and bump-hunting by the penalized likelihood method exemplified by scattering and meteorite data," with discussion, J. Amer. Statist. Assoc. 75:42-73.

[12] Badhwar, G.G., J.G. Carnes, and W.W. Austin (1982), "Use of Landsat-derived temporal profiles for corn-soybean feature extraction and classification," Remote Sensing of Environment 12:57-59.

[13] Friedman, J.H. and J.W. Tukey (1974), "A projection pursuit algorithm for exploratory data analysis," IEEE Trans. Comp. C-23, 881-890.

[14] Tukey, P.A. and J.W. Tukey (1981), "Preparation; prechosen sequences of views," in Interpreting Multivariate Data, V. Barnett, editor, John Wiley and Sons: New York, 189-213.

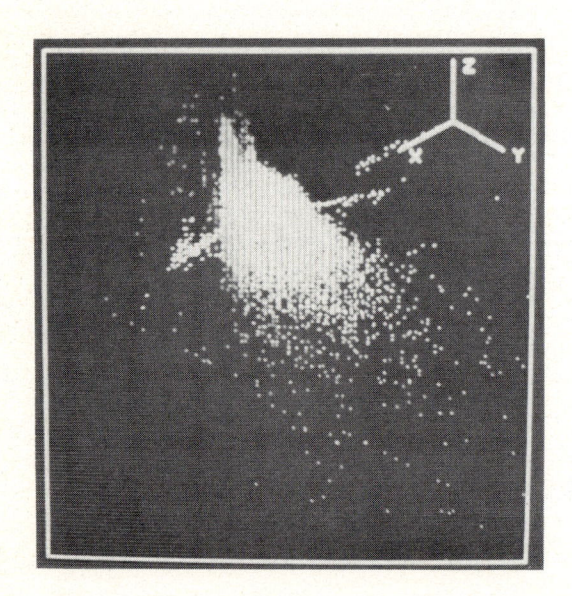

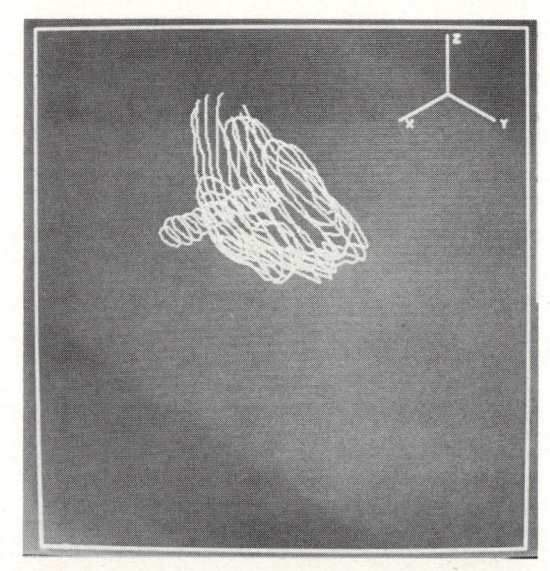

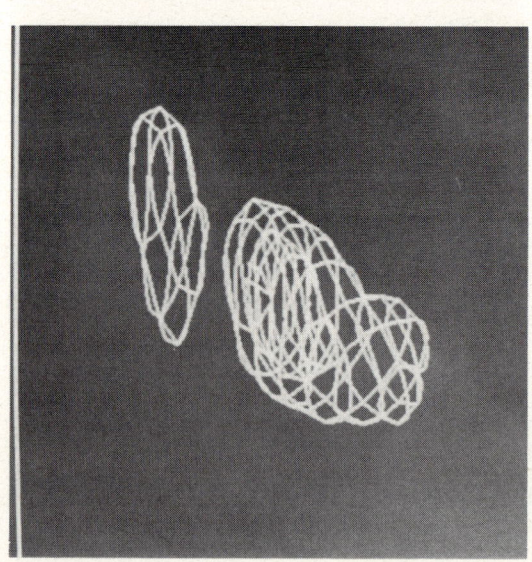

Figure 3 (above). Scattergram of remotely sensed agricultural data

Figure 4 (above right). ASH 1 per cent contours of remotely sensed agricultural data

Figure 5 (right). Zoomed view of ASH 20 per cent and 50 per cent contours of remotely sensed data

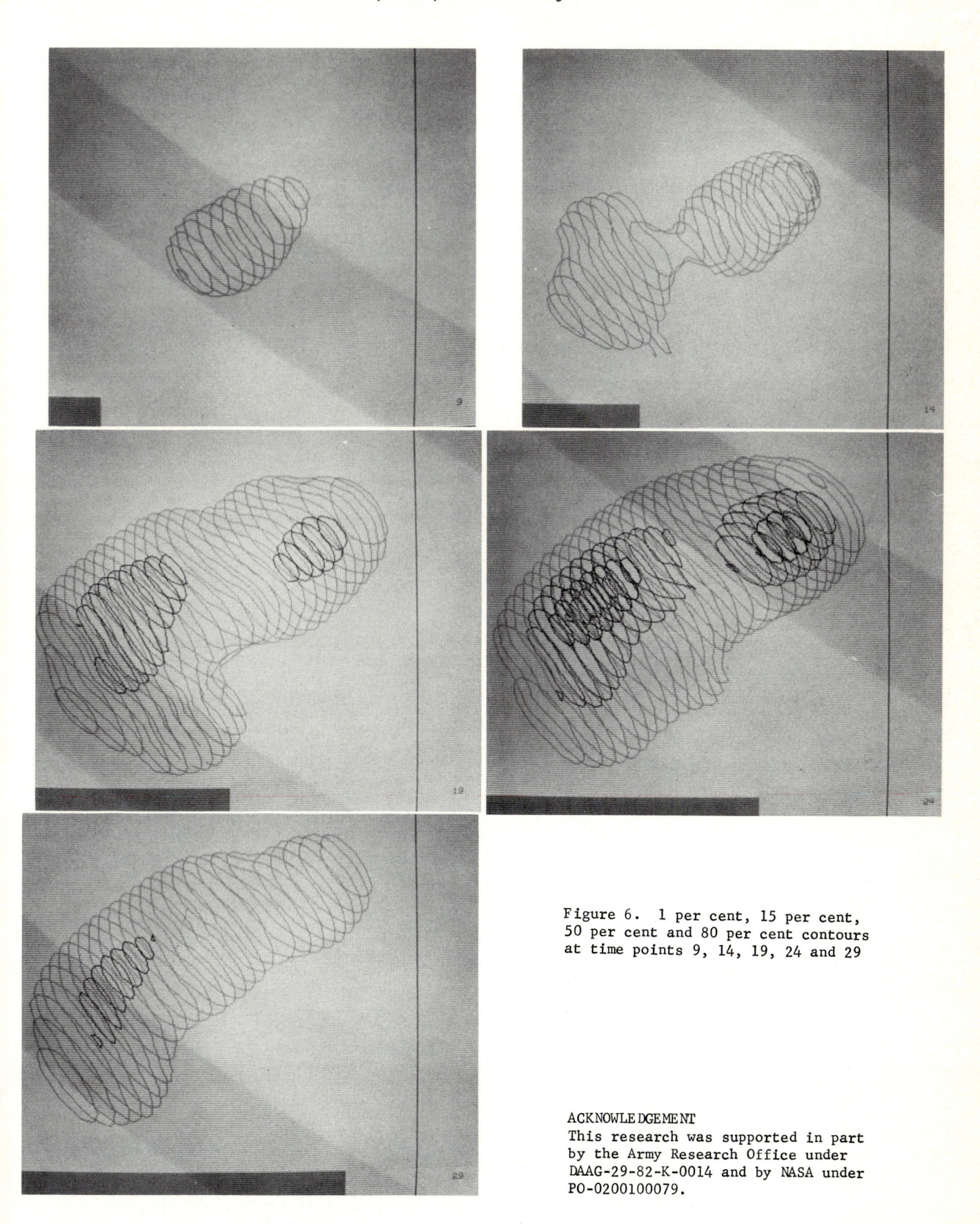

Figure 6. 1 per cent, 15 per cent, 50 per cent and 80 per cent contours at time points 9, 14, 19, 24 and 29

ACKNOWLEDGEMENT
This research was supported in part by the Army Research Office under DAAG-29-82-K-0014 and by NASA under PO-0200100079.

PATTERN RECOGNITION APPLICATIONS TO RIVER FLOW ANALYSIS

S.J. Yakowitz

University of Arizona
Tucson, Arizona

T.E. Unny and A. Wong

University of Waterloo
Waterloo, Ontario

The problem of deciding whether to issue a flood warning on the basis of rapidly rising river flows and a historical record resembles the radar warning problem. A critical distinction is that in the case of river flows, the independence assumption is unreasonable. Moreover, the task of finding suitable time series models for such sequences appears formidable. In response to these complications, we have extended a nonparametric regression procedure, namely the nearest neighbor method, to more general time series than the usual i.i.d. case. Results of application of this pattern recognition technique to actual river data are reviewed.

INTRODUCTION

The authors have separately, and now together, investigated ways to apply pattern recognition methodology to various practical problems involving physical data sets. The first author most recently has extended the kernel method to Markov sequences (Yakowitz, 1982) and applied this approach to actual river flows (Yakowitz, 1983). Yakowitz (1979 a,b) also gives methodology and riverflow applications for an alternative nonparametric time series estimation procedure. The second author and his colleagues have used feature selection ideas for quantization of flow patterns (Panu et al, (1978) and Panu and Unny (1980)). The third author is a specialist in pattern recognition applications and has a long record of experience in bringing such techniques to bear on biomedical problems (e.g., Wong and Ghahraman, 1980, Wong and Wang, 1979). Aside from our works, we are not aware of other studies which explicitly attempt to bring pattern recognition and nonparametric estimation methodology to bear on river flow analysis.

In the present study, we take a new direction, namely that of applying nearest neighbor (NN) classification techniques to the flood warning problem (FWP). The objective of this brief report is to offer a skeletal explanation of the NN approach to i.i.d. sequences, extend the NN theory to Markov sequences, and then apply these developments to a prototypical discrimination problem which we call the "flood warning problem" (FWP). Our algorithm has been coded and applied to data from the St. John River in Canada.

As far as we know, this represents the first attempt to systematically analyze and apply NN methodology in a time-series context. However, as referenced and explained in Yakowitz (1982), kernel density estimation application to such sequences dates back to a foundational study by Rosenblatt (1970) over a decade ago. Also we are aware of the work by Kassom and Thomas (1975) which brings classical nonparametric hypothesis testing ideas to bear on the pattern recognition problem in the context of a dependent process. While it is now known that under broad conditions, the convergence rate of the kernel method cannot be improved upon by alternate methodology, pattern recognition specialists (e.g., T. Wagner (1975)) have often argued that for moderate samples, the NN approach has intuitive appeal over the kernel inference approach, in that only the former effectively allows the "bandwidth" parameter to depend on the conditioning point and the "learning sample". Thus the present authors consider extension of the NN technique to the FWP a worthy exercise.

A QWIK INTRODUCTION TO NEAREST NEIGHBOR METHODOLOGY AND RESULTS

By our glib section title, we mean to warn the reader that due to limitations in space and objective here, the discussion to follow is heuristic and imprecise. Details are to be found in our citations.

Let $\{X(i),Y(i)\}$ be an i.i.d. sequence of pairs with $X(i)$ itself a random vector of dimension d. For the "pattern classification" (or discrimination") problem, the universe of the $Y(i)$'s is presumed to be the finite set $\{1,2,\ldots,M\}$ of "patterns". Intuitively, one supposes that at each decision epoch i a pattern $Y(i)$ is selected from the pattern set according to a stationary random mechanism. The decision maker observes the vector $X(i)$, often called the "feature vector", the distribution of which depends on the unobserved pattern $Y(i)$. Here and in a great deal of the pattern recognition literature, it is presumed that the conditional probability density function $f(x/y(i))$ is not known to the decision maker

(DM); in fact, the DM does not even have a parametric family available for inferring $f(x/y(i))$. The basis for the decision is a "training set" $\{(x(i),y(i)):1< i < n\}$ of "classified" (i.e., $y(i)$ is given) samples, and a loss function $L(a,y)$; $a,y \in \{1,\ldots,M\}$, reflecting the cost if the DM declares a to be the pattern when, in fact, it was really y.

A prototypical "NN" method for pattern classification on the basis of the training set and the loss function is the following: let k be a positive integer less than n, the training set sample size. Let x be an unclassified observation. The k nearest neighbors to x are those k learning samples $x(i)$ which are closest to x in Euclidean distance. Let $Q(y,x,n)$ denote the number of these k nearest neighbors which belong to pattern y, $1 < y < M$. The k nearest neighbor (k-NN) rule has us select as the guess for the pattern y associated with a feature vector x, that pattern a which minimizes

$$R_n(x,a) = \sum_{y=1}^{M} L(y,a)Q(y,x,n)/k. \quad (1)$$

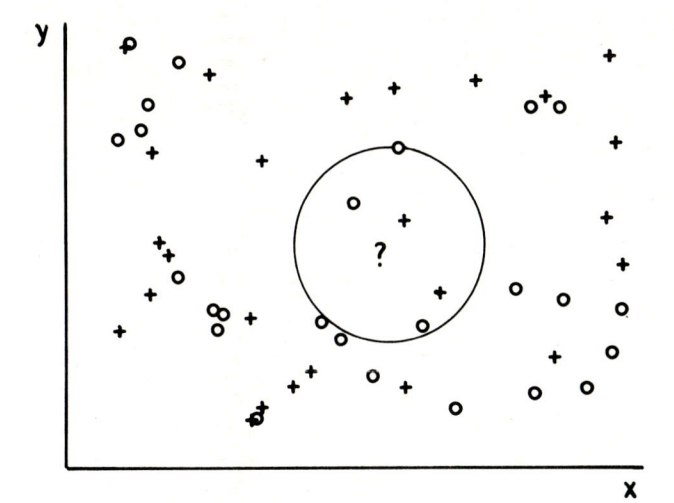

exact (but unknown) Bayes risk for choosing action a is given by

$$R(x,a) = \sum_{y=1}^{M} L(y,a)P(y/x). \quad (2)$$

The NN idea was introduced by Fix and Hodges (1961), who proved that if $k=k(n)$ increases with the number n of learning samples so that $k(n) \to \infty$, but $k(n)/n \to 0$, then in fact $R_n(x,a)$ is a consistent estimate of $R(x,a)$. Cover and Hart (1967) and Cover (1968) showed that if k remains fixed while the size n of the training sample grows, that $R_n(x,a)$ converges to a number not exceeding $2 R(x,a)$. This observation has practical merit because the factor of 2 may not be a large price to pay for avoiding the sorting problem associated with large k and n. Stone (1977) abstracted the nearest neighbor idea by allowing that each neighbor be weighted to reflect the distance from the unclassified sample as well as the locations of the other training samples. Thus the estimate of the risk takes the form

$$R_n(x,a) = \sum_{j=1}^{n} L(y(j),a)W(x,\{x(i)\},x(j)) \quad (3)$$

Stone (1977) provided necessary and sufficient conditions for these weights so that the resulting estimate $R_n(x,a)$ be L_p consistent if the pth moment of $L(Y,a)$ exists. Stone (1980) proves that regardless of the nonparametric scheme, for risk functions $R(x,a)$ being j times differentiable, an upper bound to the convergence of the square error is given by

$$E[(R(x,a)-R_n(x,a))^2] = O(n^{-r}) \quad (4)$$

with $r=2j/(2j+d)$. He shows that this upper bound is attained with a certain NN-like scheme. From Mack (1981), we have that the optimal rate is achieved by (1) in the case that the order j of differentiability is 2 and $k(n)$ grows as n^r. Also, Mack shows that the NN estimates are asymptotically normal, and provides an informative table comparing biases and variances of the kernel method with those of the NN-algorithm for the case $d=1$.

EXTENSION OF THE NN METHOD TO MARKOV SEQUENCES

For reasons explained in Yakowitz (1979b,1983), a (possibly higher-order) Markov sequence appears to be a sensible model for seasonalized river flows and other time series of interest. Thus in this section, we presume that the object $\{X(t)\}$ of analysis is a Markov series which is somehow known to be stationary and ergodic, and to have a continuously differentiable, invariant probability density function $p_i(x)$. Let

$$m(x) = E[h(X(n+1))/X(n)=x] \quad (5)$$

○ PATTERN CLASS 1
+ PATTERN CLASS 2

Figure 1: Illustration of k-NN Method

Figure 1 illustrates use of the k-NN rule on some randomly chosen data points. Here it is presumed that $d=2$, $k=5$, and $M=2$. The location of 50 classified data points are given in the figure. There are two pattern classes. The unclassified point is located by a "?". If the loss function is 1 for misclassification and 0 otherwise, then (1) implies that we should choose the class which has the most training samples among the 5 nearest neighbors. Hence, 1 would be the 5-NN choice. Of course, the denote a regression function to be inferred. (In the context of the flood warning problem, we would probably want to take $h(x)=L(x,a)$, in order to represent the risk associated with action a). Assume that $v(x)=Var(h(X(i+1))/X(i)=x)$ exists and is continuous. Yakowitz (1983) shows that, under the above assumptions, and for $B(n)$ denoting the smallest ball centered at x and containing all k nearest neighbors to x, among $\{X(i): 1 \le i < n\}$,

$$m_n(x) = \Sigma h(x(i+1))/k(n) \quad (6)$$
$$x(i) \varepsilon B(n)$$

is a consistent estimator of $m(x)$, provided only that x is in the support of $p_i()$.

However, in view of Rosenblatt (1970), one would hope that under the mixing condition he calls G, stronger results could be obtained. The purpose of the present section is to show that this hope is fulfilled: Under G_2, the k-NN estimator converges at the same rate as in the i.i.d. case. Loose ends we leave to future work include evaluation of the bias constant, and demonstration of asymptotic normality. In view of space limitations, the discussion to follow is seriously abridged; our wish is that the interested reader will be able to recover missing links by reading our citations and verifying for himself some of our assertions.

We make a slight modification in the definition of the k-NN method. The reader will be able to confirm that for our choice of $\{k(n)\}$, beyond a certain time n, our rule will coincide with the standard rule (6). Let R0 be a positive number and define

$$\tilde{B}(n) = \{x(i): ||x(i)-x|| < R0\}$$

and redefine $m_n(x)$ in (6) so that the sum is over $\tilde{B}(n)$. (If $\tilde{B}(n)$ is empty, define $m_n(x)=0$.)

The central claim is the following:

Proposition: If $\{X(i)\}$ is a G_2, stationary Markov series and x is a point in the support of the continuously differentiable invariant density $p_i()$, then, in the notation of (5) and (6), with $m(x)$ presumed twice continuously differentiable, we have that

$$E[(m_n(x)-m(x))^2] = 0(n^{-R}), \quad (7)$$

where d is the dimension of the state space of the chain, and $R=4/(d+4)$.

We may write,

$$E[(m_n(x)-m(x))^2] = Var + (bias)^2 \quad (8)$$

where, in view of the Markov property, it is clear that provided $k(n)/n \to 0$,

$$Var = v(x)/k(n) + o(1/k(n)). \quad (9)$$

Letting $\sim$ signify "proportional to", we must choose $k(n)$ so that

$$k(n) \sim n^R \quad (10)$$

if there is to be any hope for the proposition. Let us assume that (10) is satisfied. Then we need only consider the bias term. From Taylor's series considerations and our assumptions about the differentiability of $m(x)$ and $p_i(x)$, the reader may check that

$$|Bias| = |E[m_n(x)-m(x)| = q\ E[R(n)^2] + o(E[R(n)^2]) \quad (11)$$

where q is a constant and $R(n)$ denotes the radius of the smallest ball about x which contains all k nearest neighbors. Our goal now, which will complete the demonstration of the proposition, is to show that

$$E[R(n)^2] \sim n^{-R} \quad (12)$$

Toward this end, we lean heavily on Rosenblatt (1970). Take $w(x)$ to be a continuous approximation to the step function on the unit d-ball, and define the kernel density estimator

$$f(x,n)=(nr(n)^d)^{-1} \sum_{i=1}^{n} w((x(i)-x)/r(n))$$
$$\simeq N(r(n),n)/(nr(n)^d c(r)) \quad (13)$$

In (13), $c(d)$ denotes the volume of the unit d-ball and $N(r,n)$ denotes the number of $X(i)$'s, $1 < i < n$, falling into the ball $B(r)$ of radius r, centered at x. We determine $r(n)$ so that

$$E[N(r(n),n)] = 2\ k(n). \quad (14)$$

It is not difficult to see that $E[N(r,n)] = n\ p_i(x)\ c(d)\ r^d + o(nr^d)$, so that, in view of (10),

$$r(n) \sim n^{-1/(4+d)} \quad (15)$$

Also, of course, for any positive sequence $\{r(n)\}$,

$$E[R(n)^2] \le r(n)^2 + R0^2 P[R(n) \ge r(n)]. \quad (16)$$

In view of (16), our demonstration is complete if we can show that

$$P[R(n) \ge r(n)] = 0(n^{-R}) \quad (17)$$

Now under G_2, it follows from Rosenblatt (1970) that $var(f(x,n) \sim 1/(nr(n)^d)$, whence

$$var(N(r(n),n)) \sim n\ r(n)^d \sim n^R. \quad (18)$$

Now

$$P[R(n) \geq r(n)] = P[N(r(n),n) \leq k(n)]$$

$$= P[|N(r(n),n) - E\, N(r(n),n)| \geq k(n)] \leq$$

$$Var(N(r(n),n))/(k(n))^2 \qquad (19)$$

the last step being Chebyschev's inequality. But in view of (18) and (10), this last term $\sim n^{-R}$, and so (17) and the proposition follows.

APPLICATION TO THE FLOOD WARNING PROBLEM

The NN approach was used on an extremely simplified prototypical version of the flood warning problem. A 35 year record of daily flows of the St. John River, Ontario, Canada, served as our data set.

We took as definition of "flood" a flow level that was so low as to occur during essentially every year of our record. Our motivation in this was that we wanted to illustrate the nature of the NN approach, and to do this we need sufficiently many learning samples for the "flood" class. If we had chosen the 10 year flood level, for example, one would expect only 3 or 4 members of this class during the entire historical record. In this regard, the authors are at a loss for scientifically based techniques for the flood warning problem when the level is high.

Let us set forth the model of our simple flood warning problem as a pattern recognition problem. The number M of patterns is 2, with pattern 1 indicating a flood follows the feature vector, and 2 corresponding to the feature vector being followed by a flow lower than the flood level. The feature vector x(i) for day i was taken to be a two-tuple, the first coordinate of which is the flow on day i, and the second the difference between the flows on days i and i-1, i.e., the first backward difference of flow on day i. A rationale for this selection will be given shortly, when we examine the flow records. The training set was assembled as follows. After seasonalization of the record

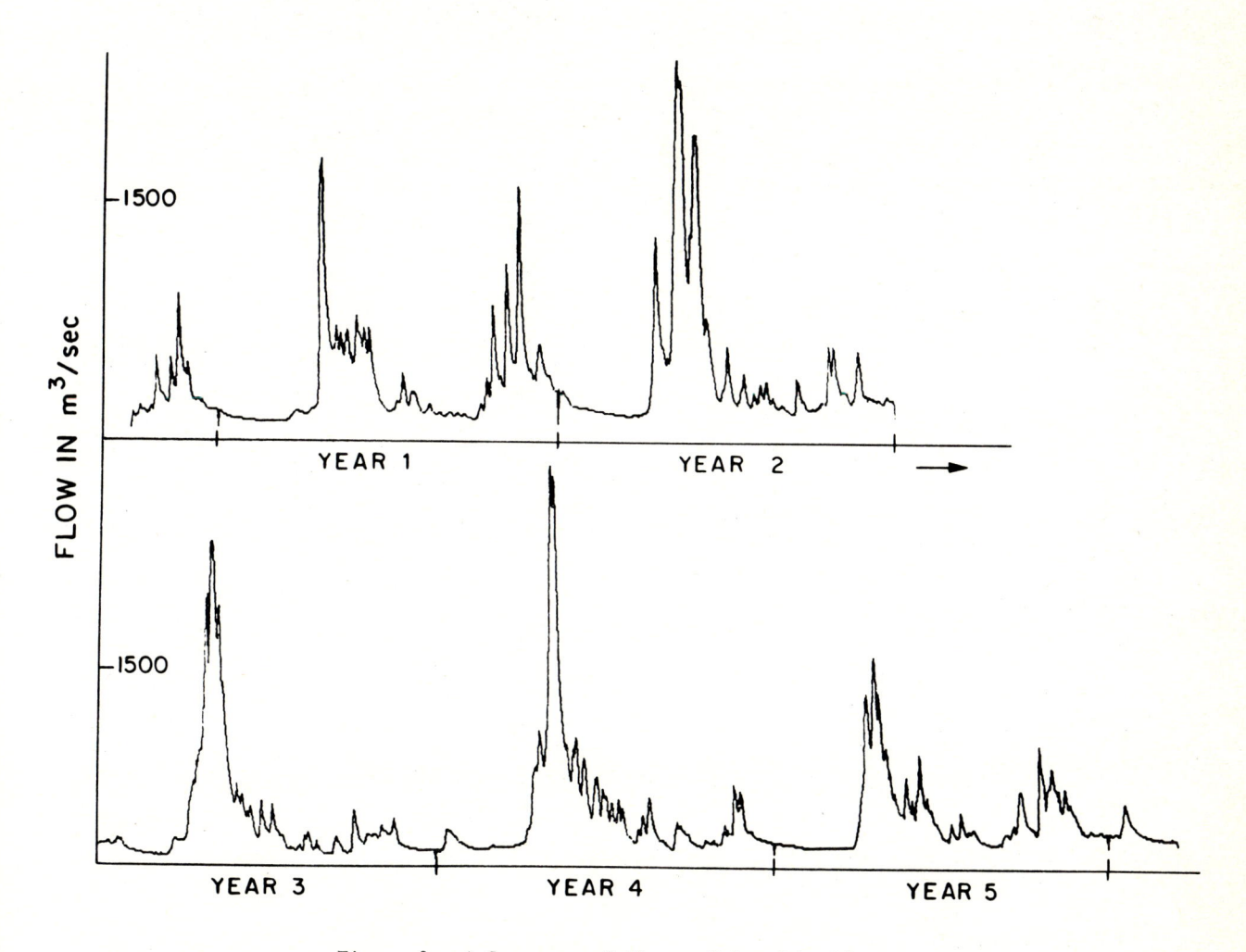

Figure 2: A Sequence of Flows of St. John River

(we will explain how this was done, presently), the record was scanned for peaks (relative high points) in the flow. If the peak was less than the flood level, the feature vector was assigned to pattern class 2, and if the peak was beyond flood level, but the preceding flows lower than the flood level, the feature vector was assigned to class 1.

Now we describe the data set. The object of study was a 35 year record of the average daily flow of the St. John River, Canada. A 2000 day sequence of this record is plotted in Figure 2. The beginning of successive years are marked by vertical lines in this figure. It is evident that the flow is not a stationary process. The tradition in river flow modelling is to subdivide the year into "seasons" during which the process is, hopefully, stationary. From examination of the record, it is apparent that the more interesting season, in terms of erratic behavior and high peak flows, is the springtime. Accordingly, we chose March through May as the stationary season of interest. In Figure 3, a 23 year record of spring season flow is presented.

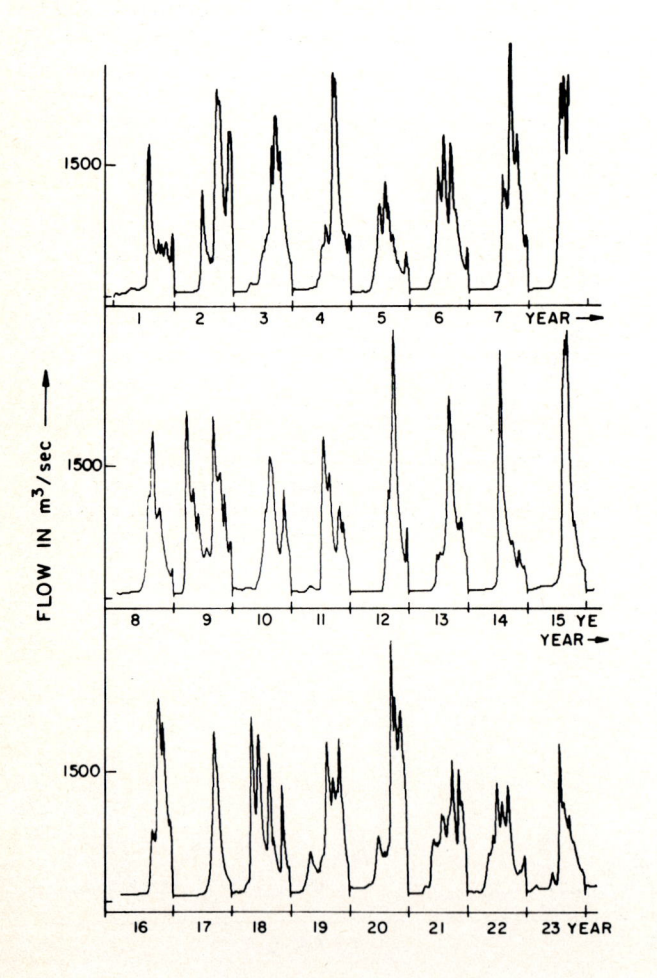

Figure 3: Spring Season Flows

Figure 4 gives the 6 flows preceding the peaks from which the training set was constructed. The solid lines are the flows associated with class 2 (no flood) and the broken line the flood class. The motivation for choosing as the feature vector the last flow and the backward difference stems from the observation that most of the information distinguishing the members of the two classes seems to occur toward the end, where the members of the flood class tend to be higher and have a sharper slope. Thus, the rationale for choosing the height and slope as the feature vector.

..... PATTERN CLASS 1 (FLOOD)
—— PATTERN CLASS 2 (NO FLOOD)

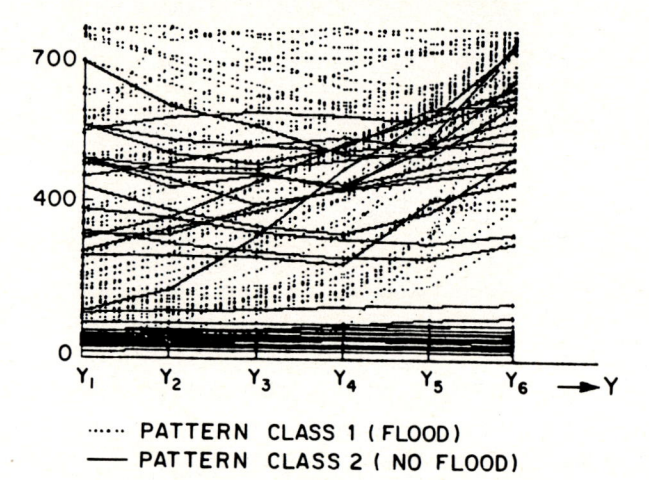

Figure 4: Sequence of Daily Flows in the Two Pattern Classes

The locations of the feature vectors are plotted in Figure 5, with 0's designating that the feature vector belongs to the flood class, and +'s, the pattern class 2. There is a substantial, but not complete, separation of the two classes. This reflects the fact that a good number of the solid lines in Figure 4 "behave" like "typical" broken lines. One can see that perfect flood forecasting will not be possible, in view of this intermingling.

The data, feature vector, pattern classes, training set, and loss function having been described, we are in a position to relate our computational findings. The number n of training samples was 79, with 34 of these being members of class 1 (flood follows), and the remaining 45 being from class 2. We performed 79 classification pattern recognition attempts, one for each training sample, by withdrawing the said training sample and trying to classify it by its k nearest neighbors. For programming simplicity, we took k to be 5, although from results we mentioned of Mack's (1981) study,

$k = n^{2j/(2j+d)} = n^{2/3} \simeq 18 \cdot$ might have given better results.

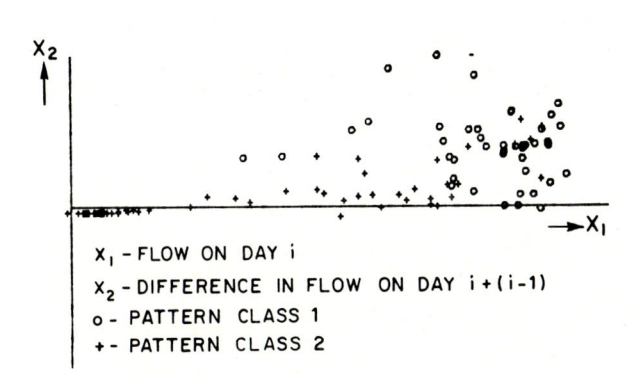

Figure 5: Plot of Training Set

REFERENCES

[1] Cover, T.M., Estimation by the nearest neighbor rule. IEEE Trans. Information Theory IT-14, 50-55, 1968.

[2] Cover, T.M. and Hart, P.E., Nearest neighbor pattern classification. IEEE Trans. Information Theory IT-13, 21-27, 1967.

[3] Fix, E. and Hodges, J.L., Jr., Discriminatory analysis, nonparametric discrimination, consistency properties. Randolph Field, Texas, Project 21-49-004, Report No. 4, 1951.

[4] Kassam, S.A. and Thomas, J.B., A class of nonparametric detectors for dependent input data, IEEE Transactions on Information Theory, Vol. IT-21, 431-437, 1975.

[5] Mack, Y.P., Local properties of k-NN regression estimates, Siam J. Alg. Disc. Meth., Vol. 2, 311-323, 1981.

[6] Panu, U. and Unny, T., Extension and application of feature prediction model for synthesis of hydrologic records, Water Resour. Res., 16(1), 77-96, 1980.

[7] Panu, U., Unny, T. and Ragade, R., A feature prediction model in synthetic hydrology based on concepts of pattern recognition, Water Resour. Res., 14(2), 335-344, 1978.

[8] Rosenblatt, M., Density Estimates and Markov sequences, in nonparametric techniques in statistical inference, M. Puri, ed., Cambridge University Press, Oxford, 1970.

[9] Stone, C., Optimal rates of convergence for nonparametric estimators, Ann. Math. Statist., 8(6), 1348-1360, 1980.

[10] Stone, C., Consistent nonparametric regression, Ann. Statist. 5(4), 595-645, 1977.

[11] Wagner, T.J., Nonparametric estimates of probability densities, IEEE Trans. on Information Theory, Vol. IT-21, 438-440, 1975.

[12] Wong, A.K.C., and Ghahraman, D.E., Random graphs: structural-contextual dichotomy, IEEE Transactions on Pattern Analysis and Machine Intelligence, Vol. PAMI-2, 341-348, 1980.

[13] Wong, A.K.C., and D.C.C. Wang, DECA: a discrete-valued data clustering algorithm, IEEE Transactions on Pattern Analysis and Machine Intelligence, Vol. PAMI-1, 343-349, 1979.

[14] Yakowitz, S., Nonparametric estimation of Markov transition functions, Ann. Statist., 7(3), 671-679, 1979a.

[15] Yakowitz, S., A nonparametric Markov model for daily river flow, Water Resour. Res., 15(5), 1035-1043, 1979b.

[16] Yakowitz, S., Nonparametric density estimation and prediction for Markov sequences, submitted for publication, 1982.

[17] Yakowitz, S., Markov flow models and the flood warning problem. Working paper, Dept. Systems and Industrial Engineering, University of Arizona, Tucson, Arizona, 1983.

METHODS FOR BANDWIDTH CHOICE IN NONPARAMETRIC KERNEL REGRESSION

John Rice

Department of Mathematics
University of California, San Diego
La Jolla, California 92093

This paper discusses several data-driven methods for bandwidth choice in nonparametric kernel regression. Several methods, including cross-validation, generalized cross validation, and Akaike's information criterion are shown to be related to a method based on unbiased risk estimation. Some numerical examples are included.

1. INTRODUCTION

This paper discusses data-based methods for the choice of a bandwidth parameter in nonparametric kernel regression. We consider cross-validation and several related criteria.

We first approach the problem in some generality in order to make clear that, although the primary focus of this paper is on kernel regression, the approach applies to other smoothing procedures as well. Suppose that Y is a random vector with

$$\underset{n \times 1}{EY} = \underset{n \times 1}{\mu}, \quad Cov(Y) = \underset{n \times n}{\sigma^2 I}.$$

A linear smoothed estimate of μ of the form $\underset{n\times 1}{\hat{Y}(\lambda)} = \underset{n\times n}{A(\lambda)} \underset{n\times 1}{Y}$ is considered, where λ is a smoothing parameter. It is desired to choose λ so that

$$R(\lambda) = E \frac{1}{n} \|\mu - A(\lambda)Y\|^2$$

is minimized. Since μ is unknown, a substitute measure based on the residual sum of squares

$$RSS(\lambda) = \frac{1}{n} \|Y - A(\lambda)Y\|^2$$

is considered. The straightforward calculation,

$$ERSS(\lambda) = R(\lambda) + \sigma^2 - \frac{2\sigma^2}{n} \operatorname{tr} A(\lambda),$$

suggests that λ be chosen to minimize an unbiased estimate of R:

$$\hat{R}(\lambda) = RSS(\lambda) - \sigma^2 + \frac{2\sigma^2}{n} \operatorname{tr} A(\lambda).$$

In order to apply this method it is necessary that σ^2 be known. We later consider modifications for unknown σ^2. This procedure is essentially that of [6]; $\hat{R}(\lambda)$ is the analogue of the C_p statistic.

We now turn to nonparametric kernel regression. We consider a special case, the data

$$y_i = f(x_i) + \varepsilon_i, \quad i = 0, \ldots, n-1$$

are equally spaced and the function f is periodic on the circle with a continuous second derivative. If the function is not periodic, the estimate must be modified at the boundary, and then will not be of a convolution form [5]. The ε_i are independent random variables with means zero and variances σ^2. A kernel estimate of f is

$$f_n(x) = \frac{1}{nb} \sum_{i=0}^{n-1} w\left(\frac{x-x_i}{b}\right) y_i.$$

Here, b is the bandwidth, and w is the kernel, a symmetric probability density with support on $[-1/2, 1/2]$, say. This estimate, evaluated at the points x_i can be written in matrix form, and, using the procedure outlined above, we consider choosing b to minimize

$$\hat{R}(b) = RSS(b) - \sigma^2 + \frac{2\sigma^2 w(0)}{nb}.$$

The basic theoretical equation to be asked is: Is $\hat{R}(b)$ a good enough estimate of $R(b)$ so that the minimizer of $\hat{R}(b)$ tends to the minimizer of R? Some results to this end are contained in [7]. There it is shown that the b minimizing $\hat{R}(b)$ (b_n, say) yields an asymptotically consistent estimate. Under the assumptions above, it is known that the asymptotically efficient choice of b is $b_n^* = cn^{-1/5}$. In [7] it is also shown that there is a (possibly local) minimizer of $\hat{R}(\lambda)$ with the property that $n^{1/5}(b_n - b_n^*) \to 0$. It is also shown that this b_n has a limiting normal distribution with standard deviation of the order $n^{3/10}$. Thus, the relative fluctuations of $(b_n - b_n^*)/b_n^*$ are of order $n^{-1/10}$ or $\sqrt{b_n^*}$.

2. OTHER METHODS

We consider some other methods that do not assume σ^2 known, and argue heuristically that they are all asymptotically equivalent to the method of the introduction. The method of cross-validation is one of the best known. It works as follows: Let $f_n^{(k)}(x)$ denote the estimated regression function with the kth point (y_k, x_k) deleted; b is chosen to minimize

$$CV(b) = \frac{1}{n} \sum_{k=0}^{n-1} (y_k - f_n^{(k)}(x_k))^2.$$

CV is easier to compute than its definition suggests. If a regression estimate for unequally spaced data [3]

$$f_n(x) = \sum (x_j - x_{j-1}) y_j \frac{1}{b} w\left(\frac{x-x_j}{b}\right),$$

is used to compute $f_n^{(k)}(x_k)$, CV can be written in closed form as

$$CV(b) = \frac{1}{n} \sum_{k=0}^{n-1} \left[y_k\left(1 + \frac{w(0)}{nb}\right) - \frac{1}{nb} w\left(\frac{1}{nb}\right) y_{k+1} - f_n(x_k) \right]^2.$$

This formula assumes equally spaced, periodic data, but similar expressions can be found in the general case. To compare CV to $\hat{R}$, the square can be expanded and the expectation of the various terms evaluated. The dominant terms are

$$\frac{1}{n} \sum (y_k - f_n(x_k))^2 + \frac{2w(0)}{n^2 b} \sum y_k(y_k - f_n(x_k))$$

and if f_n is a consistent estimate of f, the expectation of this expression is approximately equal to $ERSS(b) + 2\sigma^2 w(0)/nb$, which differs from $R(b)$ only by a constant that does not involve b.

A variety of other criteria have been proposed as well. We list some of them below. To translate to the kernel-regression case, use trace $A(\lambda) = w(0)/b$.

1. Akaike's Information Criterion [2]

    $\exp AIC(\lambda) = RSS(\lambda) \exp[2 \operatorname{tr} A(\lambda)/n]$.

2. Generalized Cross Validation [4]

    $$GCV(\lambda) = \frac{RSS(\lambda)}{\left(1 - \frac{trA(\lambda)}{n}\right)^2}.$$

3. Finite Prediction Error [1]

    $FPE(\lambda) = \frac{1 + trA(\lambda)/n}{1 - trA(\lambda)/n} RSS(\lambda)$.

4. A criterion mentioned by Shibata [8]

    $S(\lambda) = RSS(\lambda)(1 + 2trA(\lambda)/n)$.

5. A criterion to be motivated below

    $$T(\lambda) = \frac{RSS(\lambda)}{1 - 2trA(\lambda)/n}.$$

6. If σ^2 is not known it may be consistently estimated locally (from differences of adjacent y's or from residuals of straight lines fit to successive triples of points) as $\hat{\sigma}^2$, yielding

$$\tilde{R}(\lambda) = RSS(\lambda) - \hat{\sigma}^2 + 2\hat{\sigma}^2 \, tr \, A(\lambda)/n \, .$$

If for example, in kernel regression, first differences are used, and we neglect the second term, which does not depend on b, the criterion to be minimized becomes

$$\frac{1}{n} \sum (y_k - f_n(x_k))^2 + \frac{w(0)}{nb} \frac{1}{n} \sum (y_k - y_{k+1})^2$$

which has some similarity to the cross-validation criterion.

If $tr \, A(\lambda)/n \to 0$, a condition usually required for consistency, Taylor series expansions show that methods 1-3 and 5 are all asymptotically equivalent to 4. To see the relation of 4 to $\hat{R}(\lambda)$ we may write

$$RSS(\lambda)\left[1 + \frac{2trA(\lambda)}{n}\right] = \left(\hat{R}(\lambda) + \sigma^2 - \frac{2\sigma^2 trA(\lambda)}{n}\right)$$

$$\cdot \left[1 + 2tr \, \frac{A(\lambda)}{n}\right]$$

$$= \hat{R}(\lambda) + \sigma^2 + 2 \, \frac{\hat{R}(\lambda) trA(\lambda)}{n}$$

$$- 4\sigma^2 \left|\frac{trA(\lambda)}{n}\right|^2 .$$

On the assumption that $tr \, A(\lambda)/n \to 0$, the third and fourth terms are of smaller order than the first, and the second term does not depend on λ.

3. BANDWIDTH SELECTION

Although these various criteria may be asymptotically equivalent, they can behave very differently in small to moderate samples, as an example will illustrate. In terms of the measure R(b), undersmoothing is the greatest danger, as is clear from Figure 1 which shows R(b) for the function $f(x) = x^3(1-x)^3$, n = 25, σ = .002. GCV, FPE, and T have singularities for small values of b, which tend to penalize undersmoothing very heavily. The rationale behind the criterion T is that it may be expressed as

$$T(\lambda) = \frac{\hat{R}(\lambda) + \sigma^2 - 2\sigma^2 trA(\lambda)/n}{(1 - 2trA(\lambda)/n)}$$

$$= \frac{\hat{R}(\lambda)}{1 - 2trA(\lambda)/n} + \sigma^2 \, .$$

Thus, the minimizer of $T(\lambda)$ is the minimizer of $\hat{R}(\lambda)/(1 - 2trA(\lambda)/n)$ and is biased toward over-smoothing. It is easy to compare analytically the extents to which several of the criteria penalize undersmoothing. T penalizes the most, then GCV, then AIC then FPE, then S. In some *limited* simulations we have observed that CV penalizes undersmoothing somewhat more than GCV and that $\hat{R}$, with first differences is similar to GCV.

The figures below illustrate the differences that can arise. Here $f(x) = x^3(1-x)^3$, n = 25; Gaussian noise with σ = .002 was added. The quartic kernel

$$w(x) = \frac{15}{8}(1 - 4x^2)^2, \quad |x| \leq .5$$

was used. The optimal bandwidth was b = .314. Bandwidths were chosen by various criteria. Since the criterion T has a singularity at $2w(0)/nb = 1$ or b = .15, b = .15 was used as a lower bound for all the criteria in order to be "fair." One realization of this simulation is shown below. AIC chose b = .15; it was quite typical that AIC undersmoothed rather drastically. GCV undersmoothed noticeably (b = .158); GCV usually led to quite reasonable smoothing, so this example is not typical. Both T (b = .380) and CV (b = .374) oversmoothed slightly, but the results differed little from the optimal amount of smoothing.

These computations were done on a VAX 11/780. IMSL subroutines were used to generate the noise and locate the minima.

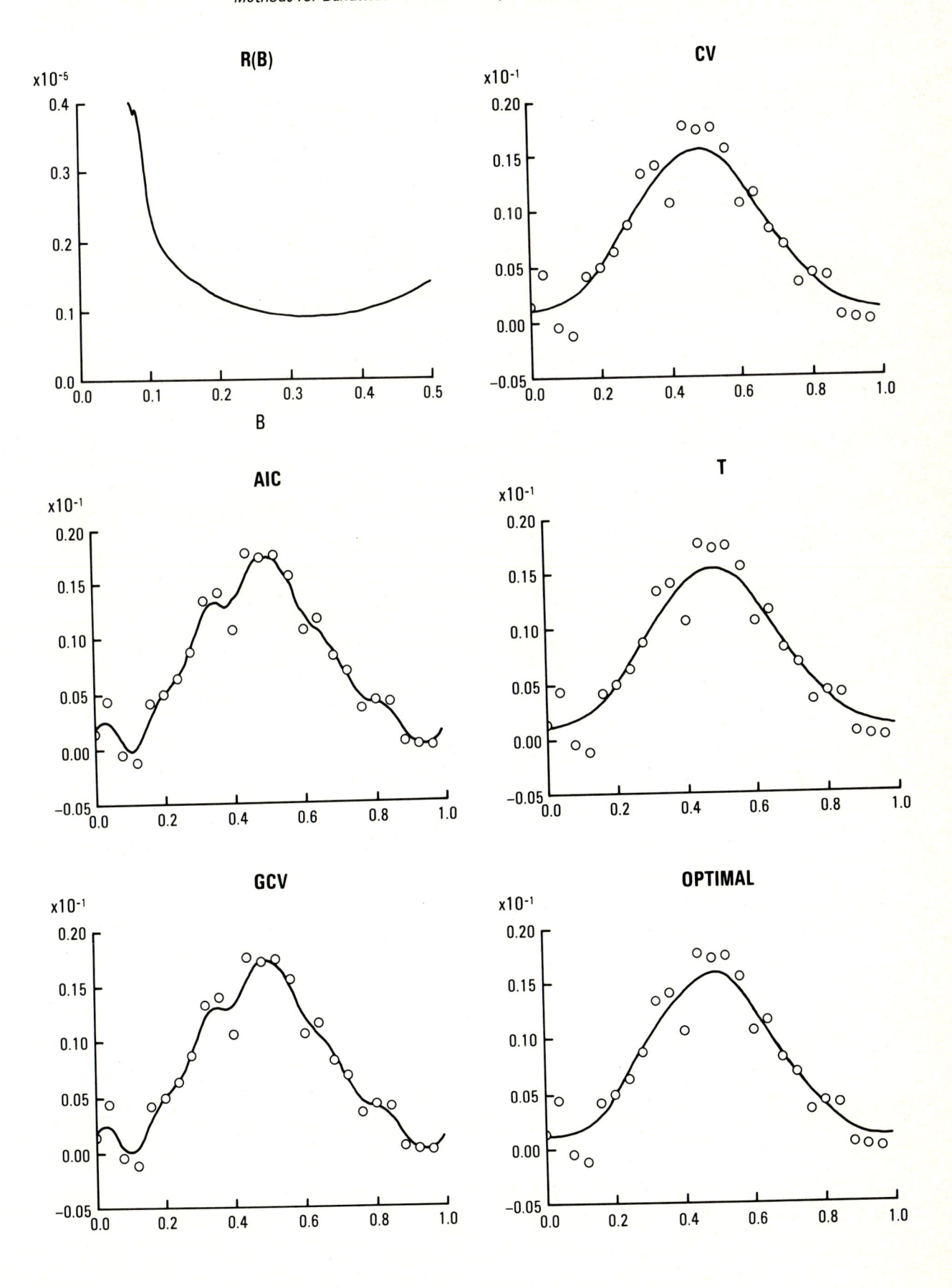

REFERENCES

[1] Akaike, H., Statistical predictor identification, Ann. Inst. Statist. Math. 22 (1970) 203-217.

[2] Akaike, H., A new look at statistical model identification, IEEE Trans. Auto Control, 19 (1974) 716-23.

[3] Benedetti, J., On the nonparametric estimation of regression functions, J. Roy. Stat. Soc. B 39 (1977) 248-253.

[4] Craven, P. and Wahba, G., Smoothing noisy data with spline functions, Num. Math. 13 (1979) 377-403.

[5] Gasser, T. and Muller, H.-G., Kernel estimation for regression functions in Gasser, T. and Rosenblatt, M. (eds.) Smoothing Techniques for Curve Estimation (lecture Notes in Mathematics 757, Springer-Verlag 1979).

[6] Mallows, C., Some comments on C_p, Technometrics 15 (1973) 661-675.

[7] Rice, J., Bandwidth choice for nonparametric kernel regression, manuscript (1982).

[8] Shibata, R., An optimal selection of regression variables, Biometrika 68 (1981) 45-54.

ON CURVE ESTIMATION AND PATTERN RECOGNITION
IN THE PRESENCE OF INCOMPLETE DATA

Michael Tarter and William Freeman

Department of Biomedical and Environmental Health Sciences
University of California at Berkeley

A nonparametric maximum likelihood estimator is obtained for the joint distribution (in the non-stationary case) from a mixed sample of complete and incomplete data. From this estimator one obtains an assessment of the degree of truncation which differs from the heuristic Bayes' Theorem estimator by a term which can be interpreted as a correction for the Gibbs phenomenon. It is shown that in the presence of truncation, conventional parametric maximum likelihood estimators of location can be obtained by trigonometric maximum likelihood procedures. Three examples of the above techniques are compared for the purpose of designing curve estimation and pattern recognition procedures for stationarity studies.

1. Introduction

This paper deals with certain problems which lie within the intersection of the following three fields: 1) curve estimation, 2) pattern recognition, and 3) survival analysis.

The last of the three topics characteristically considers data forms which are referred to as being "incomplete". For example, almost all cancer studies gather data throughout a fixed period that extends from a study beginning date, ω, to a study termination date Ω. For all such studies that involve *human* subjects, date of diagnosis X is a random variate from a density truncated to lie within the interval $[\omega,\Omega]$.

If the symbol t represents a value attained by the survival time variate T associated with a date of diagnosis $X=x$, then it is obvious that for the value t to be known, $x+t$ must be less than Ω. Data associated with a typical survival study are shown in Figure 1. The dotted curve represents an estimator $\hat{f}$ of the density of $T=t$ given date of diagnosis $X=x$ while the solid curve represents an estimator of the marginal density of the date of diagnosis variate X. The joint density of X and T is represented by the equiprobability contours shown in Figure 1.

Both the estimated conditional density curve $\hat{f}(t|x)$ and the joint density equiprobability contours are usually intersected by the study completion line $x+t=\Omega$. the fact that exact positions of any data points such that $x+t>\Omega$ cannot be ascertained is represented by the switch from solid to dotted contours over the half-plane $x+t>\Omega$. As shown in Figure 1, the degree of truncation of the density $\hat{f}(t|x)$ over the $x+t>\Omega$ half-plane is designated by the symbol p.

In this paper, the estimation of patterns and curves in the context of the above mixed incomplete-complete data will be approached from several points of view. In the next two sections nonparametric maximum likelihood methods will be applied to the estimation of joint and conditional densities which involve both the variates T and X. The method described in section 3 makes it possible to develop a curve estimation procedure based on the estimated median in the presence of incomplete data. This method combines an estimator of degree of truncation, p, with a conventional curve estimator $\hat{f}$ to estimate the median of the conditional density of T given any value of X.

In order to obtain alternatives to the above median-based procedure, in section 4 it will be shown that procedures for obtaining classical *parametric* maximum likelihood estimators from the sample trigonometric moments can be generalized to deal with truncated densities. In sections 5 and 6, the methods described in sections 2 and 4 will be combined to enable a curve estimate to be obtained in the context of incomplete data where the conditional of a variate, e.g., $\log T$, can be assumed to be normally distributed. This latter assumption makes allowance for the possibility that the location parameter μ_x, scale parameter σ_x, degree of truncation p_x, and point of truncation Ω_x are functionally related for any value of x.

The last section describes a study which uses simulated data designed to compare pattern recognition procedures based on the median with the techniques based on the normal assumption. It is demonstrated that the nonparametric median based procedure is superior to the two alternatives.

2. Description of Basic Problem

Suppose the conditional density $f(t|x)$ represents the probability that an individual who is diagnosed as having a certain disease at date $X = x$ will have a survival time $T = t$, i.e., the individual will die at exactly date $x+t$. Alternatively, suppose $f(t|x)$ represents the chance that a machine which is put into service at date $X = x$ will fail at date $x+t$. It is important to distinguish the function f from the hazard function λ or instantaneous specific death rate q. Both λ and q are conditioned not only on date $X = x$ but on survival up to but of course not including date $x+t$. In general,

$$\lambda(t|x) = \frac{f(t|x)}{S(t|x)} \qquad (2.1)$$

where

$$S(t|x) = 1 - F(t|x)$$

and

$$F(t|x) = \int_0^t f(y|x)dy$$

and thus $\lambda(t|x)$ is proportionate to the probability that, given diagnosis time x and survival through the interval $[x, x+t)$, the machine will fail or the patient will die at date $x+t$. The function S is usually referred to as a survival curve. As is well known, upon multiplying both sides of expression 2.1 by $S(t|x)$ one finds, as expected, that given date of diagnosis x the probability of death at date $x+t$ equals the product of $\lambda(t|x)$, the probability of death at date $x+t$ *given survival through the interval* $[x, x+t)$, times $S(t|x)$, the *probability of survival through the interval* $[x, x+t)$.

Cox-model and life-table procedures approach the problem of both estimating $S(t|x)$ and assessing the relationship

ILLUSTRATION OF NONPARAMETRIC REGRESSION WITH INCOMPLETE DATA

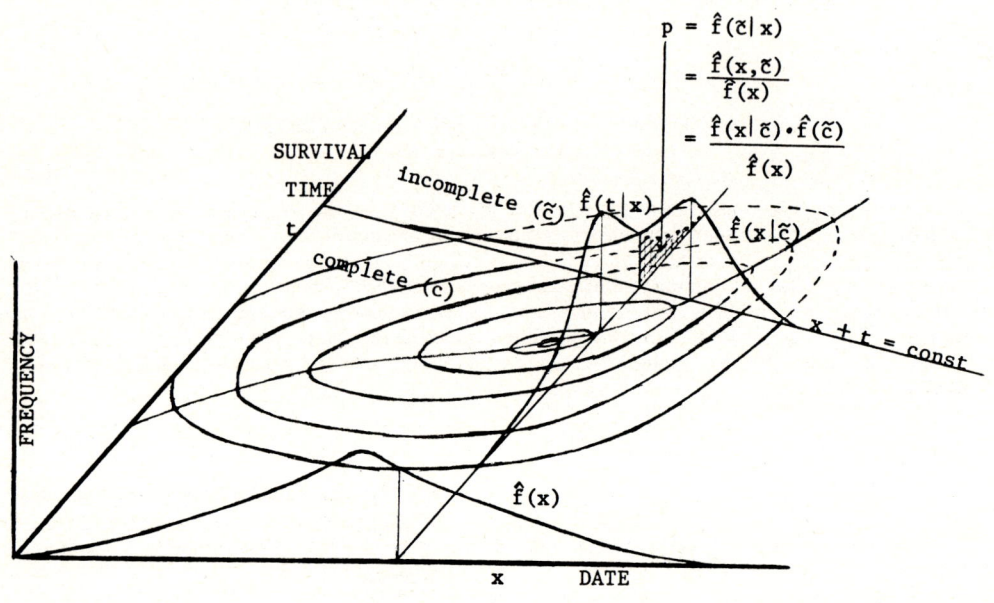

$$p = \hat{f}(\tilde{c}|x) = \frac{\hat{f}(x,\tilde{c})}{\hat{f}(x)} = \frac{\hat{f}(x|\tilde{c}) \cdot \hat{f}(\tilde{c})}{\hat{f}(x)}$$

Figure 1

between $T = t$ and a vector of concomitant variables by formulating either a model or an algorithm in terms of the function $\lambda(t|x)$. The Cox-model is a particularly useful approach since it enables the relationship between T and certain concomitant variables to be assessed unambiguously for a wide variety of underlying models (see Prentice and Kalbfleisch, 1979, and Lee, 1980, Chapter 10). However, to obtain unambiguous results the Cox-model does require a substitution of what is called conditional likelihood for the likelihood approach used elsewhere in statistics. Although Efron (1977) has shown that the conditional likelihood approach will usually provide a satisfactory approximation to the usual likelihood method, we will still base the procedures described in this section on a conventional likelihood approach based upon the density $f(t|x)$ and not the hazard function $\lambda(t|x)$.

As mentioned in the previous section the data which is available for the estimation of $f(t|x)$ is comprised of at least two forms of information. Suppose for reasons of practicality that date $x = \Omega$ is the maximum value which can be attained by the variable X and that if $x+t$ is greater than Ω then it is known that the individual has survived at least $\Omega - x$ units of time but the exact value of t, where $t \geq \Omega - x$, is unknown. In practical applications Ω is often a study (or data-gathering phase of a study) completion date. The value $\tilde{t} = \Omega - x$ is usually referred to by the term "incomplete measurement."

Data can be incomplete when an individual lives past or a machine functions beyond Ω. Alternatively, knowledge of survival up to but lack of knowledge of survival beyond a given date can be due to other causes besides the existence of a unique maximum data gathering date Ω. Often what is known as subject withdrawal takes place. A patient can be "lost to follow up" due to a variety of factors, among which the most common is probably related to change of address.

The fact that we have begun this discussion of survival analysis by introducing the variable survival time T and the variable date of diagnosis X was done purposely. Suppose density $f(t|x)$ represents the probability that an individual diagnosed at date $X = x$ will die at date $x+t$. The hypothesis or assumption that $f(t|x)$ is functionally independent of x, i.e., $H^{(1)}: f(t|x) = f(t)$ is called the *stationarity* hypothesis.

Now suppose a population of survival data, where temporarily we will assume $\Omega = \infty$, is known to contain two components. For members of the first component, both the value of T and the value of X are known. For members of the second component, X is known and a value $\tilde{t} \leq t$ is also known, where $T = t$ represents the true but unknown survival time. Let $f(t|x)$ represent the conditional density associated with members of the first population component and $f_w(t|x)$ represent the conditional density associated with members of the second population component. Here the subscript w represents the word "withdrawal." There are three important hypotheses or assumptions which involve f_w and f. These are:

$H^{(2)}: f_w(t|x) = f(t|x)$, representative withdrawal

$H^{(3)}: f_w(t|x) = f_w(t)$, stationary withdrawal

$H^{(4)}: f_w(t|x) = f(t)$, representative-stationary withdrawal

Hypothesis $H^{(2)}$ tends to apply in situations where the events which determine withdrawal are unrelated to either date of diagnosis or extent of illness. If patients who are diagnosed when they are already terminally ill tend to be lost because of travel to distant locations in search of a "miracle cure", and if this tendency to loss of information is unrelated to date of diagnosis, $H^{(3)}$ might be accepted correctly while $H^{(2)}$ and $H^{(4)}$ should be rejected. Hypothesis $H^{(4)}$ is a composite of $H^{(2)}$ and $H^{(1)}$, and as such is the most unlikely in terms of validity, but the easiest assumption to work with mathematically.

If $H^{(1)}$ and $H^{(2)}$ are both true, then for $\Omega < \infty$ there is no essential mathematical difference between two observations, one incomplete due to survival beyond date Ω and the other incomplete due to withdrawal at date $\tilde{t} < \Omega$. However, up to this point we have not discussed the relationship between date $x+t$ and survival time t.

3. Survival Analysis in the Nonstationary Case

We will now assume that $H^{(1)}$ is not valid but that all subjects for whom incomplete measurements are available were alive at date of study completion Ω. Let sample component $\{(T_j, X_j)\}$, $j=1,\ldots,n_1$ consist of complete measurements and component $\{\tilde{X}_j\}$, $j=1,\ldots,n_2$ consist of the dates of diagnosis of incomplete measurements. If each $(T_j, X_j) \sim f$, where f

has a uniformly convergent Fourier expansion whose (k, l)-th coefficient is $B_{k,l}$, and if each $\{\tilde{X}_j\} \sim g$ where density g has a uniformly convergent Fourier expansion whose k-th coefficient is $\tilde{B}_k$, then the log likelihood of a random sample of $\{(T_j, X_j)\}$ and $\{\tilde{X}_j\}$, where all X_j and $\tilde{X}_j$ are i.i.d., equals

$$\log(L) = n_1 \sum_{(u,v) \in M} \hat{B}_{u,v} \int_0^1 \int_0^1 e^{2\pi i(ut+vx)} \quad (3.1)$$

$$\log\left(\sum_{(k,l) \in M} B_{k,l} e^{2\pi i(kt+lx)}\right) dx dt +$$

$$n_2 \sum_v \tilde{B}_v \int_0^1 \left\{ e^{2\pi ivx} \log\left[(x+1-\Omega)\sum_l B_{0,l} e^{2\pi ilx} - \sum_{k \neq 0}\sum_l \left(\frac{B_{k,l}}{2\pi ik}\right)(e^{2\pi i[x(l-k)+k\Omega]} - e^{2\pi ilk})\right]\right\} dx$$

where

$$\hat{B}_{u,v} = \frac{1}{n_1}\sum_{j=1}^{n_1} e^{-2\pi i(uT_j+vX_j)} \quad (3.2)$$

and

$$\tilde{B}_v = \frac{1}{n_2}\sum_{j=1}^{n_2} e^{-2\pi iv\tilde{X}_j} \quad (3.3)$$

Hence for all l and $k \neq 0$,

$$\frac{\partial \log(L)}{\partial B_{k,l}} = n_1 \int_0^1 \int_0^1 \left[\left(\frac{\sum_u \sum_v \hat{B}_{u,v} e^{2\pi i(ut+vx)}}{\sum_u \sum_v B_{u,v} e^{2\pi i(ut+vx)}}\right) e^{2\pi i(kt+lx)}\right] dt dx -$$

$$n_2 \int_0^1 \left[\frac{\sum_v \tilde{B}_v e^{2\pi ivx}}{g(x) - \int_0^{\Omega-x} f(t,x)dt}\right] \left[\frac{e^{2\pi i[x(l-k)+k\Omega]} - e^{2\pi ilx}}{2\pi ik}\right] dx \quad (3.4)$$

Also, for all $l \neq 0$,

$$\frac{\partial \log(L)}{\partial B_{0,l}} = n_1 \int_0^1 \int_0^1 \left[\frac{\sum_u \sum_v \hat{B}_{u,v} e^{2\pi i(ut+vx)}}{\sum_u \sum_v B_{u,v} e^{2\pi i(ut+vx)}}\right] e^{2\pi ilx} dt dx -$$

$$n_2 \int_0^1 (\Omega - x - 1)\left[\frac{\sum_v \tilde{B}_v e^{2\pi ivx}}{g(x) - \int_0^{\Omega-x} f(t,x)dt}\right] e^{2\pi ilx} dx \quad (3.5)$$

The Fourier series uniqueness theorem (Edwards 1967, p. 40) implies that when for each value of $l = \pm 1, \pm 2, \cdots$ expression 3.5 is set equal to zero, with the possible exception of some set of x values of measure zero,

$$\int_0^1 \frac{\sum_u \sum_v \hat{B}_{u,v} e^{2\pi i(ut+vx)}}{\sum_u \sum_v B_{u,v} e^{2\pi i(ut+vx)}} dt = \quad (3.6)$$

$$\frac{n_2}{n_1}\left[\frac{\sum_v \tilde{B}_v e^{2\pi ivx}}{g(x) - \int_0^{\Omega-x} f(t,x)dt}\right](\Omega - x - 1) + K$$

where K takes on a constant value for all x. For any $k \neq 0$, the relationships $\partial \log(L)/\partial B_{k,l} = 0$ for all l (including $l = 0$) imply that

$$\int_0^1 \left[\frac{\sum_u \sum_v \hat{B}_{u,v} e^{2\pi i(ut+vx)}}{\sum_u \sum_v B_{u,v} e^{2\pi i(ut+vx)}}\right] e^{2\pi ikt} dt = \quad (3.7)$$

$$\frac{n_2}{n_1}\left[\frac{\sum_v \tilde{B}_v e^{2\pi ivx}}{g(x) - \int_0^{\Omega-x} f(t,x)dt}\right]\left[\frac{e^{2\pi ik(\Omega-x)} - 1}{2\pi ik}\right]$$

Since equation 3.7 specifies the Fourier coefficients of the above bracketed quantity, one finds that for some function $h(x)$

$$\frac{\sum_u \sum_v \hat{B}_{u,v} e^{2\pi i(ut+vx)}}{\sum_u \sum_v B_{u,v} e^{2\pi i(ut+vx)}} = \quad (3.8)$$

$$h(x) + \frac{n_2}{n_1}\left[\frac{\sum_v \tilde{B}_v e^{2\pi ivx}}{g(x) - \int_0^{\Omega-x} f(t,x)dt}\right]\left(\sum_{k \neq 0}\left[\frac{e^{2\pi ik(\Omega-x)} - 1}{2\pi ik}\right] e^{-2\pi ikt}\right)$$

However, since the definite integral of the right side of the above expression between zero and one equals $h(x)$ and the definite integral of the left side is specified by expression 3.6 one finds that

$$h(x) = \frac{n_2}{n_1}\left[\frac{\sum_v \tilde{B}_v e^{2\pi ivx}}{g(x) - \int_0^{\Omega-x} f(t,x)dt}\right](\Omega - x - 1) + K$$

which when substituted into expression 3.8 implies

$$\frac{\sum_u \sum_v \hat{B}_{u,v} e^{2\pi i(ut+vx)}}{\sum_u \sum_v B_{u,v} e^{2\pi i(ut+vx)}} = K + \frac{n_2}{n_1}\left[\frac{\sum_v \tilde{B}_v e^{2\pi ivx}}{g(x) - \int_0^{\Omega-x} f(t,x)dt}\right] * \quad (3.9)$$

$$\left[(\Omega - x - 1) + \sum_{k \neq 0}\left[\frac{e^{2\pi ik(\Omega-x)} - 1}{2\pi ik}\right] e^{-2\pi ikt}\right]$$

From Jolley (1961), p. 96, one finds that

$$\sum_{k \neq 0} \frac{e^{2\pi ikm}}{2\pi ik} = \sum_{k=1}^{\infty} \frac{\sin(2\pi km)}{\pi k}$$

$$= \begin{cases} \frac{1}{2} - m & \text{if } m > 0 \\ -\frac{1}{2} - m & \text{if } m < 0 \end{cases}$$

Consequently, the square bracketed quantity which appears on the right side of expression 3.9 equals 0 if $x + t < \Omega$ and -1 otherwise. Hence, one obtains the ratio

$$\frac{\hat{f}(t,x)}{\tilde{f}(t,x)} = K - \frac{n_2}{n_1} I_{[x+t > \Omega]}(t,x)\left[\frac{\sum_v \tilde{B}_v e^{2\pi ivx}}{g(x) - \int_0^{\Omega-x} \tilde{f}(t,x)dt}\right] \quad (3.10)$$

of

$$\hat{f}(t,x) = \sum_{(u,v) \in M} \hat{B}_{u,v} e^{2\pi i(ut+vx)} \quad (3.11)$$

to the maximum likelihood estimator represented by $\tilde{f}$ where $I_{[x+t > \Omega]}(t,x)$ equals one for any point (t,x) where $x + t > \Omega$ and zero otherwise, and $\tilde{g}$ is the X variate marginal of $\tilde{f}$. Upon multiplication of both sides of expression 3.10 by $\tilde{f}(t,x)$, one finds that constraining $\tilde{f}$ to integrate to one over the unit square implies that $K = (1 + n_2/n_1)$ and in turn

$$\hat{f}(t,x) = (1+n_2/n_1)\tilde{f}(t,x) + (n_2/n_1)\tilde{f}(t,x) * \quad (3.12)$$
$$I_{[x+t>\Omega]}(t,x)\left[\sum_v \tilde{B}_v \exp(2\pi ivx)/\int_{\Omega-x}^1 \tilde{f}(t,x)\,dt\right].$$

If one integrates expression 3.12 from $\Omega-x$ to 1, one finds that
$$\tilde{p} = [n_1\hat{p} + n_2\sum_v \tilde{B}_v \exp(2\pi ivx)]\,n^{-1} \quad (3.13)$$

where
$$\tilde{p} = \int_{\Omega-x}^1 \tilde{f}(t,x)\,dt$$

and
$$\hat{p} = \int_{\Omega-x}^1 \hat{f}(t,x)\,dt \quad \text{and } n = n_1 + n_2.$$

Finally, one obtains the maximum likelihood estimator
$$\tilde{f}(t,x) = \frac{n_1}{n}\hat{f}(t,x) * \quad (3.14)$$
$$\left[1 - \frac{I_{[x+t>\Omega]}(t,x)\sum_v \tilde{B}_v \exp(2\pi ivx)}{\sum_v \tilde{B}_v \exp(2\pi ivx) + n_1\hat{p}/n_2}\right]^{-1}$$

The utility of expression 3.13 is demonstrated by Figure 1. In particular let p represent the probability $f(\tilde{c}|x)$ of the event $\tilde{c}$ that a given subject will yield an incomplete measurement for a given value of the concommitant X variate $X=x$. By Bayes' Theorem
$$p = f_1(x|\tilde{c})\,f_2(\tilde{c})/f_3(x)$$
where f_1 is the density of $X=x$ for the subpopulation of incomplete measurements, f_2 is the marginal probability that a measurement will be incomplete and f_3 is the probability that $X=x$ for the composit population of complete and incomplete observations.

If $\hat{f}_3$ represents a nonparametric estimator of f_3 one can estimate p by the ratio $\tilde{p}/\hat{f}_3$. Here n_2/n can be interpreted as an estimator of f_2 while $\sum_v \tilde{B}_v \exp(2\pi ivx)$ can be interpreted as an estimator of f_1 while the $n_1\hat{p}/n$ term of $\tilde{p}$ can be interpreted as a correction for the Gibbs' phenomenon associated with the discontinuity at (x,t) point $(x,\Omega-x)$.

4. Trigonometric Maximum Likelihood Estimation in the Presence of Truncation

In Tarter (1979) it was shown that almost all classical location parameter estimators, e.g., $\bar{x}$ and the median, can be computed from the sample trigonometric moments. As an example, the underlying model
$$f_1(z) = K\exp[-c_1^2 z^2 - c_2|z|]\,I_{[0,1]}(z) \quad (4.1)$$
was considered and it was shown that as long as all data points used to estimate location parameter μ of the density $f_1(x-\mu)$ are transformed within an interval of length one-half, the new maximum likelihood estimators will yield the sample mean (when $c_2=0$) and the sample median (when $c_1=0$) *exactly*, even though the density f_1 as defined by expression 4.1 is both right and left tail truncated.

It will now be shown that the procedures described in Tarter (1979) can in almost all situations be extended to truncated densities. This result will allow the new nonparametric procedures to be applied in the next section to composit samples of complete and incomplete survival data.

Let $f(x|\mu)$ represent a density right truncated at point Ω such that $\log f$ has a uniformly convergent Fourier series on the half-open interval $[0,\Omega)$. Let the set $\{X_j\}\,j=1,\ldots,n$ consist of n points such that $0 < X_j < \Omega$ for $j=1,\ldots,n$. Now define the function
$$\tilde{f}(x|\mu) = f(x|\mu) + K\,I_{[\Omega,1]}(x) \quad (4.2)$$
where $K>0$ and I is the indicator function of the interval $[\Omega,1]$. If $\delta(z)$ represents the Dirichlet kernel over the unit interval, i.e.,
$$\delta(z) = \sum_{k=-m}^{m} \exp(2\pi ikz) \quad (4.3)$$
then
$$\sum_{j=1}^{n}\int_0^1 \ln\tilde{f}(x|\mu) = \quad (4.4)$$
$$n\sum_{k=-m}^{m}\hat{B}_k\int_0^1 \ln\tilde{f}(x|\mu)\exp(2\pi ikx)\,dx$$
where
$$\hat{B}_k = \sum_{j=1}^{n}\exp(-2\pi ikX_j)$$

By Carleson's Theorem (see Edwards 1967, p. 162) as $m\to\infty$ the left hand side of expression 4.4 approaches $\sum_{j=1}^{n}\ln\tilde{f}(X_j|\mu)$ which, since all X_j were defined to lie within the interval $[0,\Omega)$, equals $\sum_{j=1}^{n}\ln f(X_j|\mu)$ which will be represented by the term $\ln L$. Thus,
$$\ln L = n\sum_{k=-\infty}^{\infty}\hat{B}_k\int_0^{\Omega}\ln f(x|\mu)\exp(2\pi ikx)\,dx + \quad (4.5)$$
$$n(\ln K)\sum_{k=-\infty}^{\infty}\hat{B}_k\int_{\Omega}^1 \exp(2\pi ikx)\,dx.$$

and the derivative
$$\frac{\delta\ln L}{\delta\mu} = n\sum_{k=-\infty}^{\infty}\hat{B}_k\frac{\delta w_k(\mu)}{\delta\mu} \quad (4.6)$$
where
$$w_k(\mu) = \int_0^{\Omega}\ln f(x|\mu)\exp(2\pi ikx)\,dx.$$

5. The Gaussian Case (Direct Use of Trigonometric Methods)

As an example of the procedure described in section 4, consider the truncated Gaussian cumulative
$$(1-p)^{-1}\Phi\left(\frac{x-\mu}{\sigma(\mu)}\right)I_{[-\infty,\Omega)}(x) \quad (5.1)$$
where Φ represents the standard Gaussian cumulative, $(1-p)^{-1}\Phi\left(\frac{\Omega-\mu}{\sigma(\mu)}\right) = 1$ and σ is functionally related to μ. For convenience let $\beta = \Phi^{-1}(1-p)$ where Φ^{-1} is the inverse of function Φ.

Given that support is over the unit interval, for the case of the truncated Gaussian
$$w_k(\mu) = \int_0^{\Omega}\ln\left\{\frac{1}{1-p}\Phi\left(\frac{x-\mu}{\sigma(\mu)}\right)\right\}\exp(2\pi ikx)\,dx = \quad (5.2)$$
$$\int_0^{\Omega}\ln\left\{\frac{1}{(1-p)\sqrt{2\pi}\sigma(\mu)}\exp\left[-\tfrac{1}{2}\left(\frac{x-\mu}{\sigma(\mu)}\right)^2\right]\right\}\exp(2\pi ikx)\,dx.$$

Since
$$\frac{1}{1-p}\Phi\left(\frac{\Omega-\mu}{\sigma(\mu)}\right) = 1$$

and

$$\sigma(\mu) = (\Omega-\mu)/\Phi^{-1}(1-p) = \frac{\Omega-\mu}{\beta},$$

therefore

$$w_k(\mu) = \int_0^\Omega \ln\left\{\frac{\beta}{(1-p)\sqrt{2\pi}\sigma(\mu)} \right. \tag{5.3}$$

$$\left. \exp\left[-\tfrac{1}{2}\left(\frac{(x-\mu)\beta}{\Omega-\mu}\right)^2\right]\right\} \exp(2\pi ikx)\, dx.$$

Since eventually the derivative of w_k with respect to μ will be taken, the above integral is broken up as follows:

$$w_k(\mu) = \ln\left(\frac{\beta^2}{(1-p)\sqrt{2\pi}}\right)\int_0^\Omega \exp(2\pi ikx)\, dx - \tag{5.4}$$

$$\ln(\Omega-\mu)\int_0^\Omega \exp(2\pi idx)\, dx -$$

$$\frac{\beta^2}{2(\Omega-\mu)^2}\int_0^\Omega (x-\mu)^2 \exp(2\pi ikx)\, dx.$$

The last integral is evaluated as follows:

$$\int_0^\Omega (x-\mu)^2 \exp(2\pi ikx)\, dx = \tag{5.5}$$

$$\int_0^\Omega (x^2 - 2\mu x + \mu^2)\exp(2\pi ikx)\, dx =$$

$$\int_0^\Omega x^2 \exp(2\pi ikx)\, dx -$$

$$2\mu\int_0^\Omega x\exp(2\pi ikx)\, dx + \mu^2\int_0^\Omega \exp(2\pi ikx)\, dx$$

which by integration by parts equals

$$\exp(2\pi ikx)\left[\frac{x^2}{2\pi ik} - \frac{2x}{(2\pi ik)^2} + \frac{2}{(2\pi ik)^3}\right]_{x=0}^{x=\Omega} - \tag{5.6}$$

$$2\mu \exp(2\pi ikx)\left[\frac{x}{2\pi ik} - \frac{1}{(2\pi ik)^2}\right]_{x=0}^{\Omega} + \mu^2 \exp\left(2\pi \frac{ikx}{2\pi ik}\right)\Big|_0^\Omega =$$

$$\exp(2\pi ik\Omega)\left[\frac{\Omega^2}{2\pi ik} - \frac{2\Omega}{(2\pi ik)^2} + \frac{2}{(2\pi ik)^3}\right] - \frac{2}{(2\pi ik)^3} -$$

$$2\mu \exp(2\pi ik\Omega)\left[\frac{\Omega}{2\pi ik} - \frac{1}{(2\pi ik)^2}\right] -$$

$$\frac{2\mu}{(2\pi ik)^2} + \frac{\mu^2}{2\pi ik}(\exp(2\pi ik\Omega) - 1) = H(\mu).$$

$$w_k(\mu) = \ln\left(\frac{\beta^2}{(1-p)\sqrt{2\pi}}\right)\frac{\exp(2\pi ik\Omega)-1}{2\pi ik} - \tag{5.7}$$

$$\ln(\Omega-\mu)\frac{\exp(2\pi ik\Omega)-1}{2\pi ik} - \frac{\beta^2}{2(\Omega-\mu)^2}H(\mu).$$

Use of expression 5.6 to maximize the log likelihood requires taking the partial derivative of w with respect to μ.

$$\frac{\delta w_k(\mu)}{\delta\mu} = 0 + \frac{\exp(2\pi ik\Omega)-1}{2\pi ik(\Omega-\mu)} - \tag{5.8}$$

$$\frac{\beta^2}{(\Omega-\mu)^3}H(\mu) - \frac{\beta^2}{2(\Omega-\mu)^2}\frac{\delta H(\mu)}{\delta\mu} =$$

$$\frac{\exp(2\pi ik\Omega)-1}{2\pi ik(\Omega-\mu)} - \frac{\beta^2}{(\Omega-\mu)^3}H(\mu) -$$

$$\frac{\beta^2}{2(\Omega-\mu)^2}\left[-2\exp(2\pi ik\Omega)\left(\frac{\Omega}{2\pi ik} - \right.\right.$$

$$\left.\left.\frac{1}{(2\pi ik)^2}\right) - \frac{2}{(2\pi ik)^2} + \frac{2\mu}{2\pi ik}\right](\exp(2\pi ik\Omega)-1)\right]$$

From equation 5.7 we have

$$w_0(\mu) = \ln\left[\frac{\beta}{(1-p)2\pi}\right]\Omega - \Omega \ln(\Omega-\mu) - \tag{5.9}$$

$$\frac{\beta^2}{2(\Omega-\mu)^2}\int_0^\Omega (x-\mu)^2\, dx;$$

the integral equals

$$\left[\frac{x^3}{3} - x^2\mu - x\mu^2\right]_{x=0}^{x=\Omega} = \tag{5.10}$$

$$\frac{\Omega^3}{3} - \Omega^2\mu - \Omega\mu^2$$

$$\frac{\delta w_0(\mu)}{\delta\mu} = \frac{\Omega}{\Omega-\mu} - \frac{\beta^2}{(\Omega-\mu)^3}\left(\frac{\Omega}{3} - \Omega^2\mu - \Omega\mu^2\right) +$$

$$\frac{\beta^2}{2(\Omega-\mu)^2}(\Omega^2 + 2\mu\Omega) =$$

$$\frac{\Omega}{\Omega-\mu} + \frac{\Omega\beta^2}{(\Omega-\mu)^3}(2\mu^2 + \tfrac{1}{2}\mu\Omega + 1/6\Omega^2).$$

Substituting the above partial derivative into equation 5.6 and setting it equal to zero enables one to solve for the maximum likelihood estimate of μ.

6. Gaussian Case
(Indirect Use of Trigonometric Methods)

The particular model considered in the previous section can lead to a second location parameter estimator based on the sample trigonometric moments. Consider that the log likelihood computed from the i.i.d. sample $\{T_j\}$ $j=1,...,n$ where each variate distributed according to expression 5.1 equals

$$n \ln[(1-p)\beta/\sqrt{2\pi}] - n \ln(\Omega-\mu) - \tag{6.1}$$

$$\frac{\beta^2}{2}\sum_{j=1}^n [(T_j-\mu)/(\Omega-\mu)]^2$$

The derivative of equation 6.1 with respect to μ is

$$\frac{n}{\Omega-\mu} - \frac{\beta^2}{2}\sum_{j=1}^n 2\left(\frac{T_j-\mu}{\Omega-\mu}\right)\frac{(\Omega-\mu)(-1)-(T_j-\mu)(-1)}{(\Omega-\mu)^2} \tag{6.2}$$

which equals

$$\frac{n}{\Omega-\mu} - \frac{\beta^2}{(\Omega-\mu)^3}\sum_{j=1}^n (T_j-\mu)(T_j-\Omega) \tag{6.3}$$

$$\sum[(T_j-\Omega)+(\Omega-\mu)](T_j-\Omega)$$

$$\sum(T_j-\Omega)^2 + \sum(\Omega-\mu)(T_j-\Omega)$$

$$\sum(T_j-\Omega)^2 + \sum(\Omega T_j - \Omega^2 - \mu T_j + \mu\Omega)$$

$$\sum(T_j-\Omega)^2 + n\Omega\bar{T} - n\Omega^2 - n\mu\bar{T} + n\mu\Omega$$

$$\sum(T_j-\Omega)^2 + n(\Omega-\mu)\bar{T} - n\Omega(\Omega-\mu)$$

$$= \frac{n}{\Omega-\mu} - n(\bar{T}-\Omega)(\Omega-\mu) \tag{6.4}$$

$$\frac{\beta^2}{(\Omega-\mu)^3} - \frac{\beta^2}{(\Omega-\mu)^3}\sum_{j=1}^n (T_j-\Omega)^2$$

When one sets the above equation equal to zero, one finds that for $\Omega \neq \mu$

$$n = \frac{n\beta^2(\bar{T}-\Omega)}{\Omega-\mu} + \frac{\beta^2}{(\Omega-\mu)^2}\sum_{j=1}^n (T_j-\Omega)^2. \tag{6.5}$$

Upon dividing by n and multiplying by $(\Omega-\mu)^2$ one obtains the equation

$$(\Omega-\mu)^2 - \beta^2(\bar{T}-\Omega)(\Omega-\mu) - \frac{\beta^2}{n}\sum_{j=1}^{n}(T_j-\Omega)^2 = 0$$

which by quadratic formula implies that the maximum likelihood estimator equals

$$\hat{\mu} = \Omega - (\beta^2(\bar{T}-\Omega) \pm \sqrt{\beta^4(\bar{T}-\Omega)^2 + 4\frac{\beta^2}{n}\sum_{j=1}^{n}(T_j-\Omega)^2})/2. \quad (6.6)$$

Note that since $\Omega-\mu$ must be positive, the positive value of $\hat{\mu}$ is the true maximum likelihood estimator.

By using expression 3.5 of Tarter (1979) and expression 3b of Tarter and Marshall (1978) one can compute both $\bar{T}$ and $\sum_{j=1}^{n}(T_j-\Omega)^2$ in terms of the sample trigonometric moments. Thus, by using expressions 2.22, 2.23, and 2.24 of Tarter (1979), one can estimate the locus of location parameter value $\mu(x_0)$ by:

1) Using expression 3.13 to estimate degree of truncation $\tilde{p}(x_0)$.

2) Calculating $\tilde{\beta}(x_0) = \Phi^{-1}(1-\tilde{p}(x_0))$.

3) Obtaining the bivariate sample trigonometric moments as defined by expression 3.2.

4) Using expression 2.22, 2.23, and 2.24 of Tarter (1979) to estimate the Fourier coefficients of the truncated conditional $f(t|x_0)$ shown in Figure 1.

5) Calculating conditional $\bar{T}(x_0)$ and $s_\Omega(x_0)$ by substitution of the above Fourier coefficients into expression 3.5 of Tarter (1979) and expression 3b of Tarter and Marshall (1978).

6) Using expression 6.6 to estimate the location parameter in the presence of truncation.

7. Comparison of Truncation Procedures

Notice that data influences equation 6.6 through the $\bar{T} = \frac{1}{n}\sum_{j=1}^{n}T_j$ and $\frac{1}{n}\sum_{j=1}^{n}(T_j-\Omega)^2$ terms. As described in the last section, one can estimate these two statistics directly from the Fourier coefficients derived from the data by equation 3.3 as follows:

$$\bar{T} = \frac{1}{2} - \sum_{k\neq 0}(-1)^k B_k e^{\pi i k}/(2\pi i k) \quad (7.1)$$

$$\frac{1}{n}\sum_{j=1}^{n}(T_j-\Omega)^2 = \quad (7.2)$$

$$1/12 + \sum_{k\neq 0}(-1)^k \hat{B}_k e^{2\pi i k \Omega}/(2\pi^2 k^2).$$

Suppose the density is symmetric and the truncated component constitutes less than half of the density. In this case the location parameter μ can be estimated by using integration to estimate the fiftieth percentile.

We conducted a simulation experiment to compare the above procedures to the methods described in section 6 involving the following steps:

1) Generate a data sample and truncate it beyond a given Ω.

2) Estimate p by calculating the ratio of the number of points in the region of truncation divided by the total number of points.

3) Calculate $\bar{T}$ and $\frac{1}{n}\sum_{j=1}^{n}(T_j-\Omega)^2$ from the data directly and from the coefficients of the Fourier density estimate.

4) Integrate the adjusted Fourier density estimate to estimate the fiftieth percentile.

The results of the above procedure for the truncated normal case are shown in Table 1 and for various non-normal cases in Table 2. Both tables show averages over five trials (a couple of cases were tried uisng 10 and 25 trials but showed no difference in the nature of the results). Since the nonparametric method does not work for cases where the truncation throws out more than half the data points, truncation values in Table 1 stop at 0.5. Since the Weibull densities are skewed toward the left, the truncation was more extensive for the trials described in Table 2.

Notice that the nonparametric method does poorly when the truncation point Ω is close to the location parameter value μ estimated. This is due primarily to the Gibbs phenomenon caused by the discontinuity at the truncation point. Methods for surmounting this limitation are currently being investigated.

Acknowledgements

This research was supported by National Cancer Institute Grant 1 R01 CA28142-01-2 and Environmental Protection Agency Contract 68-01-6166. The authors would like to thank Ai-Chu Wu and Seemin Qayum for their technical assistance.

References

Edwards, R. (1967). *Fourier Series: A Modern Introduction*, New York: Holt, Rinehart, and Winston, Inc.

Efron, B. (1975). "The efficiency of Cox's likelihood function for censored data," *Journal of the American Statistical Association*, Vol. 72, No. 359, 557-565.

Jolley, L. (1961). *Summation of Series*, New York: Dover Publications, Inc.

Lee, E. (1980). *Statistical Methods for Survival Data Analysis*, Belmont, California: Lifetime Learning Publications.

Prentice, R. and Kalbfleisch, J. (1980). *The Statistical Analysis of Failure Time Data*, New York: John Wiley & Sons.

Tarter, M. and Marshall, J. (1978). "A new nonparametric procedure designed for simulation studies," *Communications in Statistics Part B*, Vol. 7, 283-293.

Tarter, M. (1979). "Trigonometric maximum likelihood estimation and application to the analysis of incomplete survival information," *Journal of the American Statistical Association*, Vol. 74, 132-139.

Table 1
Comparison of the Three Methods of Section Seven
in the Truncated Gaussian Case

True Mean = 2.0

Sample Size	Truncation[1]					
	0.9	0.8	0.7	0.6	0.55	0.5
50	2.15	2.00	2.11	1.79	1.65	1.49[2]
	2.24	2.25	2.22	1.86	1.69	1.50[3]
	2.14	2.14	2.19	2.76	2.76	2.76[4]
75	2.16	2.09	2.05	1.86	1.78	1.74
	2.15	2.19	2.18	1.92	1.83	1.73
	2.05	2.07	2.05	2.40	2.37	2.42
100	2.10	2.05	2.04	1.98	1.90	1.75
	2.05	2.12	2.14	2.04	1.93	1.77
	2.03	2.03	2.02	2.04	2.57	3.13
150	2.11	2.10	2.05	2.03	1.91	1.67
	2.05	2.20	2.16	2.07	1.93	1.68
	2.07	2.08	2.06	2.08	3.22	3.81
200	2.12	2.07	2.05	2.01	1.90	1.67
	2.06	2.17	2.16	2.06	1.92	1.68
	2.05	2.05	2.05	2.06	3.20	3.23
300	2.11	2.07	2.05	2.02	1.95	1.72
	1.97	2.14	2.15	2.07	1.96	1.73
	2.06	2.07	2.07	2.05	3.21	3.25

1) interval over which data is retained if the non-truncated data sample is transformed to the interval (0,1)
2) standard maximum likelihood estimate of the mean (see equation 6.6)
3) Fourier estimated mean derived from substituting equations 7.1 and 7.2 into equation 6.6
4) nonparametric estimate based on median

Table 2
Test of Robustness for Various Weibull Distributions[2]

Mean	Truncation[1]							
	0.9	0.8	0.7	0.6	0.5	0.4	0.3	0.2
$\gamma=4.0$	2.00	2.04	1.99	1.94	1.85	1.73	1.35	0.81[4]
1.85[3]	1.53	1.74	1.82	1.87	1.82	1.73	1.35	0.82[5]
$\gamma=3.0$	1.93	1.97	1.90	1.81	1.72	1.60	1.36	0.83
1.71	1.21	1.60	1.66	1.69	1.67	1.59	1.36	0.83
$\gamma=2.0$	1.77	1.77	1.74	1.68	1.57	1.42	1.25	0.90
1.51	0.80	1.14	1.36	1.48	1.47	1.39	1.24	0.90
$\gamma=1.5$	1.43	1.43	1.44	1.40	1.35	1.23	1.06	0.87
1.23	0.30	0.49	0.91	1.10	1.19	1.17	1.05	0.87
$\gamma=1.0$	1.19	1.16	1.08	1.05	1.01	0.93	0.83	0.68
0.91	-0.13	0.41	0.41	0.72	0.76	0.83	0.80	0.68
$\gamma=0.5$	0.70	0.66	0.62	0.58	0.54	0.48	0.46	0.39
0.47	0.57	0.32	0.18	0.20	0.25	0.35	0.42	0.38
Median								
$\gamma=4.0$	1.64	1.62	1.62	1.63	1.63	1.57	2.68	2.34
1.55[6]								
$\gamma=3.0$	1.41	1.41	1.41	1.43	1.42	1.42	1.93	2.18
1.35								
$\gamma=2.0$	1.14	1.13	1.14	1.14	1.15	1.15	1.14	1.99
1.10								
$\gamma=1.5$	0.89	0.89	0.89	0.89	0.89	0.89	0.92	1.08
0.85								
$\gamma=1.0$	0.63	0.63	0.63	0.63	0.63	0.63	0.63	0.63
0.58								
$\gamma=0.5$	0.32	0.32	0.31	0.31	0.32	0.32	0.31	0.32
0.30								

1) The degree of truncation refers to the interval over which the data is retained if the non-truncated data sample were transformed to the interval (0,1).
2) $f(x) = \lambda\gamma(\lambda x)^{\gamma-1} e^{-(\lambda x)^\gamma}$ where λ is the scale parameter (set to 1 for all trials) and γ is the shape parameter
3) the mean of the non-truncated sample averaged over the number of trials
4) This row shows the estimated mean from the standard maximum likelihood method.
5) This row shows the estimated mean from substituting Fourier estimates given in equations 7.1 and 7.2 into equation 6.6.
6) median of non-truncated sample averaged over the number of trials

NASA'S FUNDAMENTAL RESEARCH PROGRAM IN MPRIA

R. P. HEYDORN
JOHNSON SPACE CENTER
HOUSTON, TEXAS

NASA is currently engaged in a Fundamental Research program to understand the utility of remote sensing for monitoring the Earth resources. In the planning phases of this program four research categories were proposed:

1. Mathematical Pattern Recognition and Image Analysis (MPRIA)
2. Scene Radiation and Atmospheric Effects Characterization (SRAEC)
3. Electromagnetic Measurements and Data Handling (EMDH)
4. Information Utilization and Evaluation (IUE).

From a series of planning workshops, research issues were identified in each category and in the summer of 1982 studies in the first two categories began. This paper discusses the MPRIA research issues and summarizes some of its current studies.

1.0 INTRODUCTION

With the advent of the Earth Resources Technology Satellite (ERTS), which later was renamed LANDSAT and became the first in a series of LANDSAT satellites, NASA began to extend what had been learned from aircraft mounted multispectral sensors to this new spacecraft setting. A number of studies were begun to apply this remotely sensed data to mapping and inventory problems. One such study was the LACIE (Large Area Crop Inventory Experiment). The LACIE was an attempt to explore the feasibility of using this remotely sensed satellite data to estimate wheat production in foreign countries (c.f. MacDonald and Hall (1)). By exercising the satellite data system over a large area and over the entire growing season of wheat, a lot was learned about the variable behavior of these data and some of the causes of the variability. In fact over the four year period in which the experiment was run, several major adjustments were made to the experiment design in an attempt to incorporate these "lessons learned in real time."

The LACIE pointed out many strengths and weakness of a remote sensing approach to a problem that may have not been placed in proper prospective from an analysis of a collection of unrelated small experiments. Indeed the LACIE demonstrated the need for more in depth research into the problems of applying the principals of remote sensing to the Earth Resource Sciences. And more importantly, it contributed significantly to our understanding of where to begin.

In 1980 the Johnson Space Center started a series of planning workshops to determine the essential research issues that needed to be considered. In all there were four sets of workshops grouped under the topics

a) Mathematical Pattern Recognition and Image Analysis (MPRIA)

b) Scene Radiation and Atmospheric Effects Characterization (SRAEC)

c) Electromagnetic Measurements and Data Handling (EMDH)

d) Information Utilization and Evaluation (IUE)

The need for this grouping was partially a consession to the scientific disciplines that would be attracted to the research problems, even though it was recognized that there were strong relationships between these groups. Indeed the measurements that constitute an image are a result of the behavior of electromagnetic absorption and scattering properties of the atmosphere and of the earth's surface materials; and, the properties of the sensor and the data processing methods used to derive these measurements. Thus an intelligent development of patterns recognition methods should depend upon SRAEC and EMDH issues. The last group, IUE, addresses problems related to the needs of potential "users" of the information from a remote sensing system and the implications of this information in forecasting and planning applications.

The findings from the MPRIA and SRAEC workshops were translated into an Announcement of Opportunity which NASA issued in the summer of 1981. Approximately 20 universities and research organizations were selected to participate with NASA scientists in this program. The findings from the EMDH and IUE workshops are still under consideration.

The purpose of this paper is to present an overview of the NASA Fundamental Research Program in

Mathematical Pattern Recognition and Image Analysis. We will discuss program content and summarize some of the ideas being pursued in the ongoing studies.

2.0 THE REMOTE SENSING DATA

Before discussing the research aspects of the program it is useful to reflect on some of the elements of a remote sensing system to get some feeling for the kinds of data that would be available for a given problem.

First of all there is the sensor system. There are over 100 sensor systems that NASA is currently using, or has been associated within the past, or has in the planning stages. A good reference is Tanner (2). Table 1 lists a few of the more popular systems. The Landsat D and AVHRR systems are currently being flown. The SEASAT-SAR is no longer flying but its data is currently being analyzed. The shuttle systems have flown and presumably will fly again in the near future. As seen from table 1, the visible, near infrared, infrared, and radar ranges of the electromagnetic spectrum are represented within this collection of sensors. These bands were selected with certain possible applications in minds some of which are listed in the last column of the table.

TABLE 1.- EXAMPLE REMOTE SENSING SYSTEMS (**)

SENSOR	SATELLITE/ VEHICLE	CHANNELS: WAVE LENGTH/FREQ. RANGES (*)	GROUND RESOLUTION	SOME POTENTIAL MAPPING USES
Multispectral Scanner (MSS)	Landsat 1-3 Landsat D	(VIS) .5 − .6 µm (VIS) .6 − .7 µm (NIR) .7 − .8 µm (NIR) .8 − 1.1 µm	80 meters	Vegetation, sea-ice fields air/ water pollution, snow cover geology
Thematic Mapper (TM)	Landsat D	(VIS) .45 − .52 µm (VIS) .52 − .60 µm (VIS) .60 − .69 µm (NIR) .76 − .90 µm (NIR) 1.55 − 1.75 µm (NIR) 2.08 − 2.35 µm (IR) 10.4 − 12.5 µm	30 meters VIS & NIR 120 meters IR	Vegetation, geology, land use, vegetation moisture stress, cloud detection, snow cover
Advanced very high resolution radiometer (AVHRR)	Tiros N NOAA-6 NOAA-8	(VIS) .55 − .9 µm (NIR) .725 − 1.3 µm (IR) 3.55 − 3.93 µm (IR) 10.5 − 11.5 µm	1.1 Kilometers	Sea surface temperature, ice, snow, clouds
Synthetic Aperature Radar (SAR)	Seasat	Radar 23.5 cm (1.275 GHz L BAND)	25 meters	Sea waves, ice, geology, hydrology, glaceology
Shuttle Imaging Radar (SIR)	Shuttle	Radar 23.5 cm (1.275 GHz L BAND) 3cm (9.6 GHz-X BAND)	10-30 meters	Linements, surface texture soil moisture, vegetation, snow, rocks, drainage networks

*VIS~VISIBLE, NIR~NEAR INFRARED, IR~INFRARED.
**Data obtained from Tanner, Shelby HANDBOOK OF SENSOR TECHNICAL CHARACTERISTICS (NASA REF. PUB. 1087, 1982).

Besides the data from these space and airborne sensors other forms of data, often called auxiliary or ancillary data, are available. Example of these data sources include geological survey maps, historical crop survey and planting records, real time crop surveys, the world meterological network, and climateological records.

Often the remotely sensed data, and often the auxiliary or ancillary data, contain certain distortions in the senses that measurements occur outside of our range of expectation. For example, since these sensors are often in spacecraft which can be approximately 500-600 miles above the earth, small errors in controlling the attitude of the sensor platform can result in large displacement errors on the ground. Moreover since the measurements are passing through the entire atmosphere which changes daily, varying amounts of distortions can be in a time record of the data. The surface materials being measured can account for large sources of variance. Plants have a way of growing in very irregular ways perhaps due to environmental stresses or human intervention.

Much of the effort in the Fundamental Research program is aimed at understanding these and other error sources and, in some cases, developing corrective procedures. However there is always uncertainty which is often lumped with random error, and so our methods are often motivated by statistical concepts.

3.0 RESEARCH PHILOSOPHY

A common tendency when first exposed to a pattern recognition problem is to fit it into a familiar problem setting. Thus for example, if one is given a collection of landsat measurements of known agricultural crops in a given area, and is asked to develop a classification rule for separating these crops in the given and in similar areas, one might explore the use of a collection of linear discriminant rules. Indeed if the individual in charge of the analysis were a statistician, very likely this might be the first attempt. While the discriminant rules might apply well in the given area where sample data with known labeling is available, the use of these rules in a different area without local calibration would probably produce far poorer results. And since the cost of acquiring ground samples to calibrate the discriminants is generally expensive, the statistician may doubt the presumed cost savings of the satellite remote sensing approach. (We assume the statistician was kept ignorant of the cost of the satellite, its sensors, the ground processing system, etc.) Actually the statistician started at the "wrong end of the problem". While the satellite measurements may contain information on the "growth variables" which can be used to discriminate crop types they also contain information on "nuisance variables" such as soil color (which can be influenced by local moisture conditions) or planting times. Changes in the soil background color or planting times can produce significant changes in the satellite measurements. Rather than just attacking the discriminant problem a more rewarding pursuit might to, in addition, consider "data models" that separate the growth phenomenon from the nuisance phenomenon.

One point of this example is to bring out the fact that there should be a strong emphasis on developing models which explain the desirable elements of the data. And to the extent possible to base these models on the physical principals which generated the remotely sensed data. Both MPRIA and SRAEC studies emphasize this point. The fond hope in SRAEC is to develop models which can be "inverted" to produce variables which characterize a given material class, from the satellite point radiance measurements. But since real data has a way of providing us with answers only to within certain levels of uncertainty, other relationships that can affect the data should be studied. Since we obtain our measurements in the form of an image, we should in the MPRIA research consider models which explain the spatial information related to the arrangement of the radiance measurements in the scene. To be complete, models which explain distortions in the data due to the sensor and data transmission methods should also be included. These are included in both the MPRIA and EMDH elements of the Fundamental Research Program.

If one were to strictly follow this modeling philosophy, one might conclude that x number of years should be spent on developing point radiance models, y number of years developing sensor designs and then sensor error models, and z number of years developing the pattern recognition approach, and, do it in that order. Clearly this is an unacceptable approach. (Indeed some of us might be out of work for a period of time until the next fellow finished his job.) It fails to appreciate the interaction between the three areas. As we all know research has a way of generating new problems. While these problems may have solutions by applying one line of reasoning, these may be clumsy when compared to possible solutions from other lines of reasoning. While one discipline can often uncover certain problems, it takes the ideas from another discipline to solve them.

Research interactions can be stimulated, to some extent by a technical goal seeking aspect of the research. The goal in the previous example was to develop an approach to discriminate crops. However technical goal seeking need not be stressed at the same level through out the duration of the research. In the beginning phases creativity is important and one would like to explore ideas suggested by related research in other programs and to introduce new ideas. Thus in the early stages the technical goals may not be as precise as they could be in the later stages. Once a technical goal is defined, we would all like to believe that, relative to a given success criterion, there is an optimal approach. But there is a risk. There is a tendency at this point makes certain assumptions that admit mathematically tractable solutions. By having a multidisciplinary program the "research interaction" can, and should, question the realism of these assumptions. A meteorologist A may not completely understand the approach offered by Statistician B but quite often A can make some pointed comments about B's assumptions. Good realistic solutions, if not optimal solutions, should be stressed and hopefully the Fundamental Research Program's structure will prove to be rich enough to make this possible.

There is yet another point. As seen in the previous section remote sensing problems often have a variety of data sources to consider. There may be more than one satellite sensor or more likely several sources of auxiliary information (weather measurements, soils data, maps, historical records etc.). Often a desire is to combine these data to obtain answers that, by some standard, are more precise that can be obtained by considering only one source of data. Statistician have, of course, been familiar with this concept and loosely speaking have explored various notions of dependence between the data sources to develop estimation methods. However, other diciplines are also concerned with this problem and in fact the computer scientists have offered, in their artificial intelligence work on expert systems and related topics, notions of "convergence of evidence". Since the MPRIA program is multidisciplinary, at least statisticians and computer scientists are involved, questions like these can be explored using a variety of approaches, and if properly done, some of the benefits and weakness of these approaches can be compared.

4.0 PROGRAM CONTENT

The research issues in the MPRIA program (c.f. reference(3)) are grouped into five categories called Preprocessing, Digital Image Representation, Object Scene Inference, Computational Structures, and Continuing Studies. The first three categories form a traditional grouping of a typical pattern recognition approach. First one prepares the data perhaps by removing known distortions. This is the preprocessing step. Next the image is described in terms of basic elements and their relationships which are representative of the desirable information in the scene but which also are invariant to the undesirable, or nuisance effects. In traditional pattern recognition terminology this is the feature extraction step. But since we have attempted to capture some of the notions embedded in the ideas of artificial intellelligence in this grouping (as well as in the next grouping) we chose to refer to this as digital image representation. In the last step or inference step, one is interested in answering some questions such as "What is in the scene?" Where is it? How much of a given material is in the scene?" or "Have certain elements of the scene changed." Along with the answers to these questions one is often interested in knowing the accuracy of any determined values.

The main research issues in the computational structure category are related to parallel processing algorithms and image data structures. These issues are not currently being addressed in the program since it was felt that the more basic issues related to the first three categories need to be understood first. Continuing studies were intended to include special problems which at this point in our planning are not well defined.

4.1 PREPROCESSING

The major emphasis in preprocessing has been placed on issues related to the registration and rectification of image data. Here registration refers to the process by which two images are brought into alignment. In the usual setting the idea is to map one image onto the other so that the same area on the ground is mapped to the same image coordinates in the two images. Since in many remote sensing applications the satellite will cover the same area on the ground many times during the year, by registering each image to some reference image one can get a "multitemporal picture" of some event on the ground. Moreover, as seen in Table 1, it is possible to view the same scene from more than one satellite sensor system, and so images from multiple sensors may need to be registered.

Rectification is the process by which an image is brought into alignment with a given map of the same area. When one is interested in relating "ground truth data" with satellite remotely sensed data then rectification is required. In theory one could rectify a sequence of images and thereby register them, but since errors are often associated with the rectification process additional registration processing may be needed.

Some of the research issues related to registration and rectification of images that have been identified in the MPRIA program are as follows:

Since the sensor is mounted on a satellite or aircraft platform that is usually attitude controlled, information related to the pointing of the sensor is generally available. By knowing these pointing errors at any given instant of time one could in theory correct the image data so that it would be rectified. Hence, it is desirable to study platform error models for purposes of estimating the extent to which such correction is possible and to suggest correction methods.

While this type of correction should be the first step, it may not reduce the residual errors to acceptable levels. Additional corrective processing can be done by first identifying certain "control points" in the image that relate to entities on the ground which are, in some sense, permanent; e.g., not related to rapid changes. More generally, the control points may refer to contours or "shapes" in the image. A cloud shadow would be a poor control point. It may be possible to consider every point in some subimage as a control point with the understanding that even though some points are poor, in the above sense, enough good control points will exist so that matching these subimages would lead to a suitable corrective procedure. It is common to refer to these subimages as "control chips".

In the registration process one image is then "warped" to fit over the other image at the control points and, to within the smallest possible error, at all intermediate points. In rectification the idea is the same except that one image is now warped to fit over a given map.

There are a number of issues related to the control point problem. If an automatic registration or rectification process is desired, then the identification and location of control points can be a pattern recognition problem. "What are the easily recognizable features that make good control points?" "Moreover, how many control points are needed and how should they be spaced throughout the image?" What transformation of the image would bring out properties of good control points?" For example generalized Hough Transforms can be used to find corresponding contours in the two images that have experienced given distortions such as translation or rotation.

Warping an image to match it to another image or a map means that the image pixels must be given new values which are generally determined through some type of interpolation of least square fitting process. This is often referred to as image resampling. Undoubtly this resampling can affect the quality of the inferences that are to be made from the image. There may be trade offs between geometric fidelity and radiometric fidelity that would favor a given inference method.

Once control points are selected, there must be some process by which a control point in one image is matched to the corresponding control point in the other image or on the ground. If control chips are used, then a commonly employed matching procedure is correlation. By determining the displacement required to bring one chip into correlation with its corresponding chip, the corresponding image coordinates that must be paired are determined. "How should one estimate the exact point of correlation, that is, the global extremum point of the correlation surface?" "If control points are contours or certain shapes in the scene, what method of correspondence should be used?"

Often registration and even rectification is done under the assumption that the terrain relief is negligible. When two images are taken of the same area but at different "look angles" terrain relief can introduce substantial registration errors if corrections are not made. This is especially true if the sensor is radar. It may therefore be desirable to develop a digital elevation model (DEM) from satellite data or at least use an existing DEM and incorporate it into the registration or rectification apporach. Optimal methods related to this modeling and use of digital terrain data are of interest.

4.2 DIGITAL IMAGE REPRESENTATION

Imagine a sequence of images over a 30 square nautical mile area obtained at 18 day intervals for a period of four months. Imagine further that 600 of these 30 square nautical mile sample segment areas are needed in an inventory problem. If the sensor is MSS (see Table 1) and the images related to the same area are registered, then each pixel in the image can be represented by a 28 dimensional vector (assuming 7 passes of the satellite) and since there are 22932 pixels in a single image this means that it would take a 28 x 22932 dimensional array to represent one sample area. While such representations are clearly within the capability of the modern computer, when considered from the inventory estimation point of view, they may be statistically intractable or at least clumsy. What is needed in this case, is a representation that leads to a simple and effective inventory estimator. In general the best representation is the one that simplifies the desired inference problem.

The inference problem in this case requires that an inventory estimate be obtained from a sample of 600 segments. A good representation of each segment is one that would have the features, a) it would lead to an accurate determination of the material amounts to be inventoried (acres of wheat, acres of forest, or miles of roads, etc) using an estimator that can be deduced from data from a small number of segments and b) it would lead to a consistent estimation rule that does not have to be adapted on a segment-by-segment basis. It is highly unlikely that a representation where each segment is treated as a 28 x 22932 array would satisfy either a) or b). The problem is that the data from a segment has been treated as one "lump". While one can certainly reconstruct the image from the segment array, and hence no information is lost, it would be extremely difficult to deduce any consistant rule without observing many times more than 600 segments. We are therefore interested in representations that break this "lump" into smaller parts with some understanding of the relationships between the parts that can lead to some useful description of the segment. And if possible, going back to our remarks in the Research Philosophy Section, to understand the physical processes that generated the parts with the belief that thus physical understanding will guide us to the right representations.

An image can be throughtout of in terms of its spectral and spatial properties. The spectral properties deal with the measurements associated with a given pixel. If we dealing with a time sequence of registered imagery then a given pixel tells us something of the dynamic behavior of some material being observed. The spatial properties deal with the interrelationships of the pixel measurements due to the position of the pixels in the image. For example knowing the measurements associated with a given pixel may lead to predictable implications about the surrounding pixels.

Since remote sensing imagery often has vector rather than just scalor data associated with a pixel, there is considerable information on a pixel and so there is an emphasis in the MPRIA research to develop "point models" of pixel behavior. Much of our thinking here is along statistical lines. "How should one go about developing a probability model that would describe the statistical behavior of classes of material in the scene?" "What dimension reducing transformations would lead to a small number of derived variables that preserves the "information" related to material classes?" This latter question could be addressed by considering either a statistical approach, based possibly on a notion of spectral class separability, or upon a physical model of the dynamic behavior of the likely materials being observed. Interactions between the SRAEC and MPRIA research should have a bearing on the development of such physical models.

In considering the spatial properties of an image we are examining both statistical and some of the more recent computer science notions. With the belief that a scene can, in some workable sense, be described in terms of textures and contours, we have considered problems related to these properties. We are interested in developing mathematical descriptions of texture and rules for segmenting a scene. Texture may convey information regarding material classes and this can be useful in developing inference rules that require this knowledge, but it may be useful along with contour information in segmenting a scene into "primitive elements". We are interested in syntatic models for describing a scene in terms of these primitives. We are also interested in texture and segmentation models based on spatial statistics methods (spatial analogs of statistical time series methods).

Recognizing that many scene properties have their origin in human interpretation concepts, we are also interested in some of the so called "evidential reasoning methods" that are being developed in the expert system and artificial intelligence literature. These approaches admit the possibility of using ancillary information in conjunction with the imagery data to infer scene structure.

There is yet another property of the image which, in part, relates to our insistence on imposing an object class structure on image elements. Since our data from the sensor is made up of resolution elements (pixels), many image pixels will contain measurements from more than one object class. In natural scenes, such as a forested area, almost every pixel could contain measurements from different species of trees, grasses, and shrubs, and from different soil types. It would be desirable to be able to deduce some object class information from these composite measurements. Hence there is an interest in developing "mixed pixel models" from which it's possible to predict given class proportions in a pixel or at least over some collection of pixels.

4.3 OBJECT SCENE INFERENCE

In a sense the main questions of interest in a remote sensing application are addressed in this section. What is on the ground? Where is it? How much is there? Has it changed over some interval of time?

The first two questions can often be addressed in terms of a classification model. Given a representation of the data in an image the problem is to sort the data into classes. These classes may be real classes in the sense that they correspond to materials on the ground. The class of all trees or water or two story houses are examples of such classes. On the other hand the classes may be spectral in nature. For example the set of all measurements which fall in a given spectral region or the population of all measurements that have a given normal distribution.

Many of the classifications approaches have been derived from mathematical concepts related to modern algebras (mathematical grammar), graph theory, statistics, and combinations of these. Classification decisions that make extensive use of the spatial properties of an image may be formulated in terms of grammar parsing rules. On the other hand classification decisions that are based on just the pixel values can often be formulated in terms of a discriminant analysis model. In general we are interested in formulations of the classification problem that can accurately sort the image data into real classes with a minimal need to adapt the classifier in a changing data environment. These classifications can at some intermediate step be spectral classes, but ultimately there should a "strong correspondence" to real classes.

The third question, "How much is there?", is considered in this program mainly from a statistical point of view. We are interested in issues related to mixture model approaches for estimating the proportion of material in an area. A mixture model is simply a model that describes the probability density of the image measurements in terms of convex combinations of class conditional densities. We are also interested in regression models or parametric Empirical Bayes methods that combine ground survey data with the remotely sensed data to arrive at an inventory estimate. The idea here is generally to use the remotely sensed data to improve the precision of an estimator based on the ground survey above.

The last question, "Has it changed over some interval of time?", could be addressed in terms of "what" or "how much" depending upon the desired emphasis. For example, "has a given area changed from a forested area to an urban area?" In this case a sequential (or compound) classification approach may apply where one is interested in the point in time where pixels go from the class "forests" to the class "urban". On the other hand one may be interested in knowing how much forest area has gone to urban area. If the change is large and the time the change occurred is known, then a time series approach using intervention analysis could possibly be used to estimate the magnitude of the change. In general we are interested in formulations of models that estimate change directly that are superior to methods that estimate a sequence of absolute quantities and as a byproduct estimate the changes.

In all of the above ideas it is desirable to estimate the accuracy (perhaps the bias and variance where appropriate) of the inference. In some cases there is a small amount of "ground truth" observations which can be used. In some cases it is desirable to use this ground truth data to train a classifier, or when it is from a random sample, to estimate a proportion of material and to use the same data to estimate the classification accuracy or the accuracy of the proportion estimates. In these cases the statistical notions related to jackknifing, cross validation, or bootstrapping may apply. In cases where no ground truth is available, we are also interested in approaches to error estimation.

5.0 CURRENT INVESTIGATIONS

At the present time investigations in each of the three areas Preprocessing, Digital Image Representation, and Object Scene Inference are underway. Table 2 indicates the participants, listed by institution, and the titles of the investigations. Without presenting a detailed discussion of approaches being followed, we will summarize some of the ideas being pursued.

5.1 PREPROCESSING

Recognizing that sensor pointing errors are a major source of image distortion, relative to the ground, sensor/platform error models are being studied in an effort to understand probable magnitudes of errors and possible correction methods that make use of sensor pointing geometry and ground control points. Yet other studies are not explicitly considering the properties of the sensor, but rather they are considering methods which depend mainly on

TABLE 2.- ON GOING MPRIA INVESTIGATIONS

RESEARCH INSTITUTE	TITLE OF THE INVESTIGATION
Purdue	Aspects of Simulation for Rectification Studies
University of Maryland	Image Matching Using Hough Transforms
LNK Corporation	Subpixel Registration Accuracy and Modeling
NASA-National Space Technology Labs	Scene-To-Map Rectification
SRI International	Approaches to Image Registration and Segmentation
University of Kansas	Textural Edge Detection and Sensitivity Analysis
Hunter College and University of California at Santa Barbara	Spatial Scene Modeling
Jet Propulsion Laboratory	Reduction and Utilization of Speckle Noise in SAR Imagery Random Field Models for Use in Scene Segmentation
Texas A & M	Fun. Stat and Statistical Image Representations-An Adaptive Technique for Fitting Landsat Data
University of Texas at Austin	A Minimix Approach to Spatial Estimation Using Affinity Matrices
University of Houston	Covariance Hypotheses for Landsat Data
Rice University	Some 3-D Density Estimates
NASA-Johnson Space Center	Estimating Proportions of Materials Using Mixture Models
	Using Ground Gathered Survey and Remotely Sensed Data to Estimate Crop Acreages.

image properties and some knowledge of ground features. In some studies physical models, such as camera models, or reflectance models, are being used in effort to predict possible appearances of features from one image to the next.

Some commonly used methods in registration depend upon the pointwise cross correlation of images or chips and since the resulting correlation surfaces can be complex, studies are underway to understand the limits of such methods for registering chips to subpixel accuracy. Extending the notions of matching by these pointwise methods to matching lines, contours, and general slopes found in the image, other studies are again considering probable accuracy limits and some new approaches to the notion of matching. One matching concept is based on the notion of a generalized Hough Transform. The original Hough Transform was simply a transformation taking a set of points $\{(x,y)\}$ to a point pair (ρ, θ) which are the polar coordinates that define a given straight line. By counting, the number of (x,y) points that map to a given (ρ, θ) and by looking for peaks in this (ρ, θ)-function it is possible to pick out sets of collinear (x,y) points in the image. The generalized Hough Transform generalizes this idea to an arbitrary contour. By taking a given contour in one image it is possible to match it with the same contours or contours that have undergone admissible distortions in other images.

5.2 DIGITAL IMAGE REPRESENTATIONS

Both the spatial and pointwise (or spectral) properties of an image are being studied. In the spatial investigations the main themes are centered around texture analysis and image segmentation. In the pointwise investigations methods to select appropriate probability models for observations from a given material class are being considered.

Many of the texture studies are based on the notion of a random field model which is essentially a time series model with time being replaced by geographic or image coordinates. By fitting certain random field models to image values it may be possible to relate texture properties to model parameter values and to find "texture groups" in the image by picking out image boundaries related to changes in the model. In some studies the random process considers just the spatial properties of the point measurements in the image but in other studies the notion of a primitive element is used in place of the point measurements. A primitive element is some shape or contour in the image that can be found repeated in some regular pattern. The effects of sensor resolution on texture determination is also being considered by examining the covariance structure of the random process.

Some of texture investigations are being related to physical concepts that might explain certain arrangements of surface material. In this regard models that have been proposed in geology, meteorology, and geophysics are being studied.

Besides studying the segmentation problem in terms of the texture properties of the scene, methods are also being considered based on spline theory. Here the idea is to fit a surface to the spectral values given the number of material classes in the scene and some estimate of their spectral means.

Methods to select a probability model for the image values have considered both graphical and theoretical approaches. The idea is to develop analysis methods for deciding upon a good functional form for the probability density of pixel values from a given class and since these pixel values are generally vector quantities, multivariate forms are of interest. As an exploratory tool graphical methods are being studied which use color along with nonparametric density estimation methods to display the data in several dimensions. Nonparametric density estimation methods based on kernel functions, quantile functions, and quantile density functions are being pursued. In addition quantile functions are being considered to test for possible properties of the class conditional distributions such as "they can all be described by a location and, or, scale parameter."

5.3 OBJECT SCENE INFERENCE

While some of the studies related to the representation of image data also consider methods for classifying the data, most of emphasis at this time is aimed at developing approaches for estimating the amount of a material in the image or estimating some quantity that the pixel values are attempting to measure.

One approach is based on the so called mixture model, (as defined in section 4.3). These models have been demonstrated to be useful in several NASA applied studies in crop acreage estimation. In these studies methods for estimating model parameters over a broad class of possible component densities are being considered. Models are also being developed which use some of the spatial properties of the data.

Other estimation approaches are using concepts related to parametric Empirical Bayes and classical regression methods. Here the emphasis is to improve upon the estimate of some quantity by combining it with an estimate made with the remotely sensed data. In some of these studies the remotely sensed data is a classification of pixel values into given material classes. In this case the design of the classifier can influence the parametric form of the estimator. Given that some form is desired (e.g., such as a linear regression estimator) the idea of these studies is to develop classifier designs which will lead to that form.

6.0 CONCLUDING REMARKS

The MPRIA research is concerned with the use of remotely sensed imagery data for measuring quantities or deducing properties of the earth's surface materials. In some cases the earth surface phenomenon must be studied over a period of time and requires measurements made over several regions in the electromagnetic spectrum. This measurement orientation perhaps sets this program apart from other pattern recognition research that makes use of satellite or aircraft borne sensors for target recognition or mapping purposes. However, these other programs are addressing some similar image understanding issues and therefore the MPRIA research should be considered as part of this larger research community. Hopefully we will not reinvent wheels, but help to keep the wheels rolling down productive research paths of mutual interest.

References

1. MacDonald, R. B., Hall, F. G., Global Crop Forecasting Science, vol. 208, (1980), 670-679.

2. "Project Plan for Fundamental Research in Mathematical Pattern Recognition and Image Analysis", Earth Resources Research Division, Space and Life Sciences Directorate, Johnson Space Center, 1982.

3. Tanner, Shelby, Handbook of Sensor Technical Characteristics (NASA ref. pub. 1087, 1982).

A MATHEMATICAL EXPERIMENT

Ulf Grenander

Division of Applied Mathematics
Brown University
Providence, Rhode Island 02912

1. Statistical inference typically starts from a mathematical model, completely specified except for some parameters. With the aid of the observations that are available one tries to estimate the data and express ones confidence in the result, or to test hypotheses, or to find out how credible is the model, and so on. This is the situation usually described in statistical textbooks.

The perspective changes drastically when we have data but no model to start from, no subject matter theory helps us to formulate a model. C.S. Peirce once coined the term abduction, a term still used especially by linguists, for the thought processes by which the scientist tries to suggest hypotheses, perhaps construct models, and, in general, to structure his observations in a systematic and meaningful manner. This is, of course, related to the data analytic tendencies in contemporary statistics.

I will sketch some work done jointly by myself and Michael Katz, an embryologist. As you will see this is abduction. Our problem was morphogenesis in embryology: what are conceivable information structures that could explain, as simply and naturally as possible, the generation and transformation of shape in living organisms.

As you may know, many scientists have tried to formulate mathematical models of morphogensis. Some of the many attempts are clever and imaginative, but the success obtained so far has been less than overwhelming.

There is no shortage of data, actually we suffered from an embarrass de richesse of empirical information, but the data were only indirect, seldom numerical but pictorial, and often of doubtful validity. Starting with the pioneering experiments of Roux, Spemann and many others at the turn of the century, embryologists have collected a wealth of information via surgical experiments in the laboratory, some of which are both ingenious, dramatic, and surprising. Of course pattern generation during normal growth has been observed and classified systematically for centuries.

In addition to this macroscopic evidence there is the immense body of microscopic knowledge concerning the structure and functioning of the cell that is available to the cytologists of today. In spite of all this information a theory of morphogenesis still appears an elusive goal.

What Katz and I set out to do was not to create such a theory. We did not search for quantitative models describing the creation of biological patterns. Instead we have tried, starting from some simple but restrictive assumptions, to formulate an abstraction, a logical mechanism that would generate forms in qualitative (but not necessarily quantitative) agreement with the ones that are seen in the real biological world and in the laboratory.

The attempt at abduction was organized as a long series of thought experiments intended to resemble what happens in actual experiments; I will be more specific later. It turned out very early, that the thought experiments were more difficult to carry out with pencil and paper than we had anticipated. This is due to the complicated interactions that occur: we are dealing with complex systems (the abstraction) whose functioning is far from easy to understand intuitively.

We were therefore forced to do the thought experiments on the computer. The sort of computing involved is quite different from that in large scale statistical processing. Instead it should be thought of as using the computer as a logic machine to deduce consequences from hypotheses.

Used in this way the computer was not just a helpful tool for us in carrying out a long series of involved thought experiments, it was indispensable!

In what follows I shall avoid the probabilistic aspects. Randomness is certainly present in these biological phenomena, and must be included at some stage. The analytical problems that then arise are exciting, requiring a sophisticated analysis of interest for its own sake, but that would draw the attention away from the abduction process I want to tell you about.

2. We start from the five following restrictions A-E when designing the thought experiments. For explanations of the pattern theoretic terminology, see Grenander's Lectures on Pattern Theory, three volumes, Springer-Verlag (1976-1981).

A. Relative time: the growth mechanism is allowed to depend upon relative but not on absolute time. For convenience time was considered discrete, $t=0,1,2,...$.

B. <u>Relative space</u>: the growth mechanism is allowed to depend upon relative but not on absolute space. Space was usually chosen as the plane or as a square lattice in the plane.

C. <u>Discrete units</u>: shape is built from discrete units, the generators, and is not considered as a continuum. The generators are abstraction of cells or, often, cell aggregates.

D. <u>Local informations</u>: the decision made by a generator at a certain time is based only on the information content of the generator itself and that of its neighbors signalled by their bond values. 'Neighbor' means with respect to the discrete topology induced by some connector.

E. <u>Decision space</u>: a generator can decide to do only one of the following 1) remain the same, 2) change its internal state, 3) move, 4) die, or 5) split in two.

These assumptions are more restrictive than may be thought at first glance and they make it harder for us to design the logic so that it exhibits the macroscopic behavior we want. They will force us to make the experiments more life-like although we would certainly not claim any high degree of realism.

Here is how we operate. We select a <u>growth task</u>, either an artifical one involving <u>artificial</u> shapes (disks, rings, spirals, ovals, etc.) or <u>real</u> ones representing a particular species among the organisms that are formed by the experimenters (worms, salamanders, cockroaches, etc.).

The growth task can take several forms:

I. Design a pattern dynamics that grows a given shape.

II. Same as I, but growth should automatically stop when the desired (mature) shape has been reached.

III. Same as II but build in regenerative power so that the abstract organism can adopt itself to specific surgical modifications (amputate a limb, transplant one part of the organism somewhere else, etc.).

IV. Same as III but make the mechanism insensitive to small arbitrary disturbances.

Of course there are no unique solutions here. Of two solutions we will prefer the simpler one, where simplicity can mean several things, most often we tried to make the generator space as small as possible.

If we can design a simple pattern dynamics for a given growth task this does not mean that it necessarily should be considered a candidate for actually modelling the behavior of growth for a given species. This is not our goal at this stage of the computer experiment.

Instead we aim for <u>unifying principles</u>. We are examining a great number of growth tasks and look for general principles that seem to be useful for the logical organization of many tasks. Therefore our procedure is <u>speculative</u>. On the other hand, we select many of the tasks from actually performed experiments, in particular those which, in the laboratory, led to strange and surprising shapes and regenerative modifications.

To be able to carry out lots of such thought experiments on the computer we developed a set of programs, sufficiently flexible to allow for the variations in the multifaceted laboratory experiments, but at the same time structured enough to make it possible for us to execute the experiments fast and efficiently in terms of our own time. Little attention was given to CPU-time efficiency.

To write all the computer code that was needed was time consuming, sometimes laborious, but easy in principle. The difficulties lie elsewhere, in the logical design of the synthesis of the desired patterns.

3. We started our long series of thought experiments with growth tasks of types I and II. It takes time and experience to build up an intuition that enables one to solve such tasks with ease and elegance. Once this has been achieved, however, it is not particularly difficult to deal with tasks of these types. Time and space do not allow me to go into details, suffice it to say that we could grow many sorts of geometric shapes as well as such that looked (qualitatively) as animal forms.

The success in doing this was felt, somewhat paradoxically, as a disappointment: we did not seem to learn as much from the experiments as we had hoped. Few general principles seemed to appear, no abduction resulted from these trials.

When we turned to growth tasks of type III and IV the situation changed dramatically. For a long time we made little progress. For example: how would one organize the dynamics such that amputation and grafting the removed part back on to the body would lead to regeneration with mirror symmetry? How would one do it so that removal of a segment would lead to regeneration of it?

Eventually we began to reach solutions, sometimes natural and simple, occasionally more contrived. To show just a single example, in the figure we display a growth sequence displaying regeneration with mirror symmetry. After amputation of the limb the stump regenerates backwards with mirror symmetry. We noticed that a set of general principles seemed to emerge, principles that repeatedly showed their usefulness for the design of pattern dynamics.

Let me mention a few of these <u>morphogenetic devices</u>:

1. Let the dynamics be governed by a criterion minimizing the degree of irregularity at any time.

2. Organize the dynamics in terms of (embryonic) fields.

3. Organize the internal dynamics of the fields with buffering against disturbances.

4. Let the fields become autonomous when the organism reaches a certain age.

5. Connect the fields with a suitable topology, simple tree connection worked surprisingly often.

6. Let the dynamics pass through a (short) sequence of stages with different logical structure.

7. Allow the generators to be polarized

8. Exploit the symmetry of the target shape.

9. Express the dynamics in the intrinsic coordinate system generated by the growth process itself.

4. This work is still in progress. Armed with these morphogenetic devices we continue to pose and solve growth tasks for surgical modifications of animal organisms. We also plan to study the related problems in plant morphology/physiology, where quite different growth phenomena can occur (at least superficially different) and where other morphogenetic devices may be needed.

Statistical Packages: Implementation Techniques

Organizer: *Thomas J. Aird*

Invited Presentations:

    Design and Implementation of MINITAB, *Barbara Ryan and Thomas A. Ryan, Jr.*

    The Implementation of PROTRAN, *Thomas J. Aird and John R. Rice*

    The Conversion of SAS^{TM} Software Products to Minicomputers, *Richard D. Langston*

    Implementation of SPSS and SPSS-X, *Jonathan B. Fry*

DESIGN AND IMPLEMENTATION OF MINITAB

Barbara Ryan and Thomas A. Ryan, Jr.

Department of Statistics
The Pennsylvania State University
University Park, PA 16802 USA

1. INTRODUCTION

Minitab is a general purpose, statistical computing program, with comptabile interactive and batch modes. The first version, developed in 1972, closely followed the design of Omnitab, a program developed at the National Bureau of Standards. Since then, Minitab has evolved in response to new computing environments (e.g., interactive processing, CRT's, micro computers), new software engineering techniques, new statistical methods (e.g., descriptive and exploratory techniques, diagnostics), new audiences (large numbers of students on a campus doing a lot of homework exercises, engineers in the lab analyzing their data on the spot), new standards of quality (better algorithms, clear output, good error messages).

To the user, Minitab consists of a worksheet in which data are stored and commands that operate on the data. The worksheet can be pictured as follows:

```
  Columns             Matrices       Scalars
C1 C2 C3 C4 C5
 x  x  x  x  x        M1              K1
 x  x  x  x  x
 x  x  x  x  x  etc.  M2              K2
 x  x  x  x  x
 x     x  x  x        M3              K3
        x  x  x
        x     x         .              .
                        .              .
                      etc.           etc.
```

The variables of one or more data sets are stored in columns. In most Minitab sessions, a data analyst will use just columns. A few special calculations may require matrices and scalars. The columns can be given names (up to 8 characters). The entire worksheet (columns and names, matrices, scalars) can be stored in a Minitab save file for use at a later date. For example, to save the worksheet in the file MYDATA, type

    SAVE 'MYDATA'

To retrieve this file at a later date, type

    RETRIEVE 'MYDATA'

SAVE and RETRIEVE are two examples of Minitab commands. Most Minitab commands are short. For example,

```
HISTOGRAM C1 C2
NAME C4 = 'YIELD' C5 = 'TEMP'
HISTOGRAM 'YIELD' 'TEMP'
LET K1 = SUM( (C1-MEAN(C1)) **2 )
TABLE 'SEX' "DRUG' 'CLINIC';
   MEAN 'BP';
   MAXIMUM 'BP'.
```

Minitab reads a command, checks for errors, executes the command, prints the output and stores the answers. In the above example, the two PLOT commands and TABLE print output. The LET command stores the sum of squares in the scalar K1. Notice the TABLE command has two subcommands, MEAN and MAXIMUM. Subcommands are used when a command has many options. TABLE, for example, can store fifteen different statistics in the cells.

2. DESIGN OBJECTIVES

As was mentioned above, Minitab as it exists today evolved over the years. And so did our design objectives. One goal that has remained constant is that easy problems should be easy to do. Most users of Minitab have neither the time nor background to "learn how to use Minitab." They often have very simple problems -- a few plots, some transformations, a regression analysis -- and they want to learn as little as possible, type as little as possible, and get their answers back as quickly as possible. Here is an example.

```
READ C1-C3
 10   4.2   5.1
 14   3.1   5.0
 17   3.8   4.6
   etc.
END
PLOT C1 C2
PLOT C1 C3
LET C11 = LOGE(C1)
PLOT C11 C2
PLOT C11 C3
REGRESS C11 2 C2 C3
```

Each command is very short. Special options (for example, user controlled scales for PLOT, storage of many diagnostics for REGRESS) are available but there are rarely needed by most users. In interactive mode, the output is printed after each command. If anything is wrong (e.g., syntax error, specified the wrong variable), the user knows immediately and can retype the command.

Although Minitab is used interactively, the user interface is through a simple language rather than questions (often called a conversational system) or a menu. This makes it possible to have a compatible batch system. But what's more important, it allows the use of command files or macros in interactive (as well as batch) mode.

The syntax for most commands is quite simple and intuitive. But if a user cannot recall the syntax of a command, for example, REGRESS, he or she can ask for help by typing

    HELP REGRESS

Minitab then gives a short description of the command. There is also a quick reference card which contains all Minitab commands.

Another goal that was a part of the original system is portability. At first this meant that Minitab should run (after a few days' work) on most brands of computers. Now, however, there are a great variety of computing environments: different computer brands; different computer sizes -- large computers in a central computer center with a large support staff to minicomputers in a department with no staff to individually owned microcomputers; batch and interactive computing; output on a lineprinter, on a paper terminal, on a screen; virtual memory machines with effectively unlimited space to the 64K byte limit of a PDP-11 computer.

We would like Minitab to

(1) look the same in all environments so a user can easily go from one to the other;

(2) look like a program that belongs on its host computer, interacts well with the operating system, editors and other processors, and exploits special features of the host computer;

(3) install in less than one hour.

Let's look at (3) first. Minitab is written in "portable" Fortran code. However, a truly useful and efficient system cannot be written in portable code. Implementation includes (1) developing an interface with the computer's file system, random number generator, clock, and method of prompting in interactive mode, (2) overlaying the code to fit in a small space on machines that do not have virtual memory (and some that do), (3) writing a program to provide direct access to the messages in HELP, (4) finding a way to dynamically expand the worhsheet anytime the user needs more space, (5) handling a number of minor glitches -- no machine ever does exactly what it claims to do. Because of all these considerations, we have special versions (usually load modules) for most major brands of computers. Even a person who knows very little about his computer (and today, many people who are not computer experts own computers) can easily install these special versions.

Points (1) and (2) are to some extent inconsistent. Point (1) says we want Minitab to work the same on all brands, whereas (2) says we want Minitab to work well in its host environment. This says output must always look the same but also must work for both line-printers and CRT's. Minitab must look like a batch program on a batch system but like an interactive program on an interactive system. Minitab must work efficiently on a computer with a small memory but fully utilize the space of a virtual memory system. The way a user specifies a Minitab save file (in SAVE and RETRIEVE) or a data file (in READ) or a command file (in STORE and EXECUTE) should be the same on all computers but on a given computer should look the same as files specified by other programs.

Universities and companies that use Minitab extensively often have it on several different computers and operating systems. For them, compatibility across brands is almost essential.

We do allow some things to vary with brand. The most obvious is space. On LSI-11 based microcomputers, the worksheet has only 50 columns and can store a maximum of 2500 numbers; on a Vax, the worksheet has 1,000 columns (and on some customized versions 10,000 or more) and can store over 4,000,000 numbers. When you move from brand to brand, you must check the space limitations, if you think it could be a problem. There are some features which are added to Minitab if the host computer allows it. For example, on the IBM VM/CMS version of Minitab there is a command called CMS. If you type this, your Minitab session is temporarily suspended and you are transferred to the CMS (actually CMS subset) operating system. You then might invoke a CMS editor, make up a data set and store it in a file. Then if you type the word RETURN, you will be back in Minitab with everything as you left it when you typed CMS. This facility of moving between a program such as Minitab and the operating system depends on the operating system of the host computer. If a given computer provides the facility then we add the feature to Minitab.

Many of our users are universities with limited budgets, so the overall cost of using Minitab is a very important consideration. When we first distributed Minitab (in the early 1970's), computer time was relatively expensive so the cost of a single Minitab job had to be kept low. Now computer time is much cheaper, but many more students use the computer in their courses. So overall cost must still be kept low. The cost involves more than just the price of a CPU second. It includes amount of memory needed, the length of printed output, number of disk accesses, the probability of an error and the consequent need to redo the work, and the need to have consultants to help people use Minitab. The cost in terms of CPU time is quite low for most common statistical procedures. For example, to compute and print descriptive statistics (mean, median, quartiles, etc.) of 25,000 observations takes less than .8 seconds and costs less than 6 cents on Penn State's IBM 3033.

All the goals we've discussed so far were a part of early versions of Minitab, although they have been broadened and modified over the years. However, advances in computer design and increased user sophistication have created at least two new design criteria: Minitab should be flexible and extendible and it should be at home in an interactive environment.

Some flexibility has come as a by-product of using commands that are small and simple and from the fact that Minitab stores data in main memory rather than on disk as many statistical packages do. This allows operations based on sorting, operations based on several passes through the data, and access to individual numbers in the worksheet. Here are some examples

```
LET K1 = SUM( ABSO( C1 - MEDIAN(C1)) )/COUNT(C1)
    calculates the mean absolute deviation
    from the median
LET C1(5) = 28.3
    puts 28.3 into the 5th location of C1
LET K2 = C1( COUNT(C1) - 1 )
    puts the next to last value of C1 into K2.
```

A few simple features have been added to extend the flexibility. For example, the CK capability, although not very elegant, is very useful. As a simple example,

```
LET K1=5
PRINT C1-CK1
```

would print C1-C5. Here is a more practical example, using the stored file capability

```
STORE 'FILE1'
  PLOT C20 CK1
  REGRESS C20 on 1 pred. CK1, store res. in C21
  PLOT C21 CK1
  LET K1 = K1 + 1
END
LET K1 = 1
EXECUTE 'FILE1' 6 times
```

First, a block of commands are stored in the file FILE1. (This file can be used in this session or in a later session.) The loop is initialized by setting K1=1 and then executed 6 times. Each time through the loop, a data plot, regression, and residual plot are printed for one predictor, starting with C1 and going up through C6.

This last example also illustrates the use of macros (i.e., stored command files) and loops.

Some users have developed macros to do fairly elaborate, non-standard analysis, stored them away, and invoke them each time they want to do the special analysis.

Currently most users of Minitab work in an interactive environment. Minitab, however, was originally developed as a batch program. But because the commands were simple and short, and because the user model would read a command, process it, read the next command, etc., it was easy to convert from a batch to an interactive program. In fact, users new to Minitab are surprised to learn it was originally a batch program. This is clearly one area where we "lucked out." There are, of course, many places where interactive programs differ from batch. For example, when a serious error is detected in batch mode, the remaining commands are not executed, while in interactive mode, you get a second chance. Note that executing command files in interactive mode is essentially the same as running a batch program in terms of error recovery. The output of Minitab has always been compact and readable. However, we have had to redesign extensively to make best use of CRT screens. These screens are usually only 24 lines tall and it is difficult to look backward and forward through the output the way you do on a paper output.

3. IMPLEMENTATION

Minitab is implemented in Fortran. The reasons for this include: (a) we knew Fortran: (b) Fortran has always been available on a wide variety of computers: (c) good, optimizing compilers are available; (d) good mathematical software is available, such as the LINPACK and EISPACK: (e) tools are available, such as the PFORT verifier, Watfiv, and more are being developed, such as TOOLPACK; (f) we have developed our own extensions to handle problems with the language, such as methods for character processing, routines to do nice output, and structured flow of control (e.g., IF-THEN-ELSE). The language we write in is transformed by simple preprocessors into a portable subset of Fortran 66 (or an essentially similar subset of Fortran 77). In the (relatively few) places where we could not write code which would run or run efficiently on all computers, we write several variations of the code suitably marked so a simple program can select the code appropriate for the target computer.

Minitab stores all data -- columns, constants and matrices -- in main memory. They are stored in one long array, which is also used for the temporary scratch work done by individual commands. In the middle of a command, this array might look like

| C1 | C3 | XXX | C2 | M8 | scratch | X | scratch |

When the command is finished, scratch space is freed and "holes" are removed. The array would then be

C1	C3	C2	M8	

On computers that have a small real memory and no virtual memory capability, the size of the worksheet is small. For example, the array for data storage is just 5,000 numbers on a PDP-11 computer. On a virtual memory machine, the worksheet size can be very large. For example, the data array is 4 million numbers on a Prime computer.

Minitab stores data one column, followed by another column, followed by another, and so on. This is often called a transposed file. Most statistics packages store data rowwise, or casewise. Most statistics packages also do not have scalar or matrices and expect all columns (variables) in one file to be of the same length. Since most statistical methods (e.g., regression, analysis of variance) use algorithms to access data casewise, the columnwise storage of data Minitab uses might be inefficient on virtual memory machines. In fact, this is not true if the number of variables used by a Minitab command is less than or equal to the total number of pages that can be held in main memory.

Here is a simple example. Suppose the data file has 10,000 observations on 50 variables and that each page can hold 250 numbers. Suppose the Minitab command is PLOT C1 versus C2. To process the command, two pages are brought into main memory -- one page contains the first 250 numbers in C1 and the other contains the first 250 numbers in C2. The next 249 data references will be to the same two pages. Then two other pages are brought in -- one contains the second 250 numbers in C1 and the other the second 250 numbers in C2. This continues until all 10,000 numbers in C1 and C2 are processed and results in a total of 80 page references. If the data were stored casewise then each page would contain 5 cases on all 50 variables. A total of 2,000 page references would be required to process all the data for PLOT. Since the most commonly used statistical procedures (such as plots, tables, regression, analysis of variance) usually use relatively few variables (less than, say, 100), paging rates are normally low. Some commands, such as HISTOGRAM and LET (algebraic expression) work columnwise and thus have low paging rates no matter how many variables are used.

THE IMPLEMENTATION OF PROTRAN

Thomas J. Aird
IMSL, Inc.
7500 Bellaire Blvd.
Houston, TX 77036

John R. Rice
Mathematical Sciences
Purdue University
W. Lafayette, IN 47907

PROTRAN is a family of problem-solving software systems designed to provide quick and convenient access to the computational power of the IMSL Library of mathematical and statistical Fortran subroutines. This paper discusses the implementation of PROTRAN. PROTRAN is an extension of Fortran which accepts problem specifications, checks them for consistency, selects the appropriate IMSL routines and generates a Fortran program. All programming requirements such as workspace assignment and error message handling are handled automatically by PROTRAN. PROTRAN adds to the problem-solving power of Fortran by adding very high level problem-solving statements and it adds to program reliability by providing extensive error checking and automating the details of using the IMSL Library.

1. THE PROTRAN LANGUAGE AND SYSTEM

We first give a brief overview of the objectives and nature of PROTRAN. PROTRAN provides a problem-solving environment in which the general type of problem is specified by a mnemonic keyword. The specific type of problem and other necessary information are indicated by additional keywords. PROTRAN checks the problem specifications and data to make sure they are consistent, selects the appropriate IMSL routines, and generates a Fortran program. All programming requirements such as workspace assignments, or library error message handling are handled automatically by PROTRAN and it provides extensive checking for user errors with clear diagnostics. The generated Fortran program goes through the usual steps of compilation, linking, and execution to produce output (i.e., the problem solution). Fortran and PROTRAN may be intermixed which provides the flexibility of Fortran and provides access to any existing software available through Fortran. PROTRAN adds to the problem-solving power of Fortran by adding very high level problem-solving statements and it adds to program reliability by providing extensive checking and automating the details of using the IMSL Library.

The primary design objective of PROTRAN is to provide a very high level, user friendly problem-solving environment based on the power of the IMSL Library. Other design objectives of PROTRAN are:

portability,
reasonable cost to the end user,
stability at the user/PROTRAN interface level,
automation of programming details, missing data, ...
computed statistics available throughout program,
no restrictions imposed on problem size,
same general data structures used everywhere
robustness and flexibility for the user,
extendability for new problem solving abilities.

The nature of PROTRAN is illustrated below by a few sample statements. For further information see [IMSL, 1982], [IMSL, 1983] and [Rice, 1983]. PROTRAN statements have the following form (statement syntax):

$\quad$ \$ $\;$ procedure_name procedure_argument
$\quad\quad$ keyword_phases

To compute the integral $E5 = \int_1^2 0.5^x dx$ one uses, for example, the PROTRAN statement

$\quad$ \$ $\;$ INTEGRAL 0.5**X
$\quad\quad$ FOR (X=1,2)
$\quad\quad$ IS E5

The procedure arguments and keyword arguments consist of constants, variables, expressions, keywords or segments of Fortran code. The naming conventions for variables and the rules governing expressions are the same as those for Fortran. A second example illustrates the syntax.

\$APPROXIMATE DATA; VS POINTS; USING SPLINES
$\quad\quad$ BY FITIT ; NBASIS = NPOINT/5+1

Here the keyword phrase arguments are: SPLINES (a keyword), FITIT (the name of the approximation - a Fortran function created by PROTRAN) and NPOINTS/5+1 (an expression for the number of spline basis functions to use).

PROTRAN provides three data structures: SCALAR, VECTOR and MATRIX, and each has a relationship with corresponding Fortran entities. SCALAR entities correspond to Fortran variables, while VECTOR and MATRIX entities correspond to Fortran one-dimensional and two-dimensional arrays, respectively. Declarations of type, data structure and storage formats are illustrated below:

```
$DECLARATIONS
 REAL MATRIX A(9,9)
 REAL VECTOR B(9),BSYM(3),SOL(9),SOLSYM(3),X(9)
 DOUBLE PRECISION SCALAR ACCUM
 REAL SYMMETRIC MATRIX ASYM (20,20)
```

Facilities are provided for assigning values to scalars, vectors and matrices as well as for printing values of these entities. An example using the preceding declarations follows

```
     $ASSIGN A = (1.0,2.0)
    +           (3.0,4.0)
            B = (0.5*COS(PI/4.),-2.5*A(2,2))
            X = B
            BSYM(I) = (I+1)/2. + COS(PI*I)
            ASYM  = (1.0               )
    +               (2.5,3.3,          )
    +               (1.7, 1.9,2.1)
     $LINSYS A*SOL = B
     $LINSYS ASYM*SOLSYM = BSYM
     $PRINT A,B,SOL,ASYM,BSYM,SOLSYM
```

PROTRAN uses range variables for vectors and matrices to distinguish between storage-size (amount of memory allocated) and working-size (actual size used in a particular calculation). Range variables may be declared explicitly to provide direct control of the working-size. If range variables are not declared explicitly, they are handled implicitly and automatically be PROTRAN. This mechanism allows each PROTRAN procedure that deals with vectors or matrices to utilize the current working-size of these entities in calculations and to check that vector and matrix sizes are compatible with their use. The following program to solve both a 2 by 2 and a 3 by 3 system of linear equations illustrates the use of range variables.

```
  $ DECLARATIONS
    REAL MATRIX A(10,10)
    REAL VECTOR B(10),X(10)
C                 2 BY 2 PROBLEM
  $ ASSIGN A =  (1.0, -2.0)
  +             (3.0,  4.0)
            B = (0.5, -2.5)
  $ LINSYS A*X = B
  $ PRINT X
C                 3 BY 3 PROBLEM
  $ ASSIGN A =  (1.0,  5.0, -3.0)
  +             (2.0, -1.0,  4.0)
  +             (0.5,  2.5,  1.0)
            B = (0.5, -1.5,  3.0)
  $ LINSYS A*X = B
  $ PRINT X
  $ END

Output:
X
   -.30000    -.40000
X
   -2.33636   1.22727   1.10000
```

There are four range variables here: row range of A, column range of A, range of B, and range of X. The range variables are all implicit and are initially set to 10 (i.e., to the storage size). The first ASSIGN changes them to 2 for A and B and this is used in the first LINSYS. Then PRINT, appropriately, only prints 2 elements of X as set by LINSYS. In the second ASSIGN, the ranges for A and B are set to 3 and the second LINSYS sets the range of X to 3 so that 3 values are printed.

PROTRAN recognizes three types of errors (preprocessor, numerical and setup) and for each type there are three possible severity levels: warning, fix and fatal. Preprocessor errors, which are due to incorrect PROTRAN statements (for example, spelling errors, use of incorrect keywords) are recognized during the preprocessing stage of the run. Numerical and setup errors are recognized during the execution stage of the run.

An important concern is that Library routines have legal and consistent input and that all exceptions that occur be handled. Thus array variables that are passed to Library routines have their range variables analyzed to see if they are within storage declaration bounds and consistent with the size of the problem being solved. The user may select options (using the OPTIONS procedure) to print only certain error messages and to terminate the run on the occurrence of certain errors.

PROTRAN provides automatic workspace allocation using the Bell Labs allocation and deallocation scheme [Fox, 1978]. Each PROTRAN procedure determines the amount and type of working storage required and requests that storage from the allocation routine. After the computations are complete, storage is returned to the stack.

The typical non-programmer statistics user of PROTRAN will never need to know that PROTRAN is an extension of Fortran, see the example program in Figure 1. The more sophisticated statistics or mathematics user will use Fortran facilities from time to time. Mixing Fortran with PROTRAN is relatively easy for an experienced Fortran programmer. The PROTRAN procedure $FORTRAN on a line by itself signals the beginning of a sequence of FORTRAN statements. The sequence is terminated by the beginning of another PROTRAN procedure reference. The beginning of Fortran code is also signaled by a $ as the last nonblank character of a PROTRAN statement. To take full advantage of PROTRAN the programmer must be aware of some details such as matrix/vector storage formats, temporary variable naming conventions and error handling. The Fortran names generated by PROTRAN avoid conflict with user variable names by having six characters and ending in V7. Statement labels are in the range 30000 to 39999. The user can change the label range and V7 characters through the OPTIONS procedure. PROTRAN Library routines and utility routines all have six character names that begin with IMSL.

A complete STAT/PROTRAN program is shown in Figure 1. The file created by $BUILDFILE can be used in other PROTRAN statements if the program were longer. For example, in a more complex program one could later have the statements

```
$    DECLARATIONS
     REAL MATRIX DATA(13,5),BHAT(5,5)
     VECTOR RES(13),WGHT(13),BOLD(5)
     ...
$    REGRESS (C5) ON (C1,C2,C3,C4);ARRAY DATA
         WEIGHTS WGTS;RESIDUALS RES;COEF BHAT$
     DMAX = 0.0
     DO 25 I=1,4
         D = (BHAT(I,1)-BOLD(I))/BOLD(I)
         DMAX = AMAX1(D,DMAX)
         BOLD(I) = BHAT(I,1)
15   CONTINUE
$    PRINT DMAX,BOLD
```

This program segment shows how PROTRAN results may be used in ordinary Fortran statements. Here C1,C2, etc. are automatic names for the columns of the array data.

2. IMPLEMENTATION

The two principal obstacles to meeting the design objectives of PROTRAN have been (a) the size of the PROTRAN system and (b) the slowness that results from the usual Fortran I/O routines. To appreciate the size problem, observe that MATH/PROTRAN has 26 statements with 65 different keyword phrases. These statements can access about 80-100 IMSL Library routines (LINSYS alone accesses one of 14 routines). The amount of Fortran code to access these routines in a normal way varies from short (10 lines or so) to quite long (over 200 lines); this requires over 10,000 lines of Fortran code to be in the PROTRAN system. Obviously, these must be

```
C    VARIOUS OTHER REGRESSION OPTIONS COULD BE INVOKED, SUCH AS RESIDUAL
C    ANALYSIS, MODEL SELECTION, PRINTING (AND/OR STORING) THE CORRELATION
C    MATRIX, AND SO ON.
C
$         BUILDFILE
          READ X1,X2,X3,X4,Y
          DATA
          ====
            7    26     6    60     78.5
            1    29    15    52     74.3
           11    56     8    20    104.3
           11    31     8    47     87.6
            7    52     6    33     95.9
           11    55     9    22    109.2
            3    71    17     6    102.7
            1    31    22    44     72.5
            2    54    18    22     93.1
           21    47     4    26    115.9
            1    40    23    34     83.8
           11    66     9    12    113.3
           10    68     8    12    109.4
          ====
$         REGRESS (Y) ON (X1,X2,X3,X4)
$         END
```

Figure 1: A simple regression example using STAT/PROTRAN.

in a random access file held in auxiliary memory. The size of the object code in the preprocessor to analyze one PROTRAN procedure_name and procedure_argument runs about 300 words; a system with 50 procedures (very possible for some PROTRAN systems) would require about 15,000 words of code. If a keyword phrase can be analyzed in 150 words of code, this would require an additional 10,000 words of preprocessor code for the current MATH/PROTRAN and perhaps 15,000 words for a larger system. This 30,000 words of code is already considerably larger than existing Fortran compilers and we have included nothing for the overall management or utilities of the preprocessor.

To appreciate the slow Fortran I/O problem recall that a good Fortran compiler can read input, compile it and write object code in about 20-30% of the time a typical Fortran system needs to read the input with FORMAT(80A1). There are two potential solutions to the speed problem: (1) use machine specific routines to replace the Fortran I/O (2) avoid formatted Fortran I/O through the use of DATA statements and embedding text as character strings in FORMAT statements. These two solutions are, of course, to be complemented by a systematic effort to reduce the amount of I/O required.

The initial PROTRAN implementation uses a Fortran preprocessor with two distinct phases: language recognition and Fortran code generation. To conserve space, these are interleaved in a special manner. The code generation is done by one or more templates for each problem solving statement. A template processor generates Fortran code by inserting user supplied or PROTRAN generated text into the template. The templates themselves are written in a fairly complete, high level programming language which includes IF-THEN-ELSE, DO, INCLUDE, assignment, list operations, file assignment, generation of new labels and variables, etc., but includes no arithmetic. The template processor knows that it is creating lines of Fortran code and thus line layout and format rules are handled automatically. There are special functions related to processing PROTRAN, for example, $$RR(A) and $$DR(A) give the row range variable and row dimension of the matrix A, respectively. All the attributes of PROTRAN variables are provided by such functions.

Three steps are used to help meet the size and speed requirements. First, once the PROTRAN procedure name is identified, then an initial template is read which contains information about the rest of the statement. One can interpret this as moving part of the preprocessor into the templates. Information obtained this way includes the list of required and admissable keywords, diagnostic messages, etc. This information is then used to finish analyzing the PROTRAN statement. This technique provides a substantial reduction in the size of the preprocessor.

The second step is to have an absolutely uniform syntax for the keyword phrases so one program can analyze them all knowing the keyword, the number and nature of the arguments (if any). This latter information is data to the keyword phrase analysis program.

The third step is to move much of the code needed to access the IMSL Library routines into the routines themselves. Thus each Library routine is embedded in a routine which has a prologue and epilogue to handle most of the checking, storage allocation and error message handling. The effect of this is to move lines of Fortran from the templates into the object codes of the Fortran library that supports PROTRAN execution. This step substantially reduces the amount of Fortran I/O needed.

The initial PROTRAN implementation meets the size requirement, but is slower than desired because of the slow Fortran I/O. In order to enhance portability, special I/O routines are not used; the only machine specific code is the random access routine for the template file. Once PROTRAN was operational, it was realized that the I/O needs could be substantially decreased by redesigning the library routines in various ways; even more Fortran code can be moved from the templates to the Fortran library that supports PROTRAN. This revision of the support library is underway, but it is unlikely to provide enough speed-up to allow PROTRAN to run as fast as one would "expect".

Three further actions are being evaluated to speed up PROTRAN. The first is to write machine specific I/O routines to avoid the inefficiency of Fortran I/O for reading the template and writing the Fortran code generated. The second is to reduce the size of the templates by compiling them into an "assembly language" form. More details are given later along with the discussion of writing templates. The third action is to eliminate templates altogether and to bring all the code into the Fortran preprocessor.

Each of these actions has costs and benefits. The first one sacrifices considerable portability; one must create and maintain a set of I/O routines for all target environments and even then one has a certain loss of speed due to the overhead of subroutine calls to do I/O.

The second action requires that the "assembly language" be defined and a compiler written for it. A preliminary design has been completed for this approach and it appears quite feasible. The compiler would be another preprocessor and can be implemented fairly rapidly with the PG (Preprocessor Generator) system of [Brophy, 1982] which is part of the TOOLPACK project. An advantage of this approach is that the "source code" need not be distributed so that even with the support library in source code one is protected from others being able to maintain or modify PROTRAN with a reasonable effort. A second advantage is that the PG system allows one to write the templates in a very high level language and to quickly modify this language as needs change.

The third action creates a PROTRAN system much larger than allowable. Thus it must be accompanied by overlaying for the preprocessor to run in "normal" sized memories. This means that portability would be lost except for the few environments which support virtual memory. Note that if this implementation is selected, then the templates become individual Fortran subroutines with the Fortran code segments embedded in them.

3. PROTRAN TEMPLATES

The simplest view of PROTRAN templates is as Fortran code with places marked where various values are to be inserted. In this view the template processor is just a simple macro processor that substitutes values, variable names or blocks of code determined by the first phase of the preprocessor. This simple template concept is inadequate for PROTRAN and a fairly complete programming language is required to make decisions, construct loops, handle lists, etc. Three approaches of using templates have been evaluated for PROTRAN: 1. Write templates, put them in a file and process them with the PROTRAN template processor (a specialized macro processor), 2. Write templates in a higher level template language, compile them to an assembly template language, put these in a file and process them with a different PROTRAN template processor, 3. Write templates in a higher level template language, preprocess them into Fortran subroutines and put these in the preprocessor. Each of these possibilities is discussed briefly.

PROTRAN is implemented with the first approach; here one must keep the template language simple to process as this is done with each use. Thus the template programming language relies on special characters and short names for operators. Writing PROTRAN templates is more complex than ordinary Fortran programming both because of language complexity (one is mixing Fortran and template programming language in odd ways) and the necessity to make multitudes of decisions. Figure 2 shows a few lines from the template for SUM to illustrate to nature of this code

The second approach allows the template programming language to be much more natural. The compilation of templates into assembly language is done at IMSL before the PROTRAN system is delivered so the compilation cost has no effect on user cost and the primary objective is to make the templates easy to create and maintain. The assembly template language is designed for maximum speed of processing and minimum length for templates. The following assembly template code segment illustrates the nature of the language.

```
=MAXO(INT(-+)/),0)
 7,7,17,18,2,19,5,21,5
7,8
IF(.EQ.0)GOTO
10,7,20,10
=
7,2,8,3
DO=1,
9,9,9,11,12,7
```

The code is shown in pairs of lines, Fortran characters are on the first line and integers on the second; the integers represent the positions, operators, values and variables to be used in creating the output line of Fortran code.

```
C                       ESTABLISH INDEX INCREMENT VALUE
      $ND(INC) = $(INC)
C                       CHECK FOR ZERO INCREMENT
      IF($ND(INC) .NE. 0.00) GO TO $L(DINCOK)
      $G(IER) = 129
      CALL IMSLMS(3,2,$G(IER),6HSUM   ,$G(LINE),
.        41,41HINCREMENT VALUE OF INDEX VARIABLE IS ZERO)
      GO TO $L(FATAL)
$(DINCOK) CONTINUE
      $NI(ITCNT) = ($ND(TERM)-$ND(INIT)+$ND(INC))/$ND(INC)+0.01D0
      $ENDIF
      $ENDIF
      $ENDIF
```

Figure 2: Code in the template language for the template processor.

The third approach also allows a much more natural template programming language. It is similar in spirit to PROTRAN and could be called TEMPLATE/PROTRAN. A Fortran preprocessor translates this language into Fortran; again this compilation need not be terribly fast and the primary objective is to have templates that are easy to write and maintain. A illustrative template is given in Figure 3.

```
      SUBROUTINE LINSFE(A,X,B,EQUATI,RHS)
C                 DECLARATIONS
      KEYWORD     A,X,B,EQUAT,RHS
      KEYWORD     NEQ,NRHS,KRR,KNAME
      INTEGER     IER,ITY,IDEF
      EXTERNAL    CHKPRE
C                 PRE-PROCESSOR TIME CHECKING
C     LINSYS HAS CHECKED THAT A IS REAL OR DP
      IER=0
      CALL CHKPRE(X, <->, ITY(A), $ARRAY,IER)
      CALL CHKPRE(EQUAT,KRR(B),$ITYANY,$EXPRESS,IER)
      IF(IER .NE. 0)           RETURN
C                 CREATE NAMES
      NEQ  = KNAME($INTEGER)
      NRHS = KNAME($INTEGER)
C     GENERATE CALL TO APPROPRIATE SUBROUTINES
      FORTRAN
         NEQ  = EQUAT
         NRHS = RHS
      END FORTRAN
      IF(ITY(A) .EQ. $REAL) THEN
         FORTRAN
            CALL IMSLL8(...)
         END FORTRAN
      ELSE
         FORTRAN
            CALL IMSLL9(...)
         END FORTRAN
      END IF
      END
```

Figure 3: Typical subtemplate in the template language that is compiled into a Fortran subroutine.

This template language allows one to call on other templates (LINSFE is called from LINSYS), invoke preprocessor facilities (CHKPRE checks the legality of the types of X,B and EQUATI) and to generate Fortran code. The template compiler translates this into an actual Fortran subroutine that carries out these actions. This template compiler can be written rather easily using the PG system.

There are eight template programming rules used to help control the nature and enhance the reliability of the PROTRAN system:

1. Every template has a standard heading defining the context.

2. All variables are checked for attribute consistancy at the start of the body of the Fortran code.

3. All error messages start with *** delimited by blanks.

4. The dimensions and ranges of all arguments to PROTRAN Library routines are checked at run time.

5. Workspace for Library routines is always provided automatically.

6. Minimize use of user supplied code, put a comment before each instance of the direct use of such code which relates to the line to the user input.

7. Clear, meaningful error messages are to be issued.

8. Every template has an associated comprehensive PROTRAN test program that is self-checking.

REFERENCES

[1] Brophy, John, PG - Preprocessor Generator, CSD-TR 405, Computer Science Department, Purdue University, (1982).

[2] Fox, P.A., Hall, A.D., and Schryer, N.L., The PORT Mathematical Subroutine Library, ACM Transactions on Mathematical Software, 4 (1978), 104-126.

[3] IMSL MATH/PROTRAN User's Manual, Edition 1, IMSL, Houston, Texas, (1982).

[4] IMSL STAT/PROTRAN User's Manual, Edition 1, IMSL, Houston, Texas, (1983).

[5] Rice, J.R., Numerical Methods, Software and Analysis, (McGraw-Hill, New York, 1983).

THE CONVERSION OF SAS™ SOFTWARE PRODUCTS TO MINICOMPUTERS

Richard D. Langston

SAS Institute

The Conversion of SAS™ Software Products to Minicomputers

This paper discusses the conversion of SAS software from the IBM mainframe environment to the world of minicomputers. I will present this by enumerating the factors involved in converting existing software from one computer environment to another distinctly different environment. Then I will expound on the actual implementation process, throughout all its phases. Finally, I will give some thoughts on where we will proceed from our current portability effort.

SAS Institute currently has approximately 7,000 installed products. The current line consists of SAS, SAS/GRAPH™, SAS/ETS™, SAS/FSP™, SAS/IMS-DLI™, with SAS/OR™ due out in April 1983.

The languages used to implement all the SAS software products were PL/I and IBM assembly language. The assembly language was used in developing the supervisor and data processing step which is responsible for reading INPUT statements and data compiling and executing the DATA step, SAS data set I/O, and data handling for the SAS procedures. Assembly language was chosen primarily because of its speed and compactness and also because it allowed SAS to take advantage of many IBM operating system features. The SAS procedures, with a few exceptions, were written in PL/I. PL/I was chosen for its wide range of applications, both as a scientific and statistical language, as well as its capability of handling based structures, structured programming techniques, string manipulations, and attractive output.

Because of the fact that IBM assembly language was a predominant factor in our implementation, SAS software ran exclusively on IBM hardware. In fact, SAS was an exclusive IBM OS product, excluding the VM/CMS and DOS/VSE operating systems, because of our heavy dependence on OS-oriented assembler code. Fortunately, we were able to overcome the OS dependencies to allow SAS to run under VM/CMS by 1980 and DOS/VSE by 1982. However, the reliance on IBM hardware still remained. Meanwhile, our existing and potential users were saying, "What about SAS on our minicomputers? Our IBM system is overloaded, but our mini is in the next room hardly being used. We want SAS on that machine."

We wanted to respond to the users' needs. The reliance on the IBM assembly language would have to be eliminated by redeveloping the supervisor in PL/I. The design and coding of this monstrous undertaking would need to be done quickly, and with sufficient staff to do the job correctly and efficiently. Also, the hardware would have to be acquired to develop the new software, dubbed "Portable SAS".

We chose PL/I as our implementation language for Portable SAS. This decision was really made for us, because a large amount of code was already written in PL/I, including almost all of our procedures. The supervisor was to be rewritten in PL/I, making the entire system in one language instead of a hybrid of several.

Other languages were considered, however. PASCAL had its good points, but it's still fairly new in the development world and there is much standardization left to be done before it could be considered acceptable to us. Besides, no one in our applications staff felt comfortable programming in the language, and IBM support did not appear sufficient. FORTRAN was also pondered. Most of our staff knew it; in fact, a few of our procedures had been implemented in FORTRAN prior to being converted to PL/I. However, FORTRAN lacked the flexibility of locate-mode data access, dynamic allocation of memory, and ease of structured programming all of which we could have with PL/I. COBOL was mentioned and we all had a good laugh.

Although our procedures were already in PL/I, they were written to be compatible with IBM's PL/I compilers. IBM's PL/I is very kind to you with respect to such things as intelligently declaring based pointers, allowing #, $, and @ in variable names, and accepting multiple target assignments such as A,B=0;. These niceties are not part of the ANSI PL/I, and especially not part of the subset "G", which is the subset commonly implemented for the PL/I compiler for non-IBM computers.

The difficulty with converting our PL/I to ANSI standard subset G was that we were also rewriting our code on the IBM mainframe, whose PL/I allowed all the features we had to avoid. Therefore, we could only determine if we had removed all the non-ANSI features if we "ported" the code to the minicomputers and found that their compilers would not accept it. This cyclical process continued for some time before all the source modules for a procedure successfully compiled on the minis.

However, since we had to go into all of our code and revise it, we saw a golden opportunity to remedy some problems that had plagued us. For example, our support routines were in assembler and did not handle character arguments in the same way that PL/I does. This problem, along with other coding difficulties, is illustrated in this example.

```
/*OLD WAY OF WRITING VARAIBLE DESCRIPTOR*/
DCL X CHAR(8);
DCL X# FIXED BIN(31)DEF(X);
DCL 1 NAMESTR,
    2 NTYPE FIXED BIN(15),
       .
       .
       .
DCL VECTOR(100)FIXED BIN(31)INIT(0);
CALL SETDSN(2);/*MAKE CURRENT DATA SET */
/*WRITE OUT DESCRIPTOR FOR X*/
CALL ONAMES(NTYPE, VECTOR(1),X#,NAMESTR);
       .
       .
       .
```

In our old way of writing variable descriptors, a VECTOR array had to be maintained by the programmer, to be used by the ONAMES routine. The descriptor information would be filled into the NAMESTR structure then written out via ONAMES. However, if the variable was character, the address had to be passed by using a numeric variable overlaid on the character data, in order to avoid dope vector passing. Also, the structure name had to be added as an extra argument to prevent the PL/I optimizing compiler from moving the call to the outside of loops.

```
/*NEW WAY*/
DCL X CHAR(8);
DCL 1 NAMESTR;
    2 NTYPE FIXED BIN(15);
    .
    .
    .
NAMESTRPTR=ADDR(NAMESTR);
    .
    .
    .
/*WRITE OUT DESCRIPTOR FOR X*/
CALL XVPUTD(XVOUTPTR, NAMESTRPTR, ADDR(X));
```

In the new way, we chose to use pointers more often. The NTYPE argument was replaced with a pointer to NAMESTR, and the data value was also passed by address, to avoid any dope vector passing. The VECTOR array was made obsolete. The XVOUTPTR argument was introduced to indicate what data set to operate on, thereby eliminating the need for the SETDSN call.

In addition to new calling sequences, we found that several operations on SAS data sets were being identically coded in many SAS procedures. These operations were made into callable routines. For example,

```
CALL SETDSN(1);
N=NOVAR(1) TYPEVARNUM=0;
DO I=1 TO N WHILE(TYPEVARNUM=0);
    CALL NAMEV(1,I,NTYPE);
    IF NNAME='_TYPE_'
        THEN TYPEVARNUM=I;
    END;
```

The NAMEV routine fills in the NAMESTR structure for variable number I of list 1 starting at NTYPE. NNAME is a variable within the NAMESTR. This code was reduced to:

```
TYPEVARNUM=XVFIND(FILEID,1,'_TYPE_');
```

The problems I have just discussed were solved by substituting less cumbersome code. However, we faced difficult problems that required complete redesigning. For example, we parsed procedure statements with programs custom-written in IBM assembly language. We needed a new design that did not rely on a language that lacked portability.

This problem was solved by merging the parsing into the procedure. Parsing was accomplished by writing a input grammars to read tokens, check for validity, and perform actions based on the tokens.

We defined all procedure grammars in a BNF-style notation, then used an LL(1) parser-generator called SYNPUT (Dunn and Waite, University of Colorado) to generate a table of parser instructions, which are formed into PL/I static array declarations. The procedure then calls a parser interpreter to perform the parsing using the generated parsing tables.

People were hired to write the compiler, interpreter, word scanner, I/O routines, and to develop testing and documentation procedures. Also, hardware was procured for the development: a DEC VAX™ 11/780, a Data General ECLIPSE™ MV/8000, and a Prime™ 550. The staff grew as individuals were chosen to support each of these systems. Additional tasks, such as host-dependent routines, were divided up among these individuals.

Concurrently, the applications group was implementing a temporary version of the support routines for the IBM environment, to allow us to convert and test the procedures in our "home" environment. This concurrent development caused a few minor problems. For example, we specified a routine called XMEMDBL that would double the number of elements of a memory segment. However, if the number of elements was indicated as 0, XMEMDBL was to set the number of elements to 10 and allocate 10 elements. This feature was not implemented by the Portable Systems Staff due to a lack of communication between us and them. It took a while to determine that mysterious abends on the VAX were being caused by XMEMDBL's expecting a non-zero size. This particular bug became known as "2 times 0 is 10."

The procedures were debugged and tested on the IBM side until they appeared to produce the same results as the current product. They were then "ported" to the VAX for recompilation and further testing. After getting past the compiler incompatibilities, the procedures were found--much to our relief and joy--to function the same as they did on the IBM side. Naturally there were many bugs along the way to contend with. For example, the use of STATIC INIT was common in our code for variables that change values. However, on the VAX --our primary development machine for Portable SAS-- all of the procedures' and supervisor's code is linked together into one executable image, so that once a STATIC variable value changes, it remains changed for the duration of the execution of the image. This problem caused bizarre abends, but once it was discovered, it was easily remedied by removing the INIT option and adding the assignment statements.

Occasionally, we realized other shortcomings in our design that would require some coding changes in almost everyone's source. This delightful period was known as "change day". An example of a feature that required a change day was when the program to produce grammar DECLARE files was written. We had maintained one version for IBM, and another for the minis. The new version of the program was portable, which meant that the grammar need only be compiled on one machine. However, the new grammar DECLARE files were incompatible with the programs using the DECLAREs meaning that all of these programs had to be recompiled in order to remain compatible. I should point out that we've only experienced two "change days" so far in our development.

Naturally, all of the work in portability should lead to the ultimate release of the product to users.

As of this writing, the portability staff is preparing to send out an alpha test tape of the system to selected sites, who are already IBM SAS users with VAX machines available to them. We have developed a bug reporting system to be included with the tape shipment to aid in determining and fixing problems. Beta test will follow shortly after the alpha test is deemed complete. The production version will follow beta, with release date dependent on the success of beta test.

After production release, which will include SAS and SAS/GRAPH, we will complete the conversion of other products: SAS/OR, SAS/ETS, and SAS/FSP. At the same time, we hope to expand our range of operating systems.

As mentioned earlier, we have a Data General MV/8000 and a Prime 550 in-house, on which portability development is following very closely behind the VAX implementation. We are currently considering other hardware. The requirements for the hardware include: a PL/I compiler, 32-bit addressing (since we rely on this all too heavily in our algorithms throughout the SAS system) and virtual storage (because of the memory requirements of SAS).

In conclusion, it should be said that "porting" a system to another environment is a very involved task that takes much planning, knowledge of both the target and the source systems, diligent efforts, willingness to compromise, and moreover, a desire to provide your user base with a high-quality product for their computers. We at SAS Institute hope that the Portable SAS system indeed illustrates the recognition of these characteristics.

Implementation of SPSS* and SPSS-X*

Jonathan B. Fry

SPSS Inc.
Chicago, Illinois USA

SPSS is a mature integrated statistics package whose implementation dates back to 1966. It is implemented primarily in IBM 360 Fortran, and is very fast and reasonably compact. It is available on a wide variety of machine types and operating systems.

SPSS-X is a new system with enhanced function intended to replace SPSS, implemented in portable Fortran using a preprocessor to enhance portability. The new implementation techniques used have important consequences to its developers, converters and users. The functional characteristics and implementation of each system is discussed.

INTRODUCTION.

The SPSS Batch System is a very mature (introduced in 1966) statistical problem solving system used at over 4000 sites worldwide. It is currently available in 36 different versions for different machine - operating system combinations, including versions on nearly all the popular mainframes sold in the United States, on many European and Japanese makes and on many of the larger minicomputers. Documentation for this system includes:

    A combination user's guide and reference manual [5]
    The Release 7-9 update manual [4]
    A pocket guide [3]
    An algorithms manual [7]
    An introductory statistics guide [6]

That system is in its last release. Its successor is SPSS-X, a brand new (first released for general use in 1983) product containing large parts of the SPSS batch system, but with a new central system, new basic syntax rules, and new file management capabilities.

This package has been sent to about 500 installations to date. It is available only for IBM 370-like machines running OS or CMS. It will be available next month for DEC's VAX machines running VMS. Numerous other versions are under way. Its documentation consists of:

    A user's guide including a reference

----

* SPSS is a registered trademark of SPSS Inc. SPSS-X is a trademark of SPSS Inc.

    card [8]
    An algorithms manual [7]

    and in preparation:
    SPSS-X Basics
    An introductory statistics guide
    An advanced statistics guide
    Data Management with SPSS-X

EXTERNAL DESCRIPTIONS

Both these systems are batch programs which read command files. In SPSS, the first line of each command contains the command's name in columns 1 to 15; continuations are written in free form in column 16 and after.

In SPSS-X, continuation lines are blank in the first column; first lines are not. Otherwise, the commands are entirely free form.

The basic syntax rules described here are fairly primitive. They prevent placing more than one command on a line, and they do not encourage meaningful indenting (although SPSS-X deals with that problem). Nonetheless, the syntax has proved successful in use. It has proved itself easy to teach to computer novices, and failure to properly delimit command boundaries is not a common syntax error.

Both systems normally operate on a rectangular data matrix, performing statistical analyses, data displays, or data management operations on it. Thus the command file usually has parts that

A. Define the original data matrix, either by describing an external data file or by recalling a data matrix defined earlier and saved.

B. Modify the data matrix by adding new columns, changing values in columns, or deleting rows.

C. Modifying the data matrix in those same ways, but temporarily, meaning only for the next procedure.

D. Finally, invoke the procedure or manipulation command the user had in mind in the first place.

In SPSS, the user may then begin again at step C specifying more temporary data transformations and procedures. In SPSS-X, he or she may begin again anywhere in the sequence, and the range of possibilities in the first three steps is much wider.

The rectangular data array is a common abstraction in statistical packages for good reason. It represents simple surveys well, and can be adapted to much experimental data. It is not quite right for time series, and complex surveys are usually better represented by more complex data structures. SPSS never dealt with anything else, and this was the weakness that lead to SPSS-X. It was retained as the basic data abstraction for SPSS-X because it is easy to understand, and because the research in relational data bases showed that, subject to certain restrictions and given appropriate ways to combine rectangular data arrays, users could always represent their data and could combine rectangular arrays to produce any needed representation. Put simply, it was clear that nothing more complex was needed.

Both these systems are targeted at researchers. They provide a great deal of flexibility, they handle large data sets with reasonable efficiency, and they do not require a great deal of computer skill from their users.

INTERNAL DESCRIPTIONS

Internally, both these systems can, at first blush, be categorized as large Fortran programs. In fact, SPSS is written in IBM Fortran IV and IBM Assembler. SPSS-X is written in a portable dialect of Fortran 66 that passes through a preprocessor for application on machine dependencies, common code, and changeable parameters. Almost all of that system exists in Fortran, although much of the Fortran is replaced with assembler code in versions licenced to customers.

Fortran was an easy choice in 1966. A high-level language was preferred for productivity, and Fortran on the IBM 360 provided well-optimized code, and the subroutines the original authors intended to use or adapt were in that language.

When SPSS-X was begun in 1979, Fortran was again chosen because it offered the best portability and the company already had 155,000 lines of it in SPSS from which it expected to use the majority. The preprocessor approach was chosen to alleviate portability problems and improve the productivity of the programming team. The approach had been proven in IMSL[1] and P-STAT[2].

On most machines, overlays are used to make both system fit into memory. Procedures overlay each other and the transformation command processing code. Dynamic loading has never been used because the parts of both systems communicate through Fortran named common sections, and no system the author knows of can resolve common section references at dynamic load time.

Converting SPSS-X is a very different problem from converting SPSS to a new machine or operating system. SPSS was written to run efficiently on IBM hardware, and little attention was given to portability in writing much of the code. The converter must pay careful attention to nearly every subroutine. SPSS-X was written to be ported. The converter must create a set of character string manipulation routines and convert a preprocessor first. Once those are done, most routines require no change whatever to run on a different machine. The converter can get the entire system running fairly quickly, then concentrate on key routines to make it run faster.

This approach was chosen to minimize the time needed to perform all the conversions to each target after the first one or two. It is expected that few of the changes in such new releases will be in the parts of the system the converter had to rewrite.

INTERNAL DATA FLOW

Transformation commands like COMPUTE are translated into code for an abstract computer; that code (referred to here as t-code) is saved in memory. In SPSS-X, the class "transformations" includes most of the commands that read data from files, as well as the true data transformations which fit in that category in SPSS.

On most machines, the t-code is executed interpretively; the IBM versions of SPSS-X compile it into machine code.

The OBSERVE subsystem within SPSS or SPSS-X is responsible for performing all the data-creation and transformation actions and providing the data matrix to procedures. It executes the transformation commands, reads data, manages the data scratch file or files, and passes the data matrix to procedures one row at a time.

SPSS has one data scratch file, on which OBSERVE writes the modified data matrix. Since it has only one, the data on it cannot be modified except in memory, hence temporarily. SPSS-X has two data scratch files, so that it can permanently modify the data as often as needed.

Procedures get their data from the OBSERVE subsystem; they interpret their own commands, and can print reports, add columns to the data matrix, and write saved system files.

A procedure typically consists of three parts:

A. A parser which uses tokens built from the procedure's command using a system facility.

B. A part which processes rows of the data matrix gotten from OBSERVE to build an internal summary appropriate to the procedure (a frequency table, cross-products table, or perhaps an internal copy of the data).

C. A set of routines which use that internal summary to print the desired tables, write the desired files, or compute new column from the data matrix.

Procedures in SPSS-X have access to system service routines for:

A. Memory management
B. Lexical analysis
C. Access to a dictionary of the data columns, containing each one's name, missing values, and labels.
D. Error reporting
E. Format conversion
F. Getting the data
G. Input and output operations
H. Character string handling.
I. Parsing common syntactic entities
J. Printing and page control
K. Computing common probability functions
L. Reading matrix input

Because the transformation commands are stored in t-code form, all the commands which are to be applied for a particular use of the data (a procedure) can be applied to each row as it is read; no extra data passes are made to apply the transformations. Since the t-code, the t-code interpreter, the procedure code, the central system code, and the memory for the procedure's intermediate data summary must all be in memory together, SPSS and especially SPSS-X require large memories. Thus memory is traded for data passes.

REFERENCES

[1] Aird, T. J., Battiste, E. L. and Gregory, W. C., Portability of mathematical software coded in Fortran, Technical report, International Mathematical and Statistical Libraries Inc., (August, 1976)

[2] Buhler, R., P-STAT portability, in Proceedings of Computer Science and Statistics: 8th Annual Symposium on the Interface (UCLA, 1975)

[3] Hull, C. H. and Nie, N. H., SPSS pocket guide (McGraw-Hill, New York, 1981)

[4] Hull, C. H. and Nie, N. H., SPSS update 7-9 (McGraw-Hill, New York, 1981)

[5] Nie, Norman H. et. al., Statistical package for the social sciences, 2nd ed. (McGraw-Hill, New York, 1975)

[6] Norusis, M. J., SPSS introductory guide: basic statistics and operations (McGraw-Hill, New York, 1982)

[7] SPSS Inc, SPSS-X statistical algorithms (SPSS Inc., Chicago, 1983)

[8] SPSS Inc., SPSS-X user's guide (McGraw-Hill, New York, 1983)

Tools for Developing Statistical Software

Organizer: Webb Miller

Invited Presentations:

    Portable Tools and Environments for Software Development, Webb Miller

PORTABLE TOOLS AND ENVIRONMENTS FOR SOFTWARE DEVELOPMENT

Webb Miller

Department of Computer Science
University of Arizona
Tucson, AZ 85721

We discuss the nature and availability of some software parts, software tools and programming environments that may be useful for developing statistical software.

1. The Software Tools Users Group

The *Software Tools Users Group* is focused on a set of licence-free, *UNIX*-like utilities and system calls, written in Ratfor and Pascal. When run in conjunction with almost any local operating system, the Tools package presents a virtual operating system interface consisting of a virtual machine (system calls or "primitives"), utility programs, and a command language, thus achieving inter-system uniformity over a variety of operating systems. Originating from Kernighan and Plauger's book *Software Tools* (Prentice-Hall, 1976), the enhanced package now includes programs for text formatting, mail systems, enhanced Ratfor preprocessors, a source code control system, command line interpreter similar to the *UNIX* shell, and many other utilities. Further information can be obtained from:

> the Software Tools Users Group
> 1259 El Camino Real, #242
> Menlo Park, CA 94025

Information about a similar collection of tools is available from the following address.

> Software Tools Distribution
> Department of Computer Science
> University of Arizona
> Tucson, AZ 85721

2. Toolpack

Toolpack is a systematized collection of software tools to facilitate the development and maintenance of Fortran programs. The collection will be a mechanism to promote the use of high quality tools in software engineering research and in the development of numerical software. The project is a collaborative effort between universities, government and industry. The cooperating organizations are Argonne National Laboratories, Bell Laboratories, Jet Propulsion Laboratory, Numerical Algorithms Group (NAG), Purdue University, the University of Arizona and the University of Colorado. Further information can be obtained from:

> Wayne Cowell, MCSD-221
> Argonne National Laboratory
> 9700 South Cass Ave.
> Argonne, IL 60439

3. PFORT Verifier

The main function of the PFORT Verifier is to check that programs are written in a subset of the 1966 Fortran standard that is known to work on many computer systems. In addition, it checks for a number of interprocedural errors that are usually not diagnosed by compilers. For information about obtaining the PFORT Verifier contact:

> Francine Glick, Rm. 2F-128
> Bell Laboratories
> 600 Mountain Ave
> Murray Hill, NJ 07974
> (201) 582-7330

4. PORT Library Framework

These three program packages can be used to provide a Fortran subroutine library with machine-dependent constants, automatic error handling, and dynamic storage allocation using a stack. For a complete description see "The PORT mathematical subroutine library" by P. A. Fox, A. D. Hall and N. L. Schryer, *ACM Transactions on Mathematical Software*, June 1978, pp. 104-126. The software can be ordered (under the name "ACM Algorithm 528") by a process explained on the last few pages of any recent issue of *TOMS*.

Nonnumeric Algorithms

Organizer: *Michael Steele*

Invited Presentations:

    Letting MACSYMA Help, *Gail Gong*

    Codata Tools: Portable Software for Self-Describing Data Files, *Deane Merrill* and *John L. McCarthy*

LETTING MACSYMA HELP

Gail Gong

Carnegie-Mellon University

ABSTRACT

To illustrate the usefulness of MACSYMA, we consider an eigenvalue problem which arises from time series. MACSYMA gives us a freedom to explore the problem. Where as a tedious calculation might otherwise prevent us from looking at the problem in a certain way, MACSYMA quickly can supply us with such calculations. In the time series problem, MACSYMA shows that the inverse of a linear transformation of the matrix of interest has a simple structure. Applying some matrix theory to this simple matrix gives a very nice interpretation for the eigenvalues.

KEY WORDS:

MACSYMA, autoregressive process, eigenvalues, perturbation theory, **power method**, Green's matrix, Jacobi matrix, tridiagonal matrix, Gerschgorin Circle Theorem, Separation Theorem, pseudo-circulant.

1. INTRODUCTION

MACSYMA is a program, produced at MIT, which performs symbolic manipulations. The power of MACSYMA and the convenience which it can afford make MACSYMA a valuable tool to the statistician. To acquire some exposure to the various facilities of MACSYMA and to illustrate its usefulness, we will consider a question from time series. MACSYMA together with some matrix-theoretic results will provide an insightful answer to this question.

2. A PROBLEM FROM TIME SERIES

An economical generalization of random walk is the first order autoregressive model, AR(1),

$$Y_0 = 0,$$
$$Y_t = \rho Y_{t-1} + Z_t, \quad t=1, \ldots, n,$$

where $Z_1, \ldots, Z_n$ are iid. Because of the natural imbedding of random walk in the AR(1) model, a problem considered by many authors is that of testing $H_0 : \rho = 1$. This has attracted particular attention in the case when Z_t is assumed to be $N(0, \sigma^2)$. The Rao statistic for testing H_0 can easily be seen to be

$$U(\rho) = -\frac{1}{\sigma^2} \sum_{t=1}^{n} Z_t Y_{t-1}.$$

We wish to obtain the distribution of $U(\rho)$ for $0 < \rho \leq 1$. This problem was introduced to us by Ahtola in his PhD thesis [1]. When $0 < \rho < 1$, $U(\rho)$ is asymptotically normal. However, when $\rho = 1$, $U(1)$ is far from normal and in fact heavily skewed with a large left tail. Therefore, it seems reasonable to believe for n moderate and ρ close to 1, $U(\rho)$ would also have a large left tail. Ahtola verifies this by calculating the exact distribution of $U(\rho)$ for various values of n and ρ. For n = 50 and ρ exceeding 0.8, the asymptotic approximation fails. For practical applications, these are exactly the values of the parameters in which we are interested. The failure of the asymptotic approximations forces us to seek the finite sample distribution of $U(\rho)$.

We may write

$$U(\rho) = -\frac{1}{2} \frac{Z^T}{\sigma} C(\rho) \frac{Z}{\sigma},$$

where

$$Z = \begin{bmatrix} Z_1 \\ Z_2 \\ Z_3 \\ \vdots \\ Z_n \end{bmatrix}, \quad C(\rho) = \begin{bmatrix} 0 & 1 & \rho & \rho^2 & \cdots & \rho^{n-2} \\ 1 & 0 & 1 & \rho & \cdots & \rho^{n-3} \\ \rho & 1 & 0 & 1 & \cdots & \rho^{n-4} \\ \vdots & & & & & \vdots \\ \rho^{n-2} & \rho^{n-3} & \rho^{n-4} & \rho^{n-5} & \cdots & 0 \end{bmatrix},$$

and since $\frac{Z}{\sigma}$ is an n-dimensional standard normal random vector,

$$-2 U(\rho) = \sum_{t=1}^{n} \gamma_t \chi_t^2(1), \qquad (1)$$

where $\chi_1^2(1), \ldots, \chi_n^2(1)$ are independent chi-square random variables each with 1 degree of freedom, and $\gamma_1, \ldots, \gamma_n$ are the eigenvalues of $C(\rho)$. Ahtola points out that if we could find $\gamma_1, \ldots, \gamma_n$, then the distribution of $U(\rho)$ can easily be calculated using a one-dimensional numerical integration. See Imhof's [4] paper. Also in this vein are Schilling's [5] beautiful results.

When $\rho = 1$, the matrix $C(1)$ has a simple form; it is just the n×n matrix of all 1's except for 0's down the diagonal, and the eigenvalues are easily shown to be $\lambda_1 = (n-1)$ and $\lambda_2 = ... = \lambda_n = -1$, and so

$$-2\, U(1) = (n-1)\, \chi^2(1) - \chi^2(n-1),$$

where $\chi^2(1)$ and $\chi^2(n-1)$ are independent chi-square random variables with indicated degrees of freedom. When $\rho = 1$, the problem of finding the distribution of $U(1)$ is essentially solved. Notice that since $\chi^2(1)$ is long-tailed and $\chi^2(n-1)$ is nearly normal, then $U(1)$ is long-tailed.

To obtain the exact distribution of $U(\rho)$ when $0 < \rho < 1$, we need to solve the eigenvalue problem for $C(\rho)$. Algorithms for computing the eigenvalues of a general matrix are of order $O(n^3)$. So if $C(\rho)$ is pushed through a standard eigenvalue routine, computation quickly becomes infeasible with growing n. Therefore, Ahtola abandons the exact distribution of $U(\rho)$ and instead obtains an approximation by fitting $U(\rho)$ to a linear combination of two chi-square random variables. This is an extremely plausible approach since $U(1)$ is a linear combination of two chi-square random variables, and we would expect $U(\rho)$ to be close to $U(1)$ when ρ is close to 1. Indeed, Ahtola's approximation offers some intuitively appealing interpretations.

We take a different tack. The matrix $C(\rho)$ has a simple structure, and it is often the case that the eigenvalue problem for structured matrices can be solved in time much less than $O(n^3)$. We call on MACSYMA together with some matrix-theoretic results to study the eigenvalues of $C(\rho)$. The reward for such an approach will be a nice interpretation of the eigenvalues.

3. SOLVING THE EIGENVALUE PROBLEM

What are the eigenvalues of $C(\rho)$ when $0 < \rho < 1$?

Since $C(\rho)$ has a simple pattern, perhaps the eigenvalues of $C(\rho)$ also follow a simple pattern. In

Figure 1:
Letting MACSYMA calculate the eigenvalues of $C_n(\rho)$ for n = 3, 4, 5, 6.

```
(C2) makematrix1(mdim):=block([k,result],
     local(m),
     for i:1 thru mdim do
         for j:1 thru mdim do (
             k:abs(i-j),
             m[i,j]: if k=0 then 0
                     else r^(k-1)
         ),
     result:genmatrix(m,mdim,mdim),
     return(result))$

(C7) makematrix1(3);
                    [ 0   1   R ]
                    [           ]
(D7)                [ 1   0   1 ]
                    [           ]
                    [ R   1   0 ]

(C8) charpoly(%,x);
                2
(D8)     - X (X  - 1) + R (R X + 1) + X + R

(C9) factor(%);
                             2
(D9)           - (X + R) (X  - R X - 2)

(C10) solve(%,x);
                         2
                R - SQRT(R  + 8)
(D10)      [X = ----------------,
                       2

                        2
                SQRT(R  + 8) + R
            X = ----------------, X = - R]
                       2
```

n	Characteristic Polynomial
2	$(X - 1)(X + 1)$
3	$-(X + R)(X^2 - RX - 2)$
4	$(X^2 - R^2 X - X - 2R - 1)$
	$(X^2 + R^2 X + X + 2R - 1)$
5	$-(X^3 + R X^2 + R X + 2 R^2 - 1)$
	$(X^3 - R^3 X^2 - R X^2 - 4 R^2 X - 3 X - 4 R)$
6	$(X^3 - R^3 X^2 - R^2 X^2 - X^2 - 4 R^2 X - R X$
	$- 2 R X - 2 X - 4 R^2 - 2 R + 1)$
	$(X^3 + R^3 X^2 + R^2 X^2 + X^2 + 4 R^2 X - R X$
	$+ 2 R X - 2 X + 4 R^2 - 2 R - 1)$

Figure 1, MACSYMA calculates the eigenvalues of $C(\rho)$ for $n = 2, ..., 6$. It was a reasonable first try, but seeing any patterns seems fairly hopeless.

Ahtola recognizes that the asymptotic theory fails when one eigenvalue heavily dominates the others, and so it is reasonable to study the largest eigenvalue. We pursue two different approaches. Our first approach recognizes that when $\rho = 1$, $C(1)$ has a largest eigenvalue $\lambda_1 = n-1$ with correponding eigenvector $V = (1, ..., 1)^T$. We would expect that when $\rho \sim 1$, the largest eigenvalue and corresponding eigenvector of $C(\rho)$ would be close to those of $C(1)$. This is the essence of perturbation theory as described by Bellman [2]. More precisely, suppose A is a symmetric matrix with known eigenvalues $a_1, ..., a_n$. If a_1 is distinct from $a_2, ..., a_n$ and has eigenvector V, then for a symmetric matrix B,

$$a_1 + \epsilon \frac{V^T B V}{V^T V} ,$$

gives an approximation to an eigenvalue of $A + \epsilon B$.

In our problem, let

$$A = C(1) ,$$
$$\epsilon = 1 ,$$
$$B = C(\rho) - C(1) .$$

Since the largest eigenvalue $a_1 = n-1$ of A is distinct from the remaining eigenvalues $a_2, ..., a_n$ and has eigenvector $V = (1, ..., 1)^T$, perturbation theory suggests that the largest eigenvalue γ_1 of $C(\rho)$ has approximation

$$\gamma_1 \sim a_1 + \frac{V^T \big(C(\rho)-C(1)\big) V}{V^T V} ,$$

and since

$$a_1 = \frac{V^T C(1) V}{V^T V} ,$$

this approximation becomes

$$\gamma_1 = \frac{V^T C(\rho) V}{V^T V} .$$

An easy calculation gives

$$\gamma_1 \sim \frac{2}{n(1-\rho)^2} \Big(n(1-\rho) - (1-\rho^n)\Big) ,$$

$$= \frac{2}{1-\rho} - \frac{2(1-\rho^n)}{n(1-\rho)^2} .$$

A second approach to studying the largest eigenvalue is the power method. See Schwarz [6]. If C is a symmetric matrix with eigenvalues $\gamma_1, ..., \gamma_n$ where $|\gamma_1| > |\gamma_j|$, $j = 2, ..., n$, and if V is an arbitrary vector, then

$$\frac{V^T C^m V}{V^T C^{m-1} V} \to \gamma_1 \text{ as } m \to \infty .$$

Notice that for our choice of A, ϵ, B, perturbation theory is the first iteration in the power method with V the eigenvector corresponding to the largest eigenvalue of $C(1)$. The second iteration gives the approximation

$$\gamma_1 \sim \frac{V^T C(\rho)^2 V}{V^T C(\rho) V} ,$$

$$= \frac{1}{(1-\rho)^2}$$
$$\times \left(\frac{(2+\rho^{n-1})n(1-\rho)^2 - 4(1-\rho^n)(1-\rho) + 1 - \rho^{2n}}{n(1-\rho) - (1-\rho^n)} \right).$$

Perturbation theory and the power method offer ways of studying the dominating eigenvalue. In what follows, we use the properties of some special matrices to get a concise description of *all* the eigenvalues.

Instead of seeking the eigenvalues of $C = C(\rho)$ directly, we seek the eigenvalues of a transformation of C. This transformation is essentially the inverse of a linear polynomial in C. Because for such transformations, the eigenvalues transform in the obvious way, solving the transformed problem is equivalent to solving the original problem.

First, let $G = I + \rho C$. It turns out that G is a Green's matrix, whose inverse is a tridiagonal or Jacobi matrix. Figure 2 shows MACSYMA inverting G for $n = 6$. We see that not only is G^{-1} tridiagonal, but it is a very simple tridiagonal:

$$G^{-1} = \frac{H}{1-\rho^2} ,$$

where

$$H = \begin{bmatrix} 1 & b & & & & \\ b & a & b & & & \\ & b & a & b & & \\ & & \ddots & \ddots & \ddots & \\ & & & b & a & b \\ & & & & b & 1 \end{bmatrix},$$

and where $a = 1+\rho^2$ and $b = -\rho$. If η is an eigenvalue of H, then $(1-\rho^2)/\eta$ is an eigenvalue of G, and so

$$\gamma = \frac{1}{\rho}\left(\frac{1-\rho^2}{\eta} - 1\right), \qquad (2)$$

is an eigenvalue of C. Therefore, solving the eigenvalue problem for the very simple matrix H is sufficient for solving the eigenvalue problem for C.

To get the eigenvalues for H, we will use two very powerful theorems from matrix theory, the Gerschgorin Circle Theorem [6] and the Separation Theorem [2].

The Gerschgorin Circle Theorem is useful for specifying a region in which all the eigenvalues must lie. For a general n×n matrix A with elements a_{ij}, define

$$r_i = \sum_{\substack{j=1 \\ j \neq i}}^{n} |a_{ij}|.$$

Then the union of the n circles with centers a_{ii} and corresponding radii r_i in the complex plane contains the n eigenvalues of A. Our matrix H being symmetric has real eigenvalues. The Gerschgorin Circle Theorem says that the eigenvalues of H lie in the union of the two intervals

$$[1-b, 1+b] = [1-\rho, 1+\rho],$$

$$[a-2b, a+2b] = [(1-\rho)^2, (1+\rho)^2],$$

Figure 2:
Letting MACSYMA calculate G^{-1} for $n = 6$.

```
(C2) makematrix2(mdim):=block([k,result],
    local(m),
    for i:1 thru mdim do
        for j:1 thru mdim do (
            k:abs(i-j),
            m[i,j]:r^k
        ),
    result:genmatrix(m,mdim,mdim),
return(result))$

(C4) makematrix2(6);

(D4)
[              2    3    4    5 ]
[ 1    R    R    R    R    R    ]
[                                ]
[              2    3    4       ]
[ R    1    R    R    R    R    ]
[                                ]
[ 2              2    3          ]
[ R    R    1    R    R    R    ]
[                                ]
[ 3    2              2          ]
[ R    R    R    1    R    R    ]
[                                ]
[ 4    3    2                    ]
[ R    R    R    R    1    R    ]
[                                ]
[ 5    4    3    2               ]
[ R    R    R    R    R    1    ]

(C5) %^^(-1);

(D5)
[   1         R                                              ]
[  -----    ------     0         0        0        0         ]
[   2         2                                              ]
[  R - 1    R - 1                                            ]
[                                                            ]
[                      2                                     ]
[   R       R + 1      R                                     ]
[  -----   - ------   ------     0        0        0         ]
[   2         2         2                                    ]
[  R - 1    R - 1     R - 1                                  ]
[                                                            ]
[                                2                           ]
[              R       R + 1     R                           ]
[   0        ------  - ------   ------    0        0         ]
[              2         2        2                          ]
[            R - 1     R - 1    R - 1                        ]
[                                                            ]
[                                         2                  ]
[                       R       R + 1     R                  ]
[   0         0       ------  - ------   ------    0         ]
[                       2         2        2                 ]
[                     R - 1     R - 1    R - 1               ]
[                                                            ]
[                                                  2         ]
[                                R      R + 1      R         ]
[   0         0         0      ------ - ------   ------      ]
[                                2        2        2         ]
[                              R - 1    R - 1    R - 1       ]
[                                                            ]
[                                         R        1         ]
[   0         0         0         0     ------  - ------     ]
[                                         2        2         ]
[                                       R - 1    R - 1       ]
```

and so the eigenvalues of H lie in the interval

$$[(1-\rho)^2, (1+\rho)^2] .\qquad(3)$$

The Separation Theorem gives a beautiful result relating the eigenvalues of the general n×n matrix A to the eigenvalues of the (n−1)×(n−1) submatrix B gotten by deleting the last row and last column of A. Let A have eigenvalues $\alpha_1 \leq ... \leq \alpha_n$ and B have eigenvalues $\beta_1 \leq ... \leq \beta_{n-1}$. Then

$$\alpha_1 \leq \beta_1 \leq \alpha_2 \leq \beta_2 \leq ... \leq \alpha_{n-1} \leq \beta_{n-1} \leq \alpha_n.$$

In our problem, let K be the (n−2)×(n−2) submatrix of H gotten by deleting the first and last rows and columns of H. Let H have eigenvalues $\eta_1 \leq ... \leq \eta_n$ and K have

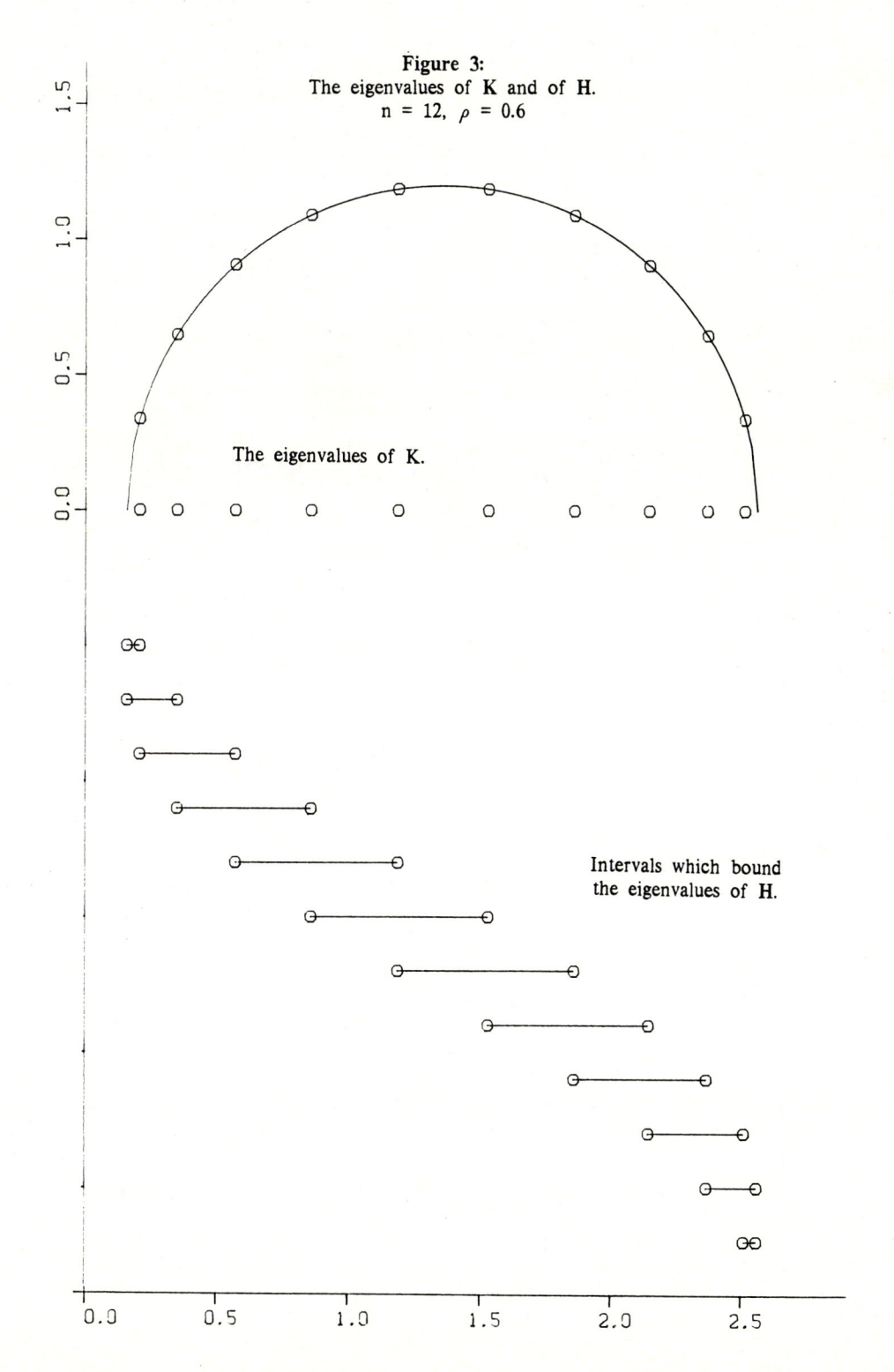

Figure 3:
The eigenvalues of K and of H.
n = 12, ρ = 0.6

The eigenvalues of K.

Intervals which bound the eigenvalues of H.

eigenvalues $\kappa_1 < ... < \kappa_{n-2}$. Then applying the separation theorem twice gives

$$\eta_1 \leq \kappa_1,$$
$$\eta_2 \leq \kappa_2,$$
$$\kappa_{j-2} \leq \eta_j \leq \kappa_j, \quad j=3, ..., n-2 \quad (4)$$
$$\kappa_{n-3} \leq \eta_{n-1},$$
$$\kappa_{n-2} \leq \eta_n.$$

Now, if we can get the eigenvalues of K, we would have a good idea of the locations of the eigenvalues of H. Hall [3] calls K a pseudo-circulant and obtains a closed-form solution for the eigenvalues:

$$\kappa_j = 1 + \rho^2 - 2\rho \cos\left(\frac{j\pi}{n-1}\right), \quad j=1, ..., n-2.$$

Figure 3 shows the construction of $\kappa_1, ..., \kappa_{n-2}$. Consider the circle with center $(1+\rho^2, 0)$ and radius 2ρ. Divide the upper half of this circle into n-1 equal arcs

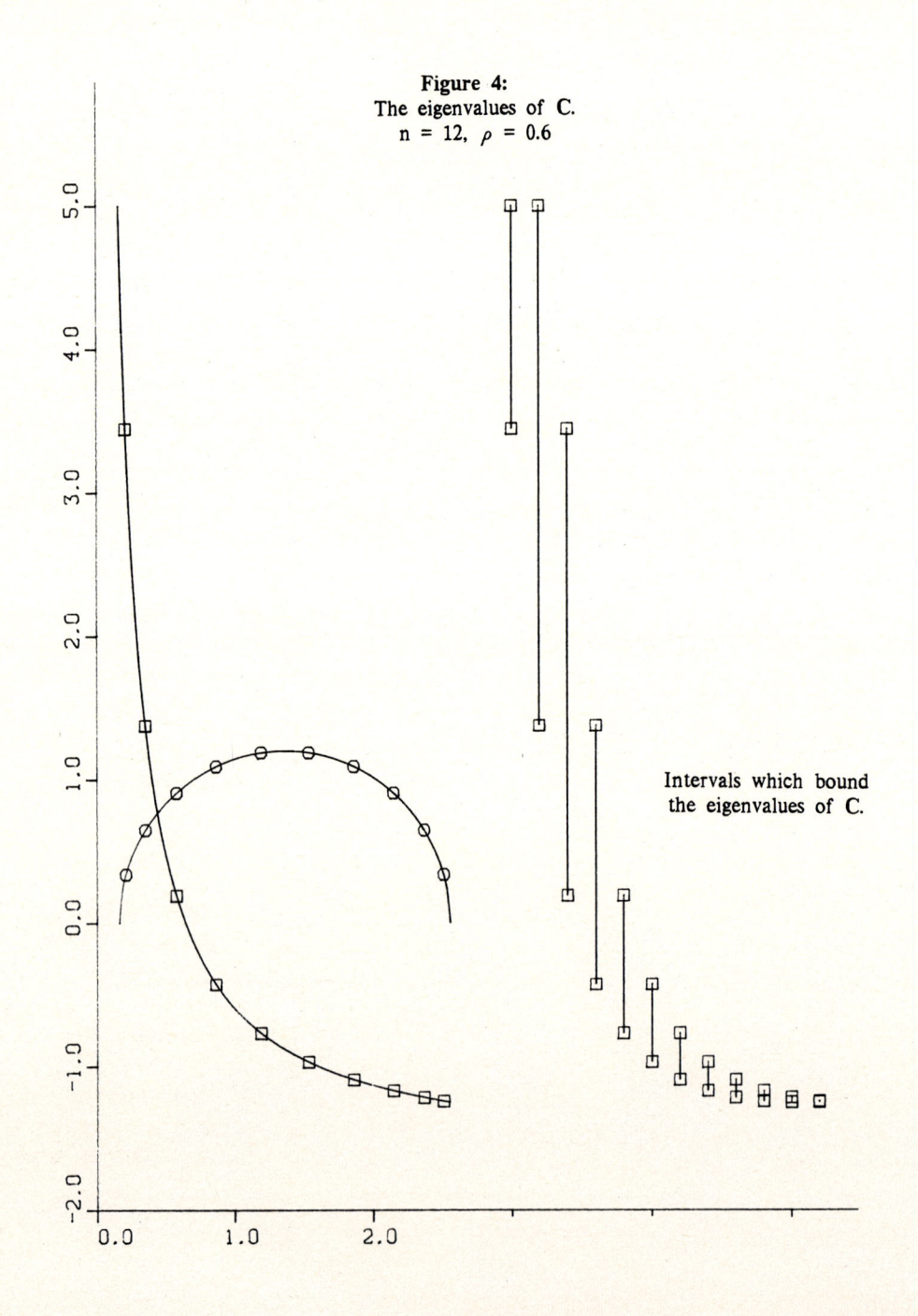

Figure 4:
The eigenvalues of C.
n = 12, ρ = 0.6

Intervals which bound the eigenvalues of C.

and project the n-2 points dividing these arcs onto the horizontal axis. These projections correspond to κ_1, ..., κ_{n-2}. Figure 3 also shows the construction of intervals which bound η_1, ..., η_n, the eigenvalues of H. These intervals result from the inequalities in (4) together with the Gershgorin bounds (3). Notice that the Gershgorin bounds are precisely the upper and lower horizontal limits of the circle above. For general n, we could not have obtained better bounds. Indeed as n grows, the Gershgorin bounds are the limits of the largest and smallest eigenvalues of K and therefore of H.

In Figure 4, we redraw the semicircle in Figure 3 and plot the function

$$\Gamma(\eta) = \frac{1}{\rho}\left(\frac{1-\rho^2}{\eta} - 1\right).$$

Notice that the scale has been reduced. Since Γ maps the eigenvalues of H to those of C by Equation (2), then Γ maps the intervals which bound η_1, ..., η_n to intervals which bound γ_1, ..., γ_n, the eigenvalues of C. These intervals are shown in Figure 4.

Figure 5:
What happens as n increases?

n = 23, ρ = 0.6

Figure 6:
What happens as ρ increases?

n = 12, ρ = 0.8

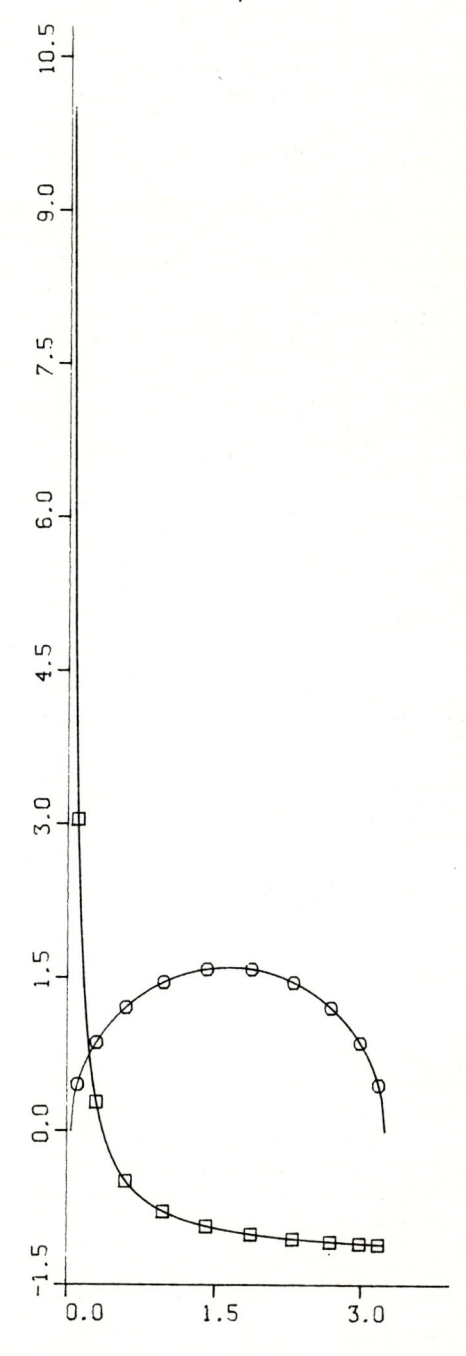

We can learn much about the behavior of the eigenvalues of C by studying the picture in Figure 4. Since $\eta_1, ..., \eta_n$ must fall between $(1-\rho)^2$ and $(1+\rho)^2$, then $\gamma_1, ..., \gamma_n$ must fall between $\Gamma((1+\rho)^2) = -2/(1+\rho)$ and $\Gamma((1-\rho)^2) = 2/(1-\rho)$. Because $\Gamma(\eta)$ has a huge negative slope when η is small and levels off quickly, then C has a few large eigenvalues and many small eigenvalues. When n is small, then the existence of one or two dominating eigenvalues precludes any hope of asymptotic approximations. What happens as n increases? Figure 5 shows the eigenvalues when the sample size is essentially doubled. For any fixed region in $[-2/(1+\rho), 2/(1-\rho)]$, notice that as n grows, the number of eigenvalues inside that fixed region also grows, and so for sufficiently large n, the normal approximation takes effect. It is just that we must take n to be very large.

Figure 4 assumes $\rho = 0.6$. Figure 6 shows a corresponding picture for $\rho = 0.8$. Because we wished to include the full range $[-2/(1+\rho), 2/(1-\rho)]$ of Γ on the vertical scale, the picture for $\rho = 0.8$ has been greatly reduced. As ρ grows toward 1, the lower horizontal limit $(1-\rho)^2$ of the circle shrinks toward 0 and the upper limit $2/(1-\rho)$ of the range of Γ grows. More importantly, the negative slope of $\Gamma(\eta)$ at small values of η becomes increasingly steep. The larger eigenvalues become more and more dominating, and asymptotic approximations require larger and larger values of n.

4. CONCLUSIONS

Using MACSYMA together with some matrix-theoretic results, we arrived at a very nice interpretation of the eigenvalues of C. We hope that this interpretation will give an increased understanding of the distribution of $U(\rho)$. This is a question of current research.

ACKNOWLEDGEMENTS

The author would like to thank J. Michael Steele for valuable motivation and assistance. Also, for the use of MACSYMA, thanks go to the Mathlab group of the Massachusetts Institute of Technology, whose work is supported, in part, by the United States Energy Research and Development Administration under Contract Number E(11-1)-3070 and by the National Aeronautics and Space Administration under Grant NSG 1323.

REFERENCES

[1] Ahtola, J. A.
Studies in asymptotic and finite sample inference of nonstationary and nearly stationary autoregressive models.
PhD thesis, University of Wisconsin, Madison, 1983.

[2] Bellman, R.
Introduction to Matrix Analysis.
McGraw-Hill Book Company, New York, 1970.

[3] Hall, G. G.
Matrices and Tensors.
Macmillan Company, New York, 1963.

[4] Imhof, J. P.
Computing the distribution of Quadratic forms in normal variables.
Biometrika 48:419-426, 1961.

[5] Schilling, M. F.
An infinite-dimensional approximation for nearest neighbor goodness of fit.
The Annals of Statistics 11:13-24, 1983.

[6] Schwarz, H.R.
Numerical Analysis of Symmetric Matrices.
Prentice-Hall, Inc., Englewood Cliffs, New Jersey, 1973.

Codata Tools: Portable Software for Self-Describing Data Files*

Deane Merrill and John L. McCarthy

Lawrence Berkeley Laboratory
University of California
Berkeley, California 94720

This paper describes the Codata tools, a set of programs which read, write, and restructure self-describing Codata (common data format) files. These tools manipulate both data and data description, so that the the output of any operation is itself a Codata file. Semantics of results and descriptions of derived Codata files are inherited from descriptions of input Codata files. Following the Software Tools philosophy, the Codata tools are modular - each tool performs a specific limited task. They follow the UNIX and Software Tools conventions of standard input and output. The output of any module can automatically serve as the input of another, and they can be "pipelined" or "chained" together. Codata tools can be used to extract specified rows and/or columns from a file, to sort a file, to perform relational joins, to perform tabulations by aggregating on common key values, and to perform other operations. The Codata tools are written in RATFOR (a transportable FORTRAN preprocessor), and can be easily adapted to run on any computer where the Software Tools have been implemented. Work is currently under way on substantial enhancements to the Codata file format and the Codata tools to provide for more efficient physical storage formats, more complex data structures, and more extensive, open-ended data description.

1. Introduction

In this paper we discuss a simple self-describing file format and associated software developed and used extensively in the Lawrence Berkeley Laboratory (LBL) Computer Science and Mathematics (CSAM) Department since early 1978. The file format is known as "Codata" (common data format) and the associated software tools are known as the "Codata Tools" [MERR81]. These developments have been closely associated with that of SEEDIS (LBL's Socio-Economic Environmental Demographic Information System), but the potential applications are much more general [COMP82, MCCA82B]. The Codata format and tools are sufficiently simple and useful that they have been used extensively outside SEEDIS. One purpose of this paper is to make them more generally known and available to other researchers.

Another purpose of this paper is to point out the need for and importance of tools for statistical data management and analysis that both *use* and *produce* self-describing data files. Statistical package programs pioneered the use of self-describing data files over the past two decades. But with the notable recent exception of System S [BECK78],

they have been deficient in several important respects:

- most of them use *internal* self-describing files, but do not *produce* such files as a routine part of any given data analysis or manipulation routine (hence it is often difficult, if not impossible to make the output of one routine easily available as input for another)

- those that do produce "savefiles" of some kind do not do so in terms of a neutral, non-procedural interchange format, but rather in terms of their own idiosyncratic procedural language.

- they do not provide facilities to easily expand or operate on the meta-data or data description portions of self-describing files

The remainder of this introductory section describes the SEEDIS Project and motivations that led to development of the Codata File Format and Codata Tools. Section 2 explains the general requirements, design principles, and implementation strategies that underlie the Codata File Format and the Codata Tools. Section 3 describes the Codata Format in some detail, while Section 4 describes the Codata Tools and summarizes the specific operations which they perform. Section 5 describes new enhancements to the Codata File Format and Codata Tools which are currently being designed and implemented. Section 6 is a concluding summary.

1.1. Background: The SEEDIS Project

In the early 1970's, LBL contracted with the Department of Labor, the Bureau of the Census, and the National Technical Information Service to produce the Urban Atlas Metropolitan Map series and a number of Manpower Planning reports for use in the Department of Labor Employment and Training Administration (then the Manpower Administration). Important resources resulting from this effort were databases from the 1970 Census of Population, geographic base map files of the 35,000 census tracts defined by the 1970 Census, and associated software for managing and displaying these data. Under supplementary funding from the Army Corps of Engineers, other socioeconomic databases such as the City-County data book and the Census of Agriculture were acquired and made available for automatic retrieval by remote users.

In 1975 the name SEEDIS (Socio-economic Environmental Demographic Information System) was invented to describe what gradually evolved into an integrated system combining information retrieval, data analysis, and display capabilities in a user-friendly environment. A prototype system was implemented on LBL's CDC computer system. Beginning in 1976, the PAREP (Populations at Risk to Environmental Pollution) project began adding environmental and health-related data bases in order to analyze the relationships between human health and air pollution [MERR82]. But further data development was hampered by the lack of adequate data management tools. Even relatively simple data manipulation tasks required the writing of special-purpose programs.

---

* This work was supported by the Office of Health and Environmental Research and the Office of Basic Energy Sciences of the U.S. Department of Energy under Contract DE-AC03-76SF00098; and the Department of Labor, Employment and Training Administration under Interagency Agreement No. 06-2063-36.

1.2. Statistical Data Management Problems

By the late 1970's it was evident that further development of SEEDIS required more systematic data management. Motivating factors included data documentation requirements, the need for different applications programs to share data without being constrained to a single physical file format, and the need to carry out simple data manipulation operations without writing special-purpose programs.

Data documentation was a primary concern. Important data files were being developed whose usefulness would outlive the projects for which they were originally acquired. "Codebooks" were frequently inadequate and idiosyncratic. People who had originally worked with given data files usually forgot important undocumented details once they stopped working with a file on a regular basis. Since documentation was almost always separate from the data, it was easily misplaced or destroyed.

Second, with continuing addition of new data files, applications programs needed to be independent of minor changes in physical file formats. Without such data independence, adding a new data field or changing the length or position of a data field might require every program using that file to be modified.

Third, a number of existing data analysis and display programs required a standard communications file format in order to communicate with one another simply and effectively. For example, CHART, used to make bar charts, line graphs, etc. and CARTE, a thematic mapping program, required different forms of input files. Even the simplest task, such as making a bar graph and a choropleth map from the same data, could not be performed without writing a program.

The situation clearly called for data standards and data management tools, but there were few statistical packages, let alone data management packages, that could run on LBL's CDC computers and the locally developed BKY operating system. Most of the available statistical and database packages lacked sufficient capacity for databases larger than a few megabytes. Statistical package programs did not provide an efficient means for local applications to communicate with one another. Available data management software was slow, complex, and expensive -- particularly for interactive applications -- and it lacked adequate facilities for data description and documentation.

2. Statistical Data Management Requirements

The problems described above did not justify acquisition or development of a full-featured database management system. Data files were archival, with very infrequent updates. Updates did not have to be made by multiple users simultaneously. Security requirements were minimal.

On the other hand, it was clear that the system had to be relatively flexible to permit continuing addition of new files. It also had to be able to handle very large numbers of records (cases), attributes (variables), and databases for different types of entities (primarily geographic areas such as states, counties, census tracts, etc.).

2.1. Design Principles for Codata Files

The solution that SEEDIS project staff adopted was to design and implement a simple, standard, self-describing file format plus a set of tools to perform basic operations on such files. Both the file format and associated tools would be as modular as possible, in order to provide flexibility and to permit further gradual and evolutionary change.

Based on experience with statistical packages, database management software, and applications programs for graphical display, project staff decided that the standard SEEDIS file format should be self-describing. That is, data files would contain essential descriptive information such as data types, missing data conventions, labels, etc. A standard library of input-output routines would be implemented to read and write such files, and applications could use such routines to access both data and data descriptions for a given file. Some additional standards were agreed upon as part of the implementation strategy, as follows:

- data and description would be stored as ASCII text in lines no longer than 132 characters, in order to permit easy inspection and modification of files with standard text-editing tools;
- descriptive items would be stored in "{keyword} = {value}" format, for ease of reading and understandability;
- data would be stored in fixed-format fields to permit easy access via conventional fixed-format read and write routines as well as via special Codata input-output subroutines.

2.2. Codata Tools Development

Beginning in 1979, the Codata Tools were originally designed to satisfy the data management requirements of the PAREP project, including sorting, row and/or column selection, and relational joins. They proved useful enough that the user interface was substantially enhanced, in imitation of the LBL Software Tools and the UNIX operating system. UNIX input-output conventions were adopted, and documentation was brought into conformity with UNIX/Software Tools formats.

At the same time, SEEDIS was being adapted to a network environment of VAX 11/780 computers running the VMS operating system. Applications modules to perform such operations as the reading and writing of computer-independent compressed data records, information retrieval (by geographic area), user-prompted data entry, on-line browsing of data dictionaries, extraction of selected variables from different databases, data transformation, table construction, and thematic mapping were linked via a menu-driven command monitor and the underlying Codata interchange file format.

Experience with the new integrated system suggested that SEEDIS itself required a rather general set of transformation routines for dealing with Codata files. The Codata Tools were further developed and incorporated into SEEDIS at points where data transformations were required, for example in combining data being extracted from more than one data source. Between 1979 and 1982, The standard Codata file format and input-output routines to read and write such files became the nucleus of a new and substantially more integrated SEEDIS system.

3. The Codata format

Codata files are fixed-format ASCII text files containing two logical parts: the data description file (DDF), and the data file (DF), as pictured schematically in Exhibit 1.

Exhibit 1: Schematic Diagram of CODATA File

```
-------------------------------------------
|                                         |
|      Data Description File (DDF) Part   |
|                                         |
-------------------------------------------
|                                         |
|         Data File (DF) Part             |
|                                         |
-------------------------------------------
```

In Codata files, both the data definition file (DDF) and the data file (DF) reside in a single physical data file, with information stored in character representation within fixed-length logical records as defined in the DDF. The DDF includes file-level information, such as the number of data records (rows), and the number of data elements (columns). It also contains information about each data element, including a short name, a descriptive label, data type (integer, floating point, or alphanumeric), field length, location in the record and other information. The data file (DF) portion of a Codata file is a simple "flat" file with fixed-length data elements (fields) in fixed-length records.

The primary virtue of the Codata format is its simplicity. Since both data and description are simple ASCII text, both can be read, written, and modified using either a text editor or formatted read and write statements from a programming language. Codata files are easy to read and understand, to transport between dissimilar computers, and to convert to other file formats. They can be easily printed, edited, or read by a user-written program. Most simple formatted data files can be converted to the Codata format simply by prepending a hand-edited DDF (data description file). Conversely, Codata tools are provided which will strip off the DDF and produce a FORTRAN format statement, for example to read a Codata file into a program like MINITAB or SPSS.

The logical view of a Codata file is that of a table (or flat file) with a fixed number of rows (records) and columns (fields), as pictured in exhibit 2.

Exhibit 2: Codata File Logical Structure

```
          =====Data Definition File (DDF)======
          | File Level Information             |
          -------------------------------------
          | DE 1 Info | DE 2 Info | DE 3 Info |
          =====Data File (DF)==================
Record 1  | Element 1 | Element 2 | Element 3 |
          -------------------------------------
Record 2  | Element 1 | Element 2 | Element 3 |
          -------------------------------------
Record 3  | Element 1 | Element 2 | Element 3 |
          -------------------------------------
Record 4  | Element 1 | Element 2 | Element 3 |
          -------------------------------------
```

Data are arranged so that each logical row of the table contains all the attributes (data elements, columns, fields, or variables) of a named entity (e.g., Alameda County, person number 2037). Data elements include keys necessary for data access and matching, a row label or "stub", and ordinary numeric or alphabetic data values. The number of logical records (rows) is equal to the number of entities in the data file, and the number of columns is equal to the number of data elements in each record.

3.1. A Simple Codata File Example

Exhibit 3 shows a small example file in Codata format. The actual data are contained in the last six lines. The remainder of the file is meta-data, i.e. descriptive information about the data that can be used by programs and people. Explanatory comments in italics at the right are not part of the Codata file but simply for purposes of this illustration.

Exhibit 3: Example Codata File

```
NDE = 6                    file level information
AREAS = 3
CARDLENGTH = 30
MISSING = -999 -991
DE = FIPS.STATE            1st data element information
  TYPE = A
  USE = K
  START = 1
  LENGTH = 2
  HEADER =#FIPS state code#
DE = STUB.GEO              2nd data element information
  TYPE = A
  USE = S
  START = 3
  LENGTH = 5
  HEADER =#state name#
DE = POP80
  TYPE = I
  USE = D
  START = 8
  LENGTH = 9
  HEADER =#1980 population#
DE = POP70
  TYPE = I
  USE = D
  START = 17
  LENGTH = 9
  HEADER =#1970 population#
DE = POP60
  TYPE = I
  USE = D
  START = 31
  LENGTH = 9
  HEADER =#1960 population#
DE = LAND.AREA             6th data element information
  TYPE = I
  USE = D
  START = 40
  LENGTH = 8
  HEADER =#land area#
  HEADER =#in square miles#
END DDF                    beginning of data (DF) part
19Iowa     2913808   2825368
   2757537      55941
48Texas   14229191  11198655
   9579677     262134
49Utah     1461037   1059273
      -999      82096
```

The first four lines constitute the global section of the DDF, and they apply to the file as a whole. The first specifies that the number of data elements (NDE) or columns in the file is six. The second specifies that the file contains three records (AREAS). CARDLENGTH specifies that physical records (lines) in the DDF and DF each contain a maximum of 30 characters of data. MISSING specifies that any numeric data value between -991 and -999 should be considered missing.

The second section of the DDF, through the "END DDF" line, contains a set of meta-data information for each data element (column) in the data file. DE is a unique identifier or name for each data element. TYPE specifies whether the data element values are alphanumeric (A), integer (I), or floating point (D). USE distinguishes between keys (K), row labels or "stubs" (S), and ordinary data elements (D). START gives the byte position for the data element (field) within each logical record. LENGTH specifies the number of bytes allocated to the data element. Each element can have one or more HEADER lines containing descriptive information, for the use of application programs such as the SEEDIS thematic mapping program.

In the example file of Exhibit 3, the six data elements (columns) are named FIPS.STATE, STUB.GEO, ... through LAND.AREA. FIPS.STATE and STUB.GEO are alphanumeric (type A); the others are integer (type I). FIPS.STATE is a key (use K); STUB.GEO is a row label or stub (use S); the others are ordinary data (use D). The column labels (headers) are indicated with each data element. FIPS.STATE starts in column 1 and occupies two characters. STUB.GEO starts in column 3 and occupies five characters. POP60 starts in column 31, which means the first position of the second line of each record (since the line length is 30 characters).

3.2. Data Definition Details

As shown in exhibit 3, the data description file (DDF) or meta-data part of a Codata file contains textual specifications in "{keyword} = {value}" format. This data description is stored in line-image form. No meta-data field can exceed the line length specified by CARDLENGTH, which is limited to a maximum of 132 characters. Blanks and upper-lower case distinctions are significant only within the text of HEADER lines. Keywords occurring before the first data element definition have global effect. That is, they hold for all data elements, unless specifically overridden by keyword definitions within the local environment of a data element definition.

3.3. Data File Details

The DF is the set of fixed length records following the "END DDF" line. As in the DDF, data fields in the DF cannot exceed the specified line length, and can never exceed 132 characters. Upper-lower case distinctions and blanks in alphanumeric fields of the DF *are* significant, unlike those of the DDF. The Codata Tools expect key elements (USE=KEY) to be alphanumeric (TYPE=A), without blanks. Missing data values (for example data suppressed for reasons of confidentiality) can be indicated by values defined in the MISSING line or by leaving the field blank (blanks are not interpreted as zeros on input, as in FORTRAN). Codata write routines automatically right-justify numeric data fields in the DF, preserving as many significant figures of information as possible. Exponential notation (e.g. 1.3E5) is not allowed.

A more complete description of the Codata format is included in [MERR 81].

3.4. DDF's for Other Data Files

The DDF syntax just described is a subset of a more general data definition language (DDL) being developed in connection with the SEEDIS project. In addition to the fixed-length, line-image, eye-readable codata files described here, the same DDL is used to describe variable-length, binary, compressed data files used for space-efficient storage of archival data. In both cases, a data set consists of two logical components -- a data definition file (DDF) and a data file (DF). Translation programs convert SEEDIS compressed files to Codata files and vice-versa. A discussion of the SEEDIS compressed data format is in preparation [GEY 83].

4. The Codata Tools

As stated in sections 1 and 2 above, development of the Codata format was primarily motivated by needs for data independence among SEEDIS applications programs. Development of the Codata Tools followed shortly thereafter due to the need for general data manipulation routines to deal with Codata files. The "tools" approach to data manipulation for SEEDIS was inspired and aided by the modular Software Tools implementation efforts that were already underway in connection with other projects at LBL,

4.1. Design Philosophy

The Codata Tools were designed and written in accordance with principles outlined by Kernighan and Plauger in *The Elements of Programming Style* [KERN74], and *Software Tools* [KERN76]. In addition, the Codata Tools utilize conventions and extensions to the original Software Tools that have been implemented at Lawrence Berkeley Laboratory [HALL81, SVEN82]. In particular, the following principles were observed:

1. Most of the Codata Tools have one input, one output, and perform a transformation on the data passing through. Such programs are known as "filters." The output from one program can be directly "pipelined" into another. For example, the command line

cocol <file1 countyname income | cosort income | lpr

extracts the variables "countyname" and "income" from file1, sorts the records in order of income, and prints the resulting file.

2. Each Codata Tool does a specific job. In the preceding example, variable extraction is performed by COCOL while sorting is performed by COSORT. The principle of modularity has been observed down through low-level subroutine calls.

3. On-line documentation in a standardized format is obtained by typing "coman <toolname>". Invoking a Codata Tool with a question mark (e.g. "cocol ?") causes a one-line reminder of the expected command syntax to be printed.

4. The Codata Tools are written in a simple extension of Fortran known as "Ratfor" (Rational Fortran). Fortran is widely available and well supported, but is a poor language for programming or describing programs. We avoid the major ideosyncrasies of Fortran, and hide the unavoidable ones in well- defined modules.

5. Input and output and low-level functions such as string manipulation are performed via standard Software Tools library subroutines.

As a result of points 4 and 5 above, the Codata Tools are easily transported to different computers and operating systems where the Software Tools primitive functions have been implemented.

4.2. Functions performed by the Codata Tools

Here we describe each of the Codata Tools individually, in order of increasing complexity. In this discussion, "{}" means "value of". For example "{file}" is an arbitrary file name, not literally "file". Arguments in square brackets "[]" are optional.

Following UNIX conventions, "<in" and ">out" indicate standard input and standard output, respectively. Including "<{file1}" in the command line causes input to be read from file {file1}; omitting it causes input to be read from the terminal. Including ">{file2}" in the command line causes output to be written to file {file2}; omitting it causes output to be written to the terminal.

4.2.1. Reformatting Functions

The first three tools perform simple reformatting functions, changing neither data values or their order in the file:

♣ **COCAT: check and reformat.**

cocat <in >out [{file}]

COCAT copies a Codata file to standard output. In general the output is not strictly identical to the input, as COCAT manipulates all data fields internally. Thus COCAT may be used to check the validity of the input file format. If {file} is not specified, COCAT reads standard input.

♣ **COCL: change line length.**

cocl <in >out [{linelength}]

COCL changes the line length of a Codata file, reading from standard input and writing to standard output. Unused space between data fields is removed, and all data elements are repositioned. If {linelength} is unspecified, the output line length is equal to the line length of the input file.

♣ **CODDF: modify DDF attributes.**

coddf <in >out {modfile}

CODDF reads a Codata file from standard input and writes a new Codata file to standard output. Changes to the DDF are specified in {modfile}. The DF portion of the output Codata file corresponds to the revised DDF.

Columns can be renamed, modified, or added; but not deleted or reordered. New columns are added at the end; the corresponding data values are written as blank if type=A, or as a missing data value if type=I or type=D. For further details, consult [MERR81].

4.2.2. Interfaces to External Applications

The next three tools can be used to provide an interface between Codata files and external applications:

♣ **COSPLIT: partition into DF and DDF.**

cosplit <in >out [{areafile} [{ddffile}]]

COSPLIT reads a Codata file from standard input and writes the DF portion to standard output. One or two file names may be specified as optional arguments. The first file, if specified, receives a copy of the "AREAS={n}" line from the DDF. The second file, if specified, receives a copy of the remainder of the Codata DDF, up through the "END DDF" line.

♣ **DDFFMT: make FORTRAN format from DDF.**

ddffmt <in >out

DDFFMT reads a Codata file, or the DDF portion of a Codata file, from standard input. A corresponding Fortran format specification is written to standard output.

Type A, I, and D fields are respectively translated to "An", "Fn.0", and "Fn.0", where n is an integer. The output format specification includes surrounding parentheses "()". A new line is begun after every 60 characters.

♣ **CONAME: list data element names.**

coname <in >out [-h][{c}] [{file}]

CONAME reads a Codata file and writes the names of the columns (data elements) to standard output. If {file} is not specified, CONAME reads from standard input.

If the "-h" flag is present, data element headers are also written. If the "-h" flag is followed by a character "{c}", header break characters (normally "#") are changed to "{c}" on output.

4.2.3. Data Manipulation Functions

The five remaining tools perform data manipulation functions:

♣ **COSORT: sort rows.**

cosort <in >out [-d] [{colname} ...]

COSORT sorts the rows of a Codata file on standard input, and writes the resulting Codata file to standard output. The command line optionally specifies the names of columns whose values comprise the sort key. If no column names are specified, the sort key is the concatenation of all columns having use=K.

Rows are put out in ascending order of the sort key unless "-d" (decreasing order) is specified. In alphabetic (type=A) data elements, blanks are significant. Upper/lower case differences are ignored.

♣ **COCOL: select columns.**

cocol <in >out {dename}
...or cocol <in >out -{colfile}

COCOL reads a Codata file from standard input and writes a Codata file to standard output. Only specified columns are copied to the output file. The command line specifies either (a) a list of column names {dename} ..., or (b) the name of a file {colfile} containing the list of column names. In mode (b), a minus sign precedes the file name. Upper/lower case differences are ignored. In the output file columns appear in the order requested.

♣ **COROW: select rows.**

corow <in >out {keyvalue}
... or corow <in >out -{keyfile}

COROW reads a Codata file from standard input and writes a Codata file to standard output. Only specified rows are copied to the standard output. The command line specifies either (a) a list of key values {keyvalue} ..., or (b) the name of a file {keyfile} containing the list of key values. In mode (b), a minus sign precedes the file name.

In both mode (a) and (b), the key value of a row is the concatenation of all data elements having use=K and type=A.

♣ **COMRG: relational join of two files.**

comrg <in >out {file2} [{colname} ...]
... or comrg <in >out {file2} [-{colfile}]

COMRG performs an asymmetric join of two Codata files which are on standard input and {file2} respectively. A resulting Codata file is written to standard output.

The rows in standard output correspond to those in standard input; row matching is performed for the keys defined in {file2}. The {file2} rows must be uniquely identified by columns having use=K and type=A. Standard input must contain corresponding columns.

Both standard input and {file2} must have been previously sorted (for example by COSORT) in ascending order of the values of these elements. (These elements need not have use=K in standard input, nor do they need to uniquely specify the rows of standard input.)

Each row of standard output corresponds to one row of standard input. All elements of standard input are copied unchanged. In addition, standard output contains additional columns corresponding to the non-key columns of {file2}. The unique keys occurring in {file2} determine which row of {file2} contributes to a row of standard output. A row of {file2} need not be used, or can be used more than once.

Optionally, the command line may specify either (a) a list of {file2} column names ({colfile} ...) or (b) the name of a file {colfile} containing a list of {file2} column names. In mode (b), a minus sign precedes the file name. If either option (a) or (b) is used, only the specified {file2} columns are written to standard output.

Other COMRG options are available to control the handling of missing data and the renaming of duplicate column labels. For further details, consult [MERR81].

• **COROAGG: aggregation on specified keys.**

coroagg <in >out [{col} ...] [-c{ctcol}] [-z]

COROAGG reads a Codata file from standard input and writes a Codata file to standard output. The output file contains the total of all the rows of the input file or (optionally) a number of subtotals.

If one or more {col} arguments are specified, these must correspond to type=A column names in standard input. In standard output, the specified {col} elements, and no others, have use=KEY.

Each row in standard output corresponds to a unique value of the specified {col} element(s). If no {col} arguments are specified, standard output contains a single row, with no use=KEY columns.

Standard input need not be sorted. Standard output is sorted in ascending order of the specified {col} elements.

If present, the optional argument {ctcol} preceded by "-c" is the name of a new column in standard output, which contains the number of input rows that contributed to the output row in question.

Numerical fields (type=I or type=D) are added. If an input data value is missing, a warning message is issued. Any missing data on input will produce a missing value in the corresponding sum, unless the "-z" option is specified. In this case missing input data values are assumed to be zero. Character (type=A) fields are copied to standard output if all contributing rows of the input file have the same value. Otherwise the corresponding output field is filled with blanks.

Other COROAGG options are available to provide weighting of each row by a specified variable, and to fill type=A fields with a character other than blanks. For further details, consult [MERR81].

4.2.4. Other Codata Tools

Other Codata tools, of a more experimental nature, are available. For complete documentation, consult [MERR81].

A notable omission is the lack of a Codata tool to calculate new data values (columns) as functions of other data values. A tool COCOCAL written to fill this need did not perform successfully. Though not following the documentation and input-output conventions of the other Codata tools, the QUERY module reads and writes Codata files and provides the required functionality in SEEDIS. QUERY uses the LANGPAC compiler generator to evaluate general algebraic and logical expressions.

4.3. Composite Example

One of the most powerful features of the Codata Tools is the idea, taken from the UNIX operating system, that output from one program can serve directly as input to another. The use of several Codata tools connected by "pipes" is illustrated in the following example. From the file listed in Exhibit 3, it is desired to list the three states in ascending order of 1980 population.

Example:

cocol <file2 stub.geo pop80 | cosort pop80 | cosplit >file20

Standard input (file2) is the same as shown in Exhibit 3.

Standard output (file20) contains:

```
Utah    1461037
Iowa    2913808
Texas  14229191
```

5. Limitations and Future Directions

As noted earlier, the primary virtue of the Codata format is its simplicity. Since 1978 it has served as the primary SEEDIS inter-module data interchange format and as a means for inexperienced users to get data in and out of SEEDIS. The Codata tools, with their extremely simple user interface, have been crucial in the development of major data files for many applications.

On the other hand, the present Codata Format and Codata Tools have inherent limitations which are being addressed in order to aid the future development of SEEDIS. Dynamic stack allocation will replace fixed-size arrays in order to efficiently handle large data files without repeated recompilation. New binary file formats are being developed in order to reduce the input-output and processing time required by the Codata tools. Input-output routines are being developed to read and write these new file formats; these will be used first in file translation utilities, and later linked directly to the major SEEDIS data manipulation and display modules. The data definition language (DDL) is being enhanced in order to provide economical description of large multi-dimensional data arrays and multiply-occurring and/or variable-length data elements [MCCA82A].

6. Summary

The Codata tools, developed in connection with the SEEDIS project at Lawrence Berkeley Laboratory, are a set of programs which read, write, and restructure self-describing Codata (common data format) files. These tools manipulate both data and data description, so that the output of any operation is itself a Codata file. The Codata tools can be used to extract specified rows and/or columns from a file, to sort a file, to perform relational joins, to perform tabulations by aggregating on common key values, and to perform other operations.

Codata files have two basic parts: data and data description. The data portion is a simple "flat" file with fixed-length data elements (fields) in fixed-length records. The data description portion consists of line images in "{name} = {value}" format. Codata files can be read, written, and modified using either a text editor or formatted READ and WRITE statements from a programming language. They are easy to read and understand, to transport between dissimilar computers, and to convert to other file formats.

Following the Software Tools philosophy, the Codata tools are modular - each tool performs a specific limited task. They follow the UNIX and Software Tools conventions of standard input and output, so that the output of one module can automatically serve as the input of another. They are written in RATFOR (a transportable FORTRAN preprocessor), and can be easily adapted to run on any computer where the Software Tools have been implemented.

Work is currently under way on substantial enhancements to the Codata file format and the Codata tools to provide for more complex data structures, more extensive, open-ended data description, and more efficient operation.

7. Acknowledgments

Carl Quong, head of the LBL Computer Science and Mathematics Department, is responsible for the stable and productive research environment in which this work was conducted. Among many others who made these results possible, the authors wish to thank especially Fred Gey and Harvard Holmes for sustained interest and valuable suggestions. Bill Hogan's QUERY module fulfilled a need not met by the other Codata tools. Linda Wong wrote the Codata tool DDFFMT. Bill Benson, Peter Wood, and Bob Healey wrote the low-level read and write routines used by all the Codata tools and SEEDIS applications modules.

This work was supported by the Office of Health and Environmental Research and the Office of Basic Energy Sciences of the U.S. Department of Energy under Contract DE-AC03-763F00098; and the Department of Labor, Employment and Training Administration under Interagency Agreement No. 06-2063-36.

8. References

BECK78 Becker, R.A., and J.M. Chambers. Design and Implementation of the S System for Interactive Data Analysis, *Proc. I.E.E.E. Compsac78*, (1978), pp. 626-629.

COMP82 Computer Science and Mathematics Department, SEEDIS Release Notes version 1.3, February, 1982; version 1.4 (preliminary), March, 1983.

GEY83 Gey, F., McCarthy, J.L. and Merrill, D.; Computer-Independent Data Compression: A Space-Efficient, Cost-Effective Storage Mechanism for Large Statistical Data Bases; Lawrence Berkeley Laboratory Report LBL-15824; submitted to the Second International Workshop on Statistical Database Management, Los Altos, California, 27-29 September 1983.

HALL81 Hall, D., Scherrer, D. and Sventek, J.; A Virtual Operating System, *Comm. ACM, Vol. 23 (1980)*, pp. 495-502.

KERN74 Kernighan, B.W., and P.J. Plauger, *The Elements of Programming Style*, New York, N.Y.: McGraw-Hill, 1974.

KERN76 Kernighan, B.W., and P.J. Plauger, *Software Tools*, Reading, Mass: Addison-Wesley, 1976.

MCCA82A McCarthy, J. L., Enhancements to the Codata Data Definition Language Lawrence Berkeley Laboratory Report LBL-14083, February, 1982.

MCCA82B McCarthy, J. L., et al., The SEEDIS Project: A Summary Overview, Lawrence Berkeley Laboratory Report PUB-424, April, 1982.

MERR81 Merrill, D., CODATA Users' Manual, Lawrence Berkeley Laboratory Internal Document LBID-021, revised August 31, 1982.

MERR82 Merrill, D. and Selvin, S.; Populations at Risk to Environmental Pollution (PAREP): Project Overview, 1976-1982; Lawrence Berkeley Laboratory Report LBL-15321, December 1982. Included in An LBL Perspective on Statistical Database Management, H. Wong, editor, Lawrence Berkeley Laboratory Report LBL-15393, December 1982.

SVEN82 Sventek, J. et al, *Software Tools Virtual Operating System User's Manual* (VMS Version), Lawrence Berkeley Laboratory internal documentation, revised 1982.

New and Innovative Ways of Looking at Data

Organizer: Robert McGill

Invited Presentations:
    Interactive Tools for Data Exploration, Sara A. Bly

INTERACTIVE TOOLS FOR DATA EXPLORATION*

Sara A. Bly

Lawrence Livermore National Laboratory
Livermore, California, USA

In complex data problems involving unknown structures and relationships, techniques which help analysts find features and characteristics of that data are particularly important. These techniques are useful predecessors to formal analysis methods. Computers have provided computational power to make a wide variety of such techniques available for data exploration.

A computing facility with locator input, color graphics output, and audio output offers the basis for interactive data presentation tools. One such tool uses visual and aural representations of data to give the analyst a variety of perspectives of the data. By combining user interactivity with various data output techniques, the analyst can dynamically explore complex data sets looking for patterns, structures, relationships, and features. With application to example data sets, this tool illustrates a means of data exploration.

1. INTRODUCTION

An interactive computer graphics facility offers capabilities to provide powerful tools for problem solving. Several such interactive graphics programs are being used in data exploration and analysis, including ORION I [7], ALDS [10], and DATAPLOT [6]. Their successes make it worthwhile to continue to search for additional ways in which interactive graphics capabilities may be used. In this paper, I would like to suggest a list of interactive graphics capabilities which are available and helpful from the system graphics perspective, to suggest that computer-generated sound be used in addition to graphical output, and to provide an example by way of a program under development.

First it is worthwhile to review a few of the needs of data exploration and second the computer capabilites which exist to serve those needs. Third the paper includes a discussion of expanding data exploration output to aural representation. Fourth, I will describe a data exploration tool under development and illustrate the use of sound in that system.

2. DATA EXPLORATION NEEDS

Typical problems in data exploration involve finding relationships among data sets and looking for patterns within a particular data set. In many cases, analytic techniques suffice to determine the differences or likenesses within and among data sets. However, particularly with multivariate data, the analyst needs to explore the data sufficiently to determine which analytic techniques are best suited to the data. It is the need to explore multivariate data that concerns the work described here.

Two factors are particularly important in exploring data -- the ability to consider the data from a variety of perspectives and the ability to manipulate the data. To consider data from a variety of perspectives implies the ability to view the data using any of several graphical representations and to transform the data using any of several analytical/mathematical tools. To manipulate the data implies the ability to alter the data by grouping the data into subsets, moving data among sets, and modifying data values.

For a variety of perspectives in exploring unknown data, one would like to have as many tools as possible at one's disposal. Several methods exist for displaying data, from 2-dimensional scatter plots to Chernoff's FACES [3] to stereo 3-dimensional output [8]. Likewise, several methods exist for transforming data to allow the analyst to view multi-dimensional data more meaningfully. Such methods may include technques such as smoothing [4] or principal components analysis. The resulting data is then viewed using any of the display methods. In all cases, the goal is to offer the analyst as much help as possible in discovering patterns and relationships within the data.

---

*Work performed under the auspices of the U.S. Department of Energy by the Lawrence Livermore National Laboratory under contract number W-7405-ENG-48.

The ability to manipulate the data also offers a great aid to exploration. Although many changes to the data itself would certainly destroy the original data features, the exploration should allow the analyst to ask "what if?". By grouping the data into distinct sets, deleting data items, and changing data values, the analyst may discover whether a few anamolies are hiding the perception of the data as a whole. Grouping the data is particularly important in looking for data clusters and often helps in simplifying a large problem. Likewise, deleting items or modifying values temporarily may help the analyst determine overall patterns without being distracted or misled by occasional outliers. It is important that the altered data be examined with various displays and analytical techniques while still retaining the ability to return easily to the original data.

Exploring multivariate data requires that the analyst be able to perceive and manipulate the data in a variety of ways. As yet, there are no well-defined steps for data exploration. The problem is one of searching for possible patterns and relationships. Tools for the analyst are techniques which provide various representations of the data.

3. COMPUTER CAPABILITIES FOR DATA EXPLORATION

The computer capabilities for providing a sophisticated data exploration tool involve both hardware and software in addition to computational power. Hardware support includes the capability to perform dynamic input and output, the capability to output both visually and aurally, and the capability to obtain hardcopy. Software characteristics include modularity and device independence in the display i/o. The use of the resulting facility comes from packaging these capabilities so that the analyst is only concerned with the data exploration, not with complicated commands, unnecessary delays, or limited flexibility in moving among functions.

Perhaps the most important characterisitic of a facility for data exploration is the dynamic i/o. The analyst concerned with discovering properties of the data must not be distracted by the workings of the system. Thus, the system must respond without hesitation and must provide updated views in a time that does not allow the analyst to lose the thread of exploration. The most fundamental tool provided by dynamic i/o is that of varying the plot of the data. The variations include rotations of the data, different renderings of the data, and displays of analytical/mathmatical functions of the data. Motion itself is one means of displaying multivariate data [5, 10].

The dynamic i/o also provides the basis for data manipulation. The user needs to indicate subsets of the data, delete outliers, alter individual characteristics, or identify unknown data items easily and naturally. Although keyboard command languages have long been the basis for interactive systems, advances in input technology provide devices for pointing and locating in more straightforward actions.

Since exploration implies a need to perceive the problem in a variety of perspectives, the available computer technology for visual displays and aural output can be utilized. We commonly use 2-dimensional pictures to describe a situation, and 2-dimensional data plots are still among the most useful tools in data exploration. In addition, color and a variety of symbols or icons allow further information to be encoded. We are accustomed to perceiving our world in three dimensions, so both stereo and motion become natural computer extensions of common visual experiences. Since sound is also a common means of obtaining information, it follows that aural output offers potential for data exploration. If we wish to offer as much information as possible about the data, both visual and aural output capabilites are important.

A useful data exploration facility will also offer hardcopy output. The output provides additional records of the data exploration, as well as comparisons among different outputs. Hopefully, the output also includes prints for reports and movies for briefings. The output may not be needed for the exploration itself, but it offers the means for communicating the resulting information.

Underlying software needs for data exploration include modularity. This aspect of software development is particularly important in insuring that the data exploration facility is not limited in the views it provides. Whenever new display techniques or algorithms are desired, it should be a trivial matter to include them. Likewise, analysis packages should be readily available and incorporated into the system so that analysis can be considered in the process of exploration.

Finally, device independence in the graphical output is important in maintaining as much display flexibility and availability as possible. Although color output or stereo views might aid a particular effort, a sophisticated facility may offer many different output devices for different purposes. Device independence also allows future addition of new display hardware.

Computer technology offers the capabilities to support data exploration. Dynamic i/o, visual

and aural output, software modularity, and
display device independence allow tools for
many alternative views and manipulations of
multivariate data.

4. SOUND FOR ADDITIONAL INFORMATION IN DATA EXPLORATION

The continuing problem of how to represent
multivariate data suggests that techniques
which enable additional dimensions of data to
be encoded are useful for data exploration.
Sound is a means of obtaining information in
our everyday world and the technology already
exists for computer-generated sound. Just as
data information is encoded in a variety of
visual outputs, the data information may be
encoded in aural output. Utilizing as many
alternative outputs as possible gives the
exploration analyst more clues.

One method of using sound is to parallel the
use of visual icons. The components of a
visual display include location (x,y,z axes),
color, size, and shape. The components of an
aural output include frequency, amplitude,
timbre, and duration. Consider a single data
item, an n-tuple, in a set of n-dimensional
data. Let each of the n values for that data
item be mapped to one of the characteristics
of sound. For example, one value might
determine a sound amplitude and another might
determine a sound frequency. For a given data
item, there now exists a corresponding
discrete sound or note. Much like graphical
icons, these notes can be used to distinguish
among data items.

Sound output need not be limited to discrete
data samples. For time-varying data, sound
offers a natural expression for highlighting
events and patterns. Other types of data may
be well suited to sound encoding. Logarithmic
data, in particular, is often difficult to
visualize. In sound, frequency and pitch are
logarithmically related and thus provide a
potential base for logarithmic data
representation.

Earlier work has shown that sound does add
information in discriminating among sets of
multivariate data [2, 9, 11]. Combined with
graphical output, sound offers the possibility
of increasing the number of dimensions which
can be represented to the user. Additionally,
sound has the advantages of an extremely high
data rate and no required physical orientation
of the listener. The very fact that sound
perception is different than visual perception
offers the user a different representation of
the data.

The work with sound for data representation
has only begun and many questions arise.
Sound, unlike graphics, is transitory and does
not offer the familiar basis for comparison of
x-y axes. Just as the perceptual issues are
important considerations in the use of visual
output, pschyoacoustics must play an integral
part in the developing use of sound for data
representation. However, in the quest to find
tools by which analysts can explore complex,
multivariate data sets, sound offers another
representation.

5. EXAMPLE OF A DATA EXPLORATION FACILITY

In the Engineering Research Division at LLNL,
we are interested in examining the ways in
which interactive techniques aid scientists
and engineers in performing their tasks. We
have put together a facility which uses a
Digital Equipment Corporation VAX 11/780, many
DEC VT100 terminals with retrographics display
capabilities and lightpen input, a Ramtek 9400
color display with lightpen and graph tablet,
and four Solid State Music boards for creating
up to four notes of sound output. A prototype
program for data exploration is intended to
suggest a few possible interactive techniques
and to illustrate the use of sound as
additional information output.

The program display consists of three types of
display areas as shown in Figure 1. The data
display area is available for any
2-dimensional projection of the data. The
variable distribution area emphasizes the
differences among sets for each given data
variable. The menu areas, when selected by
the user, determine program action.

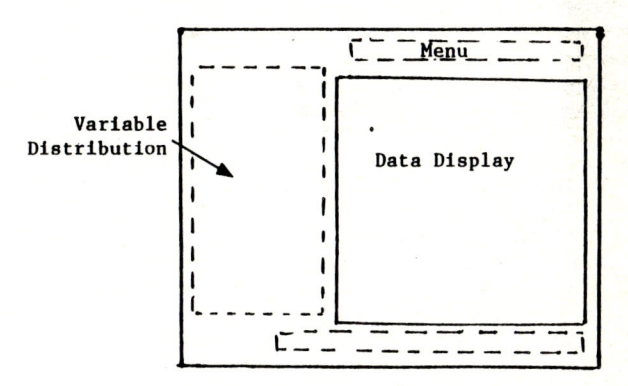

Figure 1: The display provides space for
plotting the data, showing the variable
distributions, and allowing user input via
menu items.

Initial user input comes from a file of
commands. This file determines the initial
data to be used, the data characteristics, and
the graphical and analytical module
parameters. If an input file for
initialization does not exist, the program
prompts the user for keyboard input.

The data display area is intended to show any of the data plots selected by the user. This might include a scatter plot of the data, a FACES representation of the data, a series of rotations in 3-dimensional space, or a plot resulting from a mathematical transformation of the data. An effort is made to allow the program to be used on more than one device. Thus, a scatter plot from three sets of data uses different symbols to identify sets on a black/white screen and uses color on a color output display.

The variable distribution area provides a different view of the data than the data plot by focusing on each of the data variables individually. Although a variety of outputs were considered, the current implementation shows the distribution for each variable in each of the data sets. The range of values for a particular variable is divided into a few subsets. For each data set and variable, a symmetrical histogram shows the number of data items whose values lie in each subset.

The variable distribution display allows one to determine visually the variable ranges and overlaps among sets. It also could provide a varying display as the user groups, deletes, or modifies the data displayed on the data plot. Currently, anytime a particular data item is considered, the values of the variables are overlayed on the distribution display. Figure 2 shows the variable distribution for three sets of 4-dimensional data.

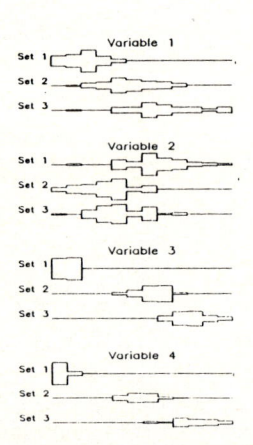

Figure 2: The variable distribution shows the spread of each variable among the sets.

The menu area consists of controls for the program which the user selects by pointing at the appropriate box with the input device. The menu allows the user to change the data plot, to manipulate the data, to listen to data items, and to save information. Not all of the functions considered useful are implemented, but Figure 3 shows those which are currently in existence. The menu button PLOT allows the user to change plots; for current examples, the PLOT merely switches between a 2-dimensional scatter plot and Andrews' functions [1]. The buttons SET 1 and SET 2 play a subset of the data from the indicated set. This allows a user to hear, as well as see, differences in data sets. The top menu, ADD TO 1 and ADD TO 2is for manipulating data items by changing the set to which they belong. The SAVE button then saves the altered data for subsequent analysis or redisplay.

Anytime an individual data item is selected, the variable distribution shows the variable values for that data item and the sound for that data item is played. Thus a fairly simple data exploration problem might be to identify an unknown data item as belonging to one of the existing known data sets. After examining the data plot, the user might wish to listen to data items near the data item to be identified. By looking at various perspectives of the data and listening to the individual items, the user chooses a set to which the unknown might belong. Figure 3 shows two sets of 4-dimensional data. Those data items (unknowns) which are not identified as belonging to one of the two sets are shown as rectangles. A circle shows an item in Set 1 and its corresponding variable values are overlaid on the distribution to the left.

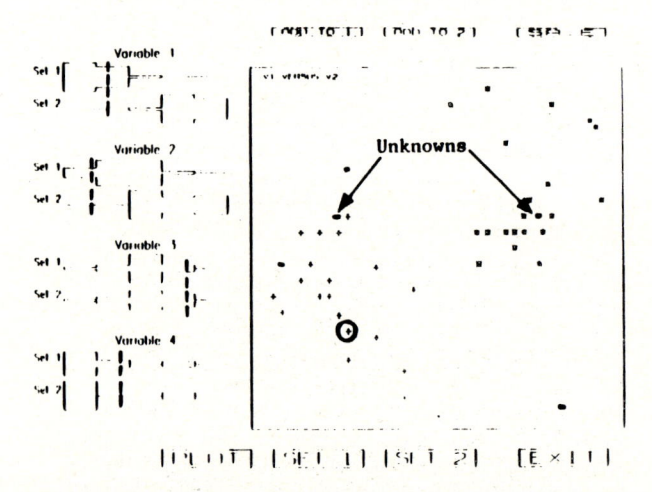

Figure 3: Two sets of 4-dimensional data are shown in a plot of variable 1 versus variable 2 with five data items whose set identifications are unknown. The variable values for the circled data item in Set 1 are indicated on the variable distribution display.

The program is intended to contain enough structure to emphasize the usefulness of interactive graphics and additional sound output. As yet, functions such as analytical tranformations, multiple window plots, and many statistical techniques have not been included. These capabilities require more sophisticated interactive menus and actions than are currently implemented. The design does show the advantages of dynamic i/o, visual and aural output, software modularity, device independence, and hardcopy output.

6. CONCLUSION

Multivariate data exploration presents a challenge to provide useful interactive graphical tools for data representation and manipulation. A facility including dynamic i/o with both visual and aural output and a wide base of software support offers a basis for a sophisticated data exploration facility. Utilizing these capabilities in both traditional and new methods of data exploration will increase our understanding of complex data sets.

7. REFERENCES

[1] Andrews, D. F., "Plots of High-dimensional Data," Biometrics, Vol. 28, p. 125, 1972.

[2] Bly, S. A., "Sound and Computer Information Presentation," Ph.D. Dissertation, University of California, Davis, 1982.

[3] Chernoff, H., "The Use of Faces to Represent Points in k-Dimensional Space Graphically," Journal of the American Statistical Association, Vol. 63, p. 361, 1973.

[4] Cleveland, W. S., "Robust Locally Weighted Regression and Smoothing Scatterplots," Journal of the American Statistical Association, Vol. 74, p. 829, 1979.

[5] Donoho, D. L., P. J. Huber, E. Ramos, and H. M. Thoma, "Kinematic Display of Multivariate Data," Proceedings of the Third Annual Conference and Exposition of the National Computer Graphics Association, Inc., Vol. 1, p. 393, 1982.

[6] Filliben, J. J., "Dataplot-An Interactive High-level Language for Graphics, Non-linear Fitting, Data Analysis, and Mathematics," Computer Graphics, Vol. 15, p. 199, 1981.

[7] Friedman, J. H., J. A. McDonald, and W. Stuetzle, "An Introduction to Real Time Graphical Techniques for Analyzing Multivariate Data," Proceedings of the Third Annual Conference and Exposition of the National Computer Graphics Association, Inc., Vol. 1, p. 421, 1982.

[8] Grotch, S. L., "Three-dimensional and Steroscopic Graphics for Scientific Data Display and Analysis," Lawrence Livermore National Laboratory, UCRL 88764, 1982.

[9] Mezrich, J. J., S. Frysinger, and R. Slivjanovski, "Dynamic Representation of Multivariate Time Series Data," Exxon Research and Engineering Company, draft, 1982.

[10] Nicholson, W. L., and R. J. Littlefield, "The Use of Color and Motion to Display Higher Dimensional Data," Proceedings of the Third Annual Conference and Exposition of the National Computer Graphics Association, Inc., Vol. 1, p. 476, 1982.

[11] Yeung, E. S., "Pattern Recognition by Audio Representation of Multivariate Analytical Data," Analytical Chemistry, Vol. 52, No. 7, p. 1120, 1980.

Developments in Optimization

Organizer: *John Dennis*

Invited Presentations:

    Constrained Maximum-Likelihood Estimation for Normal Mixtures, *Richard J. Hathaway*

CONSTRAINED MAXIMUM - LIKELIHOOD ESTIMATION FOR NORMAL MIXTURES

Richard J. Hathaway

Department of Mathematics and Statistics
University of South Carolina
Columbia, South Carolina 29208

A particular formulation of constrained maximum-likelihood estimation for a mixture of univariate normal distributions is presented. A consistency result is given for the constrained formulation, along with a corresponding constrained version of the popular EM algorithm. Results of numerical tests, demonstrating the effectiveness of the constrained approach, are included.

1. INTRODUCTION

Mixtures of univariate normal distributions are important modelling tools in many diverse fields of science. An excellent up-to-date survey of mixture distributions and their applications may be found in [9]. Beginning with Pearson in 1894 and continuing today, a major goal of research in this area has been the development of good estimation techniques. The current method of choice for normal mixtures is maximum-likelihood estimation, although such an approach gives rise to certain theoretical and computational difficulties. A constrained approach, designed to circumvent these difficulties, is presented which yields an optimizationally and statistically well-posed problem; a constrained global maximizer of the likelihood function exists and is strongly consistent for the true parameter.

2. THE CONSTRAINED PROBLEM

2.1 A Straightforward Approach

The problems inherent in a straightforward application of maximum likelihood techniques to normal mixtures are discussed after the introduction of the necessary notation.

Throughout, let $p(x;\mu,\sigma)$ denote a univariate normal density with mean μ and standard deviation σ, and let $p_m(x;\theta) = \sum_{i=1}^{m} \alpha_i p(x; \mu_i; \sigma_i)$ denote a mixture of m univariate normal densities for the parameter $\theta = (\alpha,\mu,\sigma)^T$ belonging to the parameter space $\Omega = \{\theta \in R^{3m} | \sum_{i=1}^{m} \alpha_i = 1, \alpha_i \geq 0, \sigma_i > 0, i=1,\ldots,m\}$. Whenever discussing estimation, the parameter corresponding to the true distribution of a random variable X will be denoted by θ^o. For any random sample $x_1,\ldots,x_n$, the log-likelihood function $L(\theta)$ is defined for $\theta \in \Omega$ by $L(\theta) = \log(\prod_{k=1}^{n} p_m(x_k; \theta))$.

The first problem associated with maximum likelihood estimation for normal mixtures is the unboundedness of $L(\theta)$ on Ω for any sample $x_1,\ldots,x_n$. This is seen by fixing $\alpha_i = \varepsilon > 0$ and $\mu_i = x_k$ for some i and k, and then letting $\sigma_i \to 0$. The straightforward maximum-likelihood approach of defining the estimate to be a global maximizer of $L(\theta)$ over Ω leads to an ill-posed optimization problem; one that does not have a solution. The unboundedness of $L(\theta)$ causes computational difficulties. It has been reported that optimization algorithms of both the EM and quasi-Newton types produce iterates that converge to singularities of $L(\theta)$, when certain initial guesses are used. [5,8]

Statistical theory [9], based on the results in [1], does guarantee that among all the local maximizers of $L(\theta)$, exactly one will be close to θ^o for large samples. (Precisely, there is a unique strongly consistent local maximizer of $L(\theta)$.) The problem with this localized approach is in deciding which local maximizer to use as an estimate. The limit points of iteration sequences produced by optimization algorithms are heavily dependent on the initial guess used [8], and some of these points may give larger $L(\theta)$ values than that corresponding to the consistent local maximizer. [5] In the view of some, this problem is substantial enough to justify the search for alternative estimation techniques, even though the local maximum-likelihood estimator has good asymptotic properties. [2,7]

2.2 A Constrained Approach

Day noted in [2] that the method of maximum likelihood worked well for the case of equal variances. In this case, it is possible to globally maximize $L(\theta)$ over the subset of Ω containing parameter points which satisfy $\sigma_i = \sigma_j$ for $i,j = 1,\ldots,m$. In the case of unequal variances, Day argued that any small set of sufficiently close data points would generate a spurious maximizer of $L(\theta)$, and make maximum-likelihood estimation difficult.

It was noted in [7] and the following discussion papers that maximum-likelihood estimation could be successfully used whenever the relationship between the standard deviations of the true parameter is known and appropriate constraints of the form $\sigma_i = c_{ij}\sigma_j$ for $i,j = 1,\ldots,m$ are added.

Since the exact knowledge of the coefficients c_{ij} in the above is not usually present, equality constraints of this type cannot generally be imposed. An alternative approach which compensates for imprecise knowledge of the coefficients c_{ij} is the imposition of inequality constraints of the form $\sigma_i \geq c_{ij}\sigma_j$ for $i,j = 1,\ldots,m$. In fact, constraints of this type (min $\frac{\sigma_i}{\sigma_j} \geq c$), have been suggested independently by James R. Thompson and E.M.L. Beale. [4]

The constraints chosen for the constrained formulation presented here are a subset of those above. Defining $\Omega_c = \{\theta \in \Omega \mid \sigma_i \geq c\sigma_{i+1},$ $i=1,\ldots,m-1, \sigma_m \geq c\sigma_1\}$, the constrained problem consists of defining the estimate, denoted by $\tilde{\theta}^n$, to be any global maximizer of $L(\theta)$ over Ω_c. Notice that the m constraints defining Ω_c bind the component standard deviations together in the same way that the $m(m-1)$ constraints of Thompson and Beale do, but implementing a constrained algorithm using the smaller number of constraints is simpler.

The following result, taken from [6], shows that the constrained problem is optimizationally and statistically well posed. A mixture of exactly m univariate normal densities is one which cannot be represented using fewer than m component densities. The fact that a mixture of exactly m univariate normal distributions is only identifiable up to permutations of the parameter components is neglected in the following; technicalities of this type are addressed in [6].

<u>Theorem</u> For any $c \in (0,1]$, a corresponding global maximizer $\tilde{\theta}^n$ of $L(\theta)$ over Ω_c exists. If $c \in (0,1]$ is such that the true parameter θ^o, corresponding to a mixture of exactly m univariate normal distributions, is in Ω_c, then $\tilde{\theta}^n$ is strongly consistent for θ^o.

Not only does the constrained approach lead to a well-posed problem, but it is possible, by making reasonable choices for c, to avoid spurious maximizers of the type alluded to by Day, which by necessity must occur at parameter values with widely differing component standard deviations.

3. THE CONSTRAINED ALGORITHM

3.1 The EM Algorithm

The EM algorithm is the current algorithm of choice for finding a local maximizer of the likelihood function corresponding to a mixture of normal distributions, and because of its good numerical properties, it is chosen for the extension to the constrained case. A discussion of the algorithm as it pertains to mixture densities can be found in [9], and [3] contains a description of the algorithm in more general settings. Only the bare essentials, necessary to understand the upcoming constrained version, are given here.

Let $L(\theta)$ be the log-likelihood function corresponding to a random sample $x_1,\ldots,x_n$. The following is a statement of the EM algorithm. Superscripts are used to denote the points in the iteration sequence.

The EM Algorithm

Step 1: Choose θ^1, an initial guess, and set $r = 1$.

Step 2: Compute the weights w_{ik}^r for $i=1,\ldots,m$ and $k=1,\ldots,n$ by

$$w_{ik}^r = \frac{\alpha_i^r p(x_k; \mu_i^r, \sigma_i^r)}{p_m(x_k; \theta^r)}.$$

Compute $\alpha_i^{r+1} = \frac{1}{n} \sum_{k=1}^{n} w_{ik}^r$ for $i=1,\ldots,m$.

Compute $\mu_i^{r+1} = \frac{\sum_{k=1}^{n} x_k w_{ik}^r}{\sum_{k=1}^{n} w_{ik}^r}$ for $i=1,\ldots,m$.

Compute $\sigma_i^{r+1} = \left(\frac{\sum_{k=1}^{n} (x_k - \mu_i^{r+1})^2 w_{ik}^r}{\sum_{k=1}^{n} w_{ik}^r}\right)^{1/2}$

for $i=1,\ldots,m$.

Step 3: If convergence is achieved, stop. Otherwise, set $r = r+1$ and continue with 2.

The algorithm is locally convergent to any maximizer satisfying the second order sufficient conditions, and is locally convergent to the consistent local maximizer with probability 1 as $n \to \infty$. The global result states that any limit point in Ω of a particular iteration sequence satisfies the first order necessary conditions for a local maximizer. Finally, the iteration sequence has the property that $L(\theta^{r+1}) > L(\theta^r)$

unless $\theta^{r+1} = \theta^r$, which happens only at a point satisfying the first order necessary conditions.

There seems to be no indication in the statement of the EM algorithm as to how the constraints defining Ω_c might be imposed. Fortunately an interpretation can be given to θ^{r+1} which leads to a very natural way of imposing the constraints. Define the model function $M_{\theta^r}(\theta)$ by $M_{\theta^r}(\theta) = \sum_{i=1}^{m}(\sum_{k=1}^{n} w_{ik}^r) \log \alpha_i + \sum_{i=1}^{m}(\sum_{k=1}^{n} w_{ik}^r p(x_k; \mu_i, \sigma_i))$. It is easily verified that θ^{r+1}, as defined in Step 2, globally maximizes $M_{\theta^r}(\theta)$ over Ω. The constraints defining Ω_c can be imposed during the global maximization of $M_{\theta^r}(\theta)$ at each iteration.

3.2 The Constrained EM Algorithm

A property desired of the constrained algorithm is the ability to converge to the consistent local maximizer from poor initial guesses. In the unconstrained case, a poor initial guess can cause a mixing proportion to become very close to zero, where it remains for all successive iterations. To prevent this in the constrained case, lower bounds on the mixing proportions, $\alpha_i \geq \epsilon$ for $i=1,\ldots,m$, are used. Defining $\Omega_c^{\epsilon} = \{\theta \in \Omega_c | \alpha_i \geq \epsilon, i=1,\ldots,m\}$, the constrained EM algorithm defines θ^{r+1} to be the global maximizer of $M_{\theta^r}(\theta)$ over Ω_c^{ϵ}. The The following is taken from [6].

The Constrained EM Algorithm

Step 1: Pick $0 < \epsilon < \frac{1}{m}$, $0 < c < 1$, and θ^1. Set $r = 1$.

Step 2.1: Compute $w_{ik}^r = \frac{\alpha_i^r p(x_k; \mu_i^r, \sigma_i^r)}{p_m(x_k; \theta^r)}$ for $i=1,\ldots,m$ and $k=1,\ldots,n$.

Compute $\alpha_i = \sum_{k=1}^{n} w_{ik}^r$ for $i=1,\ldots,m$.

Set $A = \{\alpha \in R^m | \sum_{i=1}^{m} \alpha_i = 1\}$.

Step 2.2: Compute $\alpha^{r+1} = \underset{\alpha \in A}{\arg\max} \sum_{i=1}^{m} a_i \log \alpha_i$.

Step 2.3: If $\alpha_i^{r+1} \geq \epsilon$, $i=1,\ldots,m$, then continue with 2.4. Otherwise, find $\alpha_j^{r+1} < \epsilon$, redefine A to be the subset of the current set A such that $\alpha_j = \epsilon$, and continue with 2.2.

Step 2.4: Compute $\mu_i^{r+1} = \frac{\sum_{k=1}^{n} x_k w_{ik}^r}{\sum_{k=1}^{n} w_{ik}^r}$

for $i=1,\ldots,m$.

Step 2.5: Compute $b_i = \sum_{k=1}^{n}(x_k - \mu_i^{r+1})^2 w_{ik}^r$

for $i=1,\ldots,m$.
Set $B = \{\sigma \in R^m | \sigma_i > 0, i=1,\ldots,m\}$.

Step 2.6: Compute $\sigma^{r+1} = \underset{\sigma \in B}{\arg\max} \sum_{i=1}^{m} a_i \log \frac{1}{\sigma_i} - \sum_{i=1}^{m} b_i \cdot \frac{1}{2(\sigma_i)^2}$.

Step 2.7: If $\sigma_i^{r+1} \geq c\, \sigma_{i+1}^{r+1}$, $i=1,\ldots,m$, then continue with 3. Otherwise, find $\sigma_j^{r+1} < c\, \sigma_{j+1}^{r+1}$, redefine B to be the subset of the current set B such that $\sigma_j = c\sigma_{j+1}$, and continue with 2.6.
(Interpret σ_{m+1} as σ_1.)

Step 3: If convergence is achieved, stop. Otherwise, set $r = r+1$ and continue with 2.1.

Since the data points are referenced no more often in the constrained version of the EM algorithm than in the unconstrained, the increase in work per constrained iteration is negligible. The constrained version has all of the convergence properties of the unconstrained, but in addition guarantees that all limit points of iteration sequences are in Ω_c; convergence to a singularity of $L(\theta)$ is impossible. [6]

The following is an example of a constrained EM iteration for a mixture of 3 normal distributions. A detailed discussion of the algorithm is in [6].

Let c and ϵ satisfy $0 < c < 1$ and $0 < \epsilon < \frac{1}{3}$. Let θ^r denote the current iterate and suppose that Steps 2.2 and 2.3 are executed. Step 2.2 first defines α^{r+1} by $\alpha_i^{r+1} = a_i / (a_1 + a_2 + a_3)$ for $i = 1,2,3$. If the constraints are not violated, then this is the final definition of α^{r+1}. Assume instead that $\alpha_1^{r+1} < \epsilon$. Then α^{r+1} is redefined in Step 2.2 by $\alpha_i^{r+1} = a_i / (a_2 + a_3)$ for $i = 2,3$, and $\alpha_1^{r+1} = \epsilon$. If α_2^{r+1} and α_3^{r+1} satisfy the constraints, then the means are calculated next, but if for example $\alpha_2^{r+1} < \epsilon$, then the final definition of α^{r+1} is $\alpha_1^{r+1} = \alpha_2^{r+1} = \epsilon$, and $\alpha_3^{r+1} = 1 - 2\epsilon$.

The new mean vector μ^{r+1} is calculated in Step 2.4.

The first definition of σ^{r+1} in Step 2.6 is $\sigma_i^{r+1} = (b_i / a_i)^{\frac{1}{2}}$ for i=1,2,3. If for example it is found in Step 2.7 that $\sigma_1^{r+1} < c\, \sigma_2^{r+1}$, then the next candidate for the final σ^{r+1} is defined by
$\sigma_2^{r+1} = ((b_1 + c^2 b_2)/(a_1 + a_2))^{\frac{1}{2}}$,
$\sigma_3^{r+1} = (b_3 / a_3)^{\frac{1}{2}}$, and $\sigma_1^{r+1} = c\, \sigma_2^{r+1}$.

Finally, if it is then found that $\sigma_2^{r+1} < c\, \sigma_3^{r+1}$, then the final definition of σ^{r+1} is $\sigma_3^{r+1} = ((b_1 + c^2 b_2 + c^4 b_3)/(a_1 + a_2 + a_3))^{\frac{1}{2}}$, $\sigma_2^{r+1} = c\, \sigma_3^{r+1}$, and $\sigma_1^{r+1} = c\, \sigma_2^{r+1}$.

4. NUMERICAL TESTS

Results of numerical tests performed by Fowlkes [5] are compared with similar tests of the constrained EM algorithm found in [6].

Fowlkes generated four samples, of size 200, distributed according to a mixture of two univariate normal densities $p_2(x; \theta^o)$ with the parameter $\theta^o = (.5,.5,0,2.5,1,1,)^T$. Fowlkes started a quasi-Newton algorithm from the initial guesses $(.25,.75,-1,1.5,.5,.5)^T$, $(.75,.25,-1,1.5,1.5,.5)^T$, $(.75,.25,1,3.5,.5,.5)^T$, $(.25,.75,1,3.5,1.5,.5)^T$, $(.75,.25,1,1.5,.5,1.5)^T$, $(.25,.75,1,1.5,1.5,1.5)^T$, $(.25,.75,-1,3.5,.5,1.5)^T$, and $(.75,.25,-1,3.5,1.5,1.5)^T$. Out of 32 trials (4 samples, 8 initial guesses), the algorithm used by Fowlkes was only successful on 10 occasions. A trial is said to be succesful if the point converged to is in fact the same point obtained when the true parameter θ^o is used as the initial guess. There was convergence in all 32 trials, but the 22 failures resulted from convergence to a spurious local maximizer or singularity of $L(\theta)$.

The experiment was repeated in [6], using ten samples and the constrained EM algorithm for several values of c and ϵ, with c and ϵ ranging from .25 to .000001 and .20 to .0001, respectively. Using the same initial guesses, sample size, and true distribution used by Fowlkes, the constrained EM algorithm was successful in all 80 trials for each choice of the constraining parameters c and ϵ.

Next, the experiment was repeated in [6] using the same samples but the following different initial guesses.

1 $(.15,.85,-3,-.5,.2,.2)^T$
2 $(.85,.15,-3,5.5,.2,.2)^T$
3 $(.85,.15,3,-.5,1.8,.2)^T$
4 $(.15,.85,3,5.5,1.8,.2)^T$
5 $(.85,.15,3,-.5,.2,1.8)^T$
6 $(.15,.85,3,5.5,.2,1.8)^T$
7 $(.15,.85,-3,-.5,1.8,1.8)^T$
8 $(.85,.15,-3,5.5,1.8,1.8)^T$

Table 1

Constrained EM: Successful Trials

		\multicolumn{8}{c}{Initial Guess}							
c	ϵ	1	2	3	4	5	6	7	8
.25	.2	9	10	8	10	10	10	10	10
.25	.1	8	10	5	10	10	10	8	10
.1	.1	8	10	5	7	10	9	6	10
.01	.05	4	10	5	0	10	8	6	10
10^{-6}	10^{-4}	0	10	5	0	10	8	6	10

Table 1 shows that from poorer initial guesses, the constraints become important. The last row of the table essentially gives the behavior of the unconstrained EM algorithm, which would be successful 16 less times than the constrained EM algorithm for the conservative choice $c = \epsilon = .1$. All the failures for the constrained EM algorithm corresponded to spurious local maximizers; convergence to a singularity is impossible. Results of more tests of the constrained EM algorithm can be found in [6], including tests of a strategy which dynamically varies c and ϵ when prior information as to their choice is not available.

5. CONCLUDING REMARKS

Maximum-likelihood estimation for a mixture of univariate normal distributions has been reformulated into a well-posed optimization problem. The corresponding constrained EM algorithm seems to be more robust against poor initial guesses than other methods. Work is currently being done in the generalization of this approach to multivariate normal mixtures.

This research was supported in part by ARO Grant DAAG 29-82-K-0014.

REFERENCES:

[1] Chanda, K.C., A note on the consistency and maxima of the roots of the likelihood equations, Biometrika 41 (1954) 56-61.

[2] Day, N.E., Estimating the components of a mixture of normal distributions, Biometrika 56 (1969) 463-474.

[3] Dempster, A.P., Laird, N.M., and Rubin, D.B., Maximum-likelihood from incomplete data via the EM algorithm, J. Royal Statist. Soc. Ser. B 39 (1977) 1-38.

[4] Dennis, J. E.,Jr., Algorithms for nonlinear fitting, Proceedings of the NATO Advanced Research Symposium, Cambridge University, Cambridge, England (July 1981).

[5] Fowlkes, E.B., Some methods for studying the mixture of two normal (lognormal) distributions, J. Amer. Statist. Assoc. 74 (1979) 561-575.

[6] Hathaway, R.J., Constrained Maximum-Likelihood Estimation for a Mixture of m Univariate Normal Distributions, Ph.D. Thesis, Dept. of Math. Sciences, Rice University (March 1983).

[7] Quandt, R.E. and Ramsey, J.B., Estimating mixtures of normal distributions and switching regressions, J. Amer. Statist. Assoc. 73 (1978) 730-738.

[8] Redner, R.A., Maximum-likelihood estimation for mixture models, NASA Tech. Memorandum (to appear).

[9] Redner, R.A. and Walker, H.F., Mixture densities, maximum likelihood, and the EM algorithm, submitted to SIAM Review.

Software Metrics and Effort Estimation

Organizer: H. E. Dunsmore

Invited Presentations:

    Defining Metrics for Ada Software Development Projects, Sylvia B. Sheppard, John W. Bailey, and Elizabeth Kruesi

    Model Evaluation in Software Metrics Research, S. M. Thebaut

    Investigation of Chunks in Complexity Measurements, John S. Davis

DEFINING METRICS FOR ADA SOFTWARE DEVELOPMENT PROJECTS

Sylvia B. Sheppard, John W. Bailey & Elizabeth Kruesi

General Electric Company
Arlington, Virginia

The Ada language is not just another programming language; Ada will be installed with a pre-defined environment that will provide programmers with a standard set of tools to perform their jobs. In addition to tools requisite for any environment (i. e., a compiler, an editor, etc.), tools may be provided for data collection. These data will provide feedback to the project's managers and programmers during the software development process and will predict the operational characteristics of the system under development. In the past, instances of data collection efforts on software projects have been erratic. However, with the commonality that will now occur for Ada environments, there is an opportunity to introduce systematic methods and procedures for measurement on a wide scale. Thus there is a need for the selection of metrics that are relevant and useful in an Ada environment.

Different metrics are useful for different purposes and at different times in the life cycle of a software development project. For example, project managers need measures to predict resources and schedules, to track costs and to locate potential problem areas. These needs differ from those of a customer or end user who would like some measure of the reliability of the system and proof of the thoroughness of the testing that has been done. Still different are the needs of a programmer who would benefit from feedback about the complexity of a just-compiled module. The problem of defining a unified set of metrics is not simple and straightforward. We must consider the needs of a variety of people who come into contact with a system over a long period of time.

Seven researchers from the University of Maryland* and from General Electric have joined in an eighteen-month collaborative effort to study this problem. Part of the effort has involved collecting measurements from an on-going Ada software development project. Our general approach has been to collect a great deal of information, to try to evaluate the different types of measures, and to select those that appear most useful. This paper will describe the software project selected, the data collection effort and the candidate metrics that are being considered.

---

*Members of the University of Maryland team are: Victor Basili, John Gannon, Elizabeth Katz, and Marvin Zelkowitz.

We chose a realistically large and complex software development project for study. It involved re-implementation of a portion of a working ground support system for a communication satellite. The original system was developed by General Electric and consisted of approximately 100,000 lines of FORTRAN and assembly code. A subset of the original system was selected for redesign and implementation in Ada. This subset is executable apart from the larger system. It includes functions to receive an operator's inputs, to perform complex computations and to display output graphically. Also included are several concurrent processes to monitor and display telemetry data from the satellite.

The project began in February 1982 with a month of formal training in Ada. Following this, the lead programmer and back-up programmer produced a requirements document describing the subsystem. The design was developed in an Ada-like program design language. Coding began in August and was completed in November, 1982. Approximately 10,000 lines of Ada source code, including comments, were produced. Some compilation has been done using the NYU Ada/Ed interpreter. Testing will be done as soon as a full Ada compiler is available.

Our approach to measurement for this project has been systematic and thorough. We began by defining goals for our data-collection effort. The goals fell naturally into three categories: those relating to software development projects in general, those relating to the use of Ada as a design and implementation language, and those relating to metrics for the APSE, the Ada Programming Support Environment. Following that, a number of specific questions related to each goal were developed. The goals and the questions relevant to each are listed in the Appendix.

Three methods of data collection were chosen to answer the questions associated with the goals. First, eight different data collection forms were adapted from those developed by Victor Basili and Marvin Zelkowitz for the NASA Software Engineering Laboratory.(1) The forms were designed to be completed by the members of the development team. The data collected on the forms provides a complete record of activities during the development process. The forms focus on three types of data: effort, changes and errors. These data will be discussed in detail below.

Second, an on-line procedure was developed for recording all versions of the design and the code for later analysis by a YACC-generated processor. The processor provides a static view of the system by counting such data as the types of Ada features used in each module, the data exchanged between modules, etc.

Third, the members of the programming team were interviewed after training and at the end of each major phase of the development cycle to determine what new concepts they had learned, what were their attitudes about Ada, and how those attitudes were changing.

In order to assure the quality of the data collected, the research team was diligent about cross-checking the data for accuracy. The results of this activity were surprising to researchers who had previously been involved in laboratory experiments. Techniques for controlled experimentation are well developed, and on-line programming experiments can be used to collect extremely precise time and error data. However, in this field study, we found a very different situation. Athough the programming team understood the necessity for recording the information precisely, there were wide variations in estimates. For example, one programmer recorded a time of one hour for a design walkthrough and another recorded a time of three hours for the same walkthrough, although each person attended the whole session. An inquiry revealed that the walkthrough on the design had been completed in an hour, but the group had remained together for an additional two hours in order to discuss methodology. Thus it was necessary to provide constant monitoring of the data collection effort at the site of the programming activity. In fact, all team meetings were attended by at least one member of the research team.

Collecting data with an on-line system would have relieved some of the problems we found with manual data collection procedures. Incomplete or inconsistent data values would have been more readily apparent. Further, on-line entry of data might have seemed more palatable to the programming team. The extensive data collection effort done here interfered to some degree with their progress of the project: filling out the forms required time and effort. On-line data collection might have been less apparent to the programming team, even though the same data would have been obtained. Finally, such a system would eliminate the need for later entry of the data into the data base. In our selection of metrics, we are considering the ease with which the data for the metrics can be collected and the degree to which it can be obtained unobtrusively.

A large number of candidate metrics have been defined and investigated. All of those that have a possibility of being useful are being evaluated. Area C, as shown in the Appendix, shows questions specifically related to defining metrics for the APSE. Some of these metrics may be useful in any software development environment, regardless of the design or implementation language used; others may be useful only for the Ada environment. Some metrics may provide enough information to make other measures superfluous. Two or more may reflect the same information, but the data for one may be more difficult to collect. Some metrics may not contribute useful information at all. Thus we are comparing the metrics in order to discard those that are not needed and to select a group that will be truly useful.

Data collection for this project has centered largely on two kinds of metrics, those dealing with the software development process itself and those dealing with the evolving product. Data collection for the process metrics include: a) the effort expended, b) the changes to the design or code, and c) the errors that occur.

For effort data, we collected the number of hours spent in various activities during the life cycle. We also recorded the names of the modules and the types of activity (e. g., design review or code reading) on which the time was expended. Because the human effort required to complete a project generally accounts for its largest single cost, effort data can be used to predict costs of future projects with similar parameters (e. g., size and application). Effort metrics are also useful for measuring productivity and for assessing the impact of new tools and techniques on productivity.

Every software development will undergo changes to previous documents (whether anticipated or not). Some development techniques, such as iterative enhancement, specifically assume that continual evolution is desirable. Others assume there will be a single pass through each phase with any further changes being carefully controlled. In either case, information about changes provides important insights into the project. Data collected about a change to our system included: the reason for the change, the time spent making the change, additional documents examined during the change, and changes to other documents because of the change. These data help determine the costs of various types of changes and assist in predicting the modifiability of the system. Change data are also useful for evaluating whether the development methodology is successful for the environment. For example, large numbers of major changes might indicate that a different development methodology should be used. In such a case, an incremental development approach might be beneficial in establishing the desirable features of the system. (2)

Data about errors included a description of the error, the activities used to detect the error, the time at which the error entered the system, and an evaluation of whether the error was related to the Ada programming language. These data provide information about the quality of the product. We can also use error data to improve our methodology. If we know which types of errors are more difficult to correct, we can put an emphasis on trying to locate and correct those types of errors during walkthroughs or other reviews of the product. For errors related to the use of Ada, error data can be used to determine where emphasis is needed in future training courses.

Product metrics are the other major type of metric we have been examining. They deal with the characteristics of the software itself and can be divided into two categories: static and dynamic. Static metrics can be collected from the software at any point in its development. Dynamic metrics are run-time measures. There are a myriad of measurements that can be collected for both categories; only a few representative ones are discussed below.

Static metrics are useful for determining the complexity and quality of the code. Collecting such metrics at several phases of the life cycle provides a view of the way in which the system changes across the phases. Static metrics can be subdivided into three types: size, control and data metrics. Size metrics are indicators of the volume of a product and the amount of work performed. They correlate well with effort and are used for estimating costs, comparing products and measuring productivity.(3) Lines of code are frequently used to measure size. However, there are alternate ways to count lines of code. Comment lines may be included or omitted, depending on the use of the metric. If we are measuring productivity, we probably want to include both code and comment lines. However, the self-documenting features of the Ada syntax may produce a smaller ratio of comment lines to code lines than is usually considered good programming practice in other languages. Further difficulties may arise when computing the size of a system which incorporates previously written components or is a modification of a previously developed system. Should productivity measures include only new development work or should they reflect all of the code that is delivered? Certain features of the Ada language (i. e., packages, generics) will make possible large libraries of reusable components that can be incorporated into a new system much the way in which circuits are now constructed with off-the-shelf chips. This new trend in software design, as well as the self-documenting capabilities of the syntax, will necessitate fresh approaches to size measures which are meaningful for Ada.

Control flow metrics measure the complexity of a product. McCabe's $v(G)$ is a count of the number of basic control path segments in a computer program.(4) This value depends on the number of decision nodes and the branches emanating from those nodes. $V(G)$ was originally developed as part of a strategy for testing software. However, it has been suggested that $v(G)$ is also useful as a measure of psychological complexity.(5) The theory is that a program with many decisions is psychologically complex: the more decisions, the more difficult a program is to understand and modify. One exception to this is a case statement where there is one path for each of several similar choices. McCabe suggests limiting $v(G)$ to 10 for any module except where $v(G)$ is inflated by a large case statement.

The current method of counting $v(G)$ may not be sufficient for the Ada language. Ada's explicit structures for exception handling alter the control flow normally found in structured languages. Additional paths can be explicitly generated to handle situations that would cause faulty execution in other languages. For example, in some other languages, an attempt to read past an end-of-file marker would produce an error message and would terminate the job stream. In Ada it is possible to specify the means to handle this possibility and to continue processing. This exception handling, however, alters the normal flow of control of the program, and it is necessary to consider alternatives for calculating $v(G)$. One method would ignore the potential occurrence of exceptions. A second method would account for all possible paths of execution as the result of the occurrence of an exception. The first alternative alters the premise that all basic paths through the module are being counted and tested. The second alternative is theoretically more appealing, but it would greatly increase the number of paths and would thus make testing very difficult.

Data metrics analyze the organization of the data structures within and between modules in an attempt to measure the ease with which they can be understood and modified. Data metrics of interest include the number of average live variables per statement, the percentage of global variables, and the number of programmer-defined types. Some data metrics, such as the information flow metrics of Henry and Kafura, focus on the interconnections between system components.(6, 7) Myers has indicated that interfaces are important because many serious, hard-to-find errors in systems result from a lack of understanding of module interdependencies and from changes made to global data areas.(8)

More work is also needed in the area of data metrics for Ada. Is the density of data flow across modules as proposed by Henry and Kafura a useful metric for the Ada language? The Ada block structure, visibility rules and packaging features all affect the way in which data is apportioned and the degree to which data is or is not visible at various locations in the code. Thus there is a need to define how to measure the number of global and local data structures and how to measure the quality with which the data structures are encapsulated in packages. These measures should be helpful in determining whether the encapsulation makes the best use of Ada's features for producing maintainable, reuseable software.

Dynamic metrics provide another method for measuring the software product. Dynamic measures can be divided into execution and test coverage metrics. Execution metrics are useful for tuning a system to make it run efficiently. Execution metrics include the number of times each statement is executed and the CPU time used.

Test coverage metrics help determine the degree and quality of the testing that has been done. McCabe's metric has already been discussed. Other candidate test coverage metrics include the number of statements executed and the number of branches executed. Because Ada has mechanisms for concurrent processing, metrics are needed to measure tasking features and usage. Further, concurrent processing presents opportunities for both starvation and deadlock; methods are needed for predicting the possibility of these occurrences during the testing phase of the life cycle.

The analysis of the data from this project and the subsequent selection of a suggested set of metrics for an Ada environment is currently underway. The systematic approach to the goal-driven data collection effort will provide a beginning methodology for subsequent efforts for monitoring software projects. The metrics selected will be useful for all of the differing needs of those associated with an Ada system throughout the life cycle. If we can automate and insert a common set of metrics into the APSE, data collection may become an integral part of software methodology, and comparisons across Ada projects of all types will become feasible. Further, quantitative indices of the progress of a project will be available for managers, procurement officers, designers and others with a need for such information.

APPENDIX

Area A: Generic goals for any software development project

Goal A1: Characterize the effort in the project.
1) How was the effort distributed over the phases of the project?
2) How was the effort for the project distributed over time?
3) How was the effort distributed across different functions in the software?
4) How are the error distributions similar to or different from other comparable software developments?

Goal A2: Characterize the changes.
1) How are the changes to the system distributed over the software development cycle?
2) How is the time for handling a change distributed? How long does it take to design and implement the change?
3) Are certain features of Ada or certain types of errors associated with particular programmers? Why?
4) Do certain programmers have problems with certain aspects of the language?
5) Do programmers want to use features available in other languages that are not available in Ada?
6) Are some features of the language overused, used incorrectly, or used inappropriately in the programmers' enthusiasm to use what they have learned?
7) Do people with no previous high level language experience have more or fewer problems with Ada than people with high level language experience?

Area B: Goals relating to Ada as a design and implementation language

Goal B1: Characterize the errors made.
1) How were the errors found? (e.g., design review, inspection of output, etc.)
2) What were the non-Ada causes of the errors? (e.g., requirements misinterpreted, mistake in computation, etc.)
3) What features of Ada are commonly involved in errors?
4) Are there features of Ada that cause problems when they are used together?
5) Are errors attributed to confusion with another language? to a lack of understanding of Ada? to a lack of experience with a feature?
6) Are the errors made when using Ada as a design language different than those made when coding?
7) Where was the information found that was needed to correct the error? (e.g., Ada Reference Manual, another programmer, etc.)
8) Is the error characteristic of the feature or of the particular application it involved?

Goal B2: Determine whether certain aspects of Ada are difficult to use for certain applications.
1) Are there certain aspects of Ada that do not apply to this type of project?
2) Are there techniques usually used for this type of application that are difficult to implement in Ada?

Goal B3: Determine which aspects of Ada contribute positively to the design and programming environment.
1) Are errors easy to find? to correct?
2) Is there a large amount of parallel development once the interfaces are defined?
3) How effective is Ada in reducing interface errors? producing software that is easy to change? reducing the development effort, especially in realtime problems?

Goal B4: Determine which combinations of Ada's features are naturally used together.
1) How fully is the language used?

2) Are there certain features of Ada that are avoided because they are difficult to learn? difficult to use? poorly implemented? error prone?

Goal B5: Determine the effect of using Ada as a PDL.
1) Does Ada PDL allow sufficient abstraction at the early stages of design?
2) Is the language really being used as a design language?
3) Does the use of Ada PDL cause a preoccupation with syntax during the design stage?
4) What is the expansion of Ada PDL to code?
5) Does Ada PDL guide the design of the project or are portions of the system primarily other language programs written in Ada syntax?
6) Is there an adequate combination of features of Ada for use as a PDL?
7) Are the most expensive errors found while using a particular set of features of Ada as a PDL?
8) Are errors uncovered at the design stage that ordinarily would have been uncovered during coding because of the use of Ada PDL?
9) What percentage of the interface errors are uncovered during the design stage?

Goal B6: Characterize the programmers and associate their background with their use of Ada.
1) What are the programmers' opinions of Ada before they begin this project? during? afterward?
2) What is each programmer's background with other languages?
3) Is there a relationship between how far into development the change was needed and how much effort was spent on the change? How many sections it affected?
4) What kind of changes were made? (e.g., error correction, planned enhancement, etc.)
5) How many components are involved in the typical change?
6) How many changes are caused by a previous change?
7) How was the need for change determined?
8) How many and what kind of interface changes need to be made?

Area C: **Goals Relating to Metrics for the APSE**

Goal C1: Select a set of static (size, control and data) metrics for the APSE

1) Are there differences in the implications of various counting measures? Are some measures more useful than others?
2) Do certain program measures provide enough information to make other measures superfluous?
3) Which static metrics can be applied throughout the design and code phases. Which cannot?
4) Which static metrics help predict run-time behavior (e.g., reliability, etc.)?
5) Which static metrics can be measured most easily?

Goal C1.1: **Develop a set of size metrics for the APSE**

1) What size metrics best predict effort?
2) What serves as a useful size metric (e.g., lines of code, modules) in Ada?
3) What constitutes a statement in Ada?
4) How should an executable statement be defined in Ada?
5) What features of Ada should be grouped when counting the number of times certain features are used?
6) How useful is Halstead's software science approach with Ada?

Goal C1.2: **Develop a set of control metrics for the APSE**

1) How can tasking and exceptions be integrated into the control metrics?
2) How useful is McCabe's cyclomatic complexity measure? How does the cyclomatic complexity compare with the essential complexity?
3) How useful are measures of nesting complexity and depth?

Goal C1.3: **Develop a set of data metrics for the APSE**

1) How can the complexity of data structures be measured?
2) What influences the number of programmer defined types?
3) How does the use of Ada influence the number of inputs to and outputs from a module?
4) How does Ada influence the use of global data?
5) How does the use of modules affect the treatment of data within a program?
6) How should the span of a variable be measured? Is there a use for the span information?
7) What do the data bindings suggest about the structure of the system?
8) Does the density of the data flow across modules provide useful feedback about the structure of the system? i.e., are information flow metrics (Henry & Kafura) useful?

Goal C2: Select a set of dynamic (test coverage and execution) metrics for the APSE

Goal C2.1: Develop a set of test coverage metrics for the APSE

1) Do any of the following measures of test coverage lead to a useful strategy for testing: number of statements executed? number of decisions executed, or number of independent paths executed?
2) Can these measures be extended to provide test coverage for concurrent processing or will new measures need to be developed? Are there measures to detect starvation, potential deadlocks, etc.?
3) Are there other features of Ada (e.g., exception handling) that require new measures for test coverage? What are those measures?

Goal C2.2: Develop a set of execution metrics for the APSE

1) What are useful execution metrics?
2) What additional information do execution statistics provide beyond what can be gained from a static view of the system?
3) Are there measures of execution complexity?
4) Are certain Ada features or combinations of features expanded into very fast or very slow code?

Goal C3: Develop a subjective evaluation system for evaluating some program and design features that are not easily or practically measured in other ways

1) Can a diverse set of experts (Ada, applications, and methodology experts) accurately evaluate the subjective aspects of the project?
2) How well do the results of these evaluations correlate with results from objective measures?
3) How well do these evaluations correlate with the opinions of the development team?
4) Can we conclude anything from the subjective results?

Acknowledgements

The categorization of the metrics used on this project was developed by Victor Basili and other members of the University of Maryland team. The software development methodology employed benefited from their direction. The authors are grateful to Victor Basili, John Gannon, Elizabeth Katz and Marvin Zelkowitz for their help. This research program is monitored by the Office of Naval Research (ONR) under contract #N00014-82-K-0225 to the University of Maryland with funding from ONR and the Ada Joint Program Office. The views expressed in this paper, however, are not necessarily those of the Office of Naval Research, the Ada Joint Program Office or the Department of Defense.

REFERENCES

[1] Basili, V. R. and Zelkowitz, M. V., Analyzing medium-scale software development, in Proceedings of the 3rd International Conference on Software Engineering (1978) 116-123.
[2] Basili, V. R., Changes and errors as measures of software development, in Basili, V. R. (ed.), Models and Metrics for Software Management and Engineering, Computer Society Press (1980) 62-64.
[3] Basili, V. R., Product metrics, in Basili, V. R. (ed.), Ibid., 214-217.
[4] McCabe, T., A complexity measure, IEEE Transactions on Software Engineering, 2 (1976) 308-320.
[5] Curtis, B., Sheppard, S. B., and Milliman, P. Third time charm: stronger prediction of programmer performance by software complexity metrics, in Proceedings of the 4th International Conference on Software Engineering (1979) 356-360.
[6] Henry, S. and Kafura, D., Software structure metrics based on information flow, IEEE Transactions on Software Engineering 5 (1981) 510-518.
[7] Kafura, D. and Henry, S., Software quality metrics based on interconnectivity, The Journal of Systems and Software 2 (1981) 121-133.
[8] Myers, G. J., Software Reliability: Principles and Practice (Wiley-Interscience, New York, 1976).

Model Evaluation in Software Metrics Research

S. M. Thebaut

Purdue University
Department of Computer Sciences
West Lafayette, Indiana 47907

Recent work has focused on the study of resource models for large-scale software development. While the goals of individual researchers differ, many have been concerned less with the development of predictive tools for project managers than with the informal testing of theories related to the development process itself. The nature of the work is such that many of the models studied are difficult to validate empirically. For example, theoretical constraints sometimes preclude the use of standard techniques in estimating unknown parameters. Furthermore, it is not always clear what criteria should be used in comparing the performance of competing models when differing and sometimes *ad hoc* estimating approaches have been employed. A case study is presented which illustrates these difficulties and some of the approaches used to deal with them.

Introduction

Several models of resource expenditure for large-scale software development have been proposed in the recent literature [1-7]. In most cases, a response (i.e., dependent) variable representing development effort or development duration is combined algebraically with one or more independent variables reflecting program size, problem complexity, etc., according to the relationship thought to exist. All such models share the characteristic of being, at best, an imperfect approximation to some enormously complex relationship involving many unknown factors.

The purposes served by these models are as varied as the models themselves. In some cases it is primarily *description*, while in others it is *prediction*. For others still, it is the (generally informal) testing of hypotheses related to the development process itself. There is, of course, considerable overlap between categories, and most models, in fact, are not easily pigeon-holed. Loosely speaking, however, those developed primarily for descriptive or predictive purposes tend to be simpler in form (e.g., linear or intrinsically linear) and are usually based on traditional estimation techniques (e.g., linear regression). The model of Walston/Felix [7] is one example. Those developed primarily for hypothesis testing, however, are often of a more specialized form and may be based on *ad hoc* or quasi-traditional estimation techniques. Examples include the models of Putnam [5], Bailey/Basili [1], and Thebaut/Shen [6].

Regardless of the motivation for development, it is usually necessary that any unknown parameters in the model be estimated on the basis of historical data. In general, however, this process is based on *accidental* - as opposed to *random* - samplings; data is collected from whatever sources happen to be available until either (a) the sample reaches some designated size, (b) the time allotted for collection expires, or (c) there are no additional sources. It is usually impossible to identify the biases, if any, introduced in such samples. As a result, what is perhaps the first-line defense against misleading results - the use of statistical tests which presuppose probability samplings - is weakened.[1]

An additional problem befalls the researcher who wishes to ascertain which of several models - some of which may carry important theoretical implications - is most consistent with reality. To begin with, it is not immediately clear what criteria should be used in ascertaining this. Some measures, for example, are closely related to specific estimation techniques (e.g., *mean square error* and *least-squares regression*), and may not be appropriate for any of several reasons. As will be seen, the problem is particularly severe if differing or non-standard estimation techniques are used.

In what follows, a case study is presented which illustrates the nature of these problems as actually encountered. A number of measures reflecting goodness-of-fit are examined with respect to their usefulness, both individually and in combination with others.

A Case Study in Model Evaluation

A recent study was undertaken to assess the effects of project team-size on the cost and duration of program development [6,9]. Project resource data obtained from various sources included the following measures:

S - program *size*, in thousands of delivered source lines

E - development *effort*, in man-months

$\bar{P}$ - average equivalent full-time programming *personnel*, in persons[2]

A major source was Boehm's *Software Engineering Economics* [3], in which statistics reflecting 63 commercial projects are provided. All illustrations are based on this data.

In the course of the study, we developed an analytic resource model, COPMO (the COoperative Programming MOdel), to isolate, in part, the effects on E of variations in $\bar{P}$. The form proposed,

$$\hat{E} = \begin{cases} b_0 + b_1 S & \bar{P} \leq k \\ b_0 + b_1 S + b_2 \bar{P}^{b_3} & \bar{P} > k \end{cases} \quad (1)$$

is based on both theoretical and empirical considerations. For example, the manner in which the expressions $b_0 + b_1 S$ and $b_2 \bar{P}^{b_3}$ are evaluated and combined corresponds to a hypothetical partitioning of development effort into components related to (a) the independent programming activity of individuals, and (b) the overhead associated with cooperative (i.e., team-oriented) programming. The expressions themselves, however, are formulated on the basis of the data and measures available.

As $\bar{P}$ is a continuous measure, the constant k represents an approximate boundary between one-person and multi-person projects. In estimating the parameters, an attempt is made to identify a least $k \geq 1$ for which a plausible relationship between S and E can be discerned for projects satisfying the condition $\bar{P} \leq k$. The relationship is then approximated using linear least-squares regression to yield parameters b_0 and b_1. Following this, projects for which $\bar{P} \leq k$ are set aside, and regression is again applied using log-transformed variables ($E - b_0 - b_1 S$) and $\bar{P}$ to yield parameters b_2 and b_3. While the details given are not critical to the discussion which follows, they do illustrate the specialized nature of the procedure used.

In order to test the reasonableness of an implicit hypothesis in COPMO, we chose to compare its performance to that of others for which alternative hypotheses are implicit. Among those considered were: (a) Boehm's Basic COCOMO (COnstructive COst MOdel) [3]; and (b) a first-order[3] model with response variable E and independent variables S and $\bar{P}$.

There are three Basic COCOMO effort estimators, one for each of three possible "modes" or levels of development constraint.[4] Each has the form

$$\hat{E} = b_0 S^{b_1} \quad (2)$$

where the parameter values depend on the mode in question. For the development data considered, the mode of each project is known. Boehm has determined the parameters for each of the three estimators using *ad hoc* techniques.

The first-order model in S and $\bar{P}$ has the form

$$\hat{E} = b_0 + b_1 S + b_2 \bar{P} \quad (3)$$

where the parameters are estimated using multiple linear regression. Of the models considered, then, it alone is based solely on standard - although, for the data considered, not necessarily appropriate - estimation techniques.

Model Comparison Criteria

We now consider the criteria to be used in comparing these models with respect to their conformability to the data in question. While formal techniques may at times be useful in testing the aptness of models, they are not always applicable.[5] We therefore focus instead on various measures of the relationship between E and $\hat{E}$, and on graphical analysis.

Several of the measures which have been used for this purpose are described below. Some have been included primarily to illustrate the more common pitfalls associated with an over-reliance on a single (and perhaps inappropriately used) measure. (In what follows, we assume some familiarity with standard statistical terminology.)

Coefficient of Multiple Determination - R^2. This measure - and its closely related counterpart, Pearson's coefficient of correlation - is among the most popular of those used. This stems, in part, from its natural association with linear least-squares regression. For the present application, it is computed as:

$$R^2 = \frac{\sum_{i=1}^{n}(E_i - \overline{E})^2 - \sum_{i=1}^{n}(E_i - f(\widehat{E}_i))^2}{\sum_{i=1}^{n}(E_i - \overline{E})^2} \quad (4.a)$$

$$= 1 - \frac{\sum_{i=1}^{n}(E_i - f(\widehat{E}_i))^2}{\sum_{i=1}^{n}(E_i - \overline{E})^2} \quad (4.b)$$

where E_i and $\widehat{E}_i$ are the ith actual and estimated values respectively, from a total of n such pairs of values, and f is that linear function of $\widehat{E}$ which maximizes the result. The naturalness of the measure when used in conjunction with the first-order model is evidenced by the fact that f, in this case, is the identity function. As will be demonstrated shortly, however, the "most apt" among competing models need not be that which maximizes R^2. This is evident for a number of reasons: (a) for models such as (1) or (2), the measure can only indicate the extent to which E and $\widehat{E}$ are *linearly* related; (b) for models such as (3), the measure depends on the assumption that a "best" linear fit between response and independent variable(s) is that which satisfies the least-squares criterion; and (c) in no case does the value of R^2 reflect how closely E and $\widehat{E}$ correspond to one another in an absolute sense. Thus, it is extremely important that the R^2 measure be considered in the context of the model and data in question.

The Spearman Rank Correlation Coefficient - r_S. Like the coefficient of multiple determination, r_S may be used to measure the degree to which values of E and $\widehat{E}$ are related. The type of relationship, however, is more general in nature: it reflects the agreement between the *rankings* of the measures involved. It is computed as:

$$r_S = 1 - \frac{6\sum_{i=1}^{n} d_i^2}{n^3 - n} \quad (5)$$

where d_i is the difference between the ranks of the measures for the ith observation. An r_S value of one corresponds to a perfect correlation between the two sets of measures while a value of zero corresponds to none. Inverse relationships are represented by negative values. The limitations of the measure - if considered in isolation - are clear in the current context. It suggests only the ordinal relationship between actual and estimated values, and therefore only the extent to which projects are correctly ranked with respect to E on the basis of $\widehat{E}$. When considered together with others, however, the measure can be a source of insight with respect to the overall performance of a model.

Average Relative Error - $\overline{RE}$. Defined here as

$$\overline{RE} = \frac{1}{n}\sum_{i=1}^{n}\frac{(E_i - \widehat{E}_i)}{E_i} \quad (6)$$

this measure is among the most widely used - and abused - of those considered. It provides essentially no information with regard to a model's precision for individual estimates since errors of opposite sign tend to cancel one another. The measure can be useful, however, as an indicator of anomalies in overall fit. A large positive/negative value of $\overline{RE}$ suggests a systematic under/over estimation of actuals by the model in question. The situation often arises, for example, when applying least-squares regression in the presence of an outlying observation.

Magnitude of Relative Error - MRE. For the ith actual and estimated values of E, the measure

$$MRE_i = \frac{|E_i - \widehat{E}_i|}{E_i} \quad (7)$$

is used in two ways: to reflect a model's *average* error in prediction for a given set of data, we use $\overline{MRE}$, the mean of MRE for the set; to reflect the model's performance at other levels of error, a "performance profile" is developed by considering the percentage of points for which $MRE \leq k$ for several values of k.

Graphical Analysis. The most direct manner for evaluating the aptness of a model involves the informal study of a scatter diagram on which the data and estimated regression function are plotted. A somewhat less direct, but often more illuminating approach involves the examination of residuals[6] plotted against either an independent variable or $\widehat{E}$. Difficulties such as the presence of outliers, inappropriateness of the model, or the violation of specific assumptions are generally apparent by inspection of these plots. Both approaches are employed in evaluating the aptness of the models considered.

Comparing the Models

Comparison-measure values for the models and data considered are given on the following page. Note the discrepancies in the rankings of the

Measure Comparisons

Model	R^2	r_S	$\overline{RE}$	$\overline{MRE}$
First-Order	.91	.93	5.25	5.92
Basic COCOMO	.44	.91	-.11	.60
COPMO	.89	.96	-.14	.43

models according to the different measures. Interestingly, no two measures show complete agreement. Based on R^2 values, for example, we see that both COPMO and the first-order model account for over twice the proportion of variance observed in E that Basic COCOMO does. Based on $\overline{RE}$ values, however, we see that the first-order model substantially under-estimates E on average, while COPMO and Basic COCOMO only slightly over-estimate E on average. This may suggest that the first-order form is inappropriate (i.e., the underlying relation is non-linear), that one or more outliers are present, or that assumptions concerning the least-squares criterion for parameter estimation have not been met.[7]

Comparing r_S and $\overline{MRE}$ values, it is evident that a strong *ordinal* relationship between E and $\hat{E}$ does not, in general, imply close agreement, on average, between actual and estimated values.

Scatter and residual plots (square root scales) for each of the models are provided at the end of the paper. For the first-order model, a large number (nearly half) of the estimates are negative. More importantly, the residuals tend to vary in a systematic fashion from positive to negative values, suggesting the need for a curvilinear regression model. Basic COCOMO provides a better fit, although some residuals are quite large. COPMO appears to provide the "tightest" fit overall.

Plots such as these generally provide a reliable, if informal, means of evaluating the aptness and overall fit of a model. Likewise, a measure such as $\overline{MRE}$ can be useful for describing, in meaningful terms, the average agreement between actual and estimated values. The *average* performance of a model, however, may not always be of greatest interest. For example, a model which performs very well for some projects and very poorly for others may be of less value, practically speaking, than one which performs only reasonably well for all. To evaluate goodness-of-fit at differing levels of accuracy, an MRE "performance profile" may be constructed as shown on the following page. The vertical axis corresponds to the percentage of projects for which $MRE \le k$. The horizontal axis corresponds to the error bound, k. Thus, while Basic COCOMO estimates are within 30% of actuals about 30% of the time, for example, COPMO estimates are with 30% of actuals nearly 50% of the time. Similarly, first-order estimates are within 80% of actuals about 30% of the time, while COPMO estimates are within 80% of actuals about 95% of the time. It is evident from the profile that the COPMO estimator is in closer agreement with E over most of the range considered than are either of the other models.

Conclusions

We have attempted to illustrate some of the problems encountered by researchers who engage in work of this type. Difficulties in estimating, evaluating, and comparing models of the type considered are at times significant and reflect both limitations in the current state of applied statistics and the relatively early stage of investigation involved.

Our experience has been that a combined use of measures, each reflecting a different facet of agreement between model and data, is generally more illuminating than the use of any one in isolation. It is critical, of course, that the measures themselves be interpreted in a way that is consistent with the model in question, the method of estimation, and the data used. Graphical analysis, while inherently subjective, has been found to be especially valuable in revealing difficulties in a model.

We have found the MRE profile to be an especially useful device in goodness-of-fit comparisons. Variations in the effort required in software development tend to increase with the size and complexity of the tasks involved. For this reason, a measure of *relative* error between estimated and actual effort is more meaningful than one of *absolute* error when applied to both small and large projects.

Acknowledgements

The author is indebted to members of the Software Metrics Research Group in the Department of Computer Sciences at Purdue University for their help in the preparation of this paper. Special thanks to Vincent Shen and Hubert Dunsmore for their many valuable suggestions and comments.

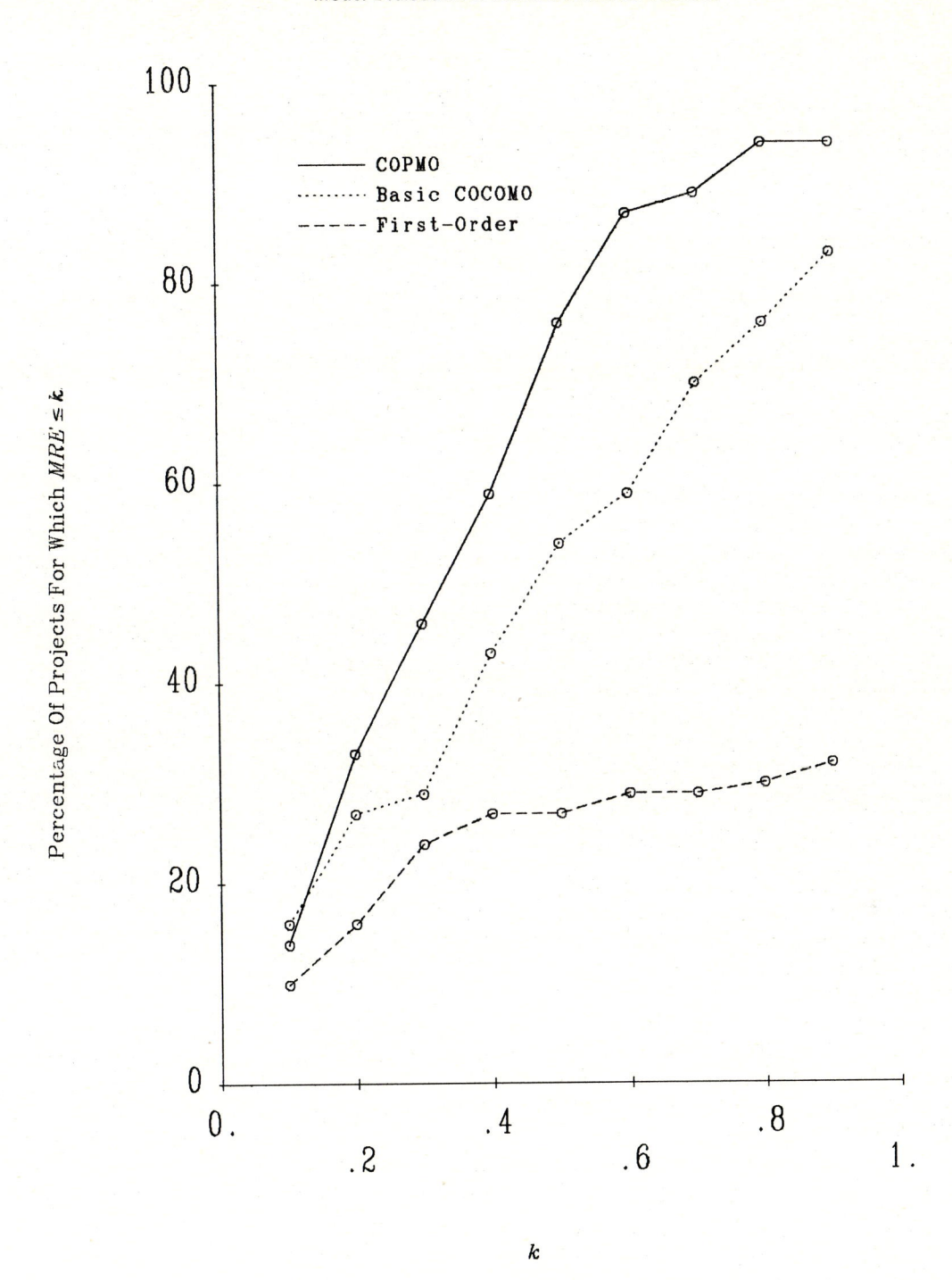

MRE Profile

Notes

1. The application of statistical procedures to accidental samples is not necessarily a meaningless practice. The issues involved, however, are both involved and controversial. An overview may be found in [8].

2. The $\bar{P}$ measure is based on the average man-months of effort expended per duration month of program development. For example, $\bar{P}$ for a project lasting 3 months and requiring 12 man-months of effort is 4 persons.

3. A *first-order* model is one which is linear in the parameters and the independent variable(s).

4. The categories used are: *organic* (least constrained), *semidetached*, and *embedded* (most constrained). The criteria used to determine which category best represents a given project are complex. See [3] for details.

5. Formal tests for determining the aptness of models depend, for example, on the availability of repeat observations at one or more levels of the independent variable(s). For an overview, see, for example, [10].

6. A *residual* is the difference between the observed value and the estimated value: $E_i - \hat{E}_i$.

7. For example, the error variance may not be constant. See, for example, [10] for a description of the various assumptions.

References

[1] J. W. Bailey and V. R. Basili, "A Meta-Model for Software Development Resource Expenditure," *Proceedings, Fifth International Conference on Software Engineering*, IEEE/ACM/NBS, March, 1981, pp. 107-116.

[2] V. R. Basili and M. V. Zelkowitz, "Analyzing Medium-Scale Software Development," *Proceedings, Third International Conference of Software Engineering*, 1978, pp. 116-123.

[3] B. W. Boehm, *Software Engineering Economics*, Prentice-Hall, Inc., Englewood Cliffs, NJ, 1981.

[4] F. N. Parr, "An Alternative to the Rayleigh Curve Model for Software Development Effort," *IEEE Transactions on Software Engineering*, May 1980, pp. 291-296.

[5] L. H. Putnam, "A General Empirical Solution to the Macro Software Sizing and Estimation Problem," *IEEE Transactions on Software Engineering*, July 1978, pp. 345-361.

[6] S. M. Thebaut and V. Y. Shen, "An Analytic Resource Model for Large-Scale Software Development," *Journal of Information Processing and Management*, 1983, to appear.

[7] C. E. Walston and C. P. Felix, "A Method of Programming Measurement and Estimation," *IBM Systems Journal*, Vol. 16, No. 1, 1977, pp. 54-73.

[8] C. Selltiz, M. Jahoda, M. Deutsch, and S. W. Cook, *Research Methods in Social Relations*, Holt, Rinehart and Winston, New York, 1965.

[9] S. M. Thebaut, "The Saturation Effect in Large-Scale Software Development: Its Impact and Control," Ph. D. Thesis, Purdue University, West Lafayette, IN, 1983.

[10] J. Neter and W. Wasserman, *Applied Linear Statistical Models*, Irwin, Inc., Homewood, IL, 1974.

RANGE OF X AXIS: -20 120
RANGE OF Y AXIS: -50 50

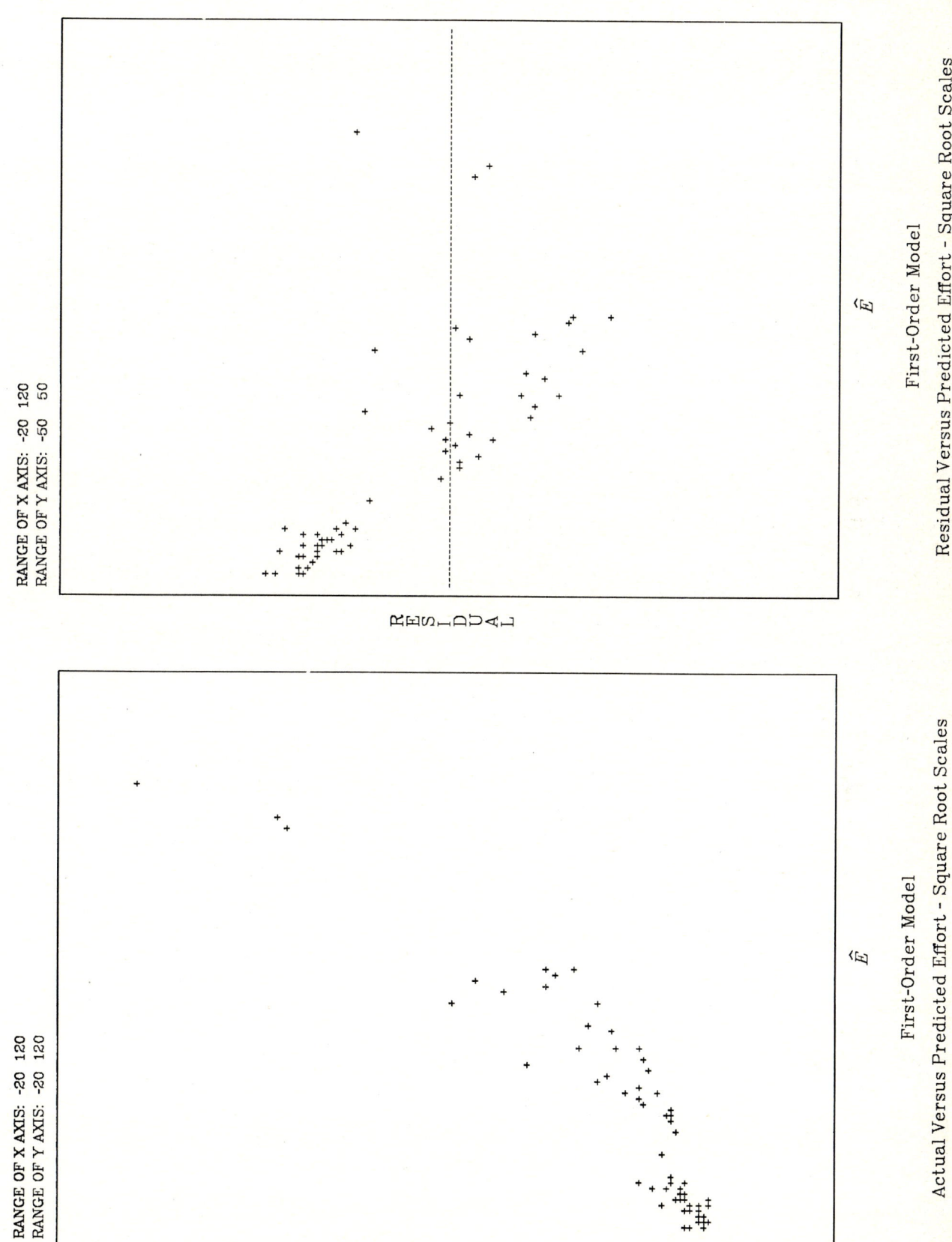

$\hat{E}$

RESIDUAL

First-Order Model
Residual Versus Predicted Effort - Square Root Scales

RANGE OF X AXIS: -20 120
RANGE OF Y AXIS: -20 120

E

$\hat{E}$

First-Order Model
Actual Versus Predicted Effort - Square Root Scales

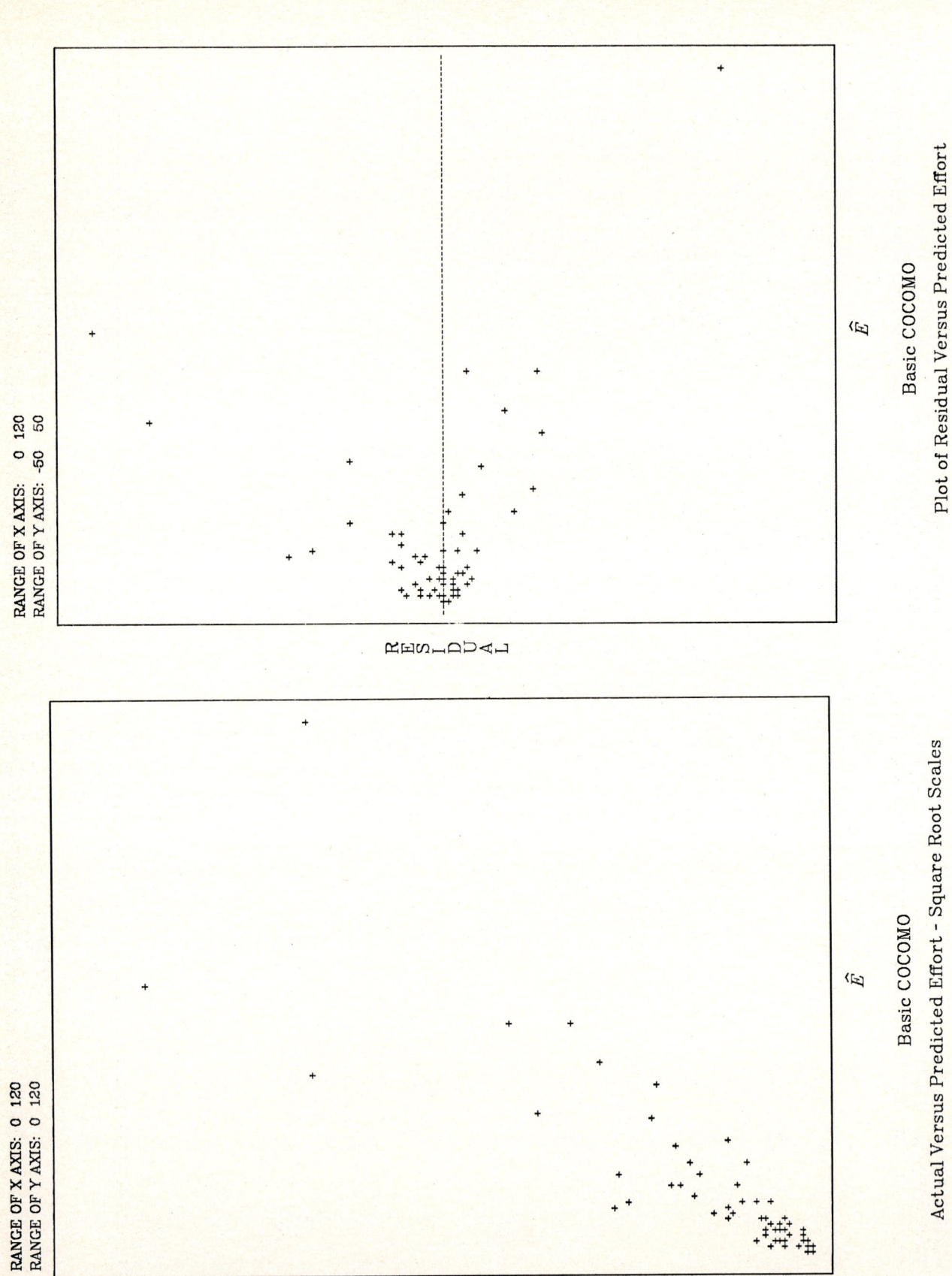

RANGE OF X AXIS: 0 120
RANGE OF Y AXIS: -50   50

$\hat{E}$

Basic COCOMO
Plot of Residual Versus Predicted Effort

RESIDUAL

RANGE OF X AXIS: 0 120
RANGE OF Y AXIS: 0 120

$\hat{E}$

Basic COCOMO
Actual Versus Predicted Effort - Square Root Scales

E

Model Evaluation in Software Metrics Research

RANGE OF X AXIS: 0 120
RANGE OF Y AXIS: -50 50

RESIDUAL

$\hat{E}$

COPMO
Residual Versus Predicted Effort - Square Root Scales

RANGE OF X AXIS: 0 120
RANGE OF Y AXIS: 0 120

E

$\hat{E}$

COPMO
Actual Versus Predicted Effort - Square Root Scales

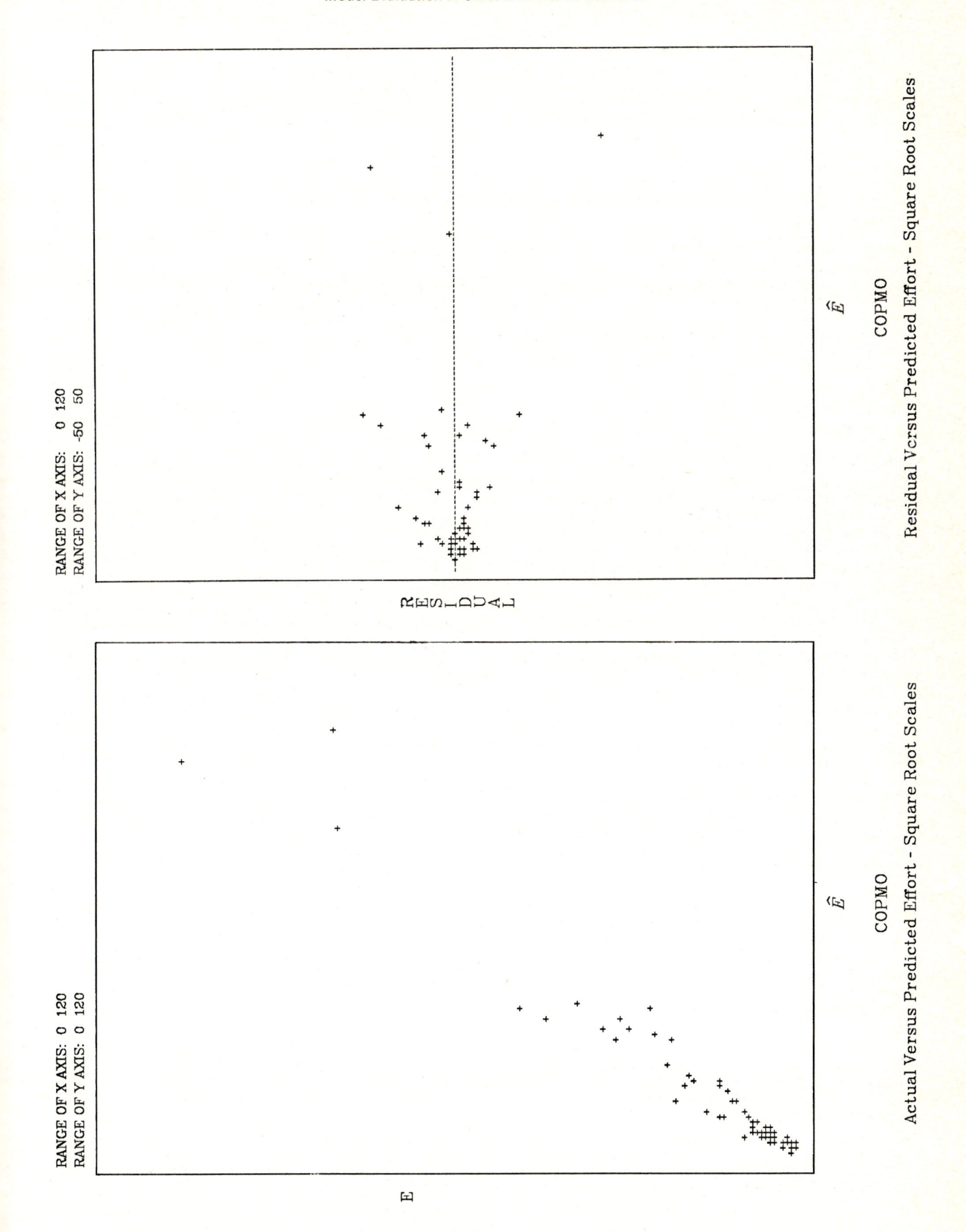

INVESTIGATION OF CHUNKS IN COMPLEXITY MEASUREMENT

John S. Davis

Army Institute for Research in Management
Information and Computer Sciences
Atlanta, Georgia

Current complexity measures are based on characteristics such as operator and operand counts and control flow paths, which are not convincing indicators of psychological complexity. This paper provides justification for using chunks in complexity measurement and presents results of an empirical study which compares performance of chunk measures with that of lines of code and McCabe's V(G). The data connections between chunks seem more significant than control connections and chunk content.

1. INTRODUCTION

The cost to develop and maintain software is dominated by labor costs, which depend significantly on the psychological complexity of the software (on how difficult the software is to understand). Complexity measures are potentially very useful in predicting the effort required in a project and in evaluating software.

Measures may have a syntactic, semantic or pragmatic basis, but the syntactic approach is the focus of this paper. A syntactic complexity measure quantifies some syntactic feature which is believed to be associated with psychological complexity.

For purposes of validating measures, we often make the assumption that psychological complexity is associated with the effort to maintain the software. We assume that certain phenomena are indicants of such effort (e.g. time to debug, number of error occurrences). A complexity measure is then judged on how well it correlates with an indicant of maintainability.

There is substantial evidence that programmers invoke a mental process called chunking to understand a program. In this process several related items, such as program statements, are formed into an entity which is remembered and understood as easily as a single item. A complexity measure should account for this process.

Most popular complexity measures are based on indicants which have not been shown to be strongly related to the process of human understanding, for example:

Measure	Indicator
McCabe's V(G) [1]	linearly independent control flow paths
Halstead's Effort [2]	operators & operands
Knots [3]	control flow path intersections
Lines of code	lines of code

Control flow paths or knots are good indicators of complexity if "playing machine" is the principal approach of programmers to understanding a program. Operators and operands or lines of code are good indicators if size is the main cause of complexity. But the results of human factors research support another view, that programmers use chunks to understand a program.

There is some evidence [4] and it is the opinion of many experienced programmers that chunks typically consist of one or more contiguous statements, but no one knows precisely what chunks are actually formed in the minds of programmers.[1] In this paper, the segments resulting from any partition of a program will be called chunks.

Perhaps the good performance of some existing measures is explained by their implicit accounting for chunks. "Lines of code" is equivalent to "number of chunks" if a line is considered a chunk.

McCabe's V(G) is based on the control flow graph of a program, defined as:

V(G) = # edges - # nodes + 2 * (# connected components).

Nodes of the graph correspond to segments of code which may be considered chunks. The knots measure similarly suggests a partition of a program into chunks. Each statement following a GO TO starts a new chunk, and each chunk will be terminated by a GO TO statement. Based on operator and operand counts, Halstead's Effort seems to be an exception. Operators and operands are too small to satisfy the role of a chunk.

2. TYPES OF CHUNKS CONSIDERED IN THIS STUDY

Explicit or implicit chunk models of program complexity may be compared on how they treat these aspects: chunk definition, chunk internal structure, and interconnection of chunks. The chunk internal structure may be related to the difficulty to form the mental concept, and the interconnections may explain the difficulty to understand how chunks work together in the whole program.

2.1 The M-Chunk

One particular definition of chunk, the M-chunk, provides a convenient basis to discuss and compare both a popular existing measure, V(G), and a proposed chunk model of complexity. Sharing a common chunk definition, the two measures have different approaches to chunk internal structure and interconnection.

The M-chunk is a sequence of one or more statements represented by a node in the control flow graph of a program. M-Chunks are the high level language equivalent of the basic block which is familiar to compiler designers: "... a sequence of operations to be executed in order, with only one entrance and one exit, the first and last operations respectively" [5]. We need only to substitute "statement" for "operation" in the definition.[2]

2.2 The S-Chunk

The second type of chunk considered in this study is the S-chunk. Every executable statement represented within M-chunks is considered an S-chunk. "Statements" here are consistent with the way M-chunks are formed, may actually be statement fragments, and are not necessarily equivalent to lines of code. For example, the line of code

    IF(X.EQ. Y)N=10

translates to two S-chunks:

    IF(X .EQ. Y)

    N = 10

The S-chunk is not an intuitive choice, but is considered primarily for contrast with the M-chunk.

3. COMPLEXITY MEASURES STUDIED

3.1 McCabe's V(G)

McCabe's V(G) essentially ignores the internal content of chunks. All nodes in the control flow graph are considered equivalent with regard to content, whether they correspond to one statement or a hundred, whether they correspond to complicated arithmetic statements or statements merely assigning a constant value to a variable.

V(G) does account for one aspect of the complexity of chunk interconnections, specifically the number of linearly independent control flow paths in the program. There is no accounting for the structure of the "data interconnections" between nodes (chunks).

3.2 Halstead's Effort

The formula used for this measure is

$$E = (n1/2)(N2/n2)(N1 + N2)\log(n1 + n2)$$

where

    n1 = number of unique operators
    n2 = number of unique operands
    N1 = total frequency of operators
    N2 = total frequency of operands

3.3 Lines of Code

Lines of code include every physical line of program text, excluding comments and continuation lines.

3.4 A Chunk Measure

Now consider a proposed measure of program complexity, originally applied to large modules [6], which here is applied to chunks. The assumption is that programmers understand a program, especially an unfamiliar one, by first scanning the entire text and then attempting to comprehend the individual chunks. When studying a particular chunk, the programmer makes reference to other related chunks when this is necessary to fully understand the function of the chunk being studied. For example, suppose the programmer is studying chunk Ki and notices the following statement:

    X = Y + 1.0

To fully understand the behavior of this chunk, he or she will refer to other chunks which affect (possibly change the value of) Y. This is an example of a data dependency between chunk Ki and, say, another chunk B which contains the statement:

    Y = 3.4

Likewise if the following statement is found in chunk Ki:

    GO TO 30

and statement 30 is in chunk B, the programmer will refer to chunk B to see "what happens" there, and will then return attention to chunk Ki. This is called a control dependency between Ki and B.

Thus a chunk may be re-read several times, but each time is easier than the last one because of residual memory. Woodfield estimated that each repeated reference (to a module) takes about 2/3 as long as the previous one [6]. This fraction is called the review constant.

The fan-in of a chunk B is the number of chunks Ki such that Ki has a data or control dependency on B. This is the number of times B must be read or re-read in the process of understanding other chunks. In accordance with the model just described, overall program complexity may be computed by the following formula from [6].[3]

$$C = \sum_{i=1}^{\#CHUNKS} \left(\sum_{m=0}^{FAN-IN_i} C_i R^m \right)$$

where

C_i = complexity of chunk i (#statements)
R = 2/3, the review constant
m : for given i, takes on values up to the "fan-in" of chunk i

Notice that this model accounts for C_i, the complexity of an individual chunk. It is not clear how this complexity should be determined, but there are two obvious alternatives: let C_i be the number of statements in chunk i, or assign a constant value (of 1) so that all chunks are considered equal.

4. ANALYSIS OF FORTRAN PROGRAMS

The author collected FORTRAN programs from the experiment of Sheppard, et. al. [8]. The data from this experiment provided a means of investigating the value of chunk approaches in explaining variations in the time required to debug. The task in the Sheppard experiment was to correct a single bug, which required changing only one line of code. Subjects were provided listings of the program, input, correct output and actual output. Nine basic programs were "stretched" into 27 by providing three different versions of each. The versions differed in length, in that certain subroutines were omitted from the listing given to subjects. In place of omitted subroutines, the experimenters provided a description of the subroutine and an assurance that it performed correctly. Each program was debugged by 6 different programmers.

Most M-chunks consist of more than one statement, with the average length 2.1 statements. It seems that there are a significant number of multi-statement M-chunks, such that if they were actually used by programmers, they would facilitate understanding the program. The distribution of chunk size, in number of statements, is shown below. Non executable statements, such as FORMAT and CONTINUE statements, were not counted in computing chunk size. Thus chunks whose only content is a CONTINUE statement have a size of zero.

#stmt	#chunks
0	97
1	405
2	317
3	131
4	58
5	41
6	11
7	9
8	0
9	3
10	0
11	0
12	6
13	6
14	0
15	0
16	0
17	3
18	0
19	0
20	3

The proposed complexity measure was computed separately for both M-chunks and S-chunks in order to gain some insight on the value of chunking, at least with respect to the particular model of programmer understanding on which the measure is based. Also varied was the choice of the type of chunk connections (dependencies) considered in the formula: control, data, or both. In the following tables, $R^{**}2$ is the coefficient of determination (in percent of variation) from a linear regression analysis, which compares measures on their ability to predict debugging time.

$R^{**}2$, Debugging Time vs. Program Complexity
(Individual chunk complexity is # statements)

Type of chunk:	Type of Chunk Dependencies Considered:		
	Control	Data	Control & Data
S-chunk	13.4	13.5	14.6
M-chunk	22.3	32.4	32.1

$R^{**}2$, Debugging Time vs. Program Complexity
(Individual chunk complexity = 1 for all chunks)

	Type of Chunk Dependencies Considered:		
	Control	Data	Control & Data
Type of chunk:			
S-chunk	18.4	20.6	21.0
M-chunk	51.9	60.4	57.3

$R^{**}2$, Debugging Time vs. Program Complexity
(M-chunks used in calculations)

	Type of Chunk Dependencies Considered:		
	Control	Data	Control & Data
Value of Ci for each chunk:			
# statements	22.3	32.4	32.1
Constant (=1)	51.9	60.4	57.3

$R^{**}2$, Debugging Time vs. # Chunks

# M-chunks:	60.3	# S-chunks:	17.7

$R^{**}2$, Debugging Time vs. Total Fan-in of All Chunks

Total Control Fan-in of M-chunks:	6.2	Total Control Fan-in of S-chunks:	6.2
Total Data Fan-in of M-chunks:	43.3	Total Data Fan-in of S-chunks:	14.7

Other Measures

Lines of code	27.0
McCabe's V(G) (Computed as McCabe recommends)	42.2
Halstead's Effort	56.2

Apparently data connections are more significant than control connections in predicting debugging time. Notice in the first three tables that $R^{**}2$ is greater in every case when data, rather than control, dependencies are accounted for, and that total data fan-in is a much better predictor of debugging time than control fan-in. This may be explained by the observation that subjects in this experiment were unfamiliar with the programs. They probably had to expend considerable effort to decipher the role played by the variables and their interrelationships. Control flow is more explicit and more easily understood. Most cases of confusing control flow behavior are acutally caused by confusing control flow <u>variable</u> behavior, which in turn depends on the data interconnections. For example, the key to understanding the behavior of the statement

    IF((A.GT.B) .OR. (C.LT.D)) GO TO 30

is to know the behavior of the variables A, B, C and D.

An unexpected result is that one of the simplest measures, number of M-chunks, performed as well as the measures which account for chunk interconnections. This seems to follow from the fact that in these programs, the number of data interconnections increases linearly with the number of chunks. The number of S-chunks did poorly because S-chunks do not account for any mental aggregation of statements, and are therefore really not chunks at all.

The results indicate that the content of the chunks is probably not as important as the number of chunks and the structure of their interconnections. The third table shows consistent improvement in $R^{**}2$ when the complexities of all chunks are considered equal. Perhaps it takes about the same mental effort to grasp the meaning of any chunk.

All variants of the chunk complexity measure improve as predictors of debugging time when the chunks are changed from the single statement level (S-chunk) to the chunk level (M-chunk). This is consistent with the hypothesis that multi-statement chunks are significant to programmers. Even when data and control connections are considered, the S-chunk does not provide a good basis for predicting debugging effort.

Halstead's Effort measure performed about as well as any of the chunk measures. The good performance here and in many other experiments remains somewhat a mystery to this author, because the measure does not seem to be based on a sound model of programmer cognition. Actually, Halstead intended that Effort be used to estimate program construction time rather than debugging effort, but he made questionable assumptions nonetheless. He imagined that humans use a binary search on a "menu" of operators and operands when they write programs. The Effort formula includes the term (n1N2/2n2), which Halstead with little justification interpreted as the number of mental discriminations per comparison during the alleged binary search.

5. CONCLUSION

Preliminary results indicate that further exploration of chunk models of program complexity is warranted. Several variants of the chunk measure outperformed existing measures in this study. The suggestion from these results that the data structure is more important than control structure in debugging seems to corroborate Shneiderman's recent finding that data structure information is more useful than control structure information in understanding programs [9].

Quantifying the variety of chunk types and interconnections may be more useful than the techniques used so far. The author intends to apply entropy to assess this variety. This idea is based on an information theory view of programmer/program interaction: the programmer is a receiver who handles the information presented by a "source" (the program) [10].

The variety in the structure of the program may be related to the difficulty of determining how chunks are related to one another, and to the difficulty of forming a mental hierarchy of chunks, a process which seems essential to understanding a program [11]. At the highest level, the meaning of the program is evident.

It is important to determine the significance of chunks and their interconnections in larger programs. It may be that interconnections become more significant as program size increases. Performance of chunk complexity measures may vary in situations where the predicted quantity is construction time or number of error occurrences, rather than debugging time.

It will be interesting to determine through further research whether programs in a given language have a rather consistent distribution of chunk size. If so, and if the number of chunks turns out to be a useful complexity measure it will then be possible to estimate the number of chunks from the lines of code in the program. This would provide a measure derived from, but superior to, lines of code.

ACKNOWLEDGMENTS

The author appreciates the experimental data furnished by S.B. Sheppard and the advice of R.A. Demillo, J. Hammer, R.J. LeBlanc, W.E. Underwood and P. Zunde.

FOOTNOTES:

1. People apparently use chunks of contiguous phrases when reading natural language texts [12]. Weiser [13] claims programmers use "slices", groups of statements which are not necessarily contiguous.

2. McCabe recommended representing statements having compound predicates by several nodes in the control graph [1]. In order to produce chunks which seem to better resemble those programmers use, we construct the graph so that compound predicates of a single statement are contained by a single node. For other reasons a single IF statement in the program may be represented by more than one node in the graph. An example is given in the description of the S-chunk.

3. A more formal description of the formula and definitions of the parameters is given in [7].

REFERENCES:

[1] McCabe, T.J., A complexity measure, IEEE Transactions on Software Engineering SE-2 (1976) 186-190.

[2] Halstead, M.H., Elements of software science (North-Holland, Amsterdam, 1977).

[3] Woodward, M., Hennell, M., and Hedley, D., A measure of control flow complexity in program text, IEEE Transactions on Software Engineering SE-5 (1979) 45-50.

[4] Norcio, A., Human memory processes for comprehending computer programs, Applied Science Department, U.S. Naval Academy (1980).

[5] Gries, D., Compiler construction for digital computers (John Wiley & Sons, NY, 1971).

[6] Woodfield, S.N., Enhanced effort estimation by extending basic programming models to include modularity factors, Ph.D. Thesis, Dept. of Comp. Sc., Purdue Univ. (December 1980).

[7] Davis, J.S., Chunks: a basis for complexity measurement, unpublished AIRMICS report (November 1982).

[8] Sheppard, S.B., Milliman, P. and Curtis, B., Factors affecting programmer performance in a debugging task, General Electric Co. (February 1979).

[9] Shneiderman, B., Control flow and data structure documentation: two experiments, CACM 25 (1982) 55-63.

[10] Chen, E.T., Program complexity and programmer productivity, IEEE Transactions on Software Engineering SE-4 (1978) 187-194.

[11] Shneiderman, B., Software psychology: factors in computer and information systems (Winthrop Publishers, 1980).

[12] Johnson, N.F., The psychological reality of phrase-structure rules, Jrnl. of Verbal Learning and Verbal Behavior 4(1965) 469-475.

[13] Weiser, M., Program slicing, Proceedings of the 5th IEEE ICSE (1981) 439-449.

Contributed Papers

The following are written versions of contributed papers that were presented as poster sessions at the 15th Interface Symposium.

COLOR ANAGLYPH STEREO SCATTERPLOTS--CONSTRUCTION DETAILS

Daniel B. Carr and Richard J. Littlefield[1]

Pacific Northwest Laboratory
Richland, Washington 99352

For those that can see stereo, stereo scatterplots provide a powerful method of displaying data that has three-dimensional structure. While not the best quality stereo presentation method, color anaglyph stereo is easily produced on raster color devices. This simplicity makes color anaglyph stereo a useful tool for exploratory interactive data analysis. Attention to several details can enhance anaglyph scatterplots. This poster paper addresses issues of perspective projection ambiguity, pixel-rounding noise, crosstalk between images, and the use of additional depth cues including overplotting, and shifts in point size, intensity, and hue.

1. INTRODUCTION

The need to display data in three or more dimensions arises frequently in exploratory data analysis.(1) One technique that has been used by several investigators is the three-dimensional (3-D) scattergram.(2,3) Three-dimensional scattergrams may be generated in many ways. True 3-D hardware has recently been developed,(4) but is not commonly available. Typically displays are made using conventional hardware and depth cueing a 2-D display.

One useful depth cue is stereo vision. Stereo presentations are commonly used in some areas, such as crystallography. In combination with motion, stereo produces quite vivid displays.(5) For scatterplots with random data, stereo by itself is not so powerful. Attention to detail is important in making such displays as clear as possible and in enlarging the audience that can actually see the depth.

2. GENERAL STEREO ISSUES

In theory, stereo images are quite simple. Two images are generated corresponding to slightly different viewpoints, and each image is presented to a different eye. Under favorable conditions, the viewer's brain fuses the two separate flat images into a single image with depth. However, there are several problems that can interfere with this process. One common source of difficult fusion is bad alignment of the images. Bad alignment of whole images is caused by images that are different sizes, vertically offset, or rotated. These causes are easily avoided. However, bad alignment of individual points can be caused by the method of computing point locations and by rounding to digital values. These point alignment problems and the problem of misalignment due to ambiguous pairing are discussed below.

Stereo pairs are usually made by selecting a single focal point, then applying a perspective transformation at each of the two eye positions.(6) The resulting images are theoretically correct only when looking at the focal point. When viewers change their focus to an arbitrary point, the point can appear badly aligned. The horizontal disparity between the eye images determines depth. At a viewing focal point, vertical disparity is noise that the brain must reject. The brain can reject this noise for familiar objects, but in scattergrams where the viewer will look directly at multiple data points, the data points far from the display focal point are likely to remain unfused or fuse incorrectly.

Vertical disparity can be removed simply by requiring that points have the same Y coordinate in both views. This is a compromise between having an exact display with a single focus and the impossible task (in a single-image display) of having an exact display for all focal points. Since the eyes tend to traverse the display in building a total mental image, the compromise is quite effective. The compromise can be thought of in terms of projection planes. The standard projection method involves two projection planes, one for each eye. In the compromise a single plane is used. This plane is orthogonal to the line from the midpoint between the eyes to the focal point.

When projection is used to determine the horizontal disparity, and the focal point is the center of the plot, a very simple algorithm can be used. It is

$$dx = k*range(x)*[max(z) - z]/range(z) \qquad (1)$$

With this approximation, the left image is plotted as $(x - dx, y)$ and the right image is plotted as $(x + dx, y)$. The selection of k depends on the physical size of the plot and the viewing distance. For our photographs the range of the data is displayed in 4.5 in. and k around 0.025 works comfortably. Straight forward modifications of (1) can be used to make the data appear in front of, behind, or on both sides of the display surface. Also the choice of having large values of Z close or far away can be made.

Using (1), of course, does not provide a complete theoretical solution to the multiple focus problem. If points in a plane at the same Z depth are displayed and the eyes shift focus to the left or to the right of the plot center, the planes could potentially appear bowed. However, this is not noticeable when the viewing distance is a few times larger than half the plot width.

A second local alignment problem with computer-generated stereo pictures is due to digital images. Suppose each data point is represented by a symbol such as an asterisk. The asterisk is represented on the screen by the coloring of selected dots called pixels. As the X coordinate for the asterisk changes slightly, the horizontal line (composed of several pixels) in the asterisk will eventually be expected to move by one pixel. However, there is no guarantee that the diagonal lines being approximated with pixels will change at the same time. Thus the actual shape of the asterisk as displayed is a function of the X (and Y) coordinate and can be different in the left and right views. Discrepancies in shape can lead to fusion difficulties. A simple solution is to round the X disparity in screen coordinates to an exact multiple of the pixel width. This will insure that the left and right images appear the same no matter how distorted they are.

Another problem related to digital images is the appearance of depth planes. For most display devices, the number of pixels in the X disparity is not large relative to the depth perception induced. A one-pixel discrepancy is observed as a different depth plane. This may not be obvious in an arbitrary scatterplot, but can be perceived as noise when stereo points are used to represent a smooth surface. In general there is no solution to this problem for a fixed-resolution device. For presentation of analytic surfaces the noise can be reduced by plotting points selected from discrete depth contours. Then the depth can be exact and the noise is transferred to X and Y values which have finer resolution.

A third type of local alignment problem is incorrect fusion due to ambiguous pairing. Ambiguous pairing occurs when stereo pairs have the same Y coordinate and are close together. In this case the brain frequently picks the wrong left and right points for fusion and has a few unmatched left and right points left over. One solution is to make slight alterations in the Y coordinate. A better solution is to use different symbols for different points. When the close points have different shapes and colors, the brain has little difficulty in matching and fusing the correct pairs. (If needed, the data can be stratified and different symbols used to distinguish otherwise ambiguous points.) However for many plots, the use of redundant depth cues (discussed below) is sufficient. That is, the size and color of the symbols can be determined by the point Z coordinate. Ambiguous pairing then occurs only for points close in all three dimensions.

Saturation can be a problem with any plot. The problem tends to be worse with anaglyphs since two symbols are used for each point and since ambiguous pairing is a problem. One alternative(7) is to have the symbol represent many points that are close together. The symbol can be chosen to indicate both the number of points (say by size) and the distribution of the points being summarized. Rocking the plot is also helpful since different views expose different points. The third approach that helps somewhat is to use symbols that have small area. We conjecture that symbols with many edges relative to their area (pixel content) are effective since the eye has good edge detection "hardware." For example an asterisk may be more effective than a dot containing the same number of pixels as long as the lines in the asterisk are thick enough to be visible.

3. COLOR ANAGLYPHS

There are many ways to produce stereo images.(8,9,10) The method discussed here uses color anaglyphs. For the present these will be described as red/green anaglyphs. In this approach, the two monochrome images of a stereo pair are displayed in red and green on the same display surface, such as a color CRT. Viewing filters in corresponding colors are used to separate the image so that they are seen only by the correct eye. No mirrors or lens systems are required. The number of simultaneous viewers can be large.

Compared to other stereo techniques, computer-generated anaglyphs are relatively straightforward, since global alignment is not a problem. However, crosstalk is a common problem with color anaglyphs. Crosstalk means that each eye sees a portion of the other eye's image. This makes it easy to obtain incorrect fusion which produces a double image without depth.

For color anaglyphs, crosstalk results when the colors and filters are not matched and light from the incorrect image leaks through the filters. Systematic experimentation can be used to address this crosstalk problem. After generating bands of increasing intensity red, green, and blue on a black background, the investigator can determine the intensity at which leakage is noticeable for each filter. For our filters, full-intensity green leaked through the red filter rather badly but blue did not. Consequently we avoided the use of full-intensity green in our blue-green data points. Full-intensity red appeared as a dull green through the cyan filter. To diminish the visibility of this leakage shadow, we added green to the background. A small amount of green in the background matches the red intensity as viewed through the cyan lens. Matching the intensity is the important factor since small hue differences are difficult to see. Unfortunately, adding green to the background reduces the contrast of the blue-green points.

There are tradeoffs between contrast and crosstalk. One of us (Littlefield) prefers less crosstalk at the cost of less contrast. He uses a yellow background with 100% red and 75% green. The points in the two images are 100% red and 80% green, respectively. The overlap is 30% green which appears as black. Of course, the contrast and crosstalk can be reduced using either a light or dark background so that choice is one of personal preference. The only requirement is that the crosstalk be negligible at the chosen viewing distance.

4. REDUNDANT CUES

Unfortunately, not all people can see in stereo, and for those that can see it, a learning period is required. To expand the audience and reduce the learning time several approaches are helpful. The use of motion has been suggested above. For static pictures the display of a familiar 3-D object helps. For static scatterplots a useful device is to redundantly code depth with other cues. If only three dimensions are being portrayed, numerous cues are available.

Symbol size is a powerful cue. In conjunction with stereo it can be used to show a fourth dimension. When only three dimensions are needed, size is an excellent redundant depth cue. The use of size helps smooth out the depth contours. It also facilitates comparison against a depth scale if one is present. The basic constraints on the size are keeping the symbol visible and avoiding saturation.

Color is a useful cue, especially intensity. Intensity is easily controlled, so the closest points can be made the brightest. Since the brain mixes the colors from the separate eyes and a cyan filter allows the use of both blue and green, hue is also a choice. Three groups of data can be distinguished by using yellow (red/green), white (red/cyan) and magenta (red/blue). As for representing depth with color, French artists have used a blue shift and Chinese artists have used gray. On our CRT the depth effect of blue lines in comparison to red lines is quite noticeable. Consequently our basic strategy is to start with an off-white (reduced green as indicated above) and diminish red and green faster than blue as a function of point distance. We have not assessed whether or not the hue shift helps. Certainly intensity is useful in calling attention to the closest points in the plot.

The implementation of color cueing may vary for different devices, and to suit personal taste. We stratify the depth into five regions. The five different red intensities are represented using three bit planes and the five blue-green intensities are represented using three other bit planes. (For a discussion of bit planes see Reference 11.) By using a write mask for the respective set of planes, left eye points can overwrite previous left eye points while providing color mixing with the right eye points and vice versa. The colors need to be defined for all pens that correspond to legitimate bit plane combinations. Two more pens are used to color the background and write text. A reasonable color choice for the text and axes is that used for overlapping points on the display surface. Whatever the choice it is important that the text/background contrast does not diminish the point/background contrast.

When using different color intensities, the problem of crosstalk must be kept in mind. The crosstalk is worst for the most intense colors. For this reason, there is some advantage in having the display behind the display surface. The brightest points are then on the surface, overlap, and have no crosstalk.

Partially hidden points are an additional depth cue. Using the above color cueing implementation, points are hidden for the separate eyes by plotting the closest points last. On our CRT this did not lead to fusion difficulties. The difficulties with photographs are discussed later.

5. IDENTIFYING COORDINATES

Exploratory analysis is typified by the search for relationships. When looking for 3-D structure, the presence of axes on the display can be a nuisance. However there are times when identifying coordinates is important. Consequently this ability needs to be assessed. Relating points to a display of all three axes is difficult. We have implemented the X disparity plotting technique using a locally modified version of the MINITAB scatterplot. That is to say, the X and Y axis are a part of the plot. X and Y values can be read directly if the eye has the same X and Y coordinates. Otherwise a small adjustment must be made to account for the plane in which the axes are drawn and the viewing angle. The depth axis (Z) is not drawn. A reference set of points at different depths can be added at the side of the plot. Here the redundant size cueing is helpful in making comparisons.

6. REFERENCE SURFACES

While analysts typically look directly at residuals, on occasion it is informative to look at both the data and a reference surface. For constructing a reference surface, a straightforward procedure is to generate uniform random X and Y points over the desired range, compute depth, and plot in stereo. However when data are added, distinguishing between surface points

and data points is unnecessarily difficult even
with moderately different shapes and colors for
the two sets of points. Using lines to represent the reference surface is a logical response
to this problem, the more obvious choices being
the use of lines in the form of contours or
crosshatching. Unfortunately long smooth lines
are not particularly effective for conveying
stereo depth. When horizontal lines are drawn,
the disparity between the eye images occurs only
at the ends of the lines. This is a poor depth
cue. For vertical lines the depth cue is
stronger, but still it is not obvious which
points in the lines match. Having distinct features to match makes image fusion easier. This
suggests that short lines and sharp angles facilitate image fusion. Jagged contour lines
work well. When the contours are smooth, the
use of textured lines helps. Simple crosshatching is not particularly effective. For the random point surfaces, two approaches are satisfactory. The first uses a short line segment as a
symbol. The direction of the segment is chosen
to be orthogonal to the line of steepest ascent.
The result looks like segments picked from contour lines. The second approach connects the
random points with a minimal spanning tree.(12).
The points themselves can be reduced in size or
eliminated to make the plot less busy. The effect is similar to viewing the cracks in a fractured glass surface. The reason the spanning
tree works well is that the connecting of nearest neighbors produces short line segments and a
variety of angles.

7. USES AND INTERPRETATION

Stereo scatterplots can be used in many ways.
Identifying clusters, suggesting 2-D models, and
viewing regression diagnostics are a few examples. However, the third dimension is relatively new to exploratory data analysts and the
experimental base for interpreting such plots is
not present. For example, a plot of studentized
residuals versus one independent variable may
reveal a wedge shape suggesting a homogeneity
problem. If the plot were made versus two independent variables, what shapes should be looked
for, a cone and cone cross sections? What
should be looked for in plots of studentized residuals versus leverage and influence? What
should be looked for in partial residual plots
for two variables? Even if verbal answers are
available, experience is needed to see the patterns. Certainly the value of 3-D scatterplots
will increase as the ability to interpret them
increases. Much remains to be learned in the
interpretation of 3-D scatterplots.

8. DIFFICULTY OF PRODUCTION

The above discussion may make it sound as if
making color anaglyph scatterplots is exceedingly difficult. Were it not for crosstalk,
making the plots would be quite simple. Once
the crosstalk problem was solved on the CRT we
found the stereo scatterplots easy to produce.
Our local implementation of a symbol drawing
language(7) has proved convenient in this regard. Basically the language interpreter takes
control just before MINITAB plots each point.
The language uses additional columns (to X and
Y) to determine the shape, separation, and color
of the anaglyphs to be drawn. After the procedure is set up, producing subsequent 3-D scatterplots is only slightly more difficult than
producing an ordinary scatterplot.

We encountered many more difficulties in going
to photographs. The colors and contrasts recorded on film were different from those on the
CRT, requiring color setting changes to reduce
crosstalk. The settings were also more difficult to determine than with the CRT because of
long film processing times and variation between
film batches. Crosstalk was not the only problem. Overplotted left-right images for different points were too intense. Each overlapping
color combination needed to be modified. Otherwise the overlapping portion can be perceived by
itself as a point in the surface of the display
rather than allowing correct fusion. The problem of new colors being produced at the edges of
high contrast regions also was noticeable. Consequently, for publication purposes, color anaglyph stereo is not nearly as practical as side-by-side stereo.

9. CONCLUSIONS

Color anaglyphs provide one of the simplest
methods for producing direct 3-D scatterplots on
CRT displays. Until superior means of producing
3-D views are readily available, color anaglyphs
will be increasingly used. The plots are particularly convenient for exploratory data analysis.
Attention to several details can enhance the
plots. However, color reproduction problems are
burdensome for film media. Consequently other
methods such as side-by-side stereo are to be
preferred for publication purposes.

10. REFERENCES

[1] Wainer, H. and Thissen, D., Graphical data
analysis, Ann. Rev. Psychol. 32 (1981) 191-241.

[2] Donoho, D., Huber, P. J. and Thoma, H. M.,
The use of kinematic displays to represent
high-dimensional data, in Eddy, W. F. (ed.),
Computer Science and Statistics: Proceedings of the 13th Symposium on the Interface
(Springer-Verlag, 1981) 274-278.

[3] Tukey, J. W., Friedman, J. H. and
Fisherkeller, M. A., PRIM-9, an interactive multidimensional data display and
analysis system, Proc. 4th International
Congress for Stereology (Gaithersburg,
Maryland, 1975).

[4] Stover, H. S., Terminal puts three-dimensional graphics on solid ground, Electronics (July 1981) 150-155.

[5] Littlefield, R. J., Stereo and motion in the display of 3-D scattergrams, in Proceedings of the 8th Annual Computer Graphics Conference (Engineering Society of Detroit, Detroit, Michigan, May 4-6, 1982).

[6] Newman, W. M. and Sproull, R. F., Principles of interactive computer graphics, (McGraw Hill, New York, New York, 1973) 239-249.

[7] Nicholson, W. L. and Littlefield, R. J., Interactive color graphics for multivariate data, in Computer Science and Statistics: Proceedings of the 14th Symposium on the Interface (in publication, 1982).

[8] desJardins, M. and Hasler, A. F., Stereographic displays of atmospheric model data, Computer Graphics 14(3) (1980) 134-139.

[9] Butterfield, J. F., Three-dimensional television, in Proceedings of the 15th Annual SPIE Symposium (1970) 3-9.

[10] Roese, J. A. and McCleary, L. E., Stereoscopic computer graphics for simulation and modeling, Computer Graphics 13(2) (1979) 41-47.

[11] Beatty, J. C., Raster graphics and color, American Statistician 37(1) (1983) 60-75.

[12] Carr, D. B., Raster color displays--examples, ideas, and principles, in Tykey Truett (ed.), Proceedings of the 1980 DOE Statistical Symposium (1981) 116-126.

---

[1] Work sponsored by the United States Department of Energy, Applied Mathematical Sciences, under Contract DE-AC06-76RLO 1830.

IMSL LIBRARIES FOR MICRO-COMPUTERS

Henry Darilek and David Kortendick

IMSL, Inc.
Houston, Texas

The development of IMSL Libraries for the Micro-computer environment is discussed.

Due to the expected proliferation of desk-top computers, and the subsequent need for basic mathematical and statistical abilities on them, IMSL has developed two subset libraries from the full IMSL LIBRARY specifically for use in a micro-computer environment. The two subsets consist of a BASIC MATHEMATICAL LIBRARY and a BASIC STATISTICAL LIBRARY.

By splitting the library and providing only the respective fundamental abilities of a MATH or STAT subroutine library, IMSL is able to provide products that are affordable to the micro user and are easier for the user to manage.

The subroutines, which are written in Fortran, are organized into groups for use in similar problems or areas of application within each library. A reference manual containing information relevant to the use of the subroutines accompanies each of the two products. Each group is documented as a separate chapter in the respective reference manuals. An attempt was made to attain a sense of completeness in a chapter by including capabilities which might be used in conjunction with the theme of the chapter. For example, the chapter titled 'Interpolation, Approximation, and Smoothing' contains abilities to interpolate, smooth, evaluate, differentiate, integrate, and plot cubic splines. Subroutines were selected for inclusion in a particular chapter based upon their contribution to the basic capabilities of the chapter, their popularity among users of the IMSL Library, and IMSL's confidence in the subroutine.

Because of space limitations on micro-computers, these chapters were made to be complete in that each group is capable of standing alone as a library module. At this time, IMSL plans to distribute the object code in library form for each chapter. The distribution media would be in the form of floppy disks. Either of the basic libraries would be available via a paidup sublicense including maintenance and consultation for one year. Support for the software will be similar to that for current products. Cost of the paidup sublicense has not been strictly set but IMSL believes the price would be around $600 per cpu and maintenance and consultation thereafter would be optional at $300 annually.

Development of the Libraries took place on an Intel MDS 800 using the 8086 numeric data processor with the 8087 numeric processor extension.[1] The numeric data formats and arithmetic operations provided by the 8087 conform to the proposed IEEE microprocessor floating point standard.[2] The 80 bit TEMPORARY REAL data type, which is provided by the Intel implementation of the standard was employed in situations where extended precision arithmetic was needed. IMSL envisions that the Libraries will also be available for machines which do not support the IEEE floating point standard. Other enhancements to the Libraries included greater use of the Basic Linear Algebra routines[3] to maximize efficiency and minimize code, and the distribution of the chapters in library object form. Respective sizes in bytes of the Libraries' Object files for the chapters on the Intel machine follows:

Basic Mathematical Library

	Single Precision	Double Precision
Differential Equations and Integration	72.6 KB	73.5 KB
Eigensystem Analysis	68.0 KB	68.9 KB
Error and Gamma Functions	34.9 KB	N/A
Interpolation Approximation, and Smoothing	60.9 KB	61.5 KB
Linear Algebraic Equations	66.5 KB	67.2 KB
Vector/Matrix Arithmetic and Sorting	97.0 KB	98.7 KB

Basic Statistical Library

	Single Precision	Double Precision
Basic Statistics	159.1 KB	159.4 KB
Regression Analysis	148.0 KB	147.2 KB
Analysis of Variance	94.7 KB	96.3 KB
Nonparametric Statistics and Tests for Goodness-of-Fit	64.8 KB	64.8 KB
Time Series Analysis and Forecasting	190.9 KB	193.9 KB
Probability Distribution Functions and Their Inverses	49.9 KB	N/A

Random Number
  Generation   41.3 KB   N/A

Descriptions of the abilities included in individual chapters within the BASIC MATHEMATICAL and BASIC STATISTICAL LIBRARIES are provided below.

In conclusion, the use of micro-computers as scientific tools is emerging and spreading widely. To enable micro-computers to be used to their full potential as scientific tools, state-of-the-art software must be made available to their users. IMSL is committed to providing state-of-the-art scientific software products for all popular Fortran computing environments. IMSL products will be made available for the micros as strong, reliable, Fortran compilers emerge in the micro arena.

BASIC MATHEMATICAL LIBRARY

Differential Equations and Integration
  DCADRE - Numerical integration of a function using cautious adaptive Romberg extrapolation
  DGEAR - Differential equation solver - Variable order Adam's predictor corrector method or Gear's method

Eigensystem Analysis
  EIGRF - Eigenvalues and (optionally) eigenvectors of a real general matrix in full storage mode
  EIGRS - Eigenvalues and (optionally) eigenvectors of a real symmetric matrix

Interpolation, Approximation, and Smoothing
  DCSEVU - Cubic spline first and second derivative evaluator
  DCSQDU - Cubic spline quadrature
  ICSCCU - Cubic spline interpolation
  ICSEVU - Evaluation of a cubic spline
  IFLSQ - Least squares approximation with user supplied functions
  USPLO - Printer plot of up to ten functions (single precision)
  USPLOD - Printer plot of up to ten functions (double precision)

Linear Algebraic Equations
  LEQOF - Linear equation solution - Full matrices (Out of core version)
  LEQT1F - Linear equation solution - Full storage mode - Space economizer solution
  LEQT1P - Linear equation solution - Positive definite matrix - Symmetric storage mode - Space economizer solution
  LEQT2F - Linear equation solution - Full storage mode - High accuracy solution
  LEQT2P - Linear equation solution - Positive definite matrix - Symmetric storage mode - High accuracy solution
  LLSQF - Solution of a linear least squares problem

Error and Gamma Functions
  ERF   - Evaluate the error function
  DERF   - Evaluate the error function of a double precision argument
  MERFI - Inverse error function
  ERFC   - Evaluate the complemented error function
  DERFC - Evaluate the complemented error function of a double precision argument
  GAMMA - Evaluate the Gamma function
  DGAMMA - Evaluate the Gamma function of a double precision argument
  ALGAMA - Evaluate the Log (Base e) of the absolute value of the Gamma function
  DLGAMA - Evaluate the Log (Base e) of the absolute value of the Gamma function of a double precision argument x

Vector-Matrix Arithmetic
  DASUM - Compute the double precision sum of absolute values
  DAXPY - Compute a constant times a vector plus a vector, all double precision
  DCOPY - Copy a vector x to a vector y, both double precision
  DDOT   - Compute double precision Dot product
  DNRM2 - Compute the Euclidean length or L2 norm of a double precision vector
  DROT   - Apply Givens plan rotation (Double precision)
  DROTG - Construct Givens plan rotation (Double precision)
  DROTM - Apply a modified Givens plan rotation (Double precision)
  DROTMG - Construct a modified Givens plane rotation (Double precision)
  DSCAL - Compute a double precision constant times a double precision vector
  DSDOT - Compute single precision Dot product using double precision accumulation
  DSWAP - Interchange vectors x and y, both double precision
  IDAMAX - Find the smallest index of the maximum magnitude of a double precision vector
  ISAMAX - Find the smallest index of the maximum magnitude of a single precision vector
  SASUM - Compute the single precision sum of absolute values
  SAXPY - Compute a constant times a vector plus a vector, all single precision
  SCOPY - Copy a vector x to a vector y, both single precision
  SDOT   - Compute a single precision Dot product
  SDSDOT - Compute a single precision Dot product and add a constant using double precision accumulation
  SNRM2 - Compute the Euclidean length or L2 norm of a single precision vector
  SROT   - Apply Givens plane rotation (Single precision)

SROTG   - Construct Givens plane rotation (Single precision)
SROTM   - Apply a modified Givens plane rotation (Single precision)
SROTMG - Construct a modified Givens plane rotation (Single precision)
SSCAL   - Compute a single precision constant times a single precision vector
SSWAP   - Interchange vectors x and y, both single precision
VMULFM - Matrix multiplication of the transpose of matrix A by matrix B (Full storage mode)
VMULFP - Matrix multiplication of matrix A by the transpose of matrix B (Full storage mode)
VSODA   - Sorting of columns of a double precision matrix in ascending order of keys in rows
VSORA   - Sorting of columns of a real matrix into ascending order of keys in rows
VSRTR   - Sorting of arrays by algebraic value - Permutations returned
VSRTRD - Sorting of double precision arrays by algebraic value - Permutations returned
VTPROF - Transpose product of matrix (Full storage mode)

BASIC STATISTICAL LIBRARY

Basic Statistics
BDCOU1 - Tally of observations into a one-way frequency table
BDCOU2 - Tally of observations into a two-way frequency table
BECORI - Estimates of means, standard deviations, and correlation coefficients (In-core version)
BECOVM - Means and variance-covariance matrix
BEGRPS - Moments estimation for grouped data with and without Sheppard's corrections
BEMMI - Estimates of means, standard deviations, correlation coefficients, and coefficients of skewness and kurtosis from a data matrix containing missing observations
BEMMO - Estimates of means, standard deviations, correlation coefficients, and coefficients of skewness and kurtosis from a data matrix containing missing observations (Out of core version)
BENSON - Mean and variance inferences using a sample from a normal population
BEPAT - Mean and variance inferences using samples from each of two normal populations with unequal variances
BEPET - Mean and variance inferences using samples from each of two normal populations with equal variances
CTRBYC - Analysis of a contingency table
USBOX - Print a boxplot (k samples)
USHST - Print a vertical histogram
USPLO - Printer plot of up to ten functions
USPLOD - Printer plot of up to ten functions (Double precision)

Regression Analysis
RLEAP - Leaps and bounds algorithm for determining a number of best regression subsets from a full regression model
RLFOR - Fit a univariate curvilinear regression model using Easy to use version
RLMUL - Multiple linear regression analysis
RLSEP - Selection of a regression model using a forward stepwise algorithm, and computation of the usual analysis of variance table entries - Easy to use version
USLEAP - Print results of the best-regressions analysis

Analysis of Variance
ACRDAN - Analysis of one-way classification design data
AFACT - Full factorial plan analysis - Easy to use version
AGLMOD - General linear model analysis
ANCOV1 - Covariance analysis for one-way classification design data
ANESTU - Analysis of completely nested design data with unequal numbers in the subclasses
ASNKMC - Student-Newman-Keuls multiple comparison test

Nonparametric Statistics and Tests for Goodness of Fit
GFIT   - Chi-Squared goodness of fit test
NAK1   - Kruskal-Wallis test for identical populations
NKS1   - Kolmogorov-Smirnov one-sample test
NKS2   - Kolmogorov-Smirnov two-sample test
NMRANK - Numerical ranking
NRWMD - Wilcoxon signed rank test
NRWRST - Wilcoxon rank-sum test (Mann-Whitney test)
VSRTR - Sorting of arrays by algebraic value - Permutations returned
VSRTRD - Sorting of double precision arrays by algebraic value - Permutations returned

Time Series Analysis and Forecasting
FTAUTO - Mean, variance, autocovariances, autocorrelations, and partial autocorrelations for a stationary time series
FTCAST - Time series forecasts and probability limits using an Arima (Box-Jenkins) model
FTCP   - Non-seasonal Arima (Box-Jenkins) Stochastic model analysis for a single time series with full parameter iteration and maximum likelihood estimation
FTCRXY - Cross-covariance of two mutually stationary time series
FTFPS - Fast fourier transform estimates of power spectra and cross spectra of time series

FTFREQ - Single of multichannel time series analysis in the time and frequency domains
FTRDIF - Transformations, differences and seasonal differences

Random Number Generation
GGAMR - One parameter gamma random deviate generator, and usable as the basis for two parameter gamma, exponential, chi-squared, chi, beta, t, and f deviate generation
GGBN - Binomial random deviate generator
GGBTR - Beta random deviate generator
GGEXN - Exponential random deviate generator
GGNPM - Normal random deviate generator via the polar method
GGNSM - Multivariate normal random deviate generator with given covariance matrix
GGPON - Poisson random deviate generator where the poisson parameter changes frequently
GGUBS - Basic uniform (0, 1) pseudo-random number generator
MDCH - Chi-squared probability distribution function
MDFDRE - F probability distribution function (Integer or fractional degrees of freedom)
MDFI - Inverse f probability distribution function
MDNOR - Normal or Gaussian probability distribution function
MDNRIS - Inverse standard normal (Gaussian) probability distribution function
MDSTI - Inverse of a Modification of Student's t probability distribution function
MDTD - Student's t probability distribution function

[1] Rash, Bill, Intel Application Note, "Getting Started With the Numeric Data Processor", AP-113, February 1981, pp. 1-4.

[2] ACM Signum Newsletter, Special Issue, October, 1979, pp. 4-12.

[3] Hanson, R. J., Krogh, F. T., and Lawson, C. L., "A Proposal for Standard Linear Algebra Subprograms", Jet Propulsion Laboratory, TM 33-660, November 1973.

Resistant Lower Rank Approximation of Matrices

K. Ruben Gabriel and Charles L. Odoroff

Department of Statistics and Division of Biostatistics
The University of Rochester
Rochester, New York 14627

Lower rank approximations to matrices are usually obtained by least squares and hence are highly sensitive to outliers. We propose several resistant methods which cycle through fits of rank one approximations and take residuals after each such fit. Our methods extend the technique of reciprocal averaging (used for least squares) by introducing medians, trimmed means and other robust averages. Several of these methods have been programmed and tried out. We have found them to be resistant to outliers in the sense that their lower rank fit is not at all influenced by a few extreme values in the matrix. They may thus be useful in providing robust methods for a variety of multivariate techniques such as principal component analysis and biplotting.

1. Lower rank approximation of matrices has a number of statistical and data analytic uses. The most obvious one is to provide a fair representation of a matrix by a smaller number of parameters. Thus, a rank r approximation of a (nxm) matrix requires (n+m-1)r numbers instead of the nm elements of the matrix itself, e.g., the rank 2 fit to a 100 x 20 matrix uses 238 numbers for approximating the matrix's 2000 elements. A related application is in modelling matrices [15,16] and their graphical display [3,5,7].

Unfortunately, if one or more elements of the matrix lie far out from a general lower dimensional pattern, they will greatly influence the least squares fitting and draw it away from the general pattern. This paper suggests a number of new procedures of lower rank fitting which are resistant to the influence of such outliers.

Additional applications of such resistant methods range over the entire field of multivariate analysis, whenever outliers are likely to influence standard statistical analyses unduly. A robust principal component analysis, for example, could be obtained by carrying it out on a lower rank approximation of the observation matrix, where the approximation - of not too much reduced rank - had served to eliminate the influence of large outliers. (Such a method would seem a priori preferable to current methods of resistant principal component analysis which smooth the cross-product elements of Y'Y rather than those of data matrix Y itself.)

2. Lower rank approximation is best discussed by formulating it in terms of a factorization. Thus, it is well-known (see, for example, [6]) that any matrix of order (nxm) can be written as a product AB' of a (nxr) matrix A and the tranpose of a (mxr) matrix B provided r is not less than the rank of the matrix. The least squares criterion of approximating a (nxm) matrix $Y = ((y_{i,j}))$ by one of rank r (or less) can therefore be written

$$\Theta = \min (||Y-AB'||^2 : A_{(nxr)}, B_{(mxr)}), \quad (1)$$

where $||\cdot||$ is the Euclidean norm. An equivalent formulation is

$$\Theta = \min (\Sigma_{i=1}^{n} \Sigma_{j=1}^{m} (y_{i,j} - \underline{a}_i'\underline{b}_j)^2 :$$

$$A_{(nxr)}, B_{(mxr)}), \quad (2)$$

where $\underline{a}_i$ and $\underline{b}_j$ denote the i-th and j-th row of A and B, respectively. The solution of this least squares problem is due to Eckart, Householder and Young - to be referred to as E-H-Y [4,6,13]. It is closely related to the eigenvalue/eigenvector problem for product matrices Y'Y and YY' and is therefore readily implemented by widely available alogrithms. It may also be computed from the solution of the equations

$$\underline{p}_u' Y = \lambda_u \underline{q}_u' \qquad u=1,\ldots,m_\wedge n, \quad (3.1)$$

$$Y\underline{q}_u = \lambda_u \underline{p}_u \qquad u=1,\ldots,m_\wedge n, \quad (3.2)$$

and

$$\underline{p}_u'\underline{p}_v = \underline{q}_u'\underline{q}_v = \delta_{u,v} \quad u,v=1,\ldots,m_\wedge n, \quad (3.3)$$

subject to inequalities

$$\lambda_1 \geq \lambda_2 \geq \cdots \geq \lambda_{m_\wedge n} \geq 0. \quad (3.4)$$

The resulting $\lambda_u, \underline{p}_u$ and $\underline{q}_u$'s are referred to as the u-th singular values, columns and rows of Y. These solutions [10] yield the rank r approximation of Y in the form

$$Y_{[r]} = \Sigma_{u=1}^{r} \lambda_u \underline{p}_u \underline{q}_u' . \quad (4)$$

In practice, one usually begins by solving equations (3.1-3.4) to obtain λ_1, $\underline{p}_1$ and $\underline{q}_1$ for the first <u>singular component</u> $Y_{[1]} = \lambda_1 \underline{p}_1 \underline{q}_1'$ which is the dyadic, i.e., rank one, approximation. Next, one subtracts this dyadic approximation from Y and solves equations (3) with Y replaced by $Y - Y_{[1]}$ to obtain λ_2, $\underline{p}_2$ and $\underline{q}_2'$ for a dyadic approximation $\lambda_2 \underline{p}_2 \underline{q}_2'$ to the residuals $Y - Y_{[1]}$. That is also the second singular component of Y whose rank 2 approximation can now be obtained by summation as $Y_{[2]} = Y_{[1]} + \lambda_2 \underline{p}_2 \underline{q}_2' = \lambda_1 \underline{p}_1 \underline{q}_1' + \lambda_2 \underline{p}_2 \underline{q}_2'$. In the third stage, one substitutes $Y - Y_{[2]}$ for Y in equations (3) and solves to obtain λ_3, $\underline{p}_3$ and $\underline{q}_3$ and thus the third singular component $\lambda_3 \underline{p}_3 \underline{q}_3'$ of Y. The rank 3 approximation to Y is then $Y_{[3]} = Y_{[2]} + \lambda_3 \underline{p}_3 \underline{q}_3' = \Sigma_{v=1}^{3} \lambda_v \underline{p}_v \underline{q}_v'$. And so one proceeds for altogether r stages.

This stagewise dyadic solution is computationally very convenient, though it is not immediately obvious that it actually yields the rank r matrix which minimizes the sum of squared properties (3.3) of the singular components.

3. Two limitations of this E-H-Y solution are (i) that it does not readily extend to weighted least squares, and (ii) that it is not resistant to the influence of isolated elements which may deviate greatly from a general lower rank pattern. The first limitation can be dealt with by introducing weighted least squares algorithms [9] - to be referred to as G-Z - which are based on the criterion

$$\Theta = \min \left\{ \Sigma_i \Sigma_j w_{i,j} (Y_{i,j} - \underline{a}_i' \underline{b}_j)^2 : A_{(n \times r)}, B_{(m \times r)} \right\}, \quad (5)$$

for any given set of weights $W_{(n \times m)} = ((w_{i,j}))$. Unfortunately, the introduction of weights destroys the orthogonality properties that the E-H-Y solution had. Hence the weighted least squares rank r fit cannot be obtained simply as the sum of stagewise dyadic fits. Instead, further iteration is needed [9] - this will be further discussed below.

4. The second limitation of the E-H-Y solution, and of least squares methods in general, is that they can be unduly influenced by a small number of "outliers." Thus, if nm-1 elements of a matrix Y lie close to a planar representation, but a single element is far away from that plane, the E-H-Y solution tends to fit that single outlier closely, at the expense of missing the plane on which all other points lie. This is analogous to the well-known tendency of a single far outlier to deflect a regression line from fitting the majority of points.

This paper proposes a number of methods of fitting lower rank approximations which are more resistant to the influence of outliers.

5. In order to develop resistant methods, we first discuss the formulation and computation of the E-H-Y solutions in some detail. We begin by describing a common iterative procedure for solving equations (3.1, 3.2) subject to (3.3) for any one value of u. The s-th step of this iteration is as follows:

(i) use a trial (nx1) vector $\underline{a}_{(s-1)}$;

(ii) multiply this into the columns of Y to yield (mx1) vector

$$\underline{b}'(s) = \underline{a}'(s-1) Y; \quad (6.1)$$

(iii) multiply that into the rows of Y and normalize to yield the new trial value

$$\underline{a}(s) = Y \underline{b}(s) / ||Y \underline{b}(s)||. \quad (6.2)$$

The iteration terminates when the $\underline{a}$'s and the $\underline{b}$'s of successive steps are sufficiently close together.

Comparing equations (6) with equations (3) one sees that the iteration's convergence is to the singular components of Y, i.e., $||\underline{b}|| = \lambda$, $\underline{b}' = \lambda \underline{q}'$ and $\underline{a} = \underline{p}$, so that $\underline{a} \underline{b}' = \lambda \underline{p} \underline{q}'$. When the procedure is applied directly to Y, it should converge to the first singular component; when Y is replaced in (6) by $Y - \lambda_1 \underline{p}_1 \underline{q}_1'$, it should converge to the second singular component, etc.

6. The iteration equations (6) can be rewritten in summation notation as

$$b_{(s)j} = \Sigma_{i=1}^{n} a_{(s-1)i} y_{i,j} \quad j = 1,\ldots,m \quad (7.1)$$

and

$$a_{(s)i} = c \Sigma_{j=1}^{m} b_{(s)j} y_{i,j} \quad i = 1,\ldots,n, \quad (7.2)$$

where the constant c (>0) is chosen to ensure that

$$\Sigma_{i=1}^{n} a_{(s)i}^2 = 1. \quad (7.3)$$

Changing the normalization at each step will not change the convergence of the procedure. Thus, we may iterate on

$$b_{(s)j} = \Sigma_{i=1}^{n} a_{(s-1)i} y_{i,j} / \Sigma_{i=1}^{n} a_{(s-1)i}^2 \quad (8.1)$$

and

$$a_{(s)i} = \Sigma_{j=1}^{m} b_{(s)j} y_{i,j} / \Sigma_{j=1}^{m} b_{(s)j}^2 \quad (8.2)$$

instead of equations (7). In this formulation, the b_j's are seen to be the regression (through the origin) coefficients of the columns of Y, i.e., the $y_{i,j}$'s (i=1,...,n), onto the a_i's; analogously, the a_i's are the regression (through the origin) coefficients of the rows of Y, i.e., the $y_{i,j}$'s (j=1,...,m), onto the b_j's. That is why this procedure has been referred to as "criss-cross regression" [18].

7. The regression formulation (8) of the fitting procedure readily lends itself to generalization for weighted least squares. Thus, for (nxm) weights matrix $W = ((w_{i,j}))$ one would iterate between the weighted regression coefficients

$$b_{(s)j} = \Sigma_i w_{i,j} a_{(s-1)i} y_{i,j} / \Sigma_i w_{i,j} a^2_{(s-1)i}$$
$$j = 1,\ldots,m \quad (9.1)$$

and

$$a_{(s)i} = \Sigma_j w_{i,j} b_{(s)j} y_{i,j} / \Sigma_j w_{i,j} b^2_{(s)j}$$
$$i = 1,\ldots,n. \quad (9.2)$$

This has been discussed by Gabriel and Zamir [9]. For rank 1 aproximations, it yields the least squares fit. For fits of higher rank, however, the sum of the successive dyadic fits to residuals from earlier fits does not necessarily lead to the overall least squares solution. (That is because the orthogonality properties of the E-H-Y case break down when the weights are not equal.) Gabriel and Zamir have considered two strategies of further iterating towards the rank r least square fit.

(i) <u>Multiple Criss-Cross Regression</u>: Replace the component-by-component regression equations of (9) by r-component multiple regression equations.

(ii) <u>Cycling Dyadic Fits</u>: Use only simple, i.e., single column or row, criss-cross regressions (9) but cycle through the r components, subtracting (r-1) fitted components at a time and refitting the remaining component dyadically.

Both strategies have been found to converge in most cases to the least squares fit.

Our development of resistant methods of lower rank approximation relies on this strategy of cycling dyadic fits. No simple resistant analogue of multiple criss-cross regression has been suggested yet. The following discussion therefore concentrates on resistant substitutes for iteration equations (9). It should be understood that for rank r(>1) approximation any of these would be incorporated in a cyclic strategy. This will not henceforth be repeated explicitly.

8. In order to develop resistant counterparts to the dyadic iterative fitting equations (8) of the E-H-Y solution, it is well to rewrite the latter in the form

$$b_{(s)j} = \Sigma_i a^2_{(s-1)i} (y_{i,j}/a_{(s-1)i})/\Sigma_i a^2_{(s-1)i}$$
$$j=1,\ldots,m \quad (10.1)$$

and

$$a_{(s)i} = \Sigma_j b^2_{(s)j} (y_{i,j}/b_{(s)j})/\Sigma_j b^2_{(s)j}$$
$$i=1,\ldots,n . \quad (10.2)$$

This shows the b_j's to be weighted means of the $(y_{i,j}/a_i)$'s ($i=1,\ldots,n$) and the a_i's to be weighted means of the $(y_{i,j}/b_j)$'s ($j=1,\ldots,m$). That may explain the term "reciprocal averaging" that has been applied to this iterative procedure [11,12].

9. This formulation of the iterative procedure by means of averages immediately suggests that resistant procedures can be obtained by introducing resistant averages. Thus, the simplest resistant procedure iterates

$$b_{(s)j} = \text{Med} \left\{ (y_{i,j}/a_{(s-1)i}) : i-1,\ldots,n \right\}$$
$$j=1,\ldots,m \quad (11.1)$$

and

$$a_{(s)i} = \text{Med} \left\{ (y_{i,j}/b_{(s)j}) : j=1,\ldots,m \right\}$$
$$i=1,\ldots,n. \quad (11.2)$$

This procedure may work well when all the $y_{i,j}$'s are positive (except, perhaps, for a few scattered zeroes). It may also work after suitable changes of signs (see Section 11 below).

Note that each $y_{i,j}/a_i$ ratio is an obvious estimate of b_j, just as each $y_{i,j}/b_j$ ratio is an estimate of a_i. A natural approach to resistant estimation [17], therefore, is to take some resistant average of these individual "obvious" estimates. That is what was done here.

10. It may be of interest to note that this method of "reciprocal medians" is equivalent to "median polish" of the matrix of logarithms $X = ((\log y_{i,j}))$ when these exist, i.e., when no elements of Y are negative. To see this, define (nx1) vector $\underline{c}_{(s)}$ by elements $c_{(s)i} = \log(a_{(s)i})$, $i = 1,\ldots,n$, and (mx1) vector $\underline{d}_{(s)}$ by elements $d_{(s)j} = \log(b_{(s)j})$, $j = 1,\ldots,m$. Then rewrite equations (11) as

$$d_{(s)j} = \text{Med} \left\{ x_{i,j} - c_{(s-1)i} : i=1,\ldots,n \right\} \quad (12.1)$$

and

$$c_{(s)i} = \text{Med} \left\{ x_{i,j} - d_{(s)j} : j=1,\ldots,m \right\}. \quad (12.2)$$

Iterative solution of these equations is equivalent to median polish of X [8].

11. Difficulties arise with the reciprocal medians procedure when the matrix Y has negative entries. The medians $a_{(s)i}$ and/or $b_{(s)j}$ calculated by (11) may be zero and then the ratios $y_{i,j}/a_{(s)i}$ and/or $y_{i,j}/b_{(s)j}$ are not

defined for the next step. The procedure therefore cannot be used. (Nor could median polish be applied to the logarithms since these are not defined for negative elements).

To obviate most difficulties of this kind, we may change the signs of any number of rows $i_1, i_2,...,i_k$, say, and any number of columns $j_1, j_2,...,j_\ell$, say, and thereby reduce the number of negative signs to one half or fewer in each row and each column. (An element $y_{i,j}$ in a row and column with changed signs would retain its original sign.) The procedure would be applied to the resulting matrix, $X^{(+)}$, say, and solutions $\underline{a}^{(+)}$ and $\underline{b}^{(+)}$ of (11) calculated. The solutions for the original matrix Y would then be obtained by changing the signs of $a_{i_1}^{(+)}$, $a_{i_2}^{(+)},..., a_{i_k}^{(+)}$ and $b_{j_1}^{(+)}, b_{j_2}^{(+)},...,b_{j_\ell}^{(+)}$. The rationale for this device is simply that if $Y_{i,j} = a_i b_j$ then $-Y_{i,j} = (-a_i) b_j = a_i (-b_j)$. (For further discussion, see [14].)

12. A further aid in dealing with negative elements and zeroes, and in avoiding the instability of the ratios $y_{i,j}/a_i$ and/or $y_{i,j}/b_j$ when a_i and/or b_j is small, might be to use a higher quartile than the median. Thus, for example, one might define $b_{(s)j}$ as the third largest of the $(y_{i,j}/a_{(s-1)i})$, $i=1,...,n$ and $a_{(s)i}$ as the third largest of the $(y_{i,j}/b_{(s)j})$, $j=1,...,m$. Also, one might use the signed absolute median (SAM), i.e., have $b_{(s)j}$ equal the algebraic value of the $y_{i,j}/a_{(s-1)i}$ ratio whose absolute value was the median of the $|y_{e,j}/a_{(s-1)i}|$, $i=1,...,n$. And similarly, $a_{(s)i}$ = SAM $\{(y_{i,j}/b_{(s)j}); j=1,...,m\}$.

Other aids could be tried out, singly or in combination, and possibly together with the sign adjustments discussed above. These need further investigation.

13. An alternative procedure of resistant fitting would use trimmed means instead of the medians in the reciprocal averaging process of (10).

14. Some of the proposed resistant procedures may lead to rapid convergence and provide good approximations. Some may depend on their initializaiton, just as median polish depends on whether it is initialized by row medians or by column medians. The indeterminacy of such methods may suggest complementing them by least squares methods with weights which reduce the influence of outliers. One such method is that of iteratively reweighted least squares which attempts to reduce the weight attached to outlying observations.

15. The method of iteratively reweighted least squares adjusts the weights in each iteration inversely to the residuals from the previous iteration's fit. Thus, if in iteration five, say, the approximation to element $y_{4,7}$ was very different from $y_{4,7}$, the large resulting residual would indicate a low weight $w_{4,7}$ for the fitting in iteration six. The idea [1] is that outliers should have large residuals and thus be allocated small weights and reduced influence.

The difficulty with this method is its initialization. If the first iteration uses the unweighted E-H-Y method, an outlier may be fitted quite closely. If this happens, the subsequent weight allocated to this outlier will be large rather than small. The method would therefore be off to a bad start and would be unlikely to reduce the outliers' weights at later iterations.

This difficulty may be overcome by using a resistant method for a preliminary fit and let its residuals determine the initial weights in the iterative reweighting procedure: that is likely to ensure the reduction of the outliers' influence.

A direct resistant method of initializing iterative reweighted least squares has been proposed elsewhere (see [9]).

16. A different approach is that of setting the minimum sum of absolute deviations as the criterion of resistant fitting. An algorithm for such fits is available [2] and could be implemented for criss-cross regression in either of two ways corresponding to those used by Gabriel and Zamir [9] for least squares fits. It could either (i) be used for criss-cross multiple regression to fit the entire reduced rank matrix in one run, or (ii) cycling dyadic fits could use criss-cross simple median regressions ([2], Section 2).

We suspect that the sum of absolute deviations is by no means as resistant to outliers as methods based on medians, quartiles or trimmed means. We therefore do not further investigate this approach.

17. We have programmed a number of techniques which use cycling iterative dyadic fits based on medians, trimmed means and other measures of location. A number of experimental computations on matrices with one or two greatly perturbed elements have shown our techniques to operate effectively, as do the iteratively reweighted least squares techniques when initialized properly. All these techniques have produced reasonably close fits to the main body of data without being influenced by large outliers. So far, we cannot see any clear advantages to one or the other of the statistics. This clearly needs further investigation.

18. Finally, we add some remarks of where we see these procedures of lower rank fitting in the context of statistical methodology. At times they may serve purposes of estimation; at others, they provide approximations to observed data. In some situations, one wishes to fit a given matrix Y by one of lower rank r, say, for some purpose such as parsimony, e.g., the need

to record and/or transmit fewer numbers. No probability model is implicit in such fitting, and the given matrix may be of interest as of itself, without any assumption that it is a single realization of some random process. If so, there is no distribution, there are no parameters in the statistical sense and there is no estimation. The criterion by which we would judge a method of fitting would simply be how well it fitted the matrix at hand.

In other situations, one assumes an underlying probability model and sees the observed Y as one realization from that "experiment." In such cases, one will obviously be concerned with estimating the parameters of the probability model rather than with mere fitting of the random Y observed. The criteria for assessing a method will in that case be standard errors, efficiency of estimation, etc. - and not the closeness of fit to the single observed Y.

REFERENCES

[1] Andrews, D.F., A robust method for multiple linear regression, Technometrics 16 (1974) 523-531.

[2] Bloomfield, P. and Steiger, W., Least absolute deviations curve fitting, SIAM J. Sci. Stat. Comp. 1 (1980) 290-301.

[3] Bradu, D. and Gabriel, K.R., The biplot as a diagnostic tool for models of two-way tables, Technometrics 20 (1978) 47-68.

[4] Eckart, C. and Young, G., The approximation of one matrix by another of lower rank, Psychometrika 1 (1936) 211-218.

[5] Gabriel, K.R., The biplot-graphic display of matrices with application to principal component analysis, Biometrika 58 (1971) 453-467.

[6] Gabriel, K.R., Least squares approximation of matrices by additive and multiplicative models, J. Roy. Statist. Soc. B, 40 (1978) 186-196.

[7] Gabriel, K.R., Biplot display of multivariate matrices for inspection of data and diagnosis, in Barnett, V. (ed.), Interpreting Multivariate Data (London, Wiley, 1981, 147-173).

[8] Gabriel, K.R., An alternative computation of median fits in two-way tables, Technical Report 83/01, Department of Statistics, University of Rochester (1983).

[9] Gabriel, K.R. and Zamir, S., Lower rank approximation of matrices by least squares with any choice of weights Technometrics 21 (1979) 489-498.

[10] Good, I.J., Some applications of the singular value decomposition of a matrix, Technometrics 11 (1969) 823-831.

[11] Greenacre, M.J., Practical Correspondence Analysis, in Barnett, V. (ed.), Interpreting Multivariate Data (London, Wiley, 1981, 119-146).

[12] Hill, M.O., Reciprocal averaging: an eigenvector method of ordination, J. Ecology, 61 (1973) 237-249.

[13] Householder, A.S. and Young, G., Matrix approximation and latent roots, Am. Math. Monthly 45 (1938) 165-171.

[14] Kester, N., Diagnosing and fitting concurrent and related models for two-way and higher-way layouts, Ph.D. Thesis, University of Rochester, Rochester, New York (1979).

[15] Mandel, J., A method of fitting empirical surfaces to physical and chemical data, Technometrics 11 (1969) 411-430.

[16] Mandel, J., A new analysis of variance model for non-additive data, Technometrics 13 (1971) 1-18.

[17] Siegel, A.F., Geometric data analysis: An interactive graphics program for shape comparison, in Launer, R.L. and Siegel, A.F. (eds.), Modern Data Analysis (New York, Academic Press, 1982, 103-122).

[18] Wold, H., Nonlinear estimation by iterative least squares procedures, in David, F.N. (ed.), Research Papers in Statistics (New York, Wiley, 1966, 411-444).

INTERACTIVE TOOLS FOR EDA

HERMAN GOLLWITZER

DREXEL UNIVERSITY
PHILADELPHIA, PA 19104

A preliminary report on APL software tools for data analysis is presented.

The recent emergence of texts pertaining to data analysis makes it important to have appropriate software tools available for students and scientists who have not had the opportunity to experiment with the techniques. Although BASIC and FORTRAN software has been given by Velleman and Hoaglin(5), no corresponding collection seems to be available with APL as the source code. Interpreters for APL are available for personal computers, and the existence of sound software for data analysis will make it simple for many people to experiment with data analysis methods.

This note is a preliminary report on efforts to develop APL functions that can be used as high level primitives in the development of data analysis routines. Nonlinear data smoothing and discrete Fourier analysis are the topics presented in the poster session.

The author is not aware of many published accounts of APL functions developed for data analysis. A good example of an earlier effort is found in a report by Heiberger(3), where various stem-and-leaf functions are developed in a modular fashion. A primer by McNeil(4) offers several useful functions, but the organization and presentation does not make it easy for the user to enter and use the functions as high level primitives. The recent text by Anscombe(1) includes an introduction to APL along with a discussion of statistical computing. In particular, he presents functions for the analysis of time series. Busby and Gollwitzer(2) presented nonlinear smoothing functions that offer an alternative to those introduced by McNeil(4).

The software developed for nonlinear smoothing includes arbitrary length smoothing with running medians; endpoint smoothing; and splitting. These primitives allow one to easily build functions for implementing smoothers given in (4),(5). The discrete Fourier analysis functions include a radix 2 FFT function that provides the user with several possible configurations. One can choose from the following list of possibilities: decimation-in-frequency or decimatio-in-time; bit-reversed inputs or outputs; direct or inverse DFT. This pivotal function is used to build a discrete convolution function and an arbitray span DFT function that is implemented via a chirp transform.

Further development and experimentation are needed to determine the strengths and weaknesses of the design. The author has used these functions to develop interactive software for courses in data analysis and statistics at Drexel University. These functions permitted the author to quickly generate application programs for these courses as needed.

REFERENCES

(1) Anscombe, F.J., Computing in Statistical Science,(Springer-Verlag 1981).

(2) Gollwitzer, H. E. and Busby, R. C., "A parallel implementation of some nonlinear data smoothing procedures" Technical Report #1-80, Department of Mathematical Sciences, Drexel University.

(3) Heiberger, R. M., "A parallel algorithm for stem and leaf displays with an implementation in APL," Technical Report No. 28, Department of Statistics, The Wharton School, University of Pennsylvania, Philadelphia, PA, 19104

(4) McNeil, D. R., Interactive Data Analysis,(J. Wiley, New York, 1977)

(5) Velleman, P. F. and Hoaglin, D. C., Applications, Basics, and Computing of Exploratory Data Analysis,(1981 Boston, MA: Dusbury Press).

On the Computation of a Class of Maximum Penalized Likelihood Estimators of the
Probability Density Function

V.K. Klonias and Stephen G. Nash

Mathematical Sciences Department
The Johns Hopkins University
Baltimore, Maryland

Using a sufficiently general roughness-penalty functional, a class of Maximum Penalized Likelihood Estimators (MPLE) of the density function can be constructed. This class includes the standard MPLE's, as well as a number of new estimators. The flexibility of the penalty functional permits the construction of splines with improved performance at the peaks and valleys of the density curves over the existing estimators. This construction leads to a large-scale, constrained optimization problem. The dimension of this problem and the denseness of the Hessian matrix make Newton's method inconvenient to use. Instead, a truncated-Newton method is suggested. This method is better able to handle the size and exploit the idiosyncracies of the optimization problem. Graphs of some density estimators are given.

1. INTRODUCTION

Let $X_1,\ldots,X_n$ be independent observations from a distribution function F with density f over $\mathbb{R}$ and let F_n denote the empirical distribution function. The nonparametric maximum penalized likelihood method of density estimation (MPLE), introduced by Good and Gaskins (1971 and 1980), produces as the estimator of f the maximizer of the log-likelihood minus a "roughness" penalty functional $\Phi(v)$ which is usually expressed in terms of the root density $v = f^{1/2}$, e.g. $\Phi_1(v) = \alpha \int (v')^2$, $\Phi_2(v) = \beta \int (v')^2 + \alpha \int (v'')^2$ with $\alpha > 0$, $\beta \geq 0$. In DeMontricher, Tapia, and Thompson (1975), the existence and uniqueness of the MPLE's were rigorously established within the framework of the Sobolev space $w^{2,m} = \{u \in L_2(\mathbb{R})$ such that $||u^{(m)}||_2 < +\infty\}$, m a positive integer, where $L_2(\mathbb{R})$ denotes the space of square integrable functions and $||u||_2^2 = \int u^2$. For a discretized version of the MPLE problem (DMPLE) see Tapia and Thompson (1978) and Scott, Tapia and Thompson (1980).

2. ON THE CONSTRUCTION OF THE ESTIMATORS

The MPLE's u of $f^{1/2}$ considered here are solutions to the following optimization problem:

(2.1) max $\{\int \log u^2(x) dF_n(x) - \lambda \int |\tilde{u}|^2 d\mu\}$, $u \in H$

subject to $u(x_i) \geq 0$, $i = 1,\ldots,n$,

where $\lambda > 0$, $\tilde{u}$ denotes the Fourier transform of $u \in L_2(\mathbb{R})$, $H = \{u \in L_2(\mathbb{R}): \int |\tilde{u}|^2 d\mu < +\infty\}$ and μ is a positive measure on $\mathbb{R}$ dominated by the Lebesgue measure. Under the condition that the Fourier transform of $(\mu')^{-1}$ exists, problem (2.1) has a unique solution which is a spline function given implicitly by

(2.2) $u(x) = \lambda^{-1} \sum u(x_i)^{-1} k_\mu(x-x_i)$, $x \in \mathbb{R}$,

where $k_\mu = (\mu')^{-1}$ and there is a unique $\lambda > 0$ such that $\int u^2 = 1$. Then f is estimated by $f_n = u^2$. For details, see Klonias (1982a).

Note that for $\mu'(t) = \sum a_j t^{2j}$ (a_0, $a_m > 0$, and $a_j \geq 0$, $j = 1,\ldots,m-1$ for an integer $m \geq 1$), problem (2.1) is equivalent to that treated in DeMontricher et al. (1975), giving for m = 1,2 the "first and second MPLE of Good and Gaskins" corresponding to ϕ_1, ϕ_2 and being generated by kernels k_μ of the form $(a/2)\exp\{-a|x|\}$ and $[4ab(a^2+b^2)]^{-1}\exp\{-|ax|\}[|b|\cos|bx| + |a|\sin|bx|]$ respectively. For m > 2, k_μ is a convolution of these two basic kernels.

The added flexibility in the choice of μ allows us to consider as kernel k_μ any symmetric density function with positive characteristic function. In particular, $k_\mu(x) = h^{-1}k(x/h)$, $h > 0$, with

(2.3) $k(x) = [2\Gamma(1+\gamma^{-1})]^{-1}\exp\{-|x|^\gamma\}$

$x \in \mathbb{R}$, $\gamma \in [1,2]$ or kernels k with finite support as long as $\tilde{k} > 0$. For example, the normal kernel corresponds to $\mu'(t) = \exp\{h^2t^2/2\}$. Under mild assumptions on f, the strong consistency in the Hellinger distance and in the $L_1(\mathbb{R})$-norm of the MPLE's with kernels given by (2.3) can be established using a generalization of the techniques in Klonias (1982b); for details see Klonias (1982a). It can also be shown that $\lambda/n \to 1$ a.s. as $n \to \infty$, so that (2.2) suggests an interesting analogy between kernel density estimators and MPLE's.

A problem common to all density estimators is that they tend to underestimate the "peaks" and overestimate the "valleys" of the density. To diminish this problem we consider penalizing the convolution of u with a strongly unimodal density $h_1^{-1}k(x/h_1)$ - so that no new modes are introduced - rather than u itself, and thus penalizing "lighter" at the "peaks" and "valleys". This corresponds to using $\mu'(t) = \tilde{k}(h_1t)^2\mu'(h_2t)$, e.g. $\tilde{k}(h_1t) = (1 + h_1^2t^2)^{-2}$, $\mu'(h_2t) = \exp\{h_2^2t^2/2\}$, leading, for $h_1=h_2$, to

(2.4) $k(x) = (2-x^2)\exp\{-x^2/2\}/(2\pi)^{1/2}$,

the kernel used in Figure 1. Nonpositive kernels like (2.4) seem to share the improved consistency properties of kernel density estimators based on nonpositive kernels, but in the case of the MPLE's $f_n = u^2$ is a true density. For the case of the "pseudo MPLE of Good and Gaskins" this is shown in Klonias (1983).

3. ON THE NUMERICAL EVALUATION OF THE MPLE'S.

In view of equation (2.2), it is enough to evaluate λ and the values of the estimator at the sample points $u_i = u(x_i)$, $i = 1,\ldots,n$. It is then sufficient to solve, given a "bandwidth" h, the following optimization problem:

(3.1) $\min A(q) = q^t \ddagger q - 2 \sum \log q_i$, $q \in \mathbb{R}^n$

subject to $q_i \geq 0$, $i = 1,\ldots,n$,

where $q_i = \{(\lambda h)^{\frac{1}{2}} u_i\}^{-1}$, $i = 1,\ldots,n$, $q^t = (q_1,\ldots,q_n)$ and $\ddagger$ is the symmetric positive definite matrix with elements $k((x_i-x_j)/h)$. Then $\lambda = q^t \ddagger * q/h$, where $\Sigma *$ is the matrix with elements $(k*k)((x_i-x_j)/h)$ and $*$ denotes convolution. Note that this approach can be used also for the numerical evaluation of the DMPLE's. For other approaches to the numerical evaluation of a number of MPLE's and DMPLE's, see Good and Gaskins (1971), Scott, Tapia and Thompson (1976), Ghorai and Rubin (1979), Hall and Klonias (1981).

4. SOLVING THE FINITE-DIMENSIONAL MINIMIZATION PROBLEM.

Many algorithms for the problem (3.1) have the following general form:

Given $q \geq 0$, an initial approximation to the solution q^*.

1. Test if q is a sufficiently good approximate solution. If so, stop.

2. Compute a search direction p such that the function A decreases locally along p.

3. Find $\alpha > 0$ such that $A(q+\alpha p) < A(q)$ and $q+\alpha p \geq 0$. (line search)

4. Set $q \leftarrow q+\alpha p$. Go to step 1.

Often step 2 is the most important, since the method used to compute the search direction p determines the class of method. Step 2 is also the most sensitive to the idiosyncracies and the size of the problem being solved.

The most effective general method for solving (3.1) is Newton's method, which takes full advantage of first- and second-derivative information about the function A. In its modern, safeguarded implementations, Newton's method provides a standard for measuring the effectiveness of other algorithms.

For Newton's method, the search direction (called the <u>Newton direction</u>) is obtained by solving

(4.1) $Gp = -g$,

where $g = g(q)$ is the gradient of $A(q)$ at the point q, and $G = G(q)$ is the Hessian matrix of second derivatives. Equation (4.1) is obtained by approximating $A(q)$ by the first three terms of its Taylor series:

(4.2) $A(q+p) \approx A(q) + p^t g + \frac{1}{2} p^t G p$.

If this approximation is accurate, $A(q+p)$ will be close to its minimum value. In this case, the step p to the solution of (3.1) can be obtained by minimizing the quadratic function (4.2) as a function of p. This is only sensible when G is positive definite (otherwise the quadratic has no finite minimum), and when q is near the solution q^* (otherwise the quadratic approximation is dubious).

Unfortunately, if the number of variables is large, or if second-derivative information is difficult to compute, Newton's method may be prohibitively expensive to use. For this reason, special methods have been developed to solve so-called "large-scale" problems. For our purposes, a problem is large-scale if the n by n Hessian matrix is difficult to manipulate.

In many applications, a large-scale problem will have a sparse Hessian matrix, i.e. the Hessian matrix will have few non-zero entries. This allows Newton's method to be extended to this problem class through the use of sparse-matrix and finite-differencing techniques. However, in this application, the number of variables will be large and the Hessian matrix will generally be dense, so these adaptations cannot be made.

Truncated-Newton methods are a family of methods that are practical for general large-scale minimization. They require only first-derivative information, do not manipulate matrices, and do not assume that the Hessian is sparse. Each step of a truncated-Newton method is an approximation (or truncation) of a step of Newton's method. Convergence theory for the family was developed by Dembo, Eisenstat, and Steihaug (1982). Nash (1982) has shown that the methods can be implemented in a stable manner, and that the methods can perform efficiently when adaptive preconditioning strategies are employed.

A truncated-Newton method computes the search direction p by approximately solving (4.1) (or equivalently by approximately minimizing (4.2)) using an iterative scheme. This results in a doubly iterative method: an "outer" iteration for minimizing the function $A(q)$ corresponding to the algorithm above, and an "inner" iteration for computing the search direction in step 2 of that algorithm. There are a number of justifications for this approach:

1. Equation (4.1) has less theoretical justification away from the solution q^*, so that a crude approximation to the Newton direction may be as effec-

tive in the early stages of the minimization.
2. The iterative algorithm for solving (4.1) can be given a variable tolerance, so that as the solution is approached, the system of equations (4.1) is solved more accurately. Thus, extensive computations are only performed when they are most useful.
3. The iterative algorithm for solving (4.1) can be designed so that second-derivative information is not required. Thus, large-scale problems can be solved.
4. A truncated-Newton algorithm requires little storage, and can be compactly programmed, thus making it suitable for small machines.

In a truncated-Newton algorithm, the equations (4.1) are often solved using the linear conjugate-gradient algorithm (Hestenes and Stiefel (1952)). This is an effective general method for solving symmetric positive-definite systems of linear equations, and can be shown to be optimal in a particular norm. It also can take advantage of a priori information about the matrix G (see Concus, Golub, and O'Leary (1976)).

The linear conjugate-gradient algorithm for solving (4.1) can be derived via the Lanczos algorithm (Lanczos (1950)) for tridiagonalizing a symmetric matrix. At stage i of the inner iteration:

(4.3a) $\quad V_i^t G V_i = T_i.$
$\quad V_i^t V_i = I,$

where V_i is an n by i orthogonal matrix, and T_i is an i by i tridiagonal matrix. The tridiagonal matrix T_i is factored into its Cholesky factors:

(4.3b) $\quad T_i = L_i D_i L_i^t,$

where D_i is diagonal with positive diagonal entries, and L_i is lower bidiagonal, with ones on the main diagonal (this factorization is only possible if T_i is positive definite). This factorization is then used to compute p_i (the i-th approximation to the solution of (4.1)):

(4.3c) $\quad T_i y_i = L_i D_i L_i^t y_i = (-V_i^t g),$
$\quad p_i = V_i y_i.$

(Paige and Saunders (1975) have derived iterative formulas for p_i based on this derivation.) At each inner iteration i, the direction p_i is tested to see if it "adequately" solves the Newton equations (4.1); if so, the inner iteration is truncated, and the search direction p is defined as p_i. (See Dembo, Eisenstat, and Steihaug (1982) for details.)

The algorithm (4.3) is well suited to the problem (3.1). At each inner iteration, a matrix/vector product of the form Gv must be formed, and this is easy to do in this particular application. However, if Newton's method were to be used, the matrix G would have to be factored at every outer iteration, and this would be very expensive if the number of data points were large. For the function (3.1), the Hessian matrix is positive definite, so there is no difficulty in applying the linear conjugate-gradient algorithm to compute a search direction. If necessary, the derivation given here permits the algorithm to be safely generalized to the non-positive-definite case.

Because a truncated-Newton algorithm is solving a sequence of equations of the form (4.1), it is possible to improve the efficiency of the method through scaling techniques. As the solution is approached, these equations will often become progressively more similar, and thus it might be hoped that information gained at one iteration might assist at the next. In the context of truncated-Newton methods, this information is generally used to form, either explicitly or implicitly, a matrix M that approximates the Hessian matrix G. Then, the better M approximates G, the better the preconditioning strategy will be (see Concus, Golub, and O'Leary (1976)). In order to use M within the linear conjugate-gradient algorithm, M must be positive-definite, and linear systems involving M should be "easy" to solve. For example, M might be diagonal or in factored form.

The formulas for most minimization algorithms can be used as preconditioning strategies since they implicitly define a positive-definite operator M. In general, if the algorithm has an efficient rate of convergence as a nonlinear minimization algorithm, the operator M will be a good preconditioning matrix (see Nash (1982)). There are other ways in which to generate a preconditioning matrix. Some of the most effective involve generating (as the function A is being minimized) an approximation to the diagonal of the Hessian (see Nash(1982)). Both these strategies have been employed to solve the density estimation problems in the numerical tests.

One final point concerns the constraints $q \geq 0$. To ensure that these constraints remain satisfied at every iteration, the search direction p will be constrained so that, if $q_i = 0$, then $p_i \geq 0$. Also, for the line search in step 3, an upper bound on α will be specified so that no constraint will be violated in the move to the new point. Loosely speaking, this is how the algorithm guarantees a non-negative solution.

5. NUMERICAL EXAMPLES

In this section we present some numerical examples concentrating exclusively on MPLE's that are not in the literature and on univariate densities, although our method can be used with multivariate (e.g. normal) kernels. For the "first and second MPLE's of Good and Gaskins" and the corresponding DMPLE's, see the references cited earlier; for MPLE's with penalties on log f see Leonard (1978) or Silverman (1982).

The estimands are the standard normal density $\phi(x)$ and the bimodal $\{\phi(x) + 2\phi(2x)\}/2$. We generated two normal samples of size 100 each using the IMSL random number generator GGNML, with DSEED's .255866175D+09 and .194929285D+10 respectively, and used the first one for the estimation of the normal, and the combined sample for the estimation of the bimodal. In the estimation of the bimodal we made an effort to estimate well the higher "peak" and the "valleys" rather than the lower "peak". We should also mention that the triangular kernel used in Figure 4 is the only kernel with finite support we have used thus far. Smoother kernels with finite support should perform well at the "valleys" of the curves.

The algorithm, initialized with $q_i = 1$, $i = 1, \ldots, n$, typically needs 26 function and gradient evaluations for n = 100, and 35 for n = 200, in order to converge with the norm of the gradient of order 10^{-12}. The rate of convergence, although theoretically only guaranteed to be superlinear, was quadratic; this indicates that the eigenvalues of the Hessian matrix of $A(q)$ are clustered. The programs were coded in FORTRAN, and double precision arithmetic was used on a VAX 11/780 computer. The CPU time is typically of the order of 10 seconds for n = 100, and 35 seconds for n = 200. In all figures, the dotted line is the true density, and the solid line is the computed estimate.

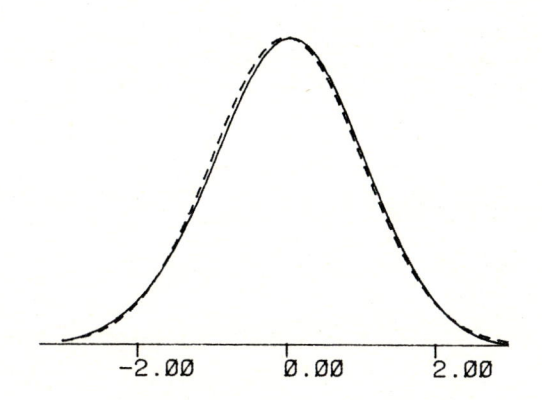

Figure 1. $k(x) = (2-x^2)\phi(x)$, h =1.275; standard normal density.

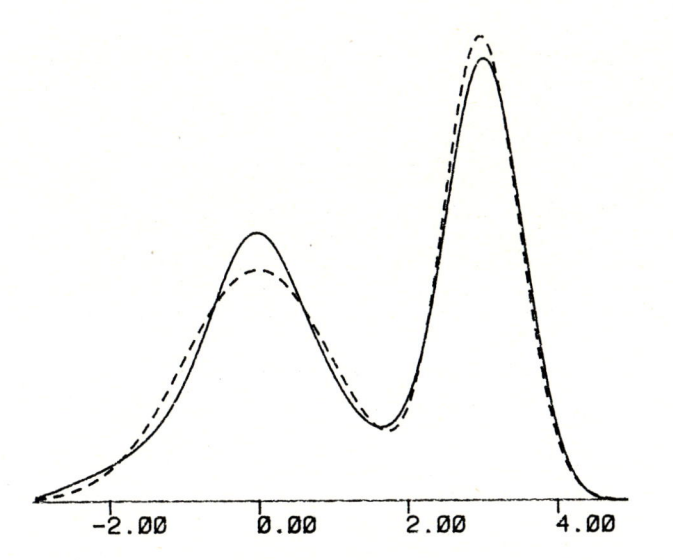

Figure 2. $k(x) = (3-x^2)\phi(x)/2$, h = .7; bimodal density.

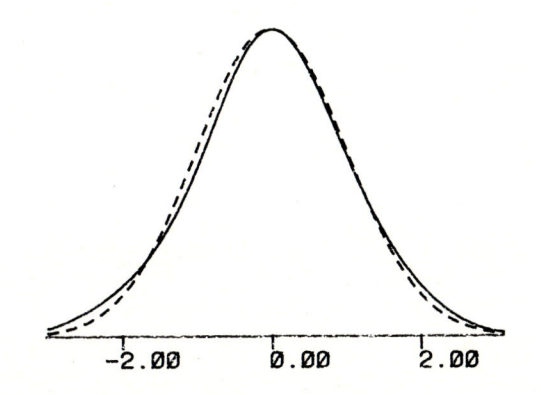

Figure 3. $k(x) = \phi(x)$, h = .72; standard normal density.

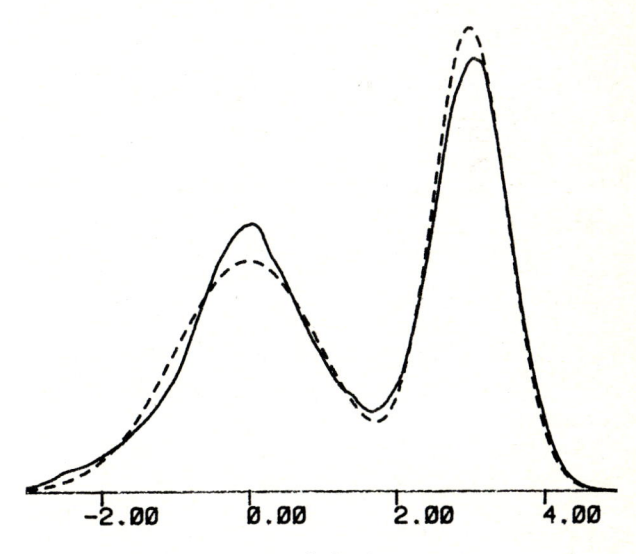

Figure 4. $k(x) = 1 - |x|$, $|x| < 1$, h = 1.1; bimodal density.

6. ACKNOWLEDGEMENTS

V.K. Klonias would like to thank Professor Richard H. Byrd for a number of stimulating discussions.

7. REFERENCES

[1] Concus, P., Golub, G.H., and O'Leary, D.P., A generalized conjugate-gradient method for the numerical solution of elliptic partial differential equations, in Bunch, J. and Rose, D. (eds.), Sparse Matrix Computations (Academic Press, New York, 1976).

[2] Dembo, R.S., Eisenstat, S.C., and Steihaug, T., Inexact Newton Methods, SIAM J. Numerical Analysis 19 (1982) 400-408.

[3] DeMontricher, G.F., Tapia, R.A., and Thompson, J.R., Nonparametric maximum likelihood estimation of probability densities by penalty function methods, Ann. Statistics 3 (1975) 1329-1348.

[4] Ghorai, J. and Rubin, H., Computational procedures for maximum penalized likelihood estimate, J. Statist. Comput. Simul. 10 (1979) 65-78.

[5] Good, I.J. and Gaskins, R.A., Nonparametric roughness penalties for probability densities, Biometrika 58 (1971) 255-277.

[6] Good, I.J. and Gaskins, R.A., Density estimation and bumphunting by the penalized likelihood method exemplified by scattering and meteorite data, (invited paper) J. Amer. Statist. Assoc. 75 (1980) 42-73.

[7] Hall, W.J. and Klonias, V.K., On the numerical evaluation of a maximum penalized likelihood estimator of the density function, Tech. Report 344, Dept. of Mathematical Sciences, The Johns Hopkins Univ. (1981).

[8] Hestenes, M. and Stiefel, E., Methods of conjugate gradients for solving linear systems, J. Res. Nat. Bur. Standards 49 (1952) 409-436.

[9] Klonias, V.K., On a class of maximum penalized likelihood estimators of the probability density function, Tech. Report 364, Dept. of Mathematical Sciences, The Johns Hopkins Univ. (1982a).

[10] Klonias, V.K., Consistency of two nonparametric maximum penalized likelihood estimators of the probability density function, Ann. Statist. 10 (1982b) 811-824.

[11] Klonias, V.K., On the consistency of the maximum penalized likelihood density estimators, Tech. Report 376, Dept. of Mathematical Sciences, The Johns Hopkins Univ. (1983).

[12] Lanczos, C., An iteration method for the solution of the eigenvalue problem of linear differential and integral operators, J. Res. Nat. Bur. Standards 45 (1950) 255-282.

[13] Leonard, T., A Bayesian method for histograms, Biometrika 60 (1978) 297-308.

[14] Nash, S.G., Preconditioning of truncated-Newton methods, Tech. Report 371, Dept. of Mathematical Sciences, The Johns Hopkins Univ. (1982).

[15] Paige, C.C. and Saunders, M.A., Solution of sparse indefinite systems of equations, SIAM J. Numerical Analysis 12 (1975) 617-629.

[16] Scott, D.W., Tapia, R.A., and Thompson, J.R., An algorithm for density estimation, presented at Computer Science and Statistics: Ninth Annual Symposium on the Interface (Harvard University, Cambridge MA, 1976).

[17] Scott, D.W., Tapia, R.A., and Thompson, J.R., Nonparametric probability density estimation by discrete maximum penalized-likelihood criteria, Ann. Statist. 8 (1980) 820-832.

[18] Silverman, B.W., On the estimation of a probability density function by the maximum penalized likelihood method, Ann. Statist. 10 (1982) 795-810.

[19] Tapia, R.A. and Thompson, J.R., Nonparametric Probability Density Estimation (The Johns Hopkins University Press, Baltimore and London, 1978).

ECONOMETRIC CALCULATIONS USING *APL*

Stephen D. Lewis
Department of Economics
University of Alberta
Edmonton, Alberta, Canada
T6G 2H4
(403) 432-5620

This paper sets forth *APL* expressions for calculations commonly required in econometric analyses. Ordinary and generalized least squares are treated in detail. Some econometric exercises are suggested. Instrumental Variables, Two-Stage Least Squares and Restricted Least Squares techniques are briefly considered. The overall discussion summaries a general treatment that has been used for other econometric problems.

In this paper, the use of *APL* [1] for econometric calculations is examined. In the first section, the equations for ordinary least squares are presented, and using a number of global variables, additional statistics and expressions are calculated. The next section examines generalized least squares. In the third section additional econometric calculations are illustrated. Finally, basic *APL* expressions for selected estimation techniques are briefly set forth.

I. Ordinary Least Squares in *APL*

Ordinary Least Squares (*OLS*) is a common econometric technique and with suitable modification for special circumstances it can accommodate many situations likely to arise in econometrics. *OLS* is viewed as a set of basic calculations that produce global variables required for additional analyses.

We start with a data matrix, D, where each column represents all observations for one variable and a column of ones for the constant term. Let V be a numeric vector such that for the equation to be estimated.

$V[1]$	Column number of D for dependent variable, Y
$1\downarrow V$	Column numbers of D for explanatory variables, X

The *OLS* estimation and calculation of basic regression results of the equation

$$Y = \underline{E} + X + . \times \underline{B}$$

where

Y	Dependent variable
X	Explanatory variables
$\underline{B}$	Unknown coefficients
$\underline{E}$	Stochastic variable

are given by the *APL* statements:

$N \leftarrow (\rho D)[1]$

$K \leftarrow (\rho V) - 1$

$Y \leftarrow D[;V[1]]$

$X \leftarrow D[;1\downarrow V]$

$B \leftarrow (\boxminus(\lozenge X)+. \times X)+. \times (\lozenge X)+. \times Y$

$E \leftarrow Y - X + . \times B$

$S2 \leftarrow (+/E*2) \div N-K$

$VCB \leftarrow S2 \times \boxminus(\lozenge X)+. \times X$

The additional variables calculated above but not previously defined are:

N	Number of observations
K	Number of explanatory variables
B	*OLS* coefficients
E	*OLS* error terms
$S2$	Estimated variance of $\underline{E}$
VCB	Variance-Covariance matrix of B

Using these basic results, additional calculations commonly required in econometric work are:

Coefficient of determination:

$R2 \leftarrow 1 - (+/E*2) \div (+/Y*2) - ((+/Y)*2) \div N$

$R2$-adjusted for degrees of freedom:

$RB2 \leftarrow R2 - ((K-1) \div N-K) \times 1 - R2$

Standard errors for B:

$SB \leftarrow (1\ 1 \lozenge VCB)*0.5$

Equation F-test that all coefficients are zero:

$FT \leftarrow (R2 \div 1 - R2) \times (N-K) \div K-1$

Durbin-Watson statistic:

$DW \leftarrow (+/((1 \downarrow E)-{}^{-}1 \downarrow E)*2) \div +/E*2$

Student t-statistics for B:

$TB \leftarrow B \div SB$

II. Generalized Least Squares in APL

OLS may be the appropriate technique when E has a homoskedastic and non-autoregressive stochastic structure. Under these circumstances the variance-covariance matrix VC for E is given by

$VC \leftarrow I \times \omega*2$

where I is an $N \times N$ identity matrix given by

$I \leftarrow (\iota N) \circ .= \iota N$

and ω is an unknown constant. When VC has a different structure than that given above, OLS estimation of $Y=E+X+.\times B$ produces inefficient estimates of B. Given a specification for VC, best linear unbiased estimates of B can be obtained from the sequence

$N \leftarrow (\rho D)[1]$

$K \leftarrow (\rho V)-1$

$Y \leftarrow D[;V[1]]$

$X \leftarrow D[;1 \downarrow V]$

$VI \leftarrow \boxdot VC \div (+/1\ 1 \varnothing VC) \div N$

$B \leftarrow (\boxdot (\varnothing X)+.\times VI+.+X)+.\times (\varnothing X)+.\times VI+.\times Y$

$E \leftarrow Y-X+.\times B$

$S2 \leftarrow E+.\times VI+.\times E \div N-K$

$VCB \leftarrow S2 \times \boxdot (\varnothing X)+.\times VI+.\times X$

The variables produced by these expressions can be used for further analyses in a manner similar to the results obtained for OLS although in most cases the expressions are a bit more complicated because VC is no longer a scalar multiple of an identity matrix.

III. Econometric Exercises Using APL

Given the results above plus a few additional calculations below it is possible to verify theoretically or empirically a number of standard results in econometric analysis.

For a particular data set, OLS estimates of $Y=E+X+.\times B$ can be calculated by the expressions given previously or more simply by

$B \leftarrow Y \boxdot X$

assuming that Y and X are as defined above. Next specify

$I \leftarrow (\iota N) \circ .= \iota N$	Identity matrix
$XP \leftarrow \varnothing X$	X-transpose
$XPY \leftarrow XP+.\times Y$	X-transpose, Y
$XPX \leftarrow XP+.\times X$	X-transpose, X
$M \leftarrow I-X+.\times (\boxdot XPX)+.\times XP$	Idempotent matrix
$YH \leftarrow X+.\times B$	Y-hats
$E \leftarrow Y-X+.\times B$	OLS error terms

Using these expressions, a number of results in OLS analysis can be verified with appropriate APL expressions using the logical dyadic primitive scalar function '='. When the expressions on each side of '=' are the same the result is 1 indicating equality. Expressions of the form $0=(APL$ numeric expression) must be written as $1=1+(APL$ numeric expression).

Important OLS expressions (left as an exercise to be verified and interpreted) are:

$\wedge/XPY=XPX+.\times B$

$(+/YH)=+/Y$

$1=1++/E$

$\wedge/1=1+XP+.\times E$

$(+/YH*2)=B+.\times XPY$

$1=1+YH+.\times E$

$1=1+B+.\times XP+.\times E$

$\wedge/\wedge/M=\varnothing M$

$\wedge/\wedge/M=M+.\times M$

$(N-K)=+/1\ 1 \varnothing M$

$\wedge/E=M+.\times Y$

When performing GLS it is necessary to specify the matrix VC. A number of APL expressions are available to greatly simplify this task. If E is heteroskedastic with variances $S \leftarrow S1, S2, \ldots, SN$ and $N=\rho S$, then the appropriate VC matrix is given by

$I \leftarrow (\iota N) \circ .= \iota N$

$H \leftarrow (N,N) \rho (,I) \backslash S$

If E is homoskedastic but has a first-order autoregressive structure with parameter R then the appropriate VC matrix is given by

$A \leftarrow (\div 1-R*2) \times |(\iota N) \circ .- \iota N$

Sometimes GLS is performed by transforming the problem into an equivalent OLS problem with an appropriate transformation matrix, P. For the heteroskedastic problem the expression for PH is obtained from

```
PH←(N,N)ρ(,I)\÷S*0.5
∧/∧/H=⌹PH+.×PH
```

The autoregressive transformation matrix PA is obtained from

```
PA←(1,(N-1)ρ0),[1](1 0↓I-(0,
    (N-1)ρ1)⌽[2]R×I)÷(1-R*2)*0.5
∧/∧/A=⌹(⌽PA)+.×PA
```

For simple numeric examples the properties of these VC and P matrices are demonstrated by:

```
      S←ι3
      N←ρS
      I←(ιN)∘.=ιN
      H←(N,N)ρ(,I)\S
      H
1 0 0
0 2 0
0 0 3
      PH
1             0              0
0             0.7071067812   0
0             0              0.5773502692
      ⌹PH+.×PH
1 0 0
0 2 0
0 0 3
      R←0.5
      A
1     0.5    0.25
0.5   1      0.5
0.25  0.5    1
      PA
¯0.8660254938  0       0
 0.5          ¯1       0
 0             0.5     1
      ⌹(⌽PA)+.×PA÷1-R*2
1     0.5    0.25
0.5   1      0.5
0.25  0.5    1
```

IV. Selected Estimation Techniques

APL statements for ordinary and generalized least squares have been discussed. To further demonstrate the usefulness of APL for econometric calculations, additional estimation techniques are briefly presented using the same format as used above.

Instrumental Variables Estimation [3]
(INST)

A common econometric problem is that some of the explanatory variables comprising X are not independent of E. In such a case, the OLS estimator is not consistent. Let Y and X be defined as before, and specify W as a data matrix of instruments for X. APL expressions for $INST$ (a consistent estimator) are given by

```
WPW←(⌽W)+.×W
WPX←(⌽W)+.×X
WPY←(⌽W)+.×Y
B←(⌹WPX)+.×WPY
E←Y-X+.×B
S2←(+/E*2)÷N-K
VCB←S2×(⌹WPX)+.×WPW+.×⌹⌽WPX
```

Two-Stage Least Squares [4]
(TSLS)

Another common problem in econometrics is the estimation of one equation from a system of simultaneous eqations. Except in special cases the OLS estimator for such an equation is inconsistent. Letting $Y1$ and $X1$ represent respectively endogenous variables and exogenous variables included within the equation to be estimated and Z system exogenous variables, then APL expressions for $TSLS$ (a consistent estimator) are given by

```
W←(Z+.×Y1⌹Z),X1
WPX←(⌽W)+.×Y1,X1
WPY←(⌽W)+.×Y
B←(⌹WPX)+.×WPY
E←Y-X+.×B
S2←(+/E*2)÷N-K
VCB←S2×⌹WPX
```

Restricted Ordinary Least Squares [5]
(ROLS)

Often OLS coefficients must satisfy linear constraints of the form $C=R+.×B$ where C and R are known matrices of appropriate dimension and rank. To take account of this prior information OLS estimators must be modified. The required APL expressions are

```
B←Y⌹X
RB←R+.×B
L←(⌹R+.×(⌹XPX)+.×⌽R)+.×C-RB
B←B+(⌹XPX)+.×(⌽R)+.×L
E←Y-X+.×B
S2←(+/E*2)÷N-K
VC←S2×⌹XPX
RV←R+.×VC
VCB←VC-(⌽RV)+.×(⌹RV+.×⌽R)+.×RV
```

V. Summary

The preceding discussion has demonstrated how econometric calculations can be formulated using *APL*. With only a little bit of knowledge about *APL* arithmetic and matrix operations, these *APL*-Econometric expressions (which are machine executable) can be easily understood and interpreted. *APL* expressions for five common econometric estimation techniques and related calculations have been presented. Given the capabilities of *APL* to operate directly on matrices and vectors, modifications can easily be made to these and other techniques. *APL*-based econometric routines provide a powerful and flexible alternative to existing econometric packages written in FORTRAN or other traditional computing languages. A computer package *APLECON* [6] has been developed with estimation capabilities that include ordinary and generalized least squares, autoregressive methods and simultaneous equation techniques. Also available are supplementary functions for residual analyses, hypotheses tests, assumptions checks, etc. Documentation and distribution details for *APLECON* can be obtained from the author.

REFERENCES:

[1] IBM Manual GC 26-3847-4, *APL* language, July 1978

[2] Kmenta, J., Elements of econometrics (The MacMillan Company, New York, 1971), chapter 12

[3] Kmenta, chapter 9

[4] Kmenta, chapter 13

[5] Theil, H., Principles of econometrics (John Wiley & Sons, New York, 1971), chapters 1, 3

[6] Lewis, S. D., *APLECON* - Econometrics with *APL*: User's Manual (Version 2), Research paper 82-11, Dept. of Econ., Univ. of Alberta (June 1982)

A SURVEY OF THE CURRENT STATUS OF STATISTICAL COMPUTING IN COLLEGES

John D. McKenzie, Jr. and David P. Kopcso

Babson College
Babson Park (Wellesley), Massachusetts

In this poster session we will present the results of a survey of approximately 40 Northeastern colleges and universities on the current state of the interface between computer science and statistics at their schools. The institutions surveyed are members of NERComP, New England Regional Computing Program. The schools range from large private universities such as Harvard and M.I.T. to small public colleges such as North Adams State College and Plymouth State College. Topics to be discussed at this poster session include the current available hardware and statistical software at each school. In addition, we will present an overview of the statistical consulting services and the courses related to the interface between computer science and statistics.

1. INTRODUCTION

Computer centers in many academic institutions are faced with the difficulty of supplying reliable hardware, meaningful software, and quality service within limited budgets. This problem is especially acute in the smaller schools. Unlike their for-profit counterparts college computer centers would usually be willing to exchange information, concerns, and queries freely if an appropriate conduit were available. NERComP, New England Regional Computing Program, is a federation of approximately 40 Northeastern colleges and universities which serves in such a capacity. This report describes the results of a recent survey of the NERComP member schools. The outcome includes descriptions of:

- the quantity, type, and manufacturer of available computer hardware.
- the types and applications of statistical software.
- the responsibility, expense, and information distribution techniques of statistical consulting.
- the course requirements, usage, and exposure in the classroom of computers and statistical software.

2. NERComP STATISTICAL COMPUTING QUESTIONNAIRE

NERComP is currently composed of five large major universities, 35 smaller universities and colleges, and one board of higher education (see Appendix A). The member schools range in size from having under 1,100 students to having over 25,000 students. Both public and private schools of varying size are represented including some of the most highly endowed colleges in the United States. Since its founding in 1967 NERComP has provided a platform for information exchange through conferences and workshops on topics that have varied from microcomputers to CENSPAC, the Census Bureau's software for accessing data from the census and other summary tape files. A member institution is linked to NERComP through an institutional representative, an appointee of the respective institution's president.

The questionnaire to access the current interface of computer science and statistics was sent to each institutional representative, who in turn circulated it to the most knowledgeable individual or group of individuals for completion. (A copy of the questionnaire is available from the authors upon request.) Before sending out the questionnaire a pretest was administered. In order to ensure a true comparable snapshot of the evolutionary computing environment, a two-week response time was requested.

An initial response of 17 of the 41 members precipitated a followup telephone call to the non-respondents. In the two weeks after this followup, nine additional questionnaires appeared. Of the 26 respondents, three were from large major universities, 22 were from the smaller universities and colleges, and one was from the board of higher education. Due to the dispersement of the computing facilities and the existence of many uncataloged systems in the large universities, only partial replies were provided. Hence we decided not to include these schools with many academic computer centers in our analysis. Since the board of higher education is a somewhat unique member its response was not included as well. The 22 remaining responses are fairly representative and form a good cross section of the majority of the NERComP membership.

Now we will report the results of these 22 questionnaires. This discussion will be organized around the five parts of the questionnaire: hardware, statistical software, statistical consulting, educational interface, and future trends.

3. DISCUSSION OF RESULTS

3.1 Hardware

The 22 NERComP members reported owning 284 machines with six schools having a total of eight mainframes, 18 schools having a total of 40 minicomputers, and 16 schools having a total of 236 microcomputers (one school owned 104 micros while the remaining 15 schools had 132 micros). Only two of the six schools with mainframes owned more than one mainframe. The mainframe manufacturers were DEC with three, Burroughs with two, and CDC, IBM, and UNIVAC with one each. The number of potential terminals on such machines averaged 50.0 with a standard deviation of 35.7. Note that one of these machines did not include on-line terminal capacity.

Eight schools had exactly one minicomputer, seven has two, two had five, and one had eight. The mini-manufacturers were again led by DEC with 28, followed by Prime with six, Data General and IBM with two each, and Texas Instruments and Wang with one each. Two of the 40 machines were strictly batch-oriented, i.e., without remote input and output. Including these two machines the minicomputers averaged 40.8 potential terminals with a standard deviation of 44.8. As these statistics show, the number of terminals varied greatly.

Excluding the one school with 104 micros the average number of micros per school was 6.3 with a standard deviation of 8.3 - still quite a spread! These manufacturers were led by Apple with 153 Apple IIs and two Apple IIIs, followed by a distinct Tandy with 22 TRS80s and six Tandy IIIs, then Commodore with 15 Pets and 1 VIC20, C3 with ten IWSs, IBM with nine PCs and the remaining manufacturers with eight different types of machines. Although the school with 104 micros has 101 Apples, Apple still dominates the NERComP microcomputer market. Many facilities had more than one type of microcomputer. While six schools had none and five schools had exactly one, four had two, four had three, one had four, and one even had six different manufacturer's micros.

Eleven of the 22 schools subscribed to an off-campus time-sharing system. Fourteen had some kind of plotting capability. The actual capability varied from hard copy separate plotting devices to graphical displays on micros.

3.2 Statistical Software

In this section of the questionnaire we first asked each facility to document what vendor software was available and on what machine it was present. Then we asked how heavily each piece of software was used in teaching, research and administrative work and in what specific way each package was used. Finally we gave questions about comprehensive packages that included a statistics submodule, in-house statistical software, and stand-alone graphics packages.

Eight examples of vendor software were explicitly mentioned in the first part of this section. BMDP, GENSTAT, IMSL, Minitab, PSTAT, SAS, and SPSS were selected since these pieces of software are heavily promoted in statistical publications and at statistical conferences. CENSPAC was mentioned since NERComP is associated with the use of this package in our region of the country. In addition to the software we presented, only one other package, TSP, was used by more than two facilities. Hence we will discuss nine examples of vendor software.

Sixteen schools had SPSS at their facility, 11 had Minitab and nine schools had BMDP. Five schools reported the availability of IMSL and TSP, while three schools had CENSPAC on their system. Only one school had SAS available while no respondents indicated the availability of GENSTAT and PSTAT. Note that the fact that SAS is currently only available on IBM equipment may explain the paucity of SAS software.

On the average the two most frequently used pieces of vendor software for teaching purposes were SPSS and Minitab. SPSS was the software most used for research purposes. Almost all social science and natural science disciplines used statistical software. Use in teaching was greater than use in research. And both of these applications surpassed use in administration work although the limited use of administrative tasks ranged from admissions to alumni records.

For the most part all of the vendor software was present on either mainframes or minicomputers. Four schools did use their time-sharing service but only two used time-sharing exclusively. Another four schools had software on microcomputers but two of these had only one package as a supplement. The other two schools depended completely on their micro-based software.

Excluding micro-based statistical software most of the 22 schools realize that there is a need for more than one piece of statistical software to handle their requirements. For, while two schools had one piece of statistical software, four schools had two pieces of such software, five had three pieces, three had four pieces, four had five pieces, one had six pieces, and one had eight pieces. In contrast Harvard, another member of NERComP with many academic computing facilities, has 40 statistical packages and programs. This major university has all pieces of statistical software discussed above except GENSTAT.

But, while most of our 22 schools feel the need for more than one piece of statistical software, only two of the schools are using a comprehensive package such as IFPS or Empire that includes statistical software as a submodule. Seven out of the 22 respondents mentioned the availability of software that was produced in-house but most of these respondents indicated there was no significant use of such software. Eight schools reported the existence of a stand-alone graphics package such as TEL-A-GRAF. Most of these packages were oriented towards DEC machines.

3.3 Statistical Consulting

Gibbons and Freund (1) mention four structures for statistical consulting at colleges and universities: subject matter department expert(s), assigned responsibility, providing consulting through computing services, and statistical center. In the first part of this section of the questionnaire we asked what structure, if any, was being used at each school. As should be expected there was a lot of variety in the way statistical consulting is performed at these 22 NERComP schools. Six schools indicated that there is no support offered for statistical software, while at five schools each statistical software package or program is informally supported by the specific subject matter discipline utilizing the software. Only one school employed discipline expert(s). Four schools provided consulting through their computer center although one school only provided limited advice on design. At two schools there was a hybrid system composed of the computer center and other departments. Two other schools indicated that all statistical software packages or programs are supported by a statistical center distinct from the computer center. Finally one school combines a statistical center with the computer center while another school combines a statistical center, the computer center, and informal support. It appears that most schools are attempting to provide some type of statistical consulting to its community.

But, while most of the responding schools indicated they were providing some form of consulting, few schools appear to have the will or resources to provide high-quality consulting. In the spotty responses to questions concerning the labor expenses of such consulting, it does not look like major support has been given to this item. For example, only eight institutions employ full-time staff for such work. And these staff members are only used a few hours a week.

Fifteen of the 16 schools that provide some form of statistical counsulting do not charge any fee for their services. The other school's respondent did not know if the distinct statistical consulting center on his campus charged a fee.

The modal response concerning access to information on statistical software was a facility that provided free and open access to the vendors's software documentation, in-house documentation, and in-house seminars, and no access to vendor's seminars. Eleven of the 15 schools responding to the vendor's software documentation question allowed such access. Out of the 15 schools responding, ten provided free and open access to in-house documentation while five schools had restricted in-house documentation. Three of these latter schools required a fee for such documentation. Eleven schools had free and open access to in-house seminars while one school provided restricted seminars with no fee, one school provided no access seminars, and nine schools did not respond to the in-house seminar question. Six schools did not allow access to vendor's seminars; the remaining schools did not answer this question. No other means of distributing statistical software information were given by the 22 respondents.

3.4 Educational Interface

Only five of the 22 respondents indicated that the school(s) serviced by their facility had a statistical literacy requirement. Nine respondents indicated that their schools have a computer literacy requirement. This requirement appeared to be more defined than the statistical requirement. One commonly mentioned computer literacy requirement was the completion of a programming course in a language such as BASIC or Pascal. Another frequently mentioned requirement was the completion of a course explaining the role of the computer in society. Except for one respondent no one seemed aware of a quantitative studies requirement as suggested by The New Liberal Arts Program of the Sloan Foundation. This respondent comes from a school that has been awarded a Sloan Foundation grant to implement this requirement.

In this next section we asked each respondent about courses which use statistical computing. For the most part we feel we asked too much of the respondents when we asked them to supply us how many courses introduce, require, and/or suggest the use of statistical software or the implementation of statistical algorithms. Many people left this part of the questionnaire blank or indicated they were estimating their answers. Still we were pleasantly surprised to see people indicating that a fair number of students are using statistical software in their course. We also were surprised to see seven schools offering at least one course that dealt with statistical algorithms.

Most schools indicate that they have facilities available for the use of computer demonstrations during class. Only two schools stated they have no such facilities while 16 schools indicated that they have at least two such facilities. Including multiple responses eight schools have a separately-equipped lab room with permanent work stations for each student, six schools have a lab room with one or more permanent monitor/receiver(s), and six schools have a lab room with one or more large projected displays, e.g., Electrohome. Thirteen schools indicated that they may bring portable work station(s) into the classroom while 11 schools indicated that they may bring monitor/receiver(s) into the classroom. Four schools have the ability to bring large projected display(s) into the classroom. Finally two schools allow classes to reserve a terminal room. It is our feeling a good deal of good experimentation is going on in order to bring the computer resource into the classroom.

3.5 Future Trends

Eighteen out of the 22 schools surveyed foresaw some dramatic change in statistical computing at their facility over the next one to five years. The most common explanation of this change from the respondents was the high probability of increased use of statistical computing. Thus one small private coeducational college is "currently looking for a broad spectrum statistical package to meet most of the needs of individual departments. We thus hope to maximize our support effort, provide the necessary services and provide our users with a uniform interface." This search is being done because their statistical computing has doubled in the last year and they expect the trend to continue. At the same time a small private college for women is anticipating a "drastic expansion of resources including acquisition of (a) new computer, software, and additional courses and training programs." But not all the change will be due to just increased use. As one respondent from another small private college for women mentioned there will also be "an increase in the level of sophistication of statistical package use. Newer faculty in the social sciences are better trained in using computers for statistical analysis." Yet with these probable increases it is important to keep in mind what the respondent from one small Catholic school stated: "Statistical packages... are production packages. What must occur at Colleges is interpretation of trends -- not statistics calculated by the packages".

4. CONCLUSION

Except for one respondent from a small university, none of the respondents anticipated some of the radical changes that we believe will be coming to statistical computing in the near future. This consultant mentioned the probable growth of the use of graphics in analyzing data and presenting results. The same type of change was foreseen by a respondent from a multi-center major university that was not included in our analysis -- "More graphical display of data and results of analysis including visual display of models..."

Another thing that is already occurring on the campuses of the surveyed schools but was not explicitly mentioned in analyzed current trend responses is the growth of microcomputers. At the present time only four schools perform statistical analyses on microcomputers. But this will probably change. As a respondent from another multi-center major university stated -- "I see that more computing will move to micros and that statistical computing will be utilizing a combination of micros and mainframes."

In addition we believe that within five years the influence of more comprehensive packages along with expert software will be felt at all of the NERComP schools. Such "user-friendly" software with its increased computing power should have a profound effect on how data are analyzed and how data analysis will be taught.

Finally all schools should be aware their students will probably be more knowledgeable in the statistical computing when they arrive on their campuses. With the continuing availability of improved technology, we believe that a large number of elementary, middle, and secondary schools,both public and private, will be adapting the recommendations of The Conference Board of the Mathematical Sciences' report "The Mathematical Sciences Curriculum K-12: What is Still Fundamental and What is Not" (2). In particular we feel the following recommendations for the secondary school curriculum will greatly affect what will be happening within the next five years.

. "That the traditional component of the secondary school curriculum be streamlined to make room for important new topics. The content, emphases, and approaches of courses in algebra, geometry, precalculus, and trigonometry need to be re-examined in light of new computer technologies".

. "That discrete mathematics, statistics and probability, and computer science now be regarded as 'fundamental' and that appropriate topics and techniques from these subjects be introduced into the curriculum. Computer programming should be included at least for college-bound students."

In this paper we have examined the current status of the interface between computer science and statistics at colleges. We have done this by means of a survey to members of the NERComP organization. Our discussion has focused on the status of statistical computing at colleges with only one major academic computing center. We hope that this profile will be helpful to the members of NERComP, other educators, and other people interested in statistical computing. In the future we hope to expand this discussion so that it includes the rest of NERComP including those schools with many academic computer centers. In addition we may prepare a sample survey for a randomly selected group of colleges throughout the United States.

REFERENCES

(1) Gibbons, Jean D. and Freund, Rudolf J., Organizations for Statistical Consulting at Colleges and Universities, The American Statistician, 34 (1980) 140-145.

(2) The Conference Board of the Mathematical Sciences, Report to NSB Commission on Pre-college Education in Mathematics, Science, and Technology - "The Mathematical Sciences Curriculum K-12: What is Still Fundamental and What is Not" (December 1, 1982)

APPENDIX A - NERComP Members

Assumption College
Babson College
Bates College
Bentley College
Bowdoin College
Central New England College
Champlain College
Colby College
Dartmouth College
Emmanuel College
Hartford Graduate Center
Harvard University
Massachusetts Institute of Technology
Merrimack College
Mount Holyoke College
New England Board of Higher Education
New Hampshire College
Nichols College
North Adams State College
Norwich University
Plymouth State College
Post College
Saint Joseph College
Salve Regina-The Newport College
Simmons College
Skidmore College
Smith College
Southeastern Massachusetts University
Suffolk University
Trinity College
U.S. Coast Guard Academy
University of Hartford
University of Massachusetts
Vermont Law School
Wellesley College
Wentworth Institute
Wesleyan University
Western New England College
Wheaton College
Worcester Polytechnic Institute
Yale University

Generating Normal Variates By Summing Three Uniforms

Eugene F. Schuster

Department of Mathematical Sciences
University of Texas at El Paso
El Paso, Texas 79968

In a high level language where one desires a relatively fast easy to program generator, a "trapezoidal" decomposition based algorithm has been recommended for sampling from the normal distribution. In this "trapezoidal" based decomposition method of sampling from the normal, one samples nearly 92% of the time from the sum of two independent uniforms. Marsaglia and Bray describe a similar decomposition procedure in which three uniform samples are required in 86.38% of all cases, two such samples are taken 11.07% of the time, and in the remaining 2.55% of all cases an acceptance-rejection technique or a tail method is used. Ahrens and Dieter report that this algorithm has performance characteristics similar to that of the "trapezoidal" algorithm. In this paper we discuss a "sum of three" decomposition based algorithm for sampling from the normal in which we show we can sample from the sum of three independent uniforms in nearly 97% of the cases.

1. INTRODUCTION AND SUMMARY

Various algorithm are known for generating normally distributed variates. Ahrens and Dieter [1] compare the most efficient ones on the basis of speed, memory requirments, and ease of programming. In a high level language like Fortran where one desires a fast, easy to program generator, these authors recommend the "trapezoidal" decomposition algorithm. In a decomposition method one takes a distribution which is hard to sample from and decomposes its probability density function, say $f(x)$, into $f(x) = \beta g(x) + (1 - \beta)h(x)$ where $g(x)$ and $h(x)$ are densities. One obtains samples from $f(x)$ by sampling from $g(x)$ or $h(x)$ with probabilities β and $1 - \beta$, respectively. If samples from $g(x)$ are easily generated then one usually wants β to be as large as possible.

In the trapezoidal decomposition method of sampling from the standard normal density $\phi(x)$ one samples from a trapezoidal density $g(x)$ with probability β. Sampling from a symmetrical trapezoid is easy: if u_1 and u_2 are two independent uniform variates, then $c_1 u_1 + c_2 u_2$ has a trapezoid density. The largest $\beta \cong 0.9195$ can be found by solving a numerical exercise: a maximum trapezoid is to be inscribed in the graph of the standard normal density $\phi(x)$. Details of the solution and a complete trapezoidal (TR) algorithm can be found in Ahrens and Dieter [1, page 879].

Marsaglia and Bray [3] describe a similar decomposition procedure in which three uniform samples are required in 86.38% of all cases, two such samples are taken 11.07% of the time, and in the remaining 2.55% of all cases an acceptance-rejection technique or a tail method is used. Ahrens and Dieter [1] report that this algorithm has performance characteristics similar to that of the trapezoidal algorithm. Although Marsaglia and Bray use a decomposition

$$(1) \qquad \phi(x) = \beta g(x) + (1 - \beta)h(x)$$

to sample from a standard normal by sampling from the density $g(x)$ of the sum of three uniforms, they do not try to maximize β. Rather, they sample the standard score

$$Z = \{Y - E(Y)\}/\sqrt{Var(Y)}$$

where Y is the sum of three independent uniforms over $(0,1)$. In this note we develop a "sum of three" (SOT) decomposition based algorithm for sampling from the normal by finding the density function $g(y)$ of the sum of three independent not necessarily identically distributed uniforms for which β in (1) is maximized. We find that we can sample from the sum of three uniforms in nearly 97% of the cases. In section 2 we indicate our solution to the maximization problem. In section 3 we discuss the SOT algorithm.

2. MAXIMIZATION PROBLEM

In this section we consider the problem of maximizing β in (1) over all densities f of the sum of three independent uniforms. We first characterize the density f. In this direction, let X_1, X_2, X_3 be independent with X_i uniformly distributed over $(-\alpha_i, \alpha_i)$ where $0 < a = \alpha_1 \leq b = \alpha_2 \leq c = \alpha_3$. Although somewhat tedious, one can use standard convolution theory to show that the density function $f = f_X$ of $X = X_1 + X_2 + X_3$ is symmetric about 0 and for $x \leq 0$

(2) $d \cdot f(x) =$
$$\begin{cases} 0 & \text{if } x < -d_4 \\ (x + d_4)^2 & \text{if } -d_4 \leq x \leq -d_3 \\ 4a(x + b + c) & \text{if } -d_3 \leq x \leq -d_2 \\ 8ab - (x + c - b - a)^2 & \text{if } -d_2 \leq x - d_1 \\ 8ab - 2x^2 - 2d_1^2 & \text{if } -d_1 \leq x \leq 0 \\ & \text{and } a + b > c \\ 8ab & \text{if } -d_1 \leq x \leq 0 \text{ and } a + b \leq c. \end{cases}$$

where $d = 16abc$, $d_1 = |a + b - c|$, $d_2 = a - b + c$, $d_3 = -a + b + c$, and $d_4 = a + b + c$. Let $\alpha = (a,b,c)$, $C = \{x \in R^3 : x_i = -1 \text{ or } x_i = +1\}$, $s(z) = (z_1 + z_2 + z_3)/2$, and $x_+ = x$ if $x \geq 0$ $x_+ = 0$ otherwise. Then another representation of f is given by

$$f(x) = \sum_{z \in C} (-1)^{s(z)}(x - z \cdot \alpha)_+^2 / d$$

can be obtained from the results in Barrow and Smith [2]. Since the derivative of $(x_+)^2$ exists everywhere and is given by $2x_+$, one can easily see that f is continuously differentiable. In the following we will write $f(x) = f(x;\alpha)$ when we want to emphasize the dependency of f on α.

For fixed $\alpha = (\alpha_1, \alpha_2, \alpha_3)$ with $0 < \alpha_1 \leq \alpha_2 \leq \alpha_3$ let $\beta(\alpha) = \sup\{z : zf(x;\alpha) \leq \phi(x) \text{ for } |x| < \alpha_1 + \alpha_2 + \alpha_3 = d_4\}$. The largest β is then $\beta^* = \sup\{\beta(\alpha) : 0 < \alpha_1 \leq \alpha_2 \leq \alpha_3\}$. Let us first consider the problem of finding $\beta = \beta(\alpha)$ for fixed α. It is not hard to show that $\beta f(x;\alpha) \leq \phi(x)$ all x with equality holding for some x. Thus $\beta(\alpha) = \min\{\phi(x)/f(x;\alpha); |x| < d_4\}$. Since $\phi(x)/f(x;\alpha)$ approaches infinity at $-d_4$ and d_4, the minimum of $H(x) = \phi(x)/f(x)$ is attained at a relative minimum. Since f is differentiable we see that

$$H'(x) = [f(x)\phi'(x) - \phi(x)f'(x)]/f^2(x)$$
$$= -\phi(x)[xf(x) + f'(x)]/f^2(x).$$

Thus $\beta(\alpha) = \phi(x)/f(x)$ for some x where

(3) $\qquad xf(x) + f'(x) = 0.$

If we examine the nature of f given in (2) we see that $xf(x) + f'(x)$ is a polynomial of degree three or less on each of the intervals $[-d_i, -d_{i-1}](d_0=0)$. Thus for $x \leq 0$ there are at most 12 solutions to (3). Let us call x a "proper solution" to (3) if x satisfies (3) for the functional form of f on $[-d_i, -d_{i-1}] = I_i$ and $x \in I_i$ some $i = 1,2,3,4$. $\beta = \beta(\alpha)$ is then

$\beta = \min\{\phi(x)/f(x), x \text{ is a proper solution}\}.$

Since the number of proper solutions is at most 12 (actually its smaller) and the zeros of a cubic can be found (at least theoretically) by radicals, one can find $\beta(\alpha)$. Three of the four cubics can be solved by factoring. The fourth required numerical methods. The four equations and solutions are given in Table 1.

We have not been able to analytically find an α which maximizes $\beta(\alpha)$. However, we have been able to show a maximizing α must lie in a bounded region as follows. First observe that $f(0) = 1/2c - k^2$ for some $k \geq 0$. If $c > 1.5$ then $f(0) < 1/3$. Thus $f(x) \leq \phi(x)$ on $[-1,1]$. This leads to $\beta(\alpha) \leq 0.82$. Now

$h(x) = d \cdot [xf(x) + f'(x)]$	Zeros of $h(x)$	Interval for $h(x)$	d Values
1. $x(x + d_4)^2 + 2(x + d_4)$	$-d_4, (-d_4 \pm \sqrt{d_4^2 - 8})/2$	$-d_4 \leq x \leq -d_3$	$d = 16abc$
2. $4ax(x + b + c) + 4a$	$[-b - c \pm \sqrt{(b + c)^2 - 4}]/2$	$-d_3 \leq x \leq -d_2$	$d_1 = |a + b - c|$
3. $x[8ab - (x + c - b - a)^2]$ $-2(x + c - b - a)$	Numerical Methods Used	$-d_2 \leq x \leq -d_1$	$d_2 = a - b + c$
4. $x[8ab - 2x^2 - 2d_1^2] - 4x$	$0, \pm\sqrt{4ab - d_1^2 - 2}$	$-d_1 \leq x \leq 0$ $(a + b > c)$	$d_3 = -a + b + c$
4' $8abx$	0	$-d_1 \leq x \leq 0$ $(a + b \leq c)$	$d_4 = a + b + c$

Table 1. Search for Proper Solutions ($0 < a \leq b \leq c$).

Marsaglia and Bray [3] show that $\beta(1,1,1) \cong 0.868$. Thus one cannot do better than at $\alpha = (1,1,1)$ if $c > 1.5$. Next observe that $f(0)$ is too large when c is small to get a large value of $\beta(\alpha)$. In this direction, if we write $f(0) = 1/2c - k^2$, then $k = 0$ when $c > a + b$ and $k = (a + b - c)^2/8abc$ otherwise. Thus in the case $k > 0$, since $0 < a \leq b \leq c$, we see that $0 \leq (a + b - c) \leq a \leq b$. Then $k^2 \leq ab/8abc = 1/8c$ or $-k^2 \geq -1/8c$, i.e., $3/8c \leq 1/2c - k^2 \leq 1/2c$. Thus for $c < 0.9$, $\beta(\alpha) \leq \phi(0)/f(0) \leq (1/\sqrt{2\pi})/(5/12) < 0.958$. Since one can show that $\beta(\sqrt{8} - 2, 1, 1) > 0.966$, we see that we can limit our search to c with $0.9 \leq c \leq 1.5$. Next, if $a + b + c \leq 2.1$, then this leads to $\beta(\alpha) < 0.965$. We first searched for the maximum value of $\beta(\alpha)$ o er a grid in the region $\{\alpha | 0 < a \leq b \leq c, 2.1 \leq a + b + c, 0.9 \leq c \leq 1.5\}$ in which each coordinate was partitioned in increments of 0.1. The maximum value of $\beta(\alpha)$ over this grid occurred at $(0.8, 1, 1)$ with $\beta \cong 0.965$. We then partitioned each coordinate in increments of 0.01 and searched the region around $(0.8, 1, 1)$ and found the maximizing value at $\alpha = (0.82, 1, 1.01)$ and $(0.83, 1, 1.01)$ with $\beta(\alpha) \cong .9671651$. If we examine Table 1 we see that $a + b + c = \sqrt{8}$ and $b + c = 2$ are critical values of $a, b,$ and c which separate the real and complex solutions to equations 1 and 2 of this table. This suggested that the best α might be $\alpha_0 = (\sqrt{8} - 2, 1, 1)$. This however did not turn out to be the case. However, because we could solve all equations in Table 1 by factoring with α_0 and since $\beta(\alpha_0)$ was <u>nearly optional</u> and had some computational advantages, we decided to use α_0 and $\beta(\alpha_0)$ in our sum of three algorithm discussed in section 3.

3. SOT ALGORITHM

Let us take $\alpha = (\sqrt{8} - 2, 1, 1)$. Then one can use Table 1 and the discussion in Section 2 to see that $\beta(\alpha) = \phi(x)/f(x, \alpha)$ where $x = -\sqrt{8\sqrt{8} - 22}$ is the proper solution corresponding to case 4 of Table 1. Thus $\beta = \beta(\alpha) = 16 \{\exp(11 - 4\sqrt{8})\}/\{(\sqrt{8} + 2)\sqrt{2\pi}\} = 0.9660119468$. Then in our "sum of three" (SOT) algorithm, we sample from the standard normal density $\phi(\alpha)$ by sampling the sum of three uniforms $X_1 + X_2 + X_3$ with density $f(x;\alpha)$ with probability β. In Figure 1 we display the graphs of $\phi(x)$ and $\beta f(x;\alpha)$. One can then sample from the residual density

$$h(x) = [\phi(x) - \beta f(x;\alpha)]/(1-\beta)$$

with probability $1 - \beta$. Although one can sample from the density h by a combination of a tail and an acceptance-rejection method, a glance at the graph of symmetric h given in Figure 2 suggests we might further decompose $h(x)$ as in Marsaglia and Bray [3] and use a modified tail acceptance-rejection method only at the third stage. Since speed is heavily dependent on the frequency $1 - \beta$ with which one must sample the residual density h, we believe that such a SOT algorithm would compare favorably with the "trapezoidal" and the Marsaglia-Bray algorithms. However, we have not carried out the final step in this study of comparing the speed of the SOT algorithm with that of the "trapezoidal" or Marsaglia-Bray algorithms. We feel an analytic solution (as opposed to the numerical solution for α presented here) can be found for the problem of finding α which maximizes $\beta(\alpha)$ of Section 2. Of course one can use the density of the sum of three independent uniforms given in Section 2 as a starting point in decomposing other densities.

4. ACKNOWLEDGEMENT

The author would like to acknowledge the assistance of W. D. Kaigh in deriving the formula for the density in (2).

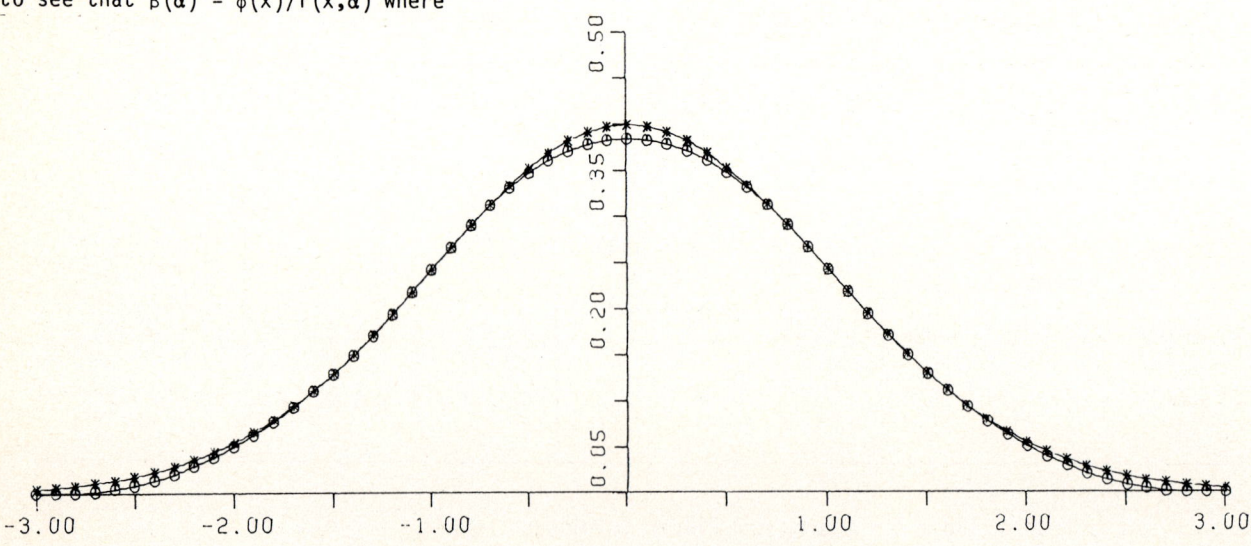

Figure 1. Comparison of $\phi(x)$ and $\beta f(x)$.

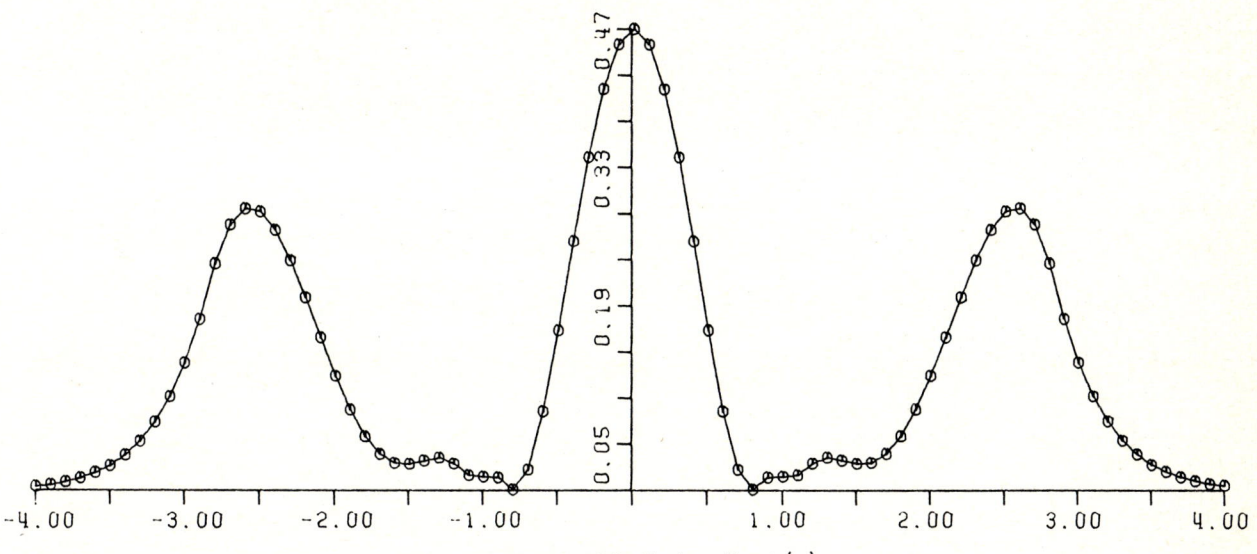

Figure 2. Residual density h(x).

REFERENCES

[1] Ahrens, J. H. and Dieter, V., Computer methods for sampling from the exponential and normal distributions, Communications of the ACM 15 (1972) 873-882.

[2] Barrow, D. L. and Smith, P.W., Spline notation applied to a volume problem, The American Mathematical Monthly, 86 (1979) 50-51.

[3] Marsaglia, G. and Bray, T. A., A convenient method for generating normal variates, SIAM Review 6 (1964) 260-263.

GRAPHPAK: INTERACTIVE GRAPHICS FOR ANALYSIS OF MULTIVARIATE DATA

Danny W. Turner[†] and Keith A. Hall[*]

[†]Department of Mathematics, Baylor University
[*]Computer Science Department, Stanford University

GRAPHPAK is a collection of FORTRAN 77 programs for generating graphical representations of multidimensional points.
GRAPHPAK produces high resolution output on Printronix 300 printers, Tektronix graphics terminals, and Calcomp pen plotters. Also, low resolution versions of all the methods allow output on any standard line printer.
The addition of a new graphical method module is relatively simple and it need not be written in FORTRAN.

1. INTRODUCTION

GRAPHPAK is a versatile set of FORTRAN 77 programs for generating graphics useful in the exploratory analysis of multivariate data. For an extensive description of GRAPHPAK, consult Turner and Hall (1983); it is essentially a user's manual for GRAPHPAK.

Section 2 contains brief descriptions for each of the graphical methods and sample output, but it is not designed to be a primer on the application of these techniques. The methods available in GRAPHPAK are used in cluster analysis, graphical representation of multivariate time series data, and outlier detection; other applications are possible. Each method represents a way to transform points in k-dimensional space to subsets of the plane.

Graphics output can be obtained on Printronix 300 printers, Tektronix graphics terminals, and Calcomp pen plotters. Also, low resolution output for all methods can be obtained on any standard line printer or hardcopy terminal.

2. DESCRIPTIONS OF THE METHODS AND SAMPLE OUTPUT

2.1 Data Set

In the following sections each of the graphical methods is described and example output is given. The emphasis is on illustrating the types of graphs that can be generated, not on presenting a complete analysis of the data. The data set used in the illustrations is described in Bruckner and Mills (1979). This data will be referred to as the oil company data. The data consists of thirteen variables measured on each of twelve cases. The variables contain information about oil leases. The cases represent oil company groups and are named Arco, Union, Getty, Mobil, Texaco, Sunoco, Chevron, Tenneco, Gulf, Amoco, Shell, and Exxon.

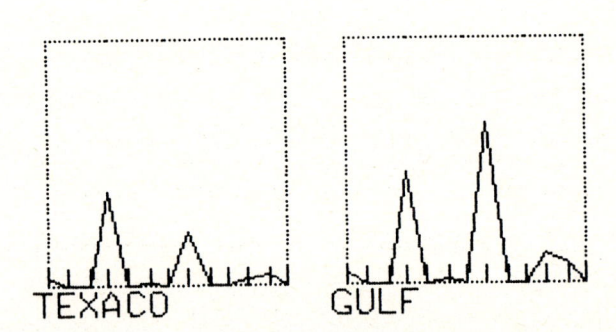

2.2 Rectangular Profiles

Suppose k variables have been selected. The k variables are given unique x-coordinates equally spaced along the interval [0,1]. (The relative variable positions are user controlled.) For each case the variable values of that case determine the corresponding y-coordinates for the rectangular profile. Thus, for each case there are k (x,y) pairs. These are plotted in rectangular coordinates and joined by line segments. Each rectangular profile is displayed in a box with tic marks along the x-axis to indicate the equally spaced x-coordinates of the variables.

Figure 1 shows rectangular profiles for selected cases in the oil data which were generated on a Printronix 300 printer. (The variable values were not scaled to the [0,1] interval.) Unless stated otherwise, assume that plots were generated on a Printronix 300. (Use of the line printer option should be obvious.)

Figure 1. Rectangular Profiles

Figure 2 shows a plot for Gulf using the line printer option. This option is available for all of the methods.

2.3 Polar Profiles (Stars)

The k variables are given unique θ coordinates equally spaced along [0,2π]. (Relative positions are user controlled.) The corresponding r-coordinates for a case are determined by its variable values. This provides k (θ,r) pairs which are plotted in polar coordinates and joined to form a k-sided polygon. Equally spaced central rays are drawn to indicate variable positions.

Figure 2. Rectangular Profile (Line Printer Option)

Figure 3 displays polar profiles for selected oil companies.

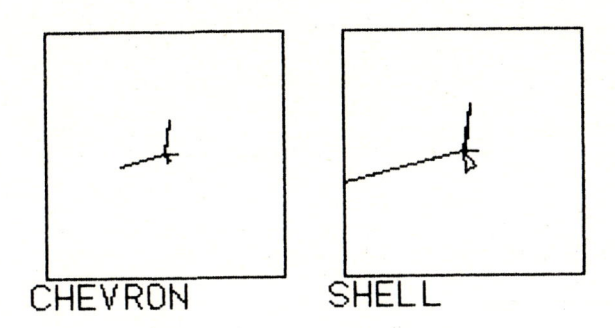

Figure 3. Polar Profiles (no [0,1] scaling)

In Figure 3 the variables were not scaled to [0,1] while Figure 4 shows selected polar profiles when the variables are scaled to [0,1].

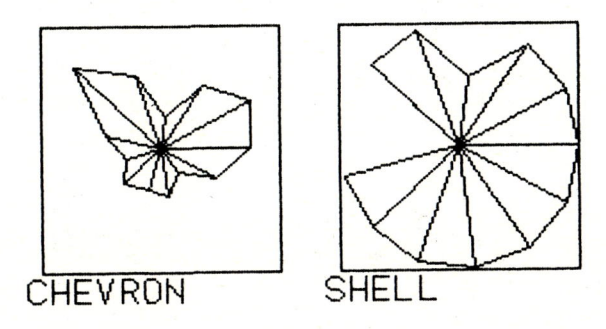

Figure 4. Polar Profiles (variables scaled linearly to [0,1]

2.4 Andrews Function Plots (Rectangular Coordinates)

Andrews (1972) devised the following scheme. Let the data for a case be represented by the vector $\underline{x} = (x_1, x_2, \ldots, x_k)$ where the x_i are variable values in some user defined order. For $-\pi \leq t \leq \pi$, define function $f_{\underline{x}}$ by the finite Fourier series $f_{\underline{x}} = \frac{x_1}{\sqrt{2}} + x_2 \sin t + x_3 \cos t + x_4 \sin 2t + x_5 \cos 2t + \ldots$, using as many terms as necessary. The graph used to represent $\underline{x}$ is obtained by plotting the graph of $f_{\underline{x}}$ in rectangular coordinates. Again the graph is enclosed in a box with cross hairs added to aid visual comparisons.

Figure 5 represents Andrews rectangular Fourier plots for selected cases in the untransformed oil company data.

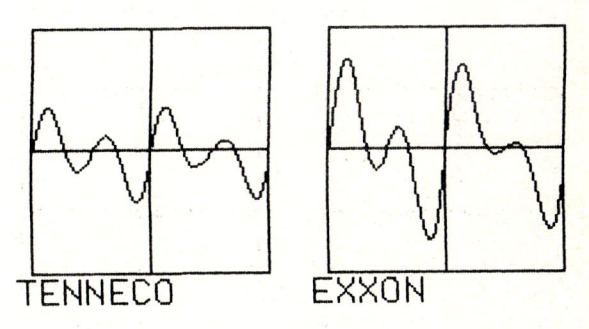

Figure 5. Andrews Rectangular Fourier Plots

Figure 6 illustrates superimposed and enlarged plots; the display box has been truncated to save space.

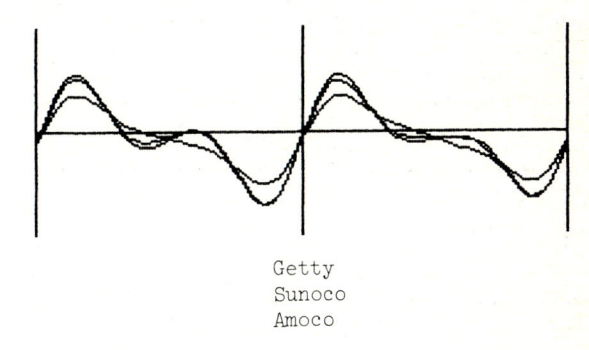

Figure 6. Superimposed and Enlarged Rectangular Fourier Plots

Figure 7 shows the same plots as Figure 5 with width = 2 and height = 1. The ability to control the width to height ratio is available for all the methods except line printer specific faces.

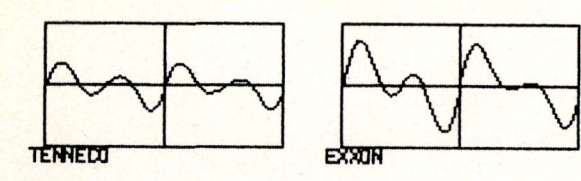

Figure 7. Andrews Rectangular Fourier Plots (2x1 format)

2.5 Andrews Function Plots (Polar Coordinates)

The situation is similar to the previous section, but the points $(t, f_x(t))$, are plotted in polar coordinates. Broken segments on a graph indicate negative f_x values.

In Figure 8 are polar Fourier plots for selected cases in the untransformed oil company data.

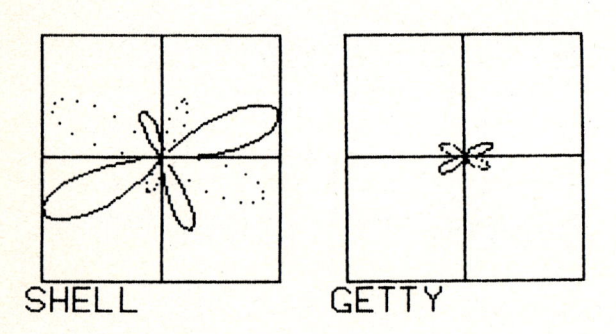

Figure 8. Polar Fourier Plots

Figure 9 shows an enlarged plot for case Getty; the display box has been truncated.

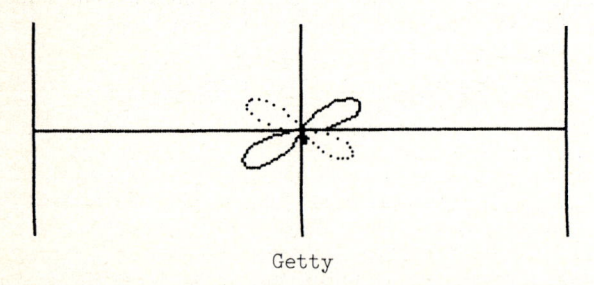

Figure 9. Enlarged Polar Fourier Plot

2.6 Printer Faces

Motivated by Chernoff (1971), Turner and Tidmore (1977) introduced a facial representation that is especially designed for output on any standard line printer or hardcopy device. Herein we refer to the Turner and Tidmore face as the line printer specific face. One innovation of Turner and Tidmore (1977) was the use of asymmetric faces. The line printer specific face has twelve features that can be controlled by the variables; these features are described in Turner and Hall (1983).

Line printer specific faces for a subset of the oil company cases are depicted in Figure 10.

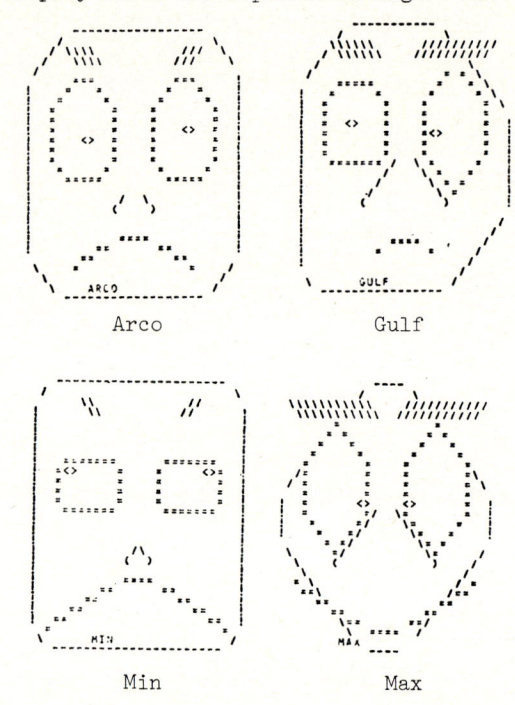

Figure 10. Line Printer Specific Faces

Figure 10 also shows two faces called MIN and MAX. MIN shows each feature mapped to its extreme low value and MAX shows each feature mapped to its extreme high value. For more on faces consult Tidmore and Turner (1983).

2.7 Modified Chernoff Faces

The modified Chernoff face in GRAPHPAK uses ideas from Chernoff (1971) and Bruckner (1978). The reader should consult Turner and Hall (1983) for more extensive geometric documentation. The modified Chernoff face is symmetric and has 20 controllable features.

Modified Chernoff faces for two of the oil companies and for the extreme faces MIN and MAX appear in Figure 11.

Figure 11. Modified Chernoff Faces

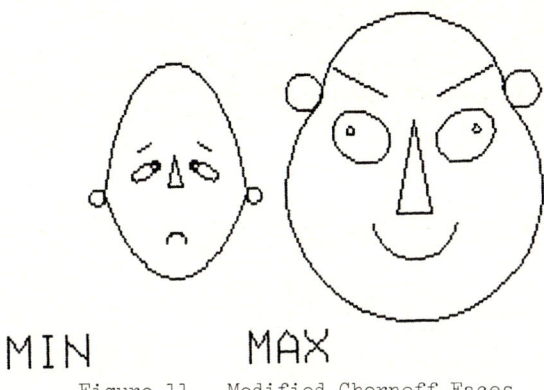

Figure 11. Modified Chernoff Faces (Continued)

Figure 12 shows MIN and MAX using the standard line printer output device option.

MIN   MAX

Figure 12. Extremes MIN and MAX for Modified Chernoff Faces Using the Standard Line Printer Output Option

2.8 Flury and Riedwyl Faces

The most recent descendant of the original Chernoff face was introduced by Flury and Riedwyl (1981). Complete geometric details are given in Flury (1980); this is the geometry upon which the corresponding GRAPHPAK program was based. Flury and Riedwyl produced the first totally asymmetric facial representation containing thirty-six features (eighteen left/right pairs); consult Turner and Hall (1983) for details.

Flury and Riedwyl faces for UNION and CHEVRON and for MIN and MAX appear in Figure 13. Cases UNION and CHEVRON were produced by a CalComp Plotter. Notice that much (perhaps too much) information has been packed into the eye group.

3. ON THE USE OF GRAPHPAK

If GRAPHPAK is installed on a VAX/VMS system, then the VAX/VMS command procedure program provided will interactively instruct the user.

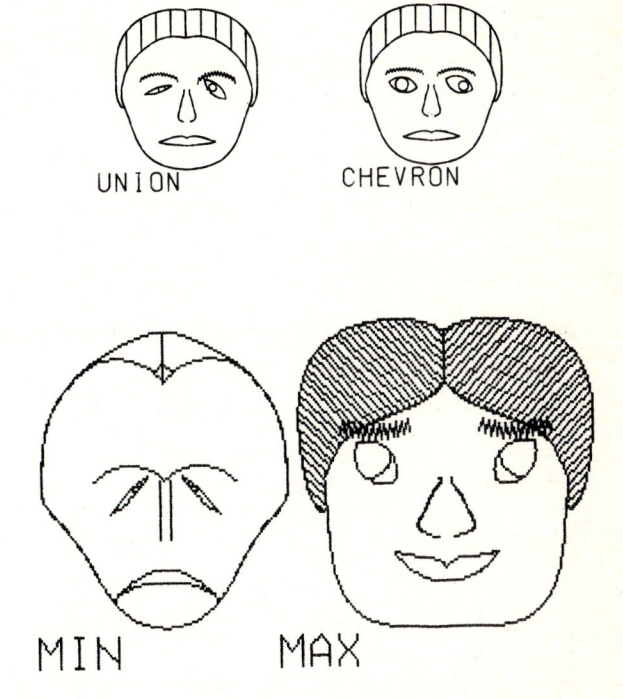

Figure 13. Flury and Riedwyl Faces. Cases UNION and CHEVRON Produced Using the CalComp Plotter Option. MIN and MAX are enlarged.

For non VAX/VMS users there are two alternatives. The first is to write a command procedure similar to the VAX/VMS command procedure program supplied. Another alternative is for the user to control the operations by running certain programs in the proper order.

Certain data files are required to implement GRAPHPAK. The basic data file contains the case names and variable values for each case. Currently the programs have a limit of 200 cases for a basic data file.

The data scaling and variable to feature assignment file, called A.Dat here, contains option selections concerning available data transformations. If transformations beyond these are required, use a statistics package such as BMDP or SAS. A.Dat also contains the variable to feature mapping. Currently, fifty variables is the maximum allowed.

The plot positioning and scaling file, called P.Dat say, gives the user control over how the final plots will appear on the page (screen). One option allows specification of a width factor and a height factor to control both the absolute size and height to width ratio of the plots. Also, the position on the page (screen) of the plot for each selected case is user specified.

The information in these files can be entered

by giving the file name when prompted, or it can be entered interactively.

4. COMMENTS ON INSTALLING GRAPHPAK

There are ten FORTRAN modules that must be operational. These programs are operational on a VAX/VMS 11/780 computer and are written (with minor exceptions) in FORTRAN 77. Adaption of the programs for another version of FORTRAN is not difficult for a FORTRAN programmer.

The only external plotting software subroutines called are Plots (initialization), Plot (straight line "pen" motion, dump plot buffer), and Symbol (plot annotation). Our graphics software libraries are set up so that these basic subroutine calls are CalComp compatible.

The VAX/VMS command procedure program interactively instructs the user in the generation of plots using the methods described in Section 2. Customization for systems similar to VAX/VMS will be routine. If this command procedure is not compatible with your system, you may write a command procedure or a job control program for your system that performs the same tasks, or let the user execute the individual programs as mentioned in Section 3.2.

5. ADDING A GRAPHICAL METHOD MODULE

The GRAPHPAK software is designed for the relatively simple addition of new or up-dated graphical method modules, which need not be written in FORTRAN. The new program must input a processed data file. This file contains the data after scaling and feature assignments. Once this information is read by the program, it must be used to generate the graphing information for the pictures that are to represent the cases. All graphs are constructed using polygons with vertices falling in the unit square. The file of graphing coordinates is used as an input file to programs that produce the final plotting information.

REFERENCES

[1] Andrews, D. F. (1972), "Plots of High Dimensional Data", Biometrics, 28, 125-136.

[2] Bruckner, L. A. (1978), "On Chernoff Faces", in Graphical Representations of Multivariate Data, Edited by P. C. C. Wang, 1978, Academic Press.

[3] Bruckner, L. A. and Mills, C. F. (1979), "The Interactive Use of Computer Drawn Faces to Study Multidimensional Data", Technical Report LA-7752-MS, Los Alamos Scientific Laboratory, Los Alamos, New Mexico, 87545.

[4] Chernoff, H. (1971), "The Use of Faces to Represent Points in n-Dimensional Space Graphically", Technical Report No. 71, Department of Statistics, Stanford University.

[5] Flury, B. (1980), "Construction of an Asymmetrical Face to Represent Multivariate Data Graphically", Technical Report No. 3, University of Berne, Department of Statistics.

[6] Flury, B. and Riedwyl, H. (1981), "Graphical Representation of Multivariate Data by Means of Asymmetric Faces", Journal of the American Statistical Association, 76, 757-765.

[7] Tidmore, F. E. and Turner, D. W. (1983), "On Clustering with Chernoff-type Faces", Communications in Statistics, A12, No. 4, 381-396.

[8] Turner, D. W. and Tidmore, F. E. (1977), "Clustering with Chernoff-Type Faces", Proceedings of the American Statistical Association, Statistical Computing Section, 372-377.

[9] Turner, D. W. and Hall, K. A. (1983), "GRAPHPAK: An Interactive Graphics System for Exploratory Analysis of Multivariate Data", Technical Report No. 8302, Department of Mathematics, Baylor University.

APPENDIX

For information on obtaining the GRAPHPAK software, write to: Danny W. Turner, Department of Mathematics, Baylor University, Waco, Texas, 76798 USA.

LP/PROTRAN, A PROBLEM SOLVING SYSTEM FOR LINEAR PROGRAMMING PROBLEMS

Granville Sewell

IMSL, Inc.
Houston, Texas

LP/PROTRAN is a member of the IMSL PROTRAN family of products and provides a user friendly environment for solving linear programming problems. Based on Hanson and Hiebert's Fortran package SPLP, it is ideally designed for large sparse problems.

1. Introduction

The IMSL PROTRAN family of products is designed to provide an environment for mathematical and statistical problem solving software which is much more user-friendly than the Fortran environment in which such software is usually found.

PROTRAN programs consist of mnemonic procedure names and keywords. Many keywords will default to reasonable values so that the basic usage of the procedures is very simple but flexibility is not sacrificed. PROTRAN diagnostics are informative and complete. For example, if a user provides an input or output array of the wrong type or data structure, or even an incorrectly dimensioned array, this error will be pinpointed by PROTRAN.

Furthermore, range variables are associated with all PROTRAN arrays so that, for example, if only the first N elements of an input array have been assigned values and more than N are needed by a procedure, an error message will be given, even though the array may have been dimensioned properly. The range variables associated with an output array will be set by PROTRAN so that, for example, if the array is to be printed, only that portion of the array which has been assigned output values will be printed.

This user-friendly environment is possible because PROTRAN execution includes a preprocessing stage in which the user's program is translated into Fortran, including Fortran calls to mathematical and statistical subroutines, many of which are adapted from the IMSL Library. Workspace arrays are set up automatically by the preprocessor and many other messy details normally involved with calls to Fortran library subroutines are handled by the preprocessor. Since PROTRAN statements are translated into Fortran, Fortran may also be added to any PROTRAN program so that all the flexibility of Fortran is available to the PROTRAN programmer.

In addition to the first two PROTRAN products, MATH/PROTRAN and STAT/PROTRAN [1], LP/PROTRAN, a problem solving system for linear programming problems, is now available from IMSL. The underlying Fortran subroutines on which LP/PROTRAN is based are adapted from R. J. Hanson and K. L. Hiebert's SPLP code [2], which is discussed in section 2. In section 3 a brief outline of the special features of LP/PROTRAN is given and some data fitting application examples are given in section 4.

2. The SPLP Program

SPLP employs a version of the revised simplex method [3] and is especially well suited for use with large LP applications. Most large linear programming applications lead to a sparse constraint coefficient matrix, that is, a matrix whose entries are mostly zeros. SPLP is designed to handle a sparse constraint matrix efficiently, and uses a version of J. K. Reid's subroutines for handling sparse linear programming bases [4].

SPLP allows upper and lower bounds on both the independent and dependent (constraint) variables. Simple bounds on the independent variables are handled implicitly, and do not increase the number of rows in the constraint matrix. Another feature useful for large problems is a save/restart feature which allows the partial results from one LP problem to be saved on a file and to be used to define a starting point for the calculations on a slightly modified problem. In applications, it is very often desirable to study the effect of varying some of the parameters and this can be very expensive if the calculations have to begin from an arbitrary feasible point each time.

3. LP/PROTRAN

LP/PROTRAN provides access to SPLP in the user-friendly PROTRAN environment. An LP/PROTRAN program may define a linear programming problem in one of two formats.

In the first format the variables and/or constraints are given names and the objective function, constraints and bounds are defined in terms of those names. In the second format the cost vector, constraint matrix and bounds vectors are assigned values in ASSIGN statements and the LP problem is defined in terms of those arrays. Each format is illustrated below.

A blending problem using the first input format:

```
        $       LPPROBLEM BLEND1
                OBJECTIVE
                ====
                0.025*BARLEY + 0.028*CORN + 0.022*SCREENINGS + 0.011*SALT
                ====
                CONSTRAINTS
                ====
                PROTEIN = 0.115*BARLEY + 0.086*CORN + 0.13*SCREENINGS
                FAT     = 0.02*BARLEY + 0.038*CORN + 0.03*SCREENINGS
                FIBER   = 0.06*BARLEY + 0.025*CORN + 0.07*SCREENINGS
                WEIGHT  = BARLEY + CORN + SCREENINGS + SALT
                ====
                BOUNDS
                ====
                WEIGHT .EQ. 2000.0
                PROTEIN .GE. 200.0
                FAT .GE. 54.0
                FIBER .LE. 90.0
                BARLEY .GE. 0.
                SCREENINGS .GE. 0.
                CORN .GE. 400.0
                CORN .LE. 1000.0
                SALT .EQ. 5.0
                ====
        $       LP MINIMIZE OBJECTIVE, PROBLEM=BLEND1
                TITLE = 'BLENDING PROBLEM'
        $       END
```

Output:

```
                          BLENDING PROBLEM

        OBJECTIVE=    .513400E+02

                                     ACTIVITY VALUES   REDUCED COSTS   BASIC?
        BARLEY                   1    .464999E+03       .000000E+00      *
        CORN                     2    .100000E+04      -.750000E-02
        SCREENINGS               3    .530001E+03       .000000E+00      *
        SALT                     4    .500000E+01      -.320000E-01

                                     CONSTRAINT VALUES    DUALS        BASIC?
        PROTEIN                  1    .208375E+03       .000000E+00      *
        FAT                      2    .632000E+02       .000000E+00      *
        FIBER                    3    .900000E+02      -.300000E+00
        WEIGHT                   4    .200000E+04       .430000E-01
```

The same problem using the second format:

```
        $       DECLARATIONS
                REAL SCALAR OBJ
                REAL VECTOR COST(4),WLOW(4),WHIGH(4),XLOW(4),XHIGH(4)
                INTEGER VECTOR XMASK2(4),WMASK1(4),WMASK2(4)
                REAL MATRIX A(4,4)
        $       ASSIGN A=(0.115,0.086,0.13,0.0)
               +         (0.02,0.038,0.03,0.0)
               +         (0.06,0.025,0.07,0.0)
               +         (1.0,1.0,1.0,1.0)
        $       ASSIGN COST=(0.025,0.028,0.022,0.011)
                       WLOW = (200,54,0,2000)
                       WHIGH = (0,0,90,2000)
                       XLOW = (0,400,0,5)
                       XHIGH = (0,1000,0,5)
                       WMASK1 = (1,1,0,1)
                       WMASK2 = (0,0,1,1)
                       XMASK2 = (0,1,0,1)
        $       LP MINIMIZE COST*X, CONSTRAINTS W=A*X,
                *   BOUNDS W.GE.WLOW.AND.WMASK1, W.LE.WHIGH.AND.WMASK2,
                *          X.GE.XLOW, X.LE.XHIGH.AND.XMASK2
                OBJECTIVE = OBJ
        $       PRINT OBJ
        $       END
```

The constraint coefficient matrix may be declared to be a normal full matrix, or if it is sparse it may be declared to be SPARSE. In this case, only about 2*NZ elements will be stored, where NZ is the number of nonzero matrix entries. There are two ways to define a sparse matrix in LP/PROTRAN but the easiest is through an ASSIGN statement of the following form:

$\quad\quad$ ASSIGN A(I,J) = G(I,J)

where G is a Fortran function subprogram which returns a new nonzero element each time it is called along with the corresponding row and column indices I and J. Function G might read a new nonzero element from a file each time it is called, for example. In the named variable format, the constraint matrix is always stored in sparse matrix storage mode, and about 3*NZ locations are used if there are NZ nonzero coefficients.

4. Examples

Although the L-one and L-infinity norm data fitting problems are normally solved using a version of the simplex method which is tailored to such problems, an LP program such as SPLP can be used to solve them reasonably efficiently.

4.1 L-one Cubic Polynomial Approximation to Data

The problem of minimizing $\sum_j |\sum_i \alpha_i \phi_i(x_j) - y_j|$ can be solved by solving the linear programming problem [5]:

$$\text{maximize} \quad y^T z$$
$$\text{with} \quad \phi z = 0$$
$$\text{and} \quad -1 \le z \le 1$$

where ϕ has elements $\phi_{ij} = \phi_i(x_j)$. The dual vector holds the coefficients α_i. In this example the approximating functions are $1, x, x^2, x^3$.

```
Input:

    $       DECLARATIONS
            REAL VECTOR X(40),Y(40)
            REAL MATRIX PHI(4,40)
    $       ASSIGN X(J)=0.025*J
    $       ASSIGN Y=(2.07458,1.78801,2.07549,1.97062,2.07776,2.05620,
            +       1.95497,1.67806,1.75925,1.30381,1.24794,1.43349,
            +       1.30241,1.13270,1.06444,1.88817,0.96055,1.01422,
            +       1.05751,0.79280,0.58909,0.88491,0.95494,0.98524,
            +       0.58597,0.62745,0.67972,0.64499,0.78238,0.88870,
            +       0.77650,0.94651,0.92542,1.06006,0.80028,0.93579,
            +       1.02410,1.27262,1.12627,0.93534)
    $       ASSIGN PHI(I,J) = X(J)**(I-1)
    $       LP MAXIMIZE Y*Z , CONSTRAINTS PHI*Z .EQ. 0 ,
    *          BOUNDS Z.GE.-1, Z.LE.1
            TITLE = 'L-ONE FIT TO DATA'
    $       END
```

4.2 L-infinity Cubic Polynomial Approximation to Data

The problem of minimizing
$$\max_j |\sum_i \alpha_i \phi_i(x_j) - y_j|$$
can be solved by solving the linear programming problem [6]:

$$\text{maximize} \quad y^T z^1 - y^T z^2$$

$$\text{with} \quad \begin{bmatrix} e^T & e^T \\ \phi & -\phi \end{bmatrix} \begin{bmatrix} z^1 \\ z^2 \end{bmatrix} = \begin{bmatrix} 1 \\ 0 \end{bmatrix}$$

$$\text{and} \quad z_j^1 \ge 0 \, , \, z_j^2 \ge 0$$

where ϕ has elements $\phi_{ij} = \phi_i(x_j)$. The dual vector holds the optimized norm immediately followed by the coefficients α_i.

In this example the approximating functions are $1, x, x^2, x^3$.

The L-one and L-infinity cubic polynomial approximations calculated by LP/PROTRAN are plotted in Figure 1.

```
        Input:
                $       DECLARATIONS
                        REAL VECTOR X(40),Y(40),C(80),B(6)
                        REAL MATRIX PHI(4,40),A(5,80)
                $       ASSIGN X(J)=0.025*J
                $       ASSIGN Y=(2.07458,1.78801,2.07549,1.97062,2.07776,2.05620,
                       +         1.95497,1.67806,1.75925,1.30381,1.24794,1.43349,
                       +         1.30241,1.13270,1.06444,1.88817,0.96055,1.01422,
                       +         1.05751,0.79280,0.58909,0.88491,0.95494,0.98524,
                       +         0.58597,0.62745,0.67972,0.64499,0.78238,0.88870,
                       +         0.77650,0.94651,0.92542,1.06006,0.80028,0.93579,
                       +         1.02410,1.27262,1.12627,0.93534)
                $       ASSIGN PHI(I,J) = X(J)**(I-1)
                $       FORTRAN
                        M = 4
                        N = 40
                        B(1) = 1.0
                        DO 1 I=1,M
                           B(I+1) = 0.0
                      1 CONTINUE
                        DO 2 J=1,N
                           C(J) = Y(J)
                           C(N+J) = -Y(J)
                           A(1,J) = 1.0
                           A(1,N+J) = 1.0
                           DO 2 I=1,M
                              A(I+1,J) = PHI(I,J)
                              A(I+1,N+J) = -PHI(I,J)
                      2 CONTINUE
                $       LP MAXIMIZE C*Z , CONSTRAINTS A*Z .EQ. B
                        TITLE = 'L-INFINITY FIT TO DATA'
                $       END
```

4.3 L-one Piecewise Linear Approximation to Data

In this example, the basis functions are piecewise linear "chapeau" functions so that the constraint coefficient matrix has only 2 nonzeros per column. LP/PROTRAN can take advantage of this sparsity as illustrated in the following example:

```
        Input:
                $       DECLARATIONS
                        REAL SCALAR ERROR,H
                        REAL VECTOR X(61),Y(61)
                C                                 A IS 51 BY 61 WITH 122 NONZEROS
                        REAL SPARSE MATRIX A(51,61,122)
                $       ASSIGN X(I) = (I-1)/60.0
                $       ASSIGN Y(I) = SIN(3.14159*X(I))
                $       FORTRAN
                        N = 60
                        DO 20 M=10,50,20
                           H = 1.0/M
                C                                 NONZERO ELEMENTS OF M+1 BY N+1
                C                                 CONSTRAINT MATRIX DEFINED HERE
                C                                 (SEE USER'S MANUAL FOR DETAILS)
                                .
                                .
                                .
```

```
         $        LP MAXIMIZE Y*Z, CONSTRAINTS A*Z.EQ. 0
                *                , BOUNDS Z.GE.-1, Z.LE.1
                  NCONSTRAINTS = M+1
                  NVARIABLES = N+1
                  NOREPORT
                  OBJECTIVE = ERROR
         $        PRINT H,ERROR
         $     FORTRAN
              20 CONTINUE
         $     END
```

Output:

```
         H
                         .10000
             ERROR
                         .13169
         H
                         .03333
             ERROR
                         .02618
         H
                         .02000
             ERROR
                         .00000
```

References

1. "PROTRAN: Problem Solving Software," T. J. Aird and J. R. Rice, IMSL Tech. Report 8201, Jan. 1982.

2. "A Sparse Linear Programming Subprogram," R. J. Hanson and K. L. Hiebert, Sandia National Labs. Tech. Report SAND81-0297, Albuquerque, NM, 1982.

3. "Linear Programming and Extensions," G. B. Dantzig, Princeton University Press, Princeton, NJ, 1963.

4. "Fortran Subroutines for Handling Sparse Linear Programming Bases," J. K. Reid, AERE-R8269, CSSD, AERE Harwell, Oxfordshire England.

5. "Minimization Techniques for Piecewise Differential Functions: The L-one Solution to an Overdetermined Linear System," R. H. Bartels, et al., SIAM J. Numer. Anal. 15, April 1978.

6. "On Cline's Direct Method for Solving Overdetermined Linear Systems in the L-infinity Sense," R. H. Bartels, et. al., SIAM J. Numer. Anal. 15, April 1978.

Figure 1

L_1 and L_∞ Cubic Polynomial Approximants to Data Set

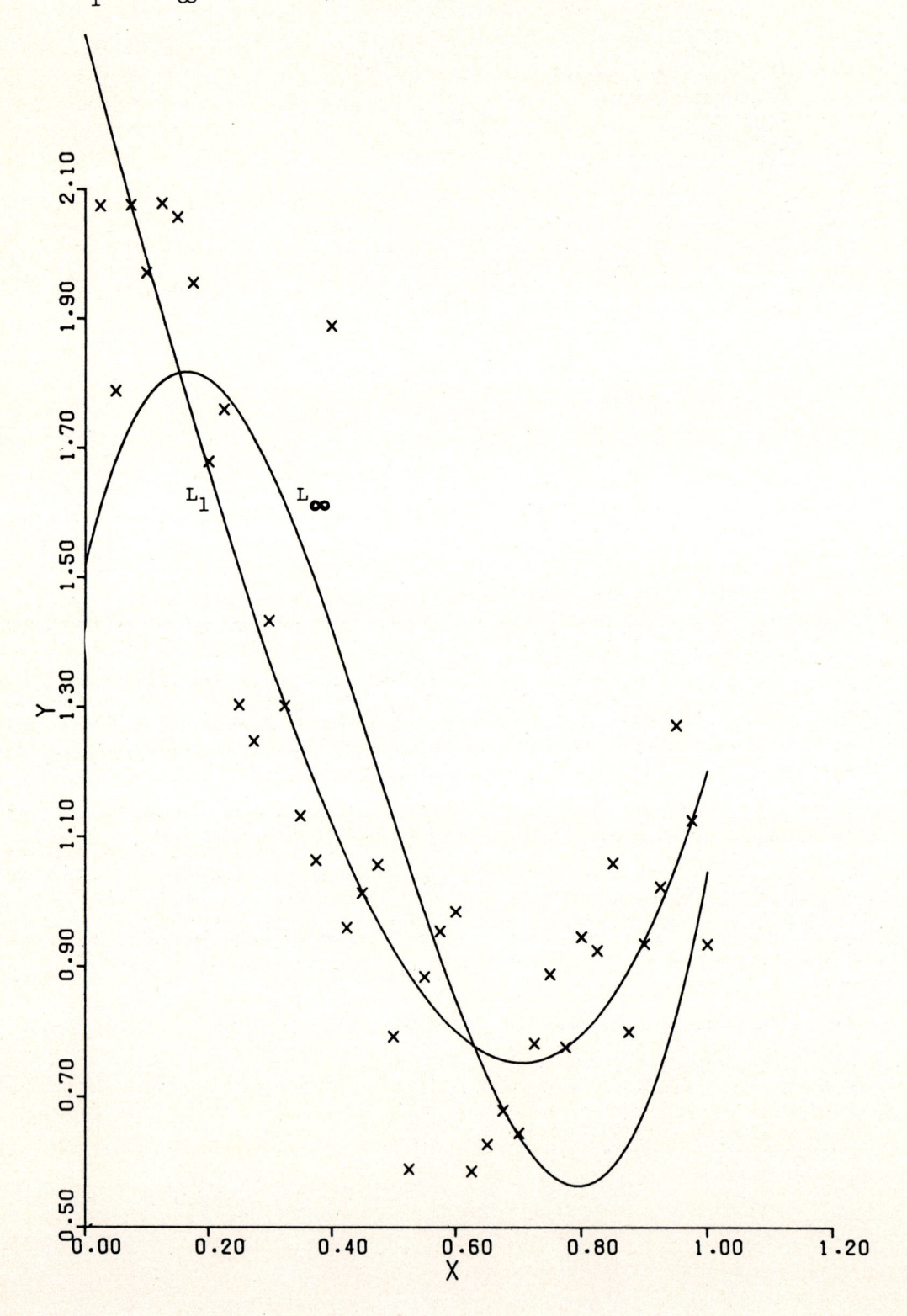

EVALUATION OF THE DOUBLY NONCENTRAL T CUMULATIVE DISTRIBUTION FUNCTION

Michele Boulanger Carey

Bell Laboratories
Holmdel, New Jersey 07733

ABSTRACT

An algorithm for the evaluation of the doubly noncentral t cumulative distribution function is described. Due to the simplicity of this new algorithm, a short routine can be written to evaluate the cdf mentionned above at any given point and for any value of the noncentrality parameters. This is a major advantage over the use of various tables that have been published for fixed (and small) values of the noncentrality parameters.

1. INTRODUCTION

There exist several algorithms in the literature to evaluate the cumulative density function of a doubly noncentral t variable (Bulgren 1968,1974; Krishnan 1968; Amos 1978). Most of them have been used to publish tables evaluating the probability integral of the doubly noncentral t distribution for fixed number of degrees of freedom and fixed (and small) values of the noncentrality parameters. The purpose of this paper is to describe a simple algorithm which will enable a small routine to be written (and stored) in order to evaluate this cdf at any given point and for any values of the noncentrality parameters and number of degrees of freedom.

In this introduction, we present some basic definitions and introduce a recursive algoritm to evaluate the mth moment of a function of a normal random variable. This algoritm simplifies greatly the evaluation of the cdf of the doubly noncentral t variable.

Let X denote a random variable which is normally distributed with mean δ and unit variance, and let U_1 be an independent random variable which follows a non-central chi-square distribution with n degrees of freedom and noncentrality parameter λ. The probability density function of X and U_1 are respectively (Searle 1971):

$$f(x) = \frac{1}{\sqrt{2\pi}} e^{-(x-\delta)^2/2} \qquad (1.1)$$

$$g_1(u_1) = e^{-\lambda} \sum_{k=0}^{\infty} \frac{\lambda^k}{k!} \frac{u_1^{n/2+k-1} e^{-u_1/2}}{2^{n/2+k} \Gamma(n/2+k)} \qquad (1.2)$$

Then:

$$T_{n,\delta,\lambda} = X/\sqrt{U_1/n} \qquad (1.3)$$

follows a doubly noncentral t distribution with n degrees of freedom and noncentrality parameters δ and λ.

Let Y follow a $N(\mu,\sigma)$; the evaluation of the mth moment of Y can be done recursively as shown in Hemmerle and Carey (1983). Similarly, $\int_0^\infty y^m f(y) dy$ can be evaluated using a recurrence formula in m as follows:

Lemma 1: Let Y follows a $N(\mu,\sigma)$; then, for any integer m:

$$\int_0^\infty y^m f(y) dy = \sigma^m B_m(\mu/\sigma) \qquad (1.4)$$

where:

$$B_m(w) = w B_{m-1}(w) + (m-1) B_{m-2}(w) \qquad (1.5)$$

$$B_0(w) = F_z(w) ;$$

$$B_1(w) = w F_z(w) + \frac{1}{\sqrt{2\pi}} e^{-w^2/2} \qquad (1.6)$$

and F_z is the cdf of the standard normal variate.

Proof:

$$\int_0^\infty y^m f(y) dy = \sigma^m \int_{z=-\mu/\sigma}^\infty \left(z+\frac{\mu}{\sigma}\right)^m \frac{1}{\sqrt{2\pi}} e^{-z^2/2} dz \qquad (1.7)$$

We will denote:

$$B_m(w) = \int_{z=-w}^\infty (z+w)^m \frac{1}{\sqrt{2\pi}} e^{-z^2/2} dz \qquad (1.8)$$

Then (1.4) is verified. If we integrate the expression (1.8) by parts, we obtain (1.5). The initial values B_0 and B_1 are simply derived.

This lemma will be used in the next chapter where the cdf of a doubly noncentral t variable is derived.

2. SERIES REPRESENTATION OF THE CDF OF A DOUBLY NONCENTRAL T VARIABLE

The evaluation of the cdf of t consists of the evaluation of $P(T \leq t)$ for all t. Three cases are studied:

2.1 t = 0

Then, according to (1.3):

$$P(T \leq 0) = P(X \leq 0) = F_z(-\delta) \text{ for all n and } \lambda \quad (2.1)$$

where F_z is the cdf of the standard normal variate Z.

2.2 t < 0

Then:

$$P(T \leq t | t < 0) = P(X < 0 \text{ and } U_1 \leq nX^2/t^2) \quad (2.2)$$

$$= \int_{x=-\infty}^{0} f(x) \int_{u_1=0}^{nx^2/t^2} g_1(u_1) du_1 dx \quad (2.3)$$

Replacing $g_1(u_1)$ by (1.2), the integral with respect to u_1 in (2.3) can be written as an infinite series, using the following series expansion of an incomplete gamma distribution (Davis 1964):

$$\int_{u=0}^{x} u^{k-1} e^{-u} du = e^{-x} \sum_{i=0}^{\infty} x^{k+i} \frac{\Gamma(k)}{\Gamma(k+i+1)} \quad (2.4)$$

So:

$$\int_{u_1=0}^{nx^2/t^2} g_1(u_1) du_1 = e^{-\lambda} \sum_{k=0}^{\infty} \frac{\lambda^k}{k!} e^{-nx^2/2t^2} \quad (2.5)$$

$$* \sum_{i=0}^{\infty} \left|\frac{nx^2}{2t^2}\right|^{n/2+k+i} \frac{1}{\Gamma(n/2+k+i+1)}$$

After checking the uniform convergence of the series in (2.5), we can switch the order of the integral and summation signs and we obtain for (2.3):

$$P(T \leq t | t < 0) = e^{-\lambda} \sum_{k=0}^{\infty} \frac{\lambda^k}{k!} \sum_{i=0}^{\infty} \left|\frac{n}{2t^2}\right|^{n/2+k+i} \quad (2.6)$$

$$* \frac{1}{\Gamma(n/2+k+i+1)} e^{-\delta^2 n/(2(n+t^2))} \frac{1}{\sqrt{1+n/t^2}}$$

$$* \int_{x=0}^{\infty} \frac{1}{\sqrt{2\pi}} \sqrt{1+n/t^2} \, x^{n+2k+2i}$$

$$* \exp\left|-\frac{1+n/t^2}{2} \left|x + \frac{t^2\delta}{t^2+n}\right|^2\right| dx$$

The integral with respect to x is now similar to $\int_0^\infty y^m f(y) dy$ where y follows a normal (μ, σ) with:

$$\mu = -t^2\delta/(t^2+n); \quad \sigma = |t| / \sqrt{t^2+n};$$

$$m = n+2k+2i \quad (2.7)$$

Using the results of Lemma 1, we obtain:

$$P(T \leq t | t < 0) = e^{-\lambda} e^{-n\delta^2/(2(n+t^2))} \frac{|t|}{\sqrt{t^2+n}} \quad (2.8)$$

$$* \sum_{k=0}^{\infty} \frac{\lambda^k}{k!} \sum_{i=0}^{\infty} \left|\frac{n}{2t^2}\right|^{n/2+k+i} \frac{1}{\Gamma(n/2+k+i+1)}$$

$$* \left|\frac{|t|}{\sqrt{t^2+n}}\right|^{n+2k+2i} B_{n+2k+2i}(w)$$

where:

$$w = \frac{|t|\delta}{\sqrt{t^2+n}} \quad (2.9)$$

After regrouping the two sums under $j=k+i$, we have:

$$P(T \leq t | t < 0) = e^{-\lambda} e^{-n\delta^2/(2(n+t^2))} \frac{|t|}{\sqrt{t^2+n}} \quad (2.10)$$

$$* \sum_{j=0}^{\infty} \left|\frac{n}{2(t^2+n)}\right|^{n/2+j} \frac{1}{\Gamma(n/2+j+1)}$$

$$* (1+\lambda+ \ldots +\frac{\lambda^j}{j!}) B_{n+2j}(w)$$

2.3 t > 0

$$P(T \leq t | t > 0) = P(X \leq 0 \text{ and } U_1 < \infty) \quad (2.11)$$

$$+ P(X > 0 \text{ and } U_1 \geq \frac{nX^2}{t^2})$$

$$= 1 - \int_{-\infty}^{0} \frac{1}{\sqrt{2\pi}} e^{-(x+\delta)^2/2} \int_{u_1=0}^{nx^2/t^2} g_1(u_1) du_1 dx \quad (2.12)$$

So:

$$P(T_{n,\delta,\lambda} \leq t | t > 0) = 1 - P(T_{n,-\delta,\lambda} \leq -t | t > 0) \quad (2.13)$$

and we can use the results of the case when t is negative.

3. COMPUTATIONS

We will restrict ourselves to $P(T \leq t | t<0)$ since the case of t positive can be treated as a particular case of t negative. The expression (2.10) can be written as:

$$P(T \leq t | t<0) = e^C \sum_{j=0}^{\infty} T_j \quad (3.1)$$

where:

$$C = -\lambda - \frac{n\delta^2}{2(n+t^2)} + \log|t| - \frac{\log(t^2+n)}{2} \quad (3.2)$$

$$T_j = \left| \frac{n}{2(t^2+n)} \right|^{n/2+j} \frac{1}{\Gamma(n/2+j+1)} \quad (3.3)$$

$$* \, (1+\ldots+\frac{\lambda^j}{j!}) \, B_{n+2j}(w)$$

We propose to evaluate T_j as a function of T_{j-1} for j greater than or equal to 1. In order to do this, we will evaluate the ratio:

$$\frac{T_j}{T_{j-1}} = \frac{n}{2(t^2+n)} \frac{1}{n/2+j} \quad (3.4)$$

$$* \left| 1+\frac{\lambda^j/j!}{1+\ldots+\lambda^{j-1}/(j-1)!} \right| \frac{B_{n+2j}(w)}{B_{n+2j-2}(w)}$$

$$= \frac{n}{2(t^2+n)} \frac{1}{n/2+j} E_j R_j \quad (3.5)$$

where:

$$R_j = B_{n+2j}(w) / B_{n+2j-2}(w) \quad (3.6)$$

$$E_j = 1 + \frac{\lambda^j/j!}{1+\ldots+\lambda^{j-1}/(j-1)!} \quad (3.7)$$

The ratio (3.4) can be easily evaluated after two remarks. Using several times the recurrence formula given in (1.5), we obtain:

$$B_{n+2j}(w) = (w^2+2n+4j-3) B_{n+2j-2}(w) \quad (3.8)$$

$$- (n+2j-2)(n+2j-3) B_{n+2j-4}(w)$$

or equivalently:

$$R_j = w^2+2n+4j-3 - (n+2j-2)(n+2j-3)/R_{j-1} \quad (3.9)$$

The second remark consists in writing E_j as a function of the previous ratio E_{j-1}:

$$E_j = 1 + \frac{\lambda \, (E_{j-1}-1)}{j \, E_{j-1}} \quad (3.10)$$

As a summary, to obtain T_j, we first obtain E_j and R_j as simple functions of E_{j-1} and R_{j-1}; then, we evaluate the ratio T_j/T_{j-1} as a function of R_j and E_j and finally obtain T_j as $T_{j-1} * (T_j/T_{j-1})$.

It remains to evaluate the initial values of these functions:

$$T_0 = \left| \frac{n}{2(t^2+n)} \right|^{n/2} \frac{1}{\Gamma(n/2+1)} B_n(w) \quad (3.11)$$

$$R_0 = B_n(w) / B_{n-2}(w) \quad (3.12)$$

$$E_0 = 1 \quad ; \quad E_1 = 1+\lambda \quad (3.13)$$

A listing of the Fortran routine we used is given at the end of this paper. We modify slightly the expression (3.1) into:

$$P(T \leq t | t<0) = \sum_{j=0}^{\infty} \exp(C+\log T_j) \quad (3.14)$$

in order to deal more efficiently with overflow problems. The program was run on a IBM 3081 D using extended precision (64 bits). We stopped the evaluation of the series when the addition of a new term contributed for less than 10^{-10} of the sum of the previous terms.

4. NUMERICAL RESULTS AND COMPARISONS

Krishnan (1968) evaluated the cdf of a doubly noncentral $T_{n,\delta,\lambda}$ variable as follows:

$$(4.1)$$

$$P(T<t) = \frac{1}{2} \sum_{j=0}^{\infty} \sum_{r=0}^{\infty} \frac{e^{-\lambda/2} \lambda^j}{2^j \Gamma(j+1)} \frac{e^{-\delta^2/2} |\delta|^r}{2^{r/2} \Gamma(r/2+1)} *$$

$$* \left| (-1)^r + \frac{H(r/2+1/2, 1-j-n/2; r/2+3/2; a) \, a^{(r+1)/2}}{\Gamma((r+3)/2) \, \Gamma(n/2+j) / \Gamma((r+n+1+2j)/2)} \right|$$

where H is the Gauss hypergeometric function.

His results of the evaluation of $P(T_{n,\delta,\lambda} \leq t)$ for n=2 (1) 20, λ=0 (2) 8, and δ=-5 (1) 5 were presented in a Table. The computations were performed on a IBM 360 using double precision. The series terminated when r and j were each run through 100 terms. Krishnan also mentioned that his series converges slowly for large values of δ and λ, and we will see that this is not necessarily the case with our algorithm.

The main difference between Krishnan algorithm (4.1) and ours (3.1) is that in the first case a doubly infinite series is evaluated while a single series is evaluated in the latter.

Similarly, Bulgren and Amos used a doubly infinite series as follows:

$$P(T_{n,delta,\lambda} < t) = 1 - F_z(\theta) + \frac{e^{(\lambda+\theta^2)/2}}{\sqrt{\pi}} \quad (4.2)$$

$$* \left| \frac{\alpha}{\sqrt{1+\alpha^2}} \sum_{m=0}^{\infty} c_m \sum_{k=0}^{\infty} a_k y_k (c=1/2) \right.$$

$$\left. - \frac{\theta}{\sqrt{2}} \sum_{m=0}^{\infty} d_m \sum_{k=0}^{\infty} b_{k+1} y_k (c=3/2) \right|$$

where θ and α are functions of t, n and δ and c_m, a_k, y_k, d_m, b_k are evaluated recursively.

Double precision was used on a IBM 7040 with a cut off point of 10^{-10}. Extensive tables using this technique were published in Selected Tables in Mathematical Statistics (1974).

Here again, a doubly infinite series of recursive terms is evaluated. The two algorithms to which we compared ours have in common that the cdf evaluation was first carried out by integrating with respect to x, then with respect to u_1:

$$P(T_{n,\delta,\lambda} \leq t) = \int_0^\infty P(X \leq t\sqrt{|u_1/n|} \mid U_1 = u_1) g(u_1) du_1 \quad (4.3)$$

while we integrated with respect to u_1 first, then with respect to x.

Another evaluation of the cdf of the noncentral t variable was done by numerical quadrature (Amos 1978). In his paper, Amos described a procedure by which cdf of symmetrical distibutions can be evaluated. The idea is to locate the mean by setting the derivative to zero and then sum quadratures for the evaluation of the cdf on intervals of length σ to the left and right of the mean until a limit of integration is reached or the truncation error is small enough. Although attractive due to its generalization to other distributions besides the noncentral t, this methodology is complicated by the search of the mean. The motivation for this search comes from the fact that the mean can vary widely with extreme parameters values as in our case with large values of δ and λ. Thus, the mean obtained and σ (which estimates the spread of the integrand) accomodate the parameters, producing meaningful results by preventing quadratures over tails which are negligible or preventing gross misjudgements of the scale of integration.

To test our algorithm, we evaluated the cdf of a noncentral t with parameters having the same large values as indicated in Amos, namely n=100, δ=100, λ=50. The results are shown in Table 1, along with N, number of items T_j that were included in the evaluation of the cdf to obtained a precision of 10^{-10}.

The results are similar to the ones of Amos up to the number of digits displayed (6) but as is indicated by the number of items in our sum, our algorithm converged very quickly in that particular case.

Table 1: cdf of a doubly noncentral t with n=100, δ=100, λ=50

P(T<t)	t	N
.050	6.4222E+01	149
.100	6.5592E+01	143
.200	6.7322E+01	135
.300	6.8621E+01	130
.400	6.9766E+01	126
.500	7.0868E+01	122
.600	7.2001E+01	118
.700	7.3248E+01	114
.800	7.4758E+01	110
.900	7.6945E+01	104
.950	7.8838E+01	100

As a summary, out of the three algorithms that we found in the literature to evaluate the cdf of a doubly noncentral t variable, two involve the evaluation of doubly infinite series of recursive terms when we only deal with one infinite series of recursive terms, and the last one is a numerical procedure which although applicable to other symmetrical distributions besides the noncentral t, is complicated by the search of the root of a complicated derivative function.

5. CONCLUSIONS

We presented a short algorithm to evaluate the probability integral of the doubly noncentral t variable. The routine listed below can be used in lieu of various tables published in the literature and offers the advantage that results can be obtained for any values of the noncentrality parameters and at any point where the cdf has to be evaluated. The same approach has been used to evaluate the cdf of a noncentral F variable with one or two noncentrality parameters and the results will be published in a following paper.

6. ACKNOWLEDGMENT

The author would like to thank Dr. William J. Hemmerle, University of Rhode Island, for his valuable comments.

REFERENCES

[1] D.E. Amos, Evaluation of some cumulative distribution function by numerical quadrature, SIAM Review 20(4) (1978), pp 778-800.

[2] W.G. Bulgren, Probability integral of the doubly noncentral t-distribution with degrees of freedom n and noncentrality parameters δ and λ, Selected Tables in Mathematical Statistics Vol 2 (1974), pp 1-12.

[3] W.G. Bulgren and D.E. Amos, A note on representations of the doubly noncentral t distribution, J. Amer. Statist. Assoc., 62.(1968), pp 1012.1019

[4] P.J. Davis, Handbook of Mathematical Functions (M. Abramovitz and I.A. Stegun, National Bureau of Standards, 5th printing) pp 252.294.

[5] W.J. Hemmerle and M.B. Carey, Some properties of generalized ridge estimators, Commun. Statist., Vol B12, No.3 (1983).

[6] N. Krishnan, Series representations of a doubly noncentral t distribution, J. Amer. Statist. Assoc., 63 (1968) pp 1004-1012.

[7] S.R. Searle, Linear Models, (Wiley 1971) pp 49-52.

ELEMENTARY METHODS FOR APPROXIMATING THE CUMULATIVE DISTRIBUTION OF
BETA, F, T, AND NORMAL VARIABLES

Dean Fearn

Statistics Department, California State University, Hayward
Hayward, California

Evaluation of the distribution of the beta random variable is by means of sums areas of rectangles and by Simpson's rule is treated. This is used as a basis for obtaining the distribution of F and the accuracy is checked against known values in tables. The evaluation of the t-distribution is done using Simpson's rule and checked against known values in the t-table. This article ends with a brief discussion of the rate of convergence for a standard continued fraction approach to obtaining the normal distribution.

In this paper the computation of the distribution of a beta random variable U, F, Student's T, and the standard normal random variable Z using rectangle quadrature and Simpson's rule will be discussed. An indication will be given as to the number of intervals required for four place accuracy. A brief treatment of the computation of the norming constants for all of the distributions will be given. A brief indication of other well-known analytical approaches to obtaining the distribution of T and Z is given.

In attempting to obtain any of the above distributions by quadrature a crucial first step is to find a very accurate approximation to the gamma function:

$$\Gamma(x) = \int_0^\infty u^{x-1} \exp(-u) du, \quad x > 0.$$

A method which gives good results along this line is to use Stirling's approximation:

$$\hat{\Gamma}(x) = \exp(-x) x^{x-.5} (2\pi)^{.5} [1 + 1/(12x) + 1/(288x^2) - 139/(51840x^3) - 571/(2488320x^4)]$$

for $x \geq 6$. For $0 < x \leq 6$ use the above formula to compute an approximation $\hat{\Gamma}(x+6)$ to $\Gamma(x+6)$. Then $\hat{\Gamma}(x) = \hat{\Gamma}(x+6)/[(x+5)(x+4)(x+3)(x+2)(x+1)x]$.

This recipe yields $\hat{\Gamma}(x)$ to about 7 or 8 significant figures. A brief account of the analysis behind this formula can be found in [1]. A more thorough account of the analysis along with a fairly thorough bibliography can be found in [3]. For example here is a table comparing $\hat{\Gamma}(x)$ to $\Gamma(x)$:

Table 1

x	$\Gamma(x)$	$\hat{\Gamma}(x)$
.5	1.7724539	1.7724537
1	1	.99999995
5	24	23.9999999
10	362880	362879.995
20	1.2164510041E17	1.21645099E17

In the case of T or F one only needs $\Gamma(x)$ for x a positive integer multiple of $1/2$. For x an odd positive multiple of $1/2$ one uses $\Gamma(x) = (x-1)\Gamma(x-1)\ldots$ when $x > \frac{1}{2}$ and $\Gamma(\frac{1}{2}) = \sqrt{\pi}$. For x an even positive multiple of $\frac{1}{2}$ $\Gamma(x) = (x-1)!$.

Now the computation of the cumulative distribution of a beta random variable U is considered. The density for U is $u^{a-1}(1-u)^{b-1}/B(a,b)$, where $B(a,b) = \Gamma(a)\Gamma(b)/\Gamma(a+b)$. The main difficulty encountered when attempting to numerically compute $F_U(x) = \int_0^x u^{a-1}(1-u)^{b-1}/B(a,b)du$ is when $0 < a < 1$. This is because $u^{a-1} \uparrow \infty$ as $u \downarrow 0$. One can avoid this problem by using the fact that

$$F_U(x) = 1 - \int_x^1 u^{a-1}(1-u)^{b-1}/B(a,b)du$$

so long as $b \geq 1$. The main case where the beta distribution with $0 < a < 1$ and $0 < b < 1$ can be in statistical applications is when $a = \frac{1}{2}$ and $b = \frac{1}{2}$ where the beta distribution is a transformation of the T distribution with 1 degree of freedom. There are continued fraction representations for $F_U(x)$ available in [1]. For the beta distribution $B(a,b)$ was approximated using Stirling's approximation for the gamma function. Call this approximation $\hat{B}(a,b)$. $F_U(x)$ was approximated using the sum of areas of rectangles:

$$\sum_{i=1}^{n} u_i^{a-1} (1-u_i)^{b-1} \, h \text{ where } h=x/n$$

and $u_i = ih - \frac{1}{2}h$. In general u_i could be defined as $ih-ch$ where $0 \leq c \leq 1$. Then

$$\hat{F}_U(x) = h/\hat{B}(a,b) \left[\sum_{i=1}^{n} u_i^{a-1} (i-u)^{b-1} \right]$$

Also $F_U(x)$ was approximated using Simpson's rule:

$$F_U(x) = (h/3\hat{B}(a,b)) \left[4\sum_{i=1}^{n} u_{2i-1}^{a-1}(1-u_{2i-1})^{b-1} + 2\sum_{i=2}^{n} u_{2i-2}^{a-1}(1-u_{2i-2})^{b-1} + u_{2n}^{a-1}(1-u_{2n})^{b-1} \right],$$

$h = x/2n$ and $u_j = jh$ for $j=1, 2, \ldots, 2n$.

Observe that the initial term $u_0^{a-1}(1-u_0)^{b-1}$ is missing since this equals zero for $u_0 = 0$. In case $a=1$, this term can be replaced by 1 with satisfactory results. Here is a little table of the results obtained:

a	b	#int.	$\hat{F}_U(.7)$ rect.	$\hat{F}_U(.7)$ Simp.
1	1	2	.700000046	
.5	.5	100	.61488477	.615214934
5	7	100	.978392691	.978380896
10	20	100	.999982994	.999982893
20	10	100	.63602216	.635995974
.5	.5	10	.580095214	.580896117
5	7	10	.979552324	.978249971
10	20	10	.999988251	.999625074
20	10	10	.638697769	.636268291
.5	.5	200	.619600475	.619837996
5	7	200	.97838384	.978380891
10	20	200	.99998292	.999982895
20	10	200	.636002501	.635995946

The F distribution with n degrees of freedom in the numerator and m degrees of freedom in the denominator was approximated using the fact that if $U=m/(m+nF)$ the U has the beta density $u^{m/2-1}(1-u)^{n/2-1}/B(\frac{m}{2},\frac{n}{2})$ whose cumulative distribution evaluated at $U=m/(m+nF)$ was approximated with the same algorithm as is indicated above. Below is a table which allows the reader to assess the accuracy of this approach and also the algorithm for the beta distribution. In the table below $F_{.05}$ is the upper 5% point of the F distribution.

n	m	#inter	$F_{0.05}$	$\hat{F}_F(F_{0.05})$ rectangles	$\hat{F}_F(F_{0.05})$ Simpson
1	1	10	161.4	.95476247	.954669352
3	7	10	4.35	.950232651	.950079922
30	20	10	2.04	.950672485	.950110189
20	40	10	1.84	.952382071	.950157245
1	1	100	161.4	.951903505	.951471487
3	7	100	4.35	.950079985	.950078448
30	20	100	2.04	.950109672	.950104001
20	40	100	1.84	.95019333	.950171228
1	1	200	161.4	.951061011	.951038343
3	7	200	4.35	.950078832	.950078448
30	20	200	2.04	.95010542	.950104002
20	40	200	1.84	.950176757	.950171232

As can be seen from the above table, there is hardly any improvement in accuracy going from 100 intervals to 200 intervals. Perhaps some improvement may be obtained by using a direct computation of the norming constants rather than Stirlings approximation. In fact this approach was taken in the approximation for the distribution of Student's T. The density for Student's T is

$$\Gamma((n+1)/2) / \left[\sqrt{n\pi}\, \Gamma(n/2)(1+\frac{t^2}{n})^{\frac{n+1}{2}} \right]$$

where n is the degrees of freedom. Here is a short table of the results:

n	#inter	$t_{.025}$	$\hat{F}_T(t_{.025})$ Simpson
1	10	12.706	.953843421
5	10	2.571	.97501084
10	10	2.228	.974992455
1	100	12.706	.974999598
5	100	2.571	.975012682
10	100	2.228	.974994113
1	200	12.706	.974999599
5	200	2.571	.975012681
10	200	2.228	.974994114

Plainly 100 intervals are enough for most practical purposes. Also it is apparent that a good approach to approximating the F-dist^n with 1 degree of freedom in the numerator is to use the fact that $1-F_F(f) = 2(1-F_T(\sqrt{f}))$ and then use Simpson's rule to approximate $F_T(\sqrt{f})$.

Although Simpson's rule would work quite well in approximating the normal distribution function a continued fraction approach

$$1-F_Z(Z) = (1/\sqrt{2\pi}) \exp(-Z^2/2) \left[\frac{1}{Z+}\, \frac{1}{Z+}\, \frac{2}{Z+}\, \frac{3}{Z+}\, \frac{4}{Z+}\cdots\right]$$

(see [2]) converges quite rapidly to the correct answer as long as $Z \geq 1$. For $Z = 1.96$, the following results were obtained using this approach

Length of continued fraction	$1-\hat{F}_Z(1.96)$
1	.0236583466
2	.0254586453
3	.0248178566
4	.025075084
5	.0249624405
10	.0249992913

For Z as low as .3 the convergence is poor. [1] has a good rendition of the algorithm for computing continued fractions.

The numbers in the above tables were produced by short programs written by the author in BASIC and run on an APPLE II plus microcumputer.

References

[1] Abramowitz, M. and Stegun, I.A. (eds), Handbook of Mathematical Functions (Dover Pub., New York, 1972).

[2] Kendall, M.G. and Stuart, A.S., The Advanced Theory of Statistics vol.1 (Griffin, London, 1963).

[3] Whittaker, E.T. and Watson, G.N., A Course of Modern Analysis 4th ed. (Cambridge Univ. Press, Cambridge, 1963)

After the decision maker assigns a value v_i to each of the bundles, the value function is estimated by ordinary least squares using the orthogonal design matrix, i.e.,

$$\hat{V} = \hat{\beta}_0 + \hat{\beta}_1 X_1 + \hat{\beta}_2 X_2 + \hat{\beta}_3 X_1 X_2 + \hat{\beta}_4 X_1^{*2} + \hat{\beta}_5 X_2^{*2} \quad (1)$$

where the quadratic terms are adjusted for orthogonality. After estimation, the coefficients are scaled such that their units are consistent with the units of the defined bundles. It is this quadratic approximation of the "true" value function that is used in the optimization procedure.

4. EXAMPLE

As an example of how the methodology might be applied, consider the blood bank inventory problem presented by Keeney [8]. The manager of a blood bank must choose an ordering policy for whole blood. Many attributes influence the managers decision, for example, "the percent or number of units of outdated blood, the percent of unsupplied demand, the age of transfused blood, and the total blood intake of the blood bank." Two salient attributes of interest are blood shortage and blood outdating. Shortage is blood that is not available in inventory when requested, and outdated blood is blood that has been in inventory longer than its legal lifetime. Define

X_1 = percent shortage of the total units stocked during a year,

X_2 = percent outdated of the total units demanded during a year.

For this problem it was determined that shortage and outdating would never exceed 10%. Therefore, the situation presented in Tables 1-3 is appropriate. Since the head nurse controls the ordering policy, the objective is to select an ordering policy that maximizes the nurse's utility as a function of the defined attributes. The nurse's utility function was represented by the following expression:

$$U(X_1, X_2) = .32(1-e^{.13X_1}) + .57(1-e^{.04X_2})$$
$$+ .107(1-e^{.13X_1})(1-e^{.04X_2}). \quad (2)$$

The function defined in (2) is assumed to be the decision maker's "true" utility function for comparison to the quadratic approximation.

The assumption is made that there exists practical lower bounds that preclude the achievement of both zero percent shortage and outdating. These lower bounds may be used to define a constraint set for the problem. For example, assume that 2% shortage and 3% outdating represent the lowest attainable attribute levels. It is reasonable to assume that outdating increases as shortage decreases since the blood bank must increase inventory to reduce shortage. If this relationship is assumed to be linear, a linear constraint set would result. In the absence of information concerning the magnitude of the lower bounds, several constraint sets were randomly generated for this example. The four problems considered are:

Maximize $\hat{U}(X_1, X_2)$ \quad (3)

Subject to: $7X_1 + X_2 \geq 4$

$3X_1 + X_2 \geq 3$

$X_1, X_2 \geq 0$,

Maximize $\hat{U}(X_1, X_2)$ \quad (4)

Subject to: $4X_1 + X_2 \geq 3$

$8X_1 + X_2 \geq 5$

$X_1, X_2 \geq 0$,

Maximize $\hat{U}(X_1, X_2)$ \quad (5)

Subject to: $2X_1 + X_2 \geq 2$

$X_1 + 2X_2 \geq 2$

$X_1, X_2 \geq 0$,

and

Maximize $\hat{U}(X_1, X_2)$ \quad (6)

Subject to: $3X_1 + X_2 \geq 3$

$5X_1 + 2X_2 \geq 4$

$.8X_1 + 4X_2 \geq 1$,

$X_1, X_2 \geq 0$

where $\hat{U}(X_1, X_2)$ is the quadratic approximation of the "true" utility function presented in (2).

Using the methodology in [12], the approximated function is estimated as

$$\hat{U}(X_1, X_2) = .00472053 - .03626965 X_1$$
$$- .02443452 X_2 + .00140472 X_1 X_2$$
$$- .00493812 X_1^2 - .00043347 X_2^2 \quad (7)$$

with the regression diagnostics presented in Table 4.

Table 4

Regression Diagnostics

Estimate	T for $H_o:\beta_i=0$	Standard Error	R^2
$\hat{\beta}_0$	-189.32	.0026	.999
$\hat{\beta}_1$	124.66	.0032	
$\hat{\beta}_2$	34.48	.0032	
$\hat{\beta}_3$	9.09	.0039	
$\hat{\beta}_4$	-22.60	.0055	
$\hat{\beta}_5$	-1.98	.0055	

Additional residual analysis confirmed model adequacy. A normal probability plot and histogram do not indicate any serious departures from normality. This was confirmed by applying the Shapiro-Wilk W-test [13] for normality ($p^*<.965$). It is difficult to make any definitive conclusions about serial correlation for small samples, but additional graphical analysis does not indicate any serious violations.

5. RESULTS

For comparison purposes the nonlinear function (2) is optimized over the four constraint sets. The complex method as presented by Box [2] is used to optimize the "true" function (2), while the complementary pivot algorithm of Lemke [10] is used to optimize the approximation (7). The results of the optimizations are presented in Table 5.

Table 5

Optimization Comparisons

Problem	Actual $U^*(X_1,X_2)$	Approximated $U^*(X_1,X_2)$
1	$U^*(.862,.426)=$ -.0475	$U^*(1,0)=$ -.0444
2	$U^*(.734,.185)=$ -.0362	$U^*(.75,0)=$ -.0327
3	$U^*(.454,.001)=$ -.0449	$U^*(.667,.667)=$ -.0441
4	$U^*(.862,.426)=$ -.0475	$U^*(.982,.054)=$ -.0448

The optimal values of the decision variables in each case are substituted into the assumed "true" utility function (2). Since the function is scaled between -1 and 0, the approximated function compares favorably in each case.

6. CONCLUSION

It must be stressed that the objective of this paper is optimization. No claims are made concerning a new methodology for assessing utility or value functions. With respect to ease of optimization, the procedure defined in this research performs extremely well.

Of course, the success of this methodology is ultimately an empirical question. The standard of comparison should be goal programming since goal programming is a linear approximation of the "true" utility or value function. Any improvement in the form of the approximation (quadratic versus linear goal programming) must be weighted against the increased difficulty in assessing the approximation.

Further testing on "real world" problems is required, but the preliminary results are optimistic.

REFERENCES:

[1] Box, G. E. P. and Wilson, K. B., On the attainment of optimal conditions, Jrnl. Royal Stat. Soc. (Series B). 13 (1951), 1-38.

[2] Box, M. J., A new method of constrained optimization and a comparison with other methods, Computer Jrnl. 8 (1965), 45-52.

[3] Charnes, A. and Cooper, W. W. Management models and the industrial applications of linear programming (John Wiley, New York, 1961).

[4] Dyer, J. S., On the relationship between goal programming and multiattribute utility theory, Discussion paper no. 69, Management science study center, Graduate school of management, University of California at Los Angeles (October, 1977).

[5] Fishburn, P. C., Approximation of two attribute utility functions, Math. Op. Res. 2 (1977), 30-44.

[6] Harrald, J., Leotta, J., Wallace, W. A., and Wendell, R. E., A Note on the limitations of goal programming as observed in resource allocation for marine environmental protection, Naval Res. Logistics Quart. 25 (1978), 733-739.

[7] Hwang, C. L. and Masud, A. S. M., Multiple objective decision making: methods and applications (Springer-Verlag, New York, 1979).

[8] Keeney, R. L., An illustrated procedure for assessing multiattributed utility functions, Sloan Mgt. Rev. 14 (1972), 37-50.

[9] Keeney, R. L. and Raiffa, H., Decisions with multiple objectives: preferences and value tradeoffs (John Wiley, New York, 1976).

[10] Lemke, C. E., Bimatrix equilibrium points and mathematical programming, Mgt. Sci. 11 (1965), 681-689.

[11] Pratt, J. W., Risk aversion in the small and in the large, Econometrica 32 (1964), 122-136.

[12] Ringuest, J. L. and Gulledge, T. R., Jr., A preemptive value-function method approach for multiobjective linear programming problems, Decision Sci. 14 (1983), 76-86.

[13] Shapiro, S. S. and Wilk, M. B., An analysis of variance for normality (complete samples), Biometrika, 52 (1965), 591-611.

[14] Shim, J. K. and Siegel, J., Quadratic preferences and goal programming, Decision Sci. 6 (1975), 662-669.

[15] von Neumann, J. and Morgenstern, O., Theory of games and economic behavior, 2nd ed. (Princeton Univ. Press, Princeton, 1947).

L_1 NORM ESTIMATES USING A DESCENT APPROACH

Lee Ann Josvanger and V. A. Sposito

Iowa State University
Ames, Iowa

Obtaining L_1 norm estimators for the general linear model, $y = X\beta + e$, are commonly formulated and solved as a linear programming problem. Barrodale and Roberts in 1973 developed an extremely fast computational-simplex approach to solve the general L_1 regression problem. Armstrong and Kung in 1978 presented a specialization of Barrodale and Roberts procedure to obtain L_1 estimates efficiently for the simple linear model. This paper will consider an alternative approach in solving the L_1 problem by considering a descent procedure.

1. INTRODUCTION

Let (x_i, y_i), $i=1,2,\ldots,n$ be n given pairs of data values. Consider the problem of determining estimates of the parameters of the simple linear model, $y = \alpha + \beta x + e$, under the criterion of L_1; i.e., under the criterion of minimizing the sum of absolute deviations. This principle can be expressed as determining α and β which minimize

$$F = \sum_{i=1}^{n} |y_i - \alpha - \beta x_i| \qquad (1)$$

It is well-known that estimates of α and β can be obtained under L_1 by reformulating (1) as a linear programming problem, as shown by Sposito (1975), Kennedy and Gentle (1980) and others. Wesolowsky (1981) derived a computational procedure to obtain estimates of (1) using a descent technique, hence solving (1) directly rather than formulating (1) as a linear programming problem.

The purpose of this paper is to consider certain modifications of Wesolowsky's procedure which will reduce the overall computational effort. A Monte Carlo study is given comparing this procedure with Armstrong and Kung's (1978) algorithm.

2. BASES OF DESCENT PROCEDURE

Wesolowsky noted that (1) is a convex function being composed of linear segments; hence the minimum will necessarily occur at the intersection of two lines. Given an initial point on any line one can determine a "better" point, which will decrease the value of F by descending to a nearby intersection. Repeating this sequence the procedure continues until a point is reached which minimizes (1).
Considering any point (x_j, y_j), Wesolowsky noted that (1) can be rewritten as

$$\underset{\beta}{\text{minimize}} \sum_{\substack{i=1 \\ x_i \neq x_j}}^{n} \left|1 - \frac{x_i}{x_j}\right| \left|\frac{y_i - (y_j x_i)/x_j}{1 - x_i/x_j} - \beta\right| \qquad (2)$$

or $\underset{\beta}{\text{minimize}} \sum_{i=1}^{n} w_i |a_i - \beta| \qquad (3)$

where w_i are positive constants and a_i are constants. Hence, (3) is equivalent to the problem of finding the weighted median of the a_i values using w_i as weights. The optimal weighted median for a given set of a_i's and w_i's are such that for an appropriate j:

$$\sum_{i=1}^{j} w_{(i)} \geq \sum_{j+1}^{n} w_{(i)} \qquad (4a)$$

and

$$\sum_{i=1}^{j-1} w_{(i)} < \sum_{i=j}^{n} w_{(i)} \qquad (4b)$$

where $w_{(i)}$ denotes the corresponding weights of the ordered a_i's.
Rather than trying to satisfy (4a) and (4b), a faster computational approach would be as follows:

(i) Consider an initial estimator, β^*

(ii) Place smaller values of a_i below β^* and likewise,

(iii) Place larger values of a_i above β^*,

(iv) Then check (4a) and (4b),

(v) If they are not satisfied then an appropriate adjustment is made; i.e., if the right hand side of (4a) is overly weighted, say, then the $w_{(i)}$ corresponding to the smallest $a_{(i)}$ is transferred to the left hand side, and the check made again.

3. MODIFICATIONS TO PROCEDURE

Minimizing $F = \sum_{i=1}^{n} |y_i - \alpha - \beta x_i|$ can be equivalently written as

$$\underset{\beta}{\text{minimize}} \sum_{i \neq j} |x_i - x_j| \left| \frac{(y_i - y_j)}{(x_i - x_j)} - \beta \right| \quad (5)$$

for a particular (x_j, y_j). Note if $x_i = x_j$ for any i then necessarily the corresponding weight is zero. With this formulation (5) is clearly equivalent to (3); hence, obtaining an alternative weighted median problem.

Another modification of Wesolowsky's procedure considers the initial estimate of β; clearly any initial estimate would suffice with a near-optimal L_1 estimate being more appropriate. Incorporating near-optimal L_1 estimates have been used previously by Sposito, et al. (1977) to reduce the overall computational effort of Barrodale and Roberts (1973) simplex procedure for general linear regression models. The near-optimal estimates used in our study include $\tilde{\alpha} = \bar{y}$ with $\tilde{\beta} = 0$, and the classical, (closed form), L_2 solution.

4. TESTING RESULTS

In 1978, Armstrong and Kung presented an extremely efficient computational procedure to determine the L_1 estimates of a simple linear model. In their study, they showed that SUBROUTINE SIMLP was considerably more efficient in terms of total computational time than those of other known computational procedures. Our enclosed study was conducted to compare SUBROUTINE SIMLP with our modifications of Wesolowsky's procedure for the two parameter linear model. One thousand repetitions on random problems were generated as follows:

(i) generate at random a solution (α, β) where α and β are iid U(-4,4)

(ii) generate an X matrix where $x_{i1} = 1$ for all i and x_{i2} are iid U(-100,100) for all i

(iii) generate the dependent variables such that: $y_i = (\alpha + \beta x)_i + e_i$ where e_i are iid U(-1,1) for all i.

Iowa State University's ITEL AS/6 was used in the study. Table I contains the summary of our study:

Table I

# of Obs.	DESLS Time (Sec)	DESL1 Time (Sec)	SIMLP Time (Sec)	Average Reduction Over SIMLP	Best
10	12.40	12.39	12.41	.00002	DESL1
25	33.59	33.58	33.62	.00004	DESL1
30	41.61	41.52	43.19	.00167	DESL1
50	69.01	69.71	94.61	.02560	DESLS
250	437.71	446.27	648.39	.11068	DESLS

Subroutine DESL1 used $\tilde{\alpha} = \bar{y}$ with $\tilde{\beta} = 0$ as an initial estimate; whereas subroutine DESLS incorporated the closed form least squares solution. For sample sizes less than 50, SUBROUTINE DESL1 is on the average slightly faster than SUBROUTINE SIMLP; for larger sample sizes SUBROUTINE DESL1 is considerably faster. Moreover, using the L_2 estimator, SUBROUTINE DESLS, is more efficient in reducing total CPU time than either of the other two subroutines. A listing of SUBROUTINE DESL1 is available from the authors.

REFERENCES

Armstrong, Ronald D. and Kung, Mabel Tam. (1978). Algorithm AS132. Least absolute value estimates for a simple linear regression problem. *Appl. Statist.*, 27, 363-366.

Arthanari, T. S. and Dodge, Y. (1981). Mathematical Programming in Statistics. John Wiley and Sons, Inc., NY.

Barrodale, I. and Roberts, F. D. K. (1973). An improved algorithm for discrete L_1 linear approximation. *Siam J. Numer. Anal.*, 10, 833-848.

Bloomfield, Peter and Steiger, William. (1980). Least Absolute deviations curve-fitting. *SIAM J. Sci. Stat. Comput.*, 1, 290-300.

Gentle, J. E. (1977). Least absolute values estimation: An introduction. *Comm. in Statist.*, Vol. B6(4), 313-328.

Karst, O. J. (1958). Linear curve fitting using least deviations. *JASA*, Vol. 53, 118-132.

Kennedy, W. J. and Gentle, J. E. (1980). Statistical Computing. Marcel Dekker, Inc., NY.

Sadovski, A. N. (1974). Algorithm AS74. L_1-norm fit of a straight line. *Appl. Statist.* 23, 244-248.

Sposito, V. A. (1975). Linear and Nonlinear Programming. Iowa State Press, Ames, Iowa.

Sposito, V. A. (1982). On Unbiasedness of L_p regression estimators. *JASA*, Vol. 77, 652-653.

Sposito, V. A. and Smith, W. C. (1976). On a sufficient condition and a necessary condition in L_1 estimation. *Appl. Statist.*, 25, 154-157.

Sposito, V. A., Hand, M. L. and McCormick, G. F. (1977). Using an approximate L_1 estimator. *Comm. in Statist.*, Vol. B6(3), 263-68.

Wesolowsky, G. O. (1981). A new descent algorithm for the least absolute value regression problem. *Comm. in Statist.*, Vol. B10(5), 479-91.

SECURITY ISSUES FOR DYNAMIC STATISTICAL DATABASES

MARY D. McLEISH
Department of Computing & Information Science
University of Guelph
Guelph, Ontario
Canada

Abstract.

A dynamic database is one in which deletions and insertions are taking place and queries are being made before and after these updates. The database studied here has been partitioned into classes in such a manner that changes are made only to individual partition sets and not across partitions.

This paper extends earlier work concerning the security of such databases, where only even-sized partition sets were allowed and changes had to be processed in pairs. Now changes to a set of records can be made involving any $n \geq 2$ records at a time, where the number of deletions plus insertions equals n, which can be either even or odd. Conditions under which the database remains secure are given, which restrict the type of records initially in the database (or in a partition set). Another alternative is also given to the earlier restriction that this number be even.

Introduction.

A statistical database is a collection of records about which queries concerning certain subsets of records may be answered, as opposed to a database which returns complete details of a record. The security problem consists of limiting the use of a database so that only statistical information is available and no sequence of queries is sufficient to derive confidential information about an individual.

A number of researchers have considered situations in which inferences from sequences of queries have lead to either partial or total compromisability of the database. c.f. [3,5,6,7,8,11,13,14]. Conditions, within the context of different models, are usually obtained under which the system is secure from the discovery of a single previously unknown individual value.

The models just referenced are not concerned with the possibility of up-dating or deleting records. This situation has been considered by F. Chin and G. Ozsoyoglu [cf.4] in the context of a partitioned database. They have found that if records are added and deleted according to certain rules and the parity of each partition set is restricted, the database may be kept secure for "mean" queries.

However, these requirements are quite restrictive (as pointed out also by L. Beck, p.318 in [1]). Certainly, in practice, partition sets cannot be expected to have prescribed parities. Techniques involving adding "dummy" records to those partition sets violating the parity restriction have been tried. However, now extraneous data is introduced and no mean queries ever return the correct values.

In this paper, the conditions on the size of the partition sets and the type of changes that can be made are generalized. A more concise, non-graphtheoretic proof of a main theorem for such a system is given. The major results are contained in Theorems 2.3.3, and 3.5.

The Dynamic Partitioned Model

Preliminary Definitions.

Let us consider a statistical database to be a subset of the Cartesian product of some attribute sets $A_1, \ldots, A_k$. Some of the attributes would normally be protected, meaning that only summary information is available (cf. Dobkin [9]; Ozsoyoglu [12], Chin [4, 15]).

A query is some statistical function, e.g. mean, median, etc., applied to a subset of the records in the database. The database becomes compromisable if one can deduce from the responses of the queries, some protected attribute values of records. (Total compromisability, in the sense used in the introduction, is not required.) Otherwise, the system is said to be secure.

For the partitioning model, we assume that there is exactly one protected attribute, whose individual values will be referred to as values v. Suppose further that at least one attribute set is partitioned into subsets.

In general, if attribute sets $A_1, \ldots A_k$ are either partitioned into or contain a finite number t_i of ordered sets or elements r_{ij} ($1 \leq j \leq t_i$), the data base may be considered to be partitioned into $\prod_{i=1}^{k} t_i$ subsets of the elements of the attribute sets exists, some artificial ordering must be imposed.) Then a record $(a_1, \ldots a_k)$ belongs to $P_{\ell_1, \ldots, \ell_k}$ if and only if a_i belongs to the ℓ_i'th subset of A_i, if A_i has been partitioned, and equals the ℓ_i'th element of A_i otherwise. (The sets $P_{\ell_1, \ldots, \ell_k}$ may be renumbered as P_j, $1 \leq j \leq \prod_{k=1}^{k} t_i$.)

Now queries will be allowed only to to return the sum and count of the protected domain values within a partition set P_j.

2.2 The change sequence.

We will allow deletions and insertions to be made within partitions, provided changes are processed in pairs. Thus, 2-tuples of records are formed which could be (r_a^i, r_b^i), or (r_a^d, r_b^d), when records r_a and r_b are either both inserted or both deleted, or of the form (r_a^i, r_b^d) for an insertion-deletion operation. The tuples are unordered (as processes within a tuple are simultaneous).

Records which are to be inserted or deleted form the change sequence. Those which appear in the change sequence are called dynamic, as opposed to static records, which belonged to the database initially and are never altered.

The assumption is made that if a record is deleted, it is not reinserted into the database, or if it is reinserted, it has an independent protected domain value.

Consider records r_a and r_b belonging to some partition set P with protected domain values v_a and v_b when they enter into a change sequence. Suppose (r_a^i, r_b^i) occurs. If P is queried before and after the change, one obtains the value of sum of v_a and v_b. Similarly a deletion-deletion tuple results in the sum and an insertion-deletion pair results in knowing the difference of the protected values involved.

Example 2.2.1

Consider the following change sequence of 2-tuples:

(r_1^i, r_2^i), (r_4^i, r_1^d), (r_3^i, r_5^i), (r_6^d, r_4^d),

(r_7^i, r_3^d), (r_8^i, r_9^d), (r_5^d, r_{10}^i).

The corresponding derivable equations can be written:

$v_1 + v_2 = c_1$ $v_4 + v_6 = c_4$

$v_1 - v_4 = c_2$ $v_3 - v_7 = c_5$

$v_3 + v_5 = c_3$ $v_8 - v_9 = c_6$

$v_5 - v_{10} = c_7$

These can be re-arranged into 3 distinct subsystems as follows: (A subsystem is any subset of equations with the property that no variable appearing in the subset, also appears elsewhere in the original system.)

$v_1 + v_2 = c_1$ $v_3 + v_5 = c_3$ $v_8 - v_9 = c_6$

$v_1 - v_4 = c_2$ $v_3 - v_7 = c_5$

$v_4 + v_6 = c_4$ $v_5 - v_{1-} = c_7$

These can further be manipulated algebraically to obtain the equivalent system, for known constants c_8, c_9, and c_{10}:

$v_1 + v_2 = c_1$ $v_3 + v_5 = c_3$ $v_8 - v_9 = c_6$

$v_2 + v_4 = c_8$ $v_5 + v_7 = c_9$

$v_4 + v_6 = c_4$ $v_7 + v_{10} = c_{10}$

It is then pointed out in [12], that this splitting of the information into sets of equations involving either sums or differences exclusively is always possible. An information graph is defined, with s or d labelled edges, depending on whether the edges represent these sum or difference equations. Active and inactive vertices are defined and several properties of the graph are proven until after considerable and lengthy effort, (taking about 6 pages) it is finally possible to conclude that the graph is 2-colorable and thus an infinite number of solutions for the system of equations may be found by adding an arbitrary number to all records which have the same colour and subtracting the same amount from all the records with the other colour.

Thus protected domain values of dynamic records are secure if no protected domain value is known initially. If the number of records in partition P to begin with is even (or zero), no protected domain value of a static record can be disclosed (by successive pairwise deletions) and the database is secure. This then proves the following theorem.

Theorem 2.2.2. (Ozsoyoglu [12]).

In a partitioned statistical database, if no protected domain value is known initially and if each partition set initially has an even number of records, then the partitioning model is secure when altered by a change sequence involving only 2-tuples of records.

2.3 An alternate proof of Theorem 2.2.2

Lemma 2.3.1. Any n x n matrix A, whole column sums are all zero is singular.

Proof. Clearly the row vectors are linearly dependent, where det(A) = 0.

Theorem 2.3.2

Suppose a change sequence results in a subsystem of n(>2) equations in n unknowns, which contains no proper subsystem of equations.
Then the system has an infinite number of solutions for each of the n variables.

Proof: For the tuples (r_a^i, r_b^i), (r_a^d, r_b^d) and (r_a^i, r_b^d), the derivable equations will now be written as:

$$v_a + v_b = c, \quad -v_a - v_b = c$$

or $v_a - v_b = c$

for some constant c. In other words, insertions will always occur with a positive sign in front of the corresponding protected domain value and deletions with a negative sign. The proof will be by induction on the number of equations in the subsystem.

Small values of n: For n = 2, the only possible such subsystem is of the form $v_1 + v_2 = c_1$,

$$-v_1 - v_2 = c_2 \ (-c_1)$$

(For n = 3, the coefficient matrix becomes:

$$\begin{pmatrix} 1 & 1 & 0 \\ 0 & -1 & 1 \\ -1 & 0 & -1 \end{pmatrix}$$

unique up to row and column interchanges.) The coefficient matrices are singular in both cases. (Recall that once a record has been deleted, it cannot be re-inserted.)

Induction Step. Assume that any system of n-1 equations in n-1 unknowns containing no proper subsystems of equations and being derived from a change sequence, is such that the column sums of the coefficient matrix are zero. (In other words, every record that is added, is also deleted and no record is deleted that was not added in an equation within this system. Thus, the columns of the coefficient matrix consists always of one, minus one and the remaining entries are zero.)

Let A be the coefficient matrix of a system of n equations coming from a change sequence and containing no proper subsystems. Suppose there is a column consisting of only one non-zero entry (which is possible as inserted or existing records need not be deleted). Then consider the row containing this entry. Let the corresponding equation be $\pm v_a \pm v_b = c$, where v_b is the value appearing only once in the system. Then the remaining n-1 equations will not contain v_b, but will contain v_a (otherwise they would form a closed subsystem within the n x n system). However, now we have a system of n-1 equations in n-1 unknowns, which contains no proper subsystem. Therefore, the columns of its coefficient matrix all contain two non-zero entries, including the column representing variable v_a. Then, in A, this column will have 3 non-zero entries, which is impossible for equations resulting from a change sequence.

Therefore A has zero-sum columns and is singular.

<u>Theorem 2.3.3</u>. The statement of Theorem 2.2.2.

<u>Proof</u>. Clearly any system of equations with more unknowns than equations and containing no proper subsystems will have no unique solutions for any of its variables (see [2, 10]). In equations resulting from a change sequence, it will never happen that there are fewer unknowns than equations. For let the number of equations be m. Then every equation involves 2 variables, so the number of unknowns is $\leq 2m$. But every variable can appear at most twice (one insertion and one deletion, therefore the number of unknowns must be $\leq \frac{2m}{2} = m$. Thus the dynamic records are secure.

3. <u>Generalization of the Dynamic Model</u>. Consider alowing a change sequence in which n-tuples, n>2, are used. For example, (r_a^i, r_b^d, r_c^d) might be a valid operation, involving the three records r_a, r_b, and r_c. Let q represent the total number of records initially in a partition set P. Assume n is fixed and any combination of deletions and insertions is allowed and that the total length of the change tuple is n. As before, we will assume deleted elements are not normally re-inserted. The major result is contained in Theorem 3.5.

The following lemma generalizes Theorem 2.3.3.

<u>Lemma 3.1</u> If q is even and n is even, the system is secure.

<u>Proof</u>. If q and n are both even, at any stage in the change sequence, the number of records remaining in the partition set P will be even. Let $v_1 + v_2 + \ldots + v_s = c$ be the result of a sum query at any stage in the change sequence, where s is then even. We will show that any equation that can subsequently be obtained by a sum query after this stage, by one even n-tuple change, could also be obtained by a sequence of 2-tuple changes. The n-tuple can be replaced by a sequence of 2-tuples. It then follows that the entire n-tuple change sequence can be replaced by a (usually longer) 2-tuple sequence and the resulting system of equations from the n-tuple system will form a subset of those generated by the 2-tuple system. As this larger system provides no unique solutions for any of the variables (by Theorem 2.3.3), neither will the subset.

Let the n-tuple in the change sequence be

$(r_{a_1}^d, \ldots r_{a_m}^d, r_{b_1}^i, \ldots, r_{b_t}^i)$

where m and t are even and sum to n. Thus
$(r_{a_1}^d, r_{a_2}^d), (r_{a_3}^d, r_{a_4}^d), \ldots, (r_{a_{m-1}}^d, r_{a_m}^d),$
$(r_{b_1}^i, r_{b_2}^i), \ldots, (r_{b_{\frac{t}{2}-1}}^i, r_{b_{\frac{t}{2}}}^i)$ can replace the n-tuple.

If m is odd and t is odd, where $t \leq m$, replace the n-tuple by
$(r_{a_1}^d, r_{b_1}^i), \ldots, (r_{a_m}^d, r_{b_m}^i), (r_{b_{m+1}}^i, r_{b_{m+2}}^i),$
$\ldots, (r_{b_{t-1}}^i, r_{b_t}^i)$, which is possible as t-m is even. If, however, t< m perform
$(r_{a_1}^d, r_{a_2}^d), \ldots, (r_{a_{m-t-1}}^d, r_{a_{m-t}}^d),$
$(r_{a_{m-t+1}}^d, r_{b_1}^i), \ldots, (r_{a_m}^d, r_{b_t}^i)$, which is again possible as m - t is even.

This theorem, although not providing any more flexibility with partition sizes, does allow speedier changes to be made, with fewer intermediate steps.

One suggestion made by Ozsoyouglu [12], is to add 'dummy' records to make the size of odd partition sets even, possibly in a statistical manner so that "on the average" their effect is minimal. However, individual queries will be inexact and other 'consistency' problems can arise.

Theorem 3.2 provides a further extension of Theorem 2.3.3 to handle situations where n and q are of any parity. Lemma 3.1 in fact proves part of this theorem. However, the general proof can be handled somewhat differently, when n > 2, and will be given in this context.

Theorem 3.2. If a fixed n is even or odd (≥ 3) and q(≥ 2) is even or odd, then the system of equations that can be derived by repeated subtraction of the sum queries obtained before and after a change has no unique solution.

Proof: When n > 2, the proof results from the fact that no subsystem of equations of any size can ever be found with fewer or an equal number of unknowns than equations. As each equation now involves n variables, m equations result in nm unknowns. As each unknown can appear at most twice (one insertion, one deletion), the total number of unknowns in any m equations must be $\geq \frac{mn}{2}$.

Since this is true for any $m \geq 1$, no set of equations arising from a change sequence can have a unique solution.

Remark 3.2.1. It is in fact possible to show that when m is even, $\frac{mn}{2}$ is obtainable and when both m and n are odd, the minimal number of unknowns possible is $\frac{mn+1}{2}$ and there is also a sequence of changes that will make this possible. For example, if m = 3 and n is odd, an optimal strategy is n additions followed by $\frac{n-1}{2}$ deletions and $n - \frac{n-1}{2} = \frac{n+1}{2}$ additions. That leaves n+1 records that could be chosen to be deleted, so an n-deletion tuple is possible, giving $n+n - \frac{n-1}{2} = \frac{3n+1}{2}$ unknown. So, for general odd m, perform pairwise n-insertions followed by n-deletions a total of $\frac{m-3}{2}$ times and then the 3 operations just described.

In the next two lemmas it is assumed that any records initially in the partition set P could be dynamic in order to produce the compromises.

Lemma 3.3. If q is even or odd and n is odd, then the database is compromisable.

Proof. Assume $n \geq 3$ as if n = 1 the proof is trivial.

If q > n, reduce the size of the partition set by successive n-deletion tuples until it equals $t \in \{1,...,n-1\}$ where $t \equiv q \pmod{n}$. If $t = 1$, clearly the database can be compromised. If $t > 1$ and is odd, let $s = \frac{t-1}{2}$ and perform s deletions and n-s additions. Then the number of remaining records will be $t - 2s + n = n + 1$. Now an n-deletion tuple will compromise the database. If t is even, perform t deletions, t' = n - t additions and proceed as with t odd replacing t by t'.

Lemma 3.4 If q is odd and n is even (≥ 2), then the database is compromisable.

Proof: Now we may always reduce the number of records to size $t \in \{1,...,n-1\}$ where t is odd, by successive n-tuple deletions and then proceed as in Lemma 3.3 for odd t.

Theorem 3.5. (i) If $q \geq 2$ is odd or even and $n \geq 3$ is odd, or if $q \geq 3$ is odd and $n \geq 4$ is even, then the dynamic statistical database is secure if and only if either no n-deletion tuples or tuples consisting of n-1 deletions and one insertion are allowed or at least two records in each partition set are kept static.

(ii) In the case when q is odd (≥ 3) and n = 2, the conditions remain the same except that now only 2-deletion tuples need to be excluded.

Proof: (i) This follows from Lemmas 3.3 and 3.4, as well as Theorem 3.2 by noting that the only way one record can be found in P is to either have n records there at some stage in the change sequence and perform n-1 deletions followed by 1 insertion, or to have n+1 records in P and perform n deletions.

(ii) Now the dynamic records are secure by Lemma 3.1 and the disallowal of a deletion-deletion pair prevents records being deleted from the partition set until only one remains. "Only if", in the case where partitions do not necessarily contain at least 2 static records, follows from Lemma 3.4.

This Theorem puts restrictions either on the change sequence or the records in P. However, it allows greater flexibility than the previously known result (Theorem 2.3.3), both for more efficient updating and for arranging of the partition sets. Of course, these results apply to any set of records subjected to a change sequence as described, whether or not they are partitioned. Essentially, the database remains secure when updating if the size of a partition set can never be reduced to one in the process and Theorem 3.5 gives the conditions to be imposed on the user to ensure that this never happens.

REFERENCES

1. Beck, L., "A security Mechanism for Statistical Databases," TODS, Vol. 5, No. 3, pp. 316-338 (Sept. 1980).

2. Campbell, H.G., "An Introduction to Matrices, Vectors and Linear Programming," 1965, Meredith, New York.

3. Chin, F.U., "Security in Statistical Databases for Queries with Small Counts," TODS, Vol. 3, No. 1, pp. 92-104 (March, 1979).

4. Chin, F. and Ozsoyoglu, G., "Security in Partitioned Dynamic Statistical Databases," Proc. IEEE, Compsac Conf., 1979, pp. 594-601.

5. Chin, F. and Ozsoyoglu, G., "Statistical Database Design," ACM TODS, Vol. 6, No. 1, pp 113-139 (March 1981).

6. DeMillo, R. and Dobkin, D., "Recent Progress in Secure Computation," Tech. Report, Georgia Institute of Technology, GIT-ICS-78.

7. Denning, D., "Secure Statistical Databases with Random Sample Queries," TODS, Vol. 5, No. 3, pp.291-315 (1980).

8. Denning, D. and Schlorer, J., "A Fast Procedure for Finding a Tracker in a Statistical Database," TODS, Vol. 5, No. 1, pp.88-102 (March 1980).

9. Dobkin, D., Jones, A.K. and Lipton, R.J., "Secure Databases: Protection Against User Influence," TODS, Vol. 4, No. 1, pp. 97-106 (March 1979).

10. Hoffman, K. and Kunze, R., "Linear Algebra," 1971, Prentice-Hall, New Jersey.

11. Kam, J. and Ullman, J., "A Model of Statistical Databases and their Security," TODS, Vol. 2, No. 1, pp. 1-10 (March 1977).

12. Ozsoyoglu, G., "Secure Statistical Database Design," Ph.D. Thesis, Dept. of Computing Science, University of Alberta, August, 1980.

13. Schlörer, J., "Security of Statistical Databases: Multidimensional Transformation," TODS, Vol. 6, No. 1, pp. 95-112 (March 1981).

14. Schwarz, M.D., Denning, D.E. and Denning, P.J., "Linear Queries in Statistical Databases," TODS, pp. 156-167 (June 1979).

15. Yu, C.T. and Chin, F., "A Study on the Protection of Statistical Databases," Proc. ACM SIGMOD Conf., 1977, pp.169-181.

On the Use of Cluster Analysis for Partitioning and Allocating Computational Objects
in Distributed Computing Systems

Lennart Pirktl

Statistisches Seminar
University of Zurich
Switzerland

The first part of this paper deals with the description of the allocation problem in distributed systems. The second part is dedicated to an alternative cluster-analytic approach to this question. Two novel methods of partitioning and allocating computational objects have been developed and are described in terms of cluster analysis. An empirical comparison of 15 procedures is followed by a method conceived to reduce dimensionality of the underlying structure.

1. INTRODUCTION

A distributed system (e.g. a banking system with a decentralized data base and a decentralized processing) contains multiple units capable of processing and interconnected by a physical network. Taking the minimization of the interprocessor communications overhead as an optimization criterion, a fast decision for the grouping of subprocesses is needed [4]. In the following text the units, or processing centers, will be referred to as nodes and independent units shall be called preferred nodes. The set V containing m nodes together with the set K of n edges define a graph $G(V,K)$. Furthermore, a subset P of V containing $1 < p < m$ preferred nodes is designated. Each edge $k(v,w)$ carries a weight $c(v,w)$ corresponding to the communication costs between units v and w if these reside in distinct logical locations. The problem to be resolved is to establish a disjoint and exhaustive partition of V into p groups (i.e. a p-cut of $G(V,K)$), containing one each of the p preferred nodes. As every possible partition of V leads to a subdivision of K into a subset of internal edges (both nodes of these edges belong to a unique group), and a complementary set of external edges (with nodes in different groups), a partition is called optimal if the sum of the weights of its external edges can be shown to be minimal.

The most obvious approach to the problem is to enumerate all the possible allocations and simply choose the most appealing one, but the number N of ways of sorting m - p nodes into p groups (including the permutations of the p preferred nodes) is a function of Stirling numbers $S(m-p,k)$ of the second kind:

$$N = \sum_{k=1}^{p} \sum_{r=1}^{k} (-1)^r \binom{k}{r} (k-r)^{m-p} \binom{p}{k} \qquad (1)$$

The problem discussed in Hässig and Jenny [3] concerning 16 edges and 13 nodes of which 3 are preferred, looks rather simple but by (1) it admits more than 50,000 solutions.

By increasing to 25 the non-preferred, and to 5 the preferred nodes, the number of possibilities is raised to about 300,000,000,000,000,000,000! This fact was the motivation for the development and application of faster tools such as the maximum spanning tree method. Nevertheless Hässig and Jenny [3, p.596] give an illustrative example showing that this graph-theoretical approach may not be optimal in certain cases. The authors introduce the PROXCUT algorithm to improve an initial partition obtained by Kruskal's maximum spanning tree. This procedure is based on switching not only single nodes but whole subsets of nodes from one group to another in order to achieve better p-cuts.

2. CLUSTERING METHODS

In this section we investigate the possible use of cluster analysis in this same field. The usual goal of this multivariate technique is the optimal partitioning of a set of objects into homogenous groups. The underlying distance-concept is derived from a set of variables describing the items. In the case at hand, objects will correspond to nodes, whereas distance (or similarity) will be a certain function of the edges' weights. As an expensive edge should preferably be internal, it is obvious that the distance between two nodes v and w linked by the edge $k(v,w)$ with weight $c(v,w) > 0$, is set to (2):

$$d(v,w) = -c(v,w) \qquad (2)$$

The problem of having to deal with negative distances does not affect the classification process. Equation (2) conforms strictly to the definition of distance given in Späth [7] as a mapping $d: V \times V \rightarrow R$, satisfying conditions (3), (4) and (5) for any pair of nodes v and w.

$$d(v,w) \geq d(0) \qquad (3)$$
$$d(v,v) = d(0) \qquad (4)$$
$$d(v,w) = d(w,v) \qquad (5)$$

The constant d(0) is an arbitrary positive or negative finite real number. Definition (2) does not, however, generate a metric, because the triangular inequality will not hold in every case. Furthermore, the following condition (6) will be violated, as V might contain distinct nodes v and w satisfying d(v,w)=d(0).

$$d(v,w) = d(0) \implies v = w \quad (6)$$

By the above definition of distance, diagonal elements of the distance matrix **D** are set to $d(0) = -(\max(c(v,w)) + 1)$. On the other hand, the distance between disconnected nodes is given a value of zero. As the set P has to be split into different groups, the distances between preferred nodes must be set to a sufficiently large number. Now clustering procedures will tend to group nodes separated by small distances (corresponding to high-valued edges) in individual ways depending on the algorithm chosen.

2.1 Hierarchical Methods

Based on this distance matrix **D**, hierarchical methods obtain the partition into p groups by n-p successive fusions of individuals or groups. At any particular stage these algorithms fuse groups of nodes which are closest. Differences between methods arise only from different ways of defining the statistical distance between groups of objects. Concepts such as Single Linkage, Complete Linkage, Average Linkage, Median, Ward, Centroid Sorting and Flexible Strategy are well known [2], but none of these techniques fulfills the implicit conditions of the problem discussed above. All these hierarchical methods may be defined by means of the single recurrence formula (7):

$$d_{pq,r} = \alpha_p d_{p,r} + \alpha_q d_{q,r} + \beta d_{p,q} + \gamma |d_{p,r} - d_{q,r}| \quad (7)$$

Concerning the problem at hand we propose to define the statistical distance between two clusters (i.e. between two sets of nodes) as the cost of not joining them. By this reasoning we obtain the parameters shown at the bottom of Table 1. This new hierarchical and agglomerative procedure, referred to here as SUM LINKAGE method, is not equivalent to McQuitty's Similarity Analysis [6]. Also, it is not reproduced by any of Lance&Williams' Flexible Strategies [5].

2.2 Iterative Methods

When thinking about clustering methods it seems quite natural to use iterative algorithms. They work much faster and are more likely to find an optimal solution. The proposed procedure starts with a random allocation respecting the independence of preferred nodes. During the following relocation scans each node w is considered in turn. Its statistical distances to all clusters are computed as the sum of the known

Procedure	α_p	α_q	β	γ
Single Linkage	1/2	1/2	0	-1/2
Complete Linkage	1/2	1/2	0	1/2
Average Linkage	$\frac{n_p}{n}$	$\frac{n_q}{n}$	0	0
Centroid	$\frac{n_p}{n}$	$\frac{n_q}{n}$	$\frac{-n_p n_q}{n^2}$	0
Median	1/2	1/2	-1/4	0
Ward	$\frac{n_p+n_r}{n+n_r}$	$\frac{n_q+n_r}{n+n_r}$	$\frac{-n_r}{n+n_r}$	0
Flexible Strategy	$\frac{1-\beta}{2}$	$\frac{1-\beta}{2}$	β	0
Sum Linkage	1	1	0	0

$(n = n_p + n_q)$

TABLE 1:
Parameters of Recurrence Formula (7)

distances between w and each of the nodes belonging to the cluster in view. If w appears to be closer to a cluster Y than to its parent cluster W, the method moves w from W to Y. The process stops either when no further amelioration is found during n-p searches, or when a maximum number of iterations is exceeded by the following search. Whether a final partition is a global optimum can not be precisely established. Therefore it is recommended to execute the procedure with different starting groupings (see the example in Table 2).

3. EMPIRICAL STUDY

To examine the quality of clustering procedures for determining optimal allocations the two proposed clustering procedures and 13 of the common hierarchical classification algorithms were tested on eight simulated networks. For some of these cases the independence restriction of preferred nodes was violated by a number of procedures. In the results of Table 2 this phenomenon is indicated by a dot. It is easy to see that Single Linkage, for example, easily agglomerates clusters containing preferred nodes because only the smallest distance (i.e. the highest weight or cost) determines fusion. Table 2 shows the obtained final sums of external edges for the 15 algorithms and eight networks.

In the case of the iterative procedure the best result out of three runs based on different random partitions was chosen. In order to obtain a reduction of these measures, every result was transformed to the percentage relative to the (local?) optimum solution concerning the respective dataset. In a next step, unweighted means of these values were calculated for each of the 15 tested methods. Table 3 shows the final result which is not a definitive indication of the relative merits of the algorithms. It indicates for each procedure the percentage by which the best solution was exceeded in average, and the number of acceptable solutions achieved.

4. WORKING WITH REDUCED GRAPHS

Depending on the individual structure of the network, there is an extremely simple way to reduce the size of the distance matrix D, and thereby the necessary CPU-time. Assuming that a provisionally (hierarchical or iterative) solution has been found, no edge whose weight is greater than the achieved sum of weights of external edges could ever be an external one for any imaginable better solution. Therefore it seems suitable to fuse all nodes connected by an internal edge with a weight superior to the obtained local optimum. While fusing two such nodes it might happen that two other edges collapse (see the example in Figure 1). It is easy to see that the resulting new edge carries the sum of the former weights. But now again the weight of the current solution might be exceeded by this sum. In this case the fusion process continues until the maximum weight becomes smaller than the obtained optimum.

Thereafter a new (and faster) partitioning step may be executed. The illustration of the proposed strategy for reduction shown in Figure 1 is based on the network examined in Hässig and Jenny [3]. Having achieved a solution with a value of 18 for the total sum of external weights from the iterative method, the graph is reduced to only 9 nodes and 11 edges. (Vichi [8] and Bruynooghe [1] propose similar methods to reduce the often prohibitive execution time required by agglomerative algorithms.)

Algorithm	Percentage above best solution	Number of valid results
Sum Linkage	6.3	8
Iterative Algorithm	8.7	8
Flex. Strat. (+0.75)	9.7	3
Single Linkage	15.0	2
Flex. Strat. (+0.25)	21.1	6
Flex. Strat. (+0.50)	26.6	5
Average Linkage	38.1	8
Flex. Strat. (0.00)	40.9	8
Flex. Strat. (-0.75)	56.0	6
Complete Linkage	79.0	8
Median	193.2	7
Centroid Sorting	194.7	7
Flex. Strat. (-0.25)	205.0	8
Ward	206.3	8
Flex. Strat. (-0.50)	210.2	8

TABLE 3:
Mean Percentage above best solution and number of admissible solutions

Algorithm	Network No.: 1	2	3	4	5	6	7	8
Sum Linkage	112	46	10	38	20	20	502	62
Iterative Algorithm	112	46	10	38	18	16	789	62
	(112	70	10	59	22	22	897	700)
	(112	49	10	45	20	20	874	76)
Single Linkage	.	.	13	.	.	.	.	62
Complete Linkage	160	60	12	54	20	21	956	279
Average Linkage	160	60	13	54	20	22	988	62
Centroid Sorting	172	79	12	82	16	23	.	718
Median	160	79	12	82	16	23	.	718
Ward	172	60	12	77	16	23	988	870
Flex. Strat. (+0.75)	.	49	.	.	.	16	.	76
Flex. Strat. (+0.50)	133	60	13	.	.	21	.	76
Flex. Strat. (+0.25)	133	60	13	38	.	20	.	76
Flex. Strat. (0.00)	160	60	13	54	20	22	988	76
Flex. Strat. (-0.25)	160	60	12	77	16	23	988	870
Flex. Strat. (-0.50)	172	60	12	77	21	23	988	870
Flex. Strat. (-0.75)	172	79	12	82	21	23	.	.

TABLE 2: Admissible solutions of 15 clustering algorithms

Original Version | Reduced Versions

critical value: 18 | critical value: 10

13 nodes | 9 nodes | 6 nodes

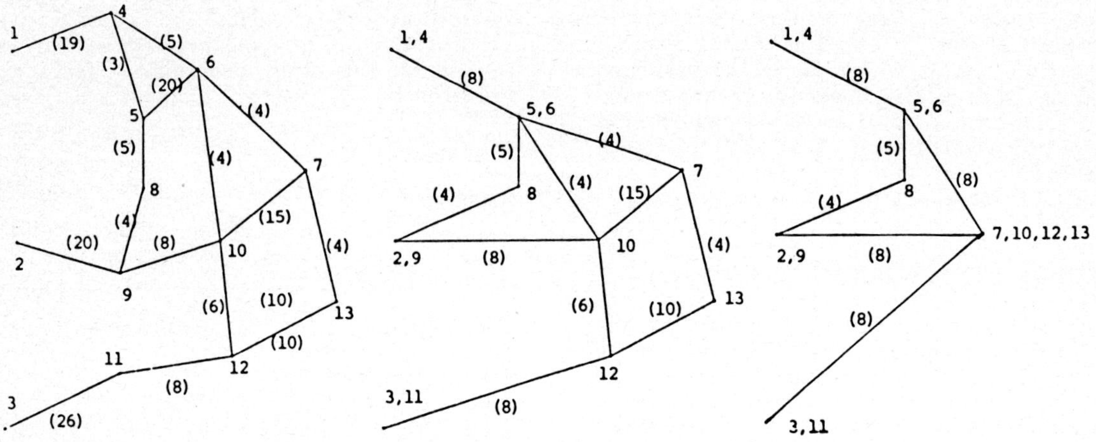

FIGURE 1: Reduction of the Network

Note: A very speculative approach to data reduction might be successful in some cases. Deciding to eliminate all weights superior or equal to 10, the original graph is reduced to a graph (which still conforms to the requirement for separation of preferred nodes) with only 6 nodes and 6 edges (see Figure 1). By (1) the number of solutions to be examined in full enumeration has been reduced from 59,049 to 27.

5. CONCLUSION

Clustering algorithms offer an alternative means of minimizing the interprocessor communications overhead. The problem of deciding the number of clusters present does not arise, as it corresponds to the given number p of preferred nodes. On the other hand, classification criteria have to be chosen carefully. Both SUM LINKAGE and its corresponding iterative algorithm prove to be most suitable among the 15 clustering procedures tested. In large problems preference should be given to the faster iterative method. Furthermore, it is pointed out that even a poor solution or a speculative guess about the optimal solution may be useful for reducing size of the network and CPU-time of further analyses.

6. ACKNOWLEDGEMENTS

I would like to thank Professor Kurt Hässig (University of Zurich) and Dr Martin Casey (Swiss Federal Institute of Technology) for helpful discussions and comments.

REFERENCES

[1] Bruynooghe, M., Classification ascendante hiérarchique des grands ensembles de données. Un algorithme rapide fondé sur la construction des voisinages reductibles, Les cahiers de l'analyse des données 3, (1978) 7-33.

[2] Everitt, B., Cluster Analysis (Wiley, New York, 1980).

[3] Hässig, K. and Jenny, C.J., Partitioning and Allocating Computational Objects in Distributed Computing Systems, Proc. IFIP Congress, Tokyo and Melbourne (1980) 593 - 598.

[4] Jenny, C.J., On the Allocation of Computational Objects in Distributed Systems (Diss, Zürich, 1980).

[5] Lance G.N. and Williams W.T., A General Theory of Classificatory Sorting Strategies 1: Hierarchical Systems, Comp. J. 9 (1967) 373-380.

[6] McQuitty, L.L., Similarity Analysis by Reciprocal Pairs for Discrete and Continuous Data, Educ. Psychol. Measur. 26 (1966) 825-831.

[7] Späth H., Cluster Analysis Algorithms for Data Reduction and Classification of Objects (Wiley, New York, 1980).

[8] Vichi E.G., Cluster Analysis: An Algorithm, More Applications, Compstat, Physica, Vienna (1982) 267-268.

EXACT FACTORIZATION OF NONNEGATIVE POLYNOMIALS AND ITS APPLICATION

Mohsen Pourahmadi

Department of Mathematical Sciences
Northern Illinois University
DeKalb, IL 60115 U.S.A.

This is a brief report on newly developed exact methods for factorization of nonnegative polynomials. An exact recursive method for finding the coefficients of the analytic factor of such polynomials is given. The recursion depends on the Fourier coefficents of the log of the polynomial. Also, a different combinatorial formula for the coefficients of the analytic factor is provided.

1. INTRODUCTION

A well-known theorem of F. Riesz and L. Féjer states that if f is a nonnegative trigonometric polynomial of degree n, that is

$$f(\lambda) = \sum_{k=-n}^{n} \gamma_k e^{ik\lambda} \geq 0, \text{ for all } \lambda \in [-\pi,\pi],$$

then there exists an analytic trigonometric polynomial $P(\lambda) = \sum_{k=0}^{n} c_k e^{ik\lambda}$ such that $f(\lambda) = |P(\lambda)|^2 = |\sum_{k=0}^{n} c_k e^{ik\lambda}|^2$. Furthermore, the coefficients c_k can be chosen in such a way that all the roots of the equation

$$P(z) = \sum_{k=0}^{n} c_k z^k = 0$$

lie outside the open unit disc $D = \{z \in \mathbb{C}; |z| < 1\}$. This polynomial $P(\lambda)$ is unique if we require $c_0 > 0$.

Nonnegative polynomials can arise as the spectral density function of a pure moving average process of order n, or as an estimate of the spectral density function of a stationary time series based on the estimated autocovariances from the data, etc. Analytic factorization of the above type plays extremely important roles in time series analysis [1,3], network and control theory [5], digital signal analysis and filtering [2]. In all these areas it is important to find the coefficients c_k in terms of γ_k's or $f(\lambda)$.

There are several iterative methods for solving this problem [2]. The most successful iterative method is that of Wilson [5], which is essentially a Newton-Raphson method for solving the system of non-linear equations

$$\sum_{j=0}^{n-k} c_j c_{j+k} = \gamma_k, \quad k = 0,..,2,...,n,$$

and therefore the convergence is quadratic. However, Wilson's method is restricted to the polynomials with no roots on the unit circle. If $\sum_{k=-n}^{n} \gamma_k z^k = 0$ has roots with modulus one or close to one, the algorithm provided by Wilson may not converge or may converge very slowly.

In this note we provide two exact methods for finding the c_k's even when some or all of the roots of $\sum_{k=-n}^{n} \gamma_k z^k = 0$ have unit modulus. In fact we show that,

$$c_{k+1} = \sum_{j=0}^{k} (1-j/k+1)a_{k+1-j}c_j,$$

$$k = 0,1,2,...,n-1, \qquad (1)$$

where

$$a_k = \int_{-\pi}^{\pi} e^{-ik\lambda} \log\left(\sum_{j=-n}^{n} \gamma_j e^{ij\lambda}\right) \frac{d\lambda}{2\pi}$$

and $c_0 = \exp\{a_0/2\}$.

It is interesting to note that c_{k+1} is expressed as a "weighted" Césaro average of the previous c_k's where the weight a_k is the kth Fourier coefficient of log of the polynomial $\sum_{k=-n}^{n} \gamma_k e^{ik\lambda}$. Thus having $a_0, a_1,...,a_n$, (1) will provide an exact method for finding $c_0,...,c_n$. However, in most applications only $f(\lambda)$ is given and it is necessary to compute the a_k's by using a numerical integration technique. Therefore the precision of the c_k's will depend very much on the precision of the numerical integration technique used in computing a_k's.

The second exact method is a combinatorial method which expresses the c_k's in terms of a_k's and c_0. This will be discussed in section 3.

In this note we shall give a brief description and proof of these methods and some of its application. The numerical details and the possible extension of these methods to nonnegative functions (spectral densities) will be discussed in separate papers.

2. Proof of formula (1).

Proof of the recursive formula (1) follows from the fact that for a nonnegative polynomial $f(\lambda) \not\equiv 0$, $\int_{-\pi}^{\pi} \log f(\lambda) \, d\lambda > -\infty$. Thus, it follows from Szegö's theorem [4, p. 53] that the analytic factor $f(\lambda)$ has the representation;

$$\sum_{k=0}^{n} c_k z^k = \exp\{\frac{1}{2} \int_{-\pi}^{\pi} \frac{e^{i\lambda}+z}{e^{i\lambda}-z} \log f(\lambda) \frac{d\lambda}{2\pi}\},$$

$$z \in D. \qquad (2)$$

It is easy to check that

$$\frac{1}{2} \int_{-\pi}^{\pi} \frac{e^{i\lambda}+z}{e^{i\lambda}-z} \log f(\lambda) \frac{d\lambda}{2\pi} = a_0/2 + \sum_{k=1}^{\infty} a_k z^k \text{ with}$$

$a_k = \int_{-\pi}^{\pi} e^{-ik\lambda} \log f(\lambda) \frac{d\lambda}{2\pi}$. By differentiating the identity (2) with respect to z and replacing $\exp\{a_0/2 + \sum_{k=1}^{\infty} a_k z^k\}$ by $\sum_{k=0}^{n} c_k z^k$ and some algebraic manipulations we obtain,

$$\sum_{k=0}^{n-1} (k+1) c_{k+1} z^k = \sum_{k=0}^{\infty} [\sum_{j=0}^{k} (k+1-j) a_{k+1-j} c_j] z^k \qquad (3)$$

Comparing like powers of z in (3) it follows that

$$c_{k+1} = \sum_{j=0}^{k} (1-j/k+1) a_{k+1-j} c_j,$$

$$k = 0,1,2,\ldots,n-1.$$

The fact that $c_0 = \exp\{a_0/2\}$ follows from (2) by letting $z = 0$.

In several problems such as deconvolution of a digital filter and autoregressive representation of a stationary time series with the spectrum $f(\lambda)$, it is necessary to find the Fourier (Taylor) coefficients d_k of $(\sum_{k=0}^{n} c_k z^k)^{-1}$. For this, usually the d_k's are found by writing $(\sum_{k=0}^{n} c_k z^k)^{-1} = \sum_{k=0}^{\infty} d_k z^k$, which for large n does not provide a reasonable method for finding the d_k's.

Observing that

$$\sum_{k=0}^{\infty} d_k z^k = (\sum_{k=0}^{n} c_k z^k)^{-1} = \exp\{-a_0/2 - \sum_{k=1}^{\infty} a_k z^k\}$$

and using the technique used in the proof of (1) the following recursion for the d_k's emerges,

$$d_{k+1} = -\sum_{j=0}^{k} (1-j/k+1) a_{k+1-j} d_j,$$

$$k = 0,1,2,\ldots, \text{ with } d_0 = \exp\{-a_0/2\}. \qquad (4)$$

In most applications $\gamma_k = \gamma_{-k}$, $k = 1,2,\ldots,n$. Thus, $f(\lambda) = \gamma_0 + 2\sum_{k=1}^{n} \gamma_k \cos k\lambda$, and the problem of finding a_k's is reduced to that of evaluating the integrals

$$\frac{1}{\pi} \int_0^{\pi} \cos k\lambda \, \log(\gamma_0 + 2\sum_{j=1}^{n} \gamma_j \cos j\lambda) d\lambda,$$

$k = 0,1,2,\ldots$. Numerical results obtained by using the IMSL subroutine DCADRE are quite satisfactory for small k and n. However, for large k and n, since $\cos k\lambda$ oscillates rapidly Filon's method (which is a certain modification of Simpson's rule) is more suitable.

3. A combinatorial solution.

It follows from the formulae (1) and (4) that the c_k and d_k can be expressed in terms of $c_0, a_1, a_2,\ldots,a_k$ through an algebraic function. For example $c_1 = a_1 c_0$, $c_2 = (a_2 + \frac{1}{2} a_1^2) c_0$, $c_3 = (a_3 + a_1 a_2 + \frac{1}{6} a_1^3) c_0$, etc. The general form of the algebraic function expressing c_k in terms of $c_0, a_1,\ldots,a_k$ can not be obtained from (1). However, by proposing a different method of finding c_k's we obtain the desired algebraic form.

To do this, we note that from the identity (2) it follows that

$$\sum_{k=0}^{n} c_k z^k = c_0 \prod_{m=1}^{\infty} \exp\{a_m z^m\} =$$

$$c_0 \prod_{m=1}^{\infty} (\sum_{j=0}^{\infty} a_m^j z^{mj}/j!), \qquad (5)$$

and the coefficient of z^k on the right hand side of (5) can be written as $\sum_{k}' \prod_{j=1}^{k} a_j^{p_j}/p_j!$, when the summation $\sum_{k}'$ extends over all k-tuples $(p_1,\ldots,p_k)$ of nonnegative integers with the property $p_1 + 2p_2 + \ldots + kp_k = k$. Therefore,

$$c_k = c_0 \sum_{k}' \prod_{j=1}^{k} a_j^{p_j}/p_j!, \qquad (6)$$

and similarly, $d_k = d_0 \sum_{k}' \prod_{j=1}^{k} (-1)^{p_j} a_j^{p_j}/p_j!$ with $d_0 = \exp\{-a_0/2\}$.

It is interesting to note that as a consequence of (1) or (6) it follows that for a nonnegative polynomial not all a_k's can be nonnegative.

REFERENCES

[1] Box, G. E. P. and Jenkins, G. M. Time Series Analysis, Forecasting and Control (Holden-Day, San Francisco, 1976).

[2] Gabel, R. A., Neuvo, Y. and Rudin, H. "Iterative spectral factorization algorithms for digital filters", in Digital Signal Processing, edited by V. Cappellini and A. G. Constantinides, Academic Press, 1980, pp. 11-19.

[3] Hannan, J. Multiple Time Series, (John Wiley, New York, 1970).

[4] Hoffman, K. Banach Spaces of Analytic Functions, (Prentic Hall, New Jersey, 1962).

[5] Wilson, G. Factorization of the covariance generating function of a pure moving average process. SIAM J. NUMER. ANAL. 6(1969), 1-7.

A COMPARISON OF ECONOMIC DESIGNS OF NON-NORMAL CONTROL CHARTS

M.A. RAHIM and R.S. LASHKARI

Department of Industrial Engineering
University of Windsor
Windsor, Ontario, Canada   N9B 3P4

This paper investigates the relative performances of one-sided $\bar{x}$-chart, $\bar{x}$-chart with warning limits, and cusum chart in controlling non-normal process means. The process is subject to a single assignable cause with exponentially distributed occurrence time. Associated with each chart is a loss-cost function, consisting of the fixed and variable costs of sampling, the cost of searching for the assignable cause when it does or does not exist, and adjustment and repair costs. The optimal values of the design parameters of each chart are obtained by minimizing the corresponding loss-cost function, using a computerized search technique. Minimum loss-cost criterion is used for each chart as a measure of its performance at various levels of shift in the process mean. In addition, the average run lengths when the process is in control and when it is out of control, are compared.

1. INTRODUCTION

In any production process, some variations in product quality are unavoidable. These variations can be divided into two categories, (i) random variations and (ii) variations due to assignable causes. If the random variations exhibit a stable pattern, the process is said to be operating under a stable system of chance-causes, or simply, to be in a state of in control. Variations that are not within the stable pattern of chance-causes are attributed to assignable causes and the process is then said to be in a state of out-of-control. It is desirable that, when there is evidence that assignable causes of variation are present, these causes be detected and removed so as to bring the process back to the in-control state. This is facilitated by the use of quality control charts.

When the product quality is measured on a continuous scale, commonly used statistical quality control charts for controlling the process means are the $\bar{x}$-chart, the $\bar{x}$-chart with warning limits, and the cumulative sum (cusum) chart. To use a control chart, the user must specify a sample size, a sampling interval and the control limits or the critical region for the chart. Selection of these parameters is called the design of the control chart.

The design of the control chart with respect to economic criteria has been a subject of interest during the last three decades. The objective of the design has always been the optimal determination of the design parameters so as to minimize the average per hour loss-cost.

Duncan [5] and Cowden [4] independently developed the economic design of $\bar{x}$-charts. The pioneering work of Duncan was then extended by others. The economic design of $\bar{x}$-charts with warning limits was first investigated by Gordon and Weindling [8] and was later extended by Chiu and Cheung [2]. The economic design of cusum charts was first introduced by Taylor [16]. Later, following Duncan's work, Goel and Wu [6] developed a single assignable cause model for the economic design of cusum charts. Chiu [1] developed a single-cause economic model of the cusum chart following the general modelling structure of both Duncan's $\bar{x}$-chart model and Taylor's cusum model.

An assumption common to the works mentioned above is that the process means are normally distributed. In many industrial situations the measurable quality characteristic is a random variable whose density function depends upon one or two parameters of the product quality and often has a non-normal distribution. Nagendra and Rai [12] and Lashkari and Rahim [10] proposed models for the economic design of $\bar{x}$-charts to control non-normal means. Rahim and Lashkari [14] proposed an economic design of $\bar{x}$-chart with warning limits to control non-normal process means. Under the non-normality assumptions, Lashkari and Rahim [11] also developed an economic design of cusum charts.

In the past, using the run length criterion under normality assumption, some authors [7,15] have compared the relative efficiencies of $\bar{x}$- and cusum charts in detecting lack of control. However, a comprehensive comparison of the performances of the $\bar{x}$-chart, the $\bar{x}$-chart with warning limits and the cusum chart for controlling non-normal process means is lacking. The present paper is an attempt to evaluate the relative performances of the three control charts at various degrees of shift in the level of the process mean.

2. PRODUCTION PROCESS

The process is assumed to start in a state of statistical control having a non-normal probability density function $f(\mu_0, \sigma^2, \beta_1, \beta_2)$ with mean μ_0, variance σ^2, the measure of skewness β_1, and the measure of kurtosis β_2. A single assignable cause may shift the process mean to $\mu_0 + \delta\sigma$, where δ is the shift parameter. The time between the occurrences of the assignable cause is exponentially distributed with the mean $1/\lambda$ hours. It is assumed that the production process is shut-down during the search for the assignable cause. A production cycle is defined as the time period from the beginning of production (adjustment) to the detection/elimination of the assignable cause. The production cycle consists of four periods: (1) the in-control period; (2) the out-of-control period; (3) the search period due to false alarms; and (4) the search and repair period due to true alarms.

3. LOSS-COST FUNCTION

Given relevant income and cost parameters associated with each period of production cycle operating under the surveillance of the control chart, the expected income per production cycle can easily be derived. In order to formulate the loss-cost function for the economic design of control charts, the following notations will be used.

T_a = random time during which the process operates under the state of control
M = number of samples taken before the process goes out of control
G = number of samples taken after the Mth sample and up to the moment the chart signals lack of control
N = number of false alarms occuring among the first M samples
n = sample size
s = sampling interval
τ_s = expected time to search for the assignable cause
K_s = expected cost of searching for the assignable cause
τ_r = expected time for adjustment and repair if assignable cause exists
K_r = expected adjustment and repair cost
R_0 = average run length (ARL) of the control chart operating at the acceptable quality level μ_0
R_1 = ARL of the control chart operating at the rejectable quality level $\mu_1 = \mu_0 + \delta\sigma$
V_0 = profit per hour from the process when operating in control
V_1 = profit per hour from the process when operating out of control
b = fixed cost of sampling
c = variable cost of sampling
L = average per hour loss-cost of the process
I = average income per hour.

The expected length of the production cycle is:

$$sE(M) + \tau_s E(N) + sE(G) + \tau_s + \tau_r \quad (1)$$

and the expected income from the production cycle is [11]:

$$V_0/\lambda + V_1 E(Ms+Gs-T_a) - E(N)K_s - (b+cn)E(M+G) - K_s - K_r \quad (2)$$

Hence, the average net income per hour is:

$$I = \text{equation}(2)/\text{equation}(1) \quad (3)$$

Defining $L = V_0 - I$, the following expression is obtained after some simplification:

$$L = \frac{\lambda UB_1 + VB_0 + \lambda W + (b+cn)(1+\lambda B_1)/s}{1 + \lambda B_1 + \tau_s B_0 + \lambda(\tau_r + \tau_s)} \quad (4)$$

where $U = V_0 - V_1$, $V = K_1 + V_0 \tau_s$,
$W = K_r + K_s^1 + V_0(\tau_s + \tau_r)$,

$B_0 = (1/s - \lambda/2 + \lambda^2 s/12) / R_0$, and

$B_1 = (R_1 - 1/2 + \lambda s/12)s$.

The expressions for R_0 and R_1 depend on the particular type of the control chart that is to be used for controlling the process means.

4. OPTIMAL DESIGN

To take into consideration the effects of non-normality on the control chart, the non-normal probability density function of the process is approximated by the first four terms of an Edgeworth series [13]. The optimum design parameters and loss-cost function of the $\bar{x}$-chart, the $\bar{x}$-chart with warning limits and the cusum chart are obtained by the methods explained below. All three charts are considered to be one-sided.

4.1 Design of $\bar{x}$-Chart

In $\bar{x}$-control chart, the control limit is set at $\mu_0 + k\sigma/\sqrt{n}$, where k is the control limit coefficient. A sample of size n is taken from the process every s hours and the sample mean is plotted on the chart. The occurrence of the sample means outside the control limit is regarded as an indication that the process is out of control and some rectifying action is to be taken. Thus, to operate an $\bar{x}$-chart, the user has to specify n, s, and k. Before minimizing equation (4) to obtain the design parameters, the values of the average run length, R_0 and R_1 are derived as follows [13]:

$$R_0 = 1/\alpha = 1/(1-\Phi(k)), \text{ and, } R_1 = 1/(1-\beta) = 1/p$$

where $p = 1 - \Phi(k-\delta\sqrt{n}) + \frac{\gamma_1}{6\sqrt{n}} \phi^{(2)}(k-\delta\sqrt{n})$

$$- \frac{\gamma_2}{24n} \phi^{(3)}(k-\delta\sqrt{n}) - \frac{\gamma_1^2}{72n} \phi^{(5)}(k-\delta\sqrt{n})$$

and, $\gamma_1 = \sqrt{\beta_1}$, $\gamma_2 = \beta_2 - 3$,
$\Phi(x)$ = the cdf of the standardized normal variate x,
$\phi(x)$ = the pdf of the standardized normal variate x,
$\phi^{(r)}(x) = (\frac{d}{dx})^r \phi(x)$.

p is the probability of true alarm, β is the probability of Type II error, and α is the probability of Type I.

The pattern search techniques of Hooke and Jeeves [9] is employed to minimize equation (4) in order to obtain the optimum design parameters of the $\bar{x}$-chart.

4.2 Design of $\bar{x}$-Chart with Warning Limit

A sample of fixed size n is taken at regular intervals of s hours and the sample mean is plotted on a one-sided $\bar{x}$-chart with warning limit. The action limit is set at $\mu_0 + k_a \sigma/\sqrt{n}$, where k_a is the control limit coefficient. The warning limit is set at $\mu_0 + k_w \sigma/\sqrt{n}$, where $0 < k_w < k_a$. A search for the assignable cuase is undertaken if the last sample mean taken falls outside the action limit, or if it completes a run length, R_c, between the warning and the action limits. The corresponding expressions for R_0 and R_1 to be used in equation (4) are derived from the following equation [14]:

$$ARL = (1 - q^{R_c})/[1 - q - p(1 - q^{R_c})] \quad (5)$$

where

$$p = \Phi(k_w - \delta\sqrt{n}) - \frac{\gamma_1}{6\sqrt{n}} \cdot \phi^{(2)}(k_w - \delta\sqrt{n}) + \frac{\gamma_2}{24n} \cdot$$
$$\phi^{(3)}(k_w - \delta\sqrt{n}) + \frac{\gamma_1 2}{72n} \cdot \phi^{(5)}(k_a - \delta\sqrt{n}) \quad (6)$$

$$q = \Phi(k_a - \delta\sqrt{n}) - \Phi(k_w - \delta\sqrt{n}) - \frac{\gamma_1}{6\sqrt{n}} \cdot$$
$$[\phi^{(2)}(k_a - \delta\sqrt{n}) - \phi^{(2)}(k_w - \delta\sqrt{n})] + \frac{\gamma_2}{24n} \cdot$$
$$[\phi^{(3)}(k_a - \delta\sqrt{n}) - \phi^{(3)}(k_w - \delta\sqrt{n})] + \frac{\gamma_1}{72n} \cdot$$
$$[\phi^{(5)}(k_a - \delta\sqrt{n}) - \phi^{(5)}(k_w - \delta\sqrt{n})] \quad (7)$$

p is now the probability that a point falls below the warning limit, and q is the probability that it falls between the warning and action limits. R_1 is obtained by substituting equations (6) and (7) into equation (5). To obtain R_0, we let $\delta = 0$ in equation (5). The direct search method of Hooke and Jeeves [9] is again employed to minimize L with respect to the variables (n, s, k_a, k_w, R_c) in order to obtain the optimal design parameters of $\bar{x}$-chart with warning limit.

4.3 Design of Cusum Chart

The operation of a cusum chart for controlling the mean of a process involves taking samples of size n at regular interval of s_r hours, and plotting the cumulative sums $S_r = \sum_{j=1}^{r}(\bar{x}_j - K)$ vs. the sample number r, where $\bar{x}_j$ is the sample mean of the jth sample, and K is a prescribed reference value. If the cumulative sum exceeds the decision interval h, it is concluded that an upward shift in the process mean has occurred. The optimum values of the design parameters n, s, h and K are obtained by minimizing the loss-cost function L.

As before, function L depends on ARL whose value, for a one-sided cusum chart for controlling non-normal process mean with horizontal boundaries at (0,H), is defined as [11]:

$$ARL = \frac{N(0)}{1 - P(0)} \quad (8)$$

P(0) and N(0) are special cases, at z = 0 of P(z), and N(z), which are defined as follows:

$$P(z) = \int_{-\infty}^{-z} g_1(x)dx + \int_{0}^{H} P(x) \cdot g_1(x-z)dx$$
$$0 \le z \le H \quad (9)$$

$$N(z) = 1 + \int_{0}^{H} N(x) \cdot g_1(x-z)dx \quad 0 \le z \le H \quad (10)$$

where $g_1(x)$ is the pdf of the standardized increment $(x_n - K)/(\sigma/\sqrt{n})$ in the cumulative sum, for the non-normal process with mean $\theta = (\mu_0 - K)/(\sigma/\sqrt{n})$ and parameters γ_1 and γ_2, and H is the standardized decision interval defined as $H = h/(\sigma/\sqrt{n})$. Equations (9) and (10) are first approximated by a system of linear equations [11] and then solved numerically to obtain the values of R_0 and R_1. These values are substituted in equation (4) to derive the loss-cost function of the cusum chart. An iterative optimization algorithm is used to minimize the resulting loss-cost function to determine the optimal values of the design parameters [11].

5. A RELATIVE COMPARISON AMONG THE THREE CHARTS

Because of its simplicity and ease of operation, $\bar{x}$-chart with action limits has been in use for about fifty years. The $\bar{x}$-chart with warning limits has become popular during the recent years since it is generally believed that it is more efficient than the $\bar{x}$-chart for detecting the shifts in the process mean. But there are certain disadvantages in the operation of $\bar{x}$-chart with warning limits. The use of warning limits, in addition to control limits, implies that a certain number, R_c, of successive sample means must fall between the warning and control limits to take action. In other words, there is no cumulative effect of the mean points that deviate from the expected value unless R_c number of points successively appear between the warning and the control limits. Now, it is certainly possible that a shift could occur in the process mean and remain undetected for a substantial period. In contrast, the cusum chart is based on all sample points rather than the last few samples. The chart deals with retrospective examination of the samples to detect the occurrence of significant changes in process mean. Due to the differing ways in which the three charts establish control over the process, it is instructive to compare, in quantitative terms, the performances of the $\bar{x}$-chart, the $\bar{x}$-chart with warning limits and the cusum chart at various degrees of shift in the process level.

Under non-normality assumption, the minimum loss-cost criterion is employed as a measure of the chart's performance. The same cost factors are used in calculating the loss-cost function in each case. The optimum design parameters

TABLE 1: Comparison of Three Minimum Cost Control Procedures

δ	x̄-Charts					x̄-Charts With Warning Limits					Cusum Charts				
	n	s	R_0	R_1	L	n	s	R_0	R_1	L	n	s	R_0	R_1	L
0.50	30	1.26	20	1.17	11.1468	30	1.24	20	1.17	11.1443	29	1.21	21	1.19	11.1447
0.75	19	1.01	43	1.12	9.4623	19	1.01	44	1.12	9.4593	19	1.01	46	1.13	9.4587
1.00	13	0.88	70	1.10	8.5272	13	0.88	73	1.10	8.5250	13	0.88	72	1.10	8.5244
1.25	10	0.80	111	1.07	7.9397	10	0.80	114	1.07	7.9391	10	0.80	115	1.08	7.9387
1.50	8	0.76	161	1.06	7.5742	8	0.74	165	1.06	7.5426	8	0.74	174	1.06	7.5227
1.75	6	0.69	175	1.06	7.2590	6	0.68	182	1.06	7.2592	6	0.68	190	1.06	7.2591
2.00	5	0.66	222	1.05	7.0449	5	0.65	230	1.05	7.0458	5	0.66	222	1.05	7.0458
2.25	5	0.66	434	1.03	6.8885	4	0.62	238	1.05	6.8915	5	0.65	441	1.03	6.8908

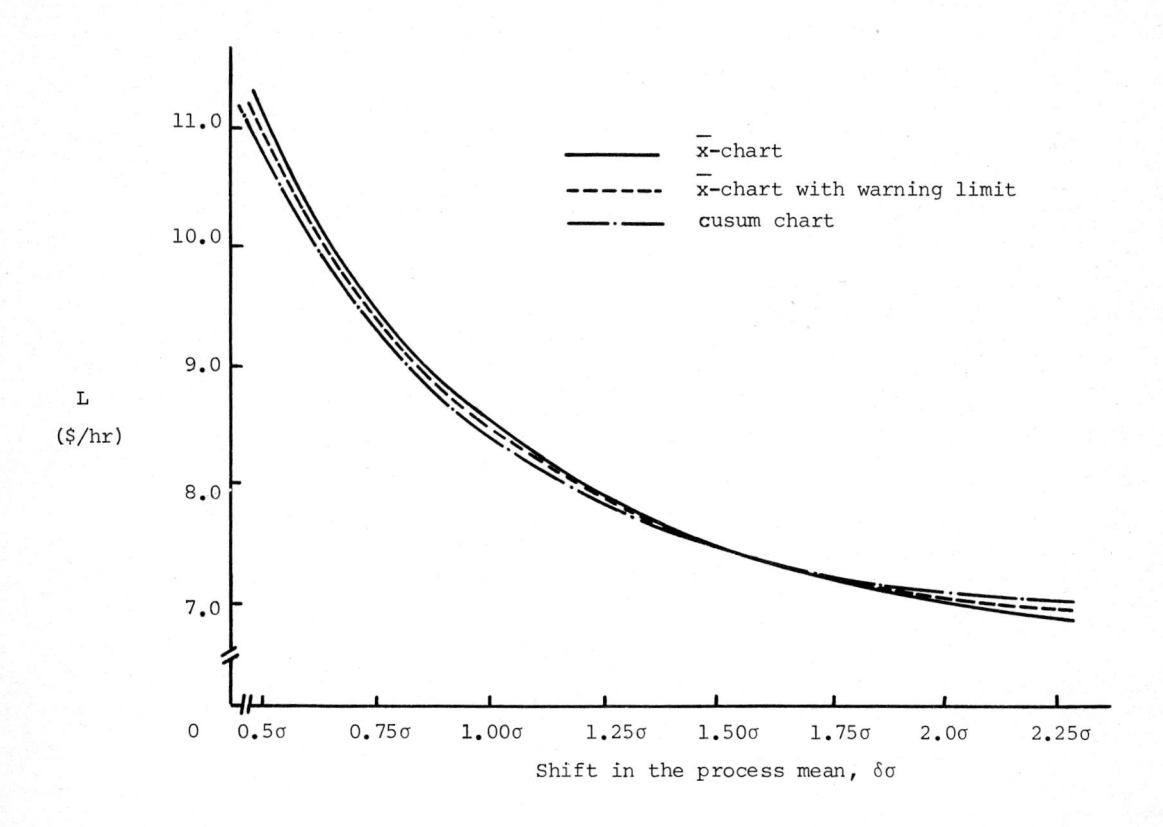

Figure 1. Loss-Cost vs. δσ for the Three Control Charts

and the loss-cost function of the charts are obtained by the methods given in sections 4.1 - 4.3, and organized in Table 1 for various values of the shift parameter. The corresponding average run lengths R_0 and R_1 are also shown.

The parameter values associated with Table 1 are:

$\lambda = 0.05$, $\mu_0 = 0.0$, $\sigma^2 = 1.0$, $\gamma_1 = 0.5$,

$\gamma_2 = 0.5$, $V_0 = 150.0$, $V_1 = 50$, $k_r = 20$,

$k_s = 10$, $\tau_r = 0.2$, $\tau_s = 0.1$, $b = 0.5$ and

$c = 0.1$.

6. DISCUSSION AND CONCLUSIONS

It can be seen from Table 1 and Figure 1 that for a shift in the process mean between 0.5σ and 1.5σ the loss-cost for the cusum chart is slightly less than that of the x-chart with warning limits, which, in turn, is less than that of the x-chart with control limits. When the shift in the process mean is above 1.5σ, the loss-cost due to the x-chart becomes less than those of the x-chart with warning limit and the cusum chart. It is observed from Table 1 that, for small shifts in the process mean, the sample size and the sampling interval are relatively large. However, for shifts greater than 1.5σ, small sample size and frequent sampling are desirable.

Quality control engineers are always concerned with the optimal selection of producer's and consumer's risks which are, respectively, analogous to the probabilities of type I and type II errors for an x-chart. It is desirable that these errors be as small as possible. This is achieved by having a large average run length R_0, when the process is in control, and a short average run length R_1, when the process is out of control. Extensive numerical studies based on Table 1 reveal that there are no appreciable differences in the R_1 values among the three charts.

ACKNOWLEDGEMENTS

Financial support for this study was provided by the Natural Sciences at Engineering Research Council of Canada. Their assistance is gratefully acknowledged.

REFERENCES

[1] Chiu, W.K., The economic design of cusum charts for controlling normal means, Applied Statistics, 23(1974) 420-433.

[2] Chiu, W.K. and Cheung, K.C., An economic study of x-charts with warning limits, Jrnl. of Qual.Technology, 9(1977) 166-171.

[3] Chiu, W.K. and Wetherill, G.B., A simplified scheme for the economic design of x-charts, Jrnl. of Qual.Technology, 6(1974) 63-69.

[4] Cowden, D.J., Statistical Methods in Quality Control (Prentice Hall, Englewood Cliffs, N.J., 1957).

[5] Duncan, A.J., The economic design of x-charts used to maintain current control of a process, Jrnl. Amer. Statis. Assoc. 51(1956) 228-242.

[6] Goel, A.L.and Wu, S.M., Economically optimal design of cusum charts, Manag. Science, 19(1973) 1271-1282.

[7] Goldsmith, P.L. and Whitefield, H., Average run lengths in cumulative chart quality control schemes, Technometrics, 3(1961) 11-20.

[8] Gordon, G.R. and Weindling, J.I., A cost model for economic design of warning limits control chart schemes, AIIE Transac. 7(1975) 319-329.

[9] Hooke, R. and Jeeves, T.A., Direct search solution of numerical statistical problems, Jrnl. of the Assoc. of Comp. Mach. 8(1961) 212-229.

[10] Lashkari, R.S. and Rahim, M.A., An economic design of x-charts for controlling non-normal process means considering the cost of process shut-down, Proc. of Computer Science and Statistics, 12th Symposium on the Interface (University of Waterloo, 1979).

[11] Lashkari, R.S. and Rahim, M.A., An economic design of cumulative sum charts to control non-normal process means, Comput. & Indus. Engng. 6(1982) 1-18.

[12] Nagendra, Y. and Rai, G., Optimum sample size and sampling interval for controlling the mean of non-normal variables, Jrnl. of Amer. Statis. Assoc. 66(1971) 637-646.

[13] Rahim, M.A., Economic design of x- and cusum charts as applied to non-normal processes, Ph.D. thesis, Dept. of Indus. Engng., University of Windsor (Sept. 1981).

[14] Rahim, M.A. and Lashkari, R.S., An economic design of x-chart with warning limits, Proc. Computer Science and Statistics, 13th Symposium on the Interface,(Springer-Verlag, N.Y., 1981).

[15] Roberts, S.W., A comparison of some control charts procedures, Technometrics, 8(1966) 411-430.

[16] Taylor, H.M., The economic design of cumulative sum control charts, Technometrics, 10 (1968) 479-488.

THE DESIGN OF UNITEST, A UNIFIED FORTRAN PROGRAM TESTING SYSTEM

Vincent L. Tang

IMSL, Inc.
Houston, Texas

It is possible to build a powerful program-testing tool by implementing various
test methods in a single system as the capabilities of these methods are complementary in nature. A Fortran 77 preprocessor that analyzes a given program is
used to extract program attributes to construct a database. The given program
is also instrumented with additional statements to update the database during a
program testing session. This paper describes some experiences in implementing
and using such an automated program testing system.

I. INTRODUCTION

It is well recognized that software is the dominant cost in the development of a computer system. In some cases, it is found that software
cost is as high as eighty percent of the total
system cost[1]. This ratio tends to increase as
computer hardware can be cheaply manufactured.
In order to ensure that a piece of expensive
computer software will work properly, there has
been a lot of research on designing and developing methodology that can be used to test and
validate computer programs. This paper describes an automated tool that aims to test and
validate computer programs.

We can conclude that a program is in error if
a test failed, but no matter how many tests suc-
ceed, we cannot be certain that the program in
question works as intended for all cases. Therefore, the goal of program testing and validation
is to expose as many errors in a piece of computer software as possible. Some of the most
commonly used methods are program structure
checks using a program graph[2], dataflow analysis
on the pattern of usage of program variables[3],
and generation of a program profile to show that
every branch in a program is traversed at least
once[4]. Unfortunately, no single method is as
powerful as the totality of all the program
testing methods. Furthermore, existing program
testing systems that are implemented on one or
two testing methods can only be used as stand-
alone tools. Since the capabilities of these
methods are complimentary in nature, it is suggested that we can treat various methods in a
unified manner and develop a new design concept
that makes it possible to implement different
methods in a single system economically and efficiently[5]. This paper describes how such a
program testing system, called Unitest, is implemented.

Unitest is designed using the unified concept
for testing and validating Fortran 77 programs[5].
It was a research project of the Computer
Science Department of the University of Houston.
The first version is run on a VAX-11 780 computer under the MVS operating system. It has
also been installed on a Data General MV/8000
system under the AOS operating system.

II. THE UNITEST SYSTEM

The Unitest system consists of a Fortran 77 preprocessor, a database generator, and a collection of utility programs. The Fortran preprocessor is essentially a static analyzer that
analyzes programs to be tested and extracts desired program attributes for the database generator. The database generator collects all
the necessary information to build a database.
A set of the Unitest utility programs are used
as real time monitor routines for performing
dynamic analysis on the programs being tested.
While the remaining set of utility programs are
used as post-processing routines that generate
reports based on the contents of the Unitest
database.

A. The Unitest Preprocessor

The Unitest preprocessor implements a technique
known as "program instrumentation". This refers
to the process in which additional statements are
inserted into the source program for the purpose
of computing the values of the program attributes. The instrumentation process is completely automated, and the inserted statements
do not alter the intended function of the input
program. A command procedure which is written
in computer command language is run on top of
the Unitest system. So that the instrumented
program can be compiled and linked to the
Unitest utility programs in response to a single
user command.

B. The Unitest Database

The Unitest database, which resides on disk,
logically consists of a symbol table, a use
table and a node table. The tables are not
operated independent from each other and they

are linked together by logical associations. A movement from one table to another causes a logically related item to be addressed. The Unitest database behaves like a network database in the sense that many of its tables are linked by pointers. The use of pointers speeds up the retrieval of related data because no additional key hashing is needed. It also provides the flexibility of inserting and deleting data elements. The values of the pointers are included as part of the program instrumentation, so that concerned areas in the database can be instantly addressed and updated. Furthermore, an index table that is kept in the main memory is used to perform direct access on the disk files that form the database.

Certain Unitest utility programs require memory workspace to perform certain program testing tasks. For example, the Unitest program that is used to perform dataflow analysis requires a table to monitor the usage pattern of all program variables during execution time. Sometimes it is preferable to have a portion of the database loaded to the main memory for faster processing time. Also, since there are a lot of input/output activities between the system and the database, improvement on the data transfer mechanism is desired. Data transfer mechanisms such as blocked input/output data transfer should be adopted.

C. The Unitest Utility Programs

Utility programs are written to perform dataflow analysis, symbolic trace, and analysis of the number of times each branch or each subroutine call is exercised. All the utility programs use the data stored in the Unitest database, but they function independently. The instrumentation provides break points at each node of the program graph, so that the utility programs can be activated. Any questionable event occurring during execution time can then be detected and reported.

The user can use the post-processing programs to retrieve program testing results stored in the Unitest database. The function of a post-processing routine should not be limited to the display of the contents of the database. It should also perform analysis on test results, as specified by the user, to reduce redundancy in the report elements. Some utility programs may be used to assist the user in searching the Unitest database for answers to the questions that arose by previous queries to the system.

III. CONCLUSION

By implementing the first as well as the experimental version of the Unitest system, experience has been gained in designing an automated program testing system using the unified concept. The unified concept is attractive because it produces a powerful system without wasting efforts in duplicating those elements that are complimentary to different program testing methods. The Unitest system is expandable. Additional utility programs can be built or tied onto the existing system to support additional program testing activities. Furthermore, a unified program testing system like Unitest is not so big and complex that makes it more difficult to be maintained in comparison to some other program testing systems. The current Unitest system has about 26,000 Fortran source statements, while Dave[3] alone has about 25,000 Fortran source statements. In the Unitest system, ninety-four subprograms are written for the Fortran 77 preprocessor and database generator. While there are thirty-six utility subprograms that are used to perform dataflow analysis, symbolic trace generation, and analysis on branch execution.

The design of a database for a unified program testing system is also an important part of the project. The database needs to be general and versatile enough that once it is implemented, no major modification is necessary for the introduction of additional program testing processes to the system. Also emphasis should be made on designing utility programs that make use of the contents of the database to generate desirable but not excessive program testing reports for the user.

[1] B. W. Boehm, CHARACTERISTICS OF SOFTWARE QUALITY, TRW report, December 1973.

[2] C. Ramamoorthy, TESTING LARGE SOFTWARE WITH AUTOMATED SOFTWARE EVALUATION SYSTEMS, Proceedings of 1975 International Conference on Reliable Software.

[3] L. J. Osterweil, SOME EXPERIENCE WITH DAVE - A FORTRAN PROGRAM ANALYZER, National Computer Conference, 1976.

[4] D. E. Knuth, AN EMPIRICAL STUDY OF FORTRAN PROGRAMS, Software-Practice and Experience, 1(2), April 1971.

[5] J. C. Huang, SOFTWARE TOOLS FOR PROGRAM VALIDATION: A UNIFIED APPROACH, Technical Report, Computer Science Department, University of Houston, 1981.

EXACT SAMPLING DISTRIBUTIONS OF THE COEFFICIENT OF KURTOSIS USING A COMPUTER

Derrick S. Tracy and William C. Conley

University of Windsor and University of Wisconsin at Green Bay

The ratio statistic k_4/k_2^2 provides the coefficient of kurtosis (peakedness) of a distribution. To find the sampling distribution of this statistic, using analytical methods, is almost impossible. The exact sampling distributions are obtained here, using computer techniques, for sampling with or without replacement from a discrete uniform distribution for various sample and population sizes.

1. INTRODUCTION

The peakedness of a distribution is measured by its coefficient of kurtosis, κ_4/κ_2^2, where κ_2 and κ_4 are its second and fourth cumulants. For a sample, Fisher's k-statistic k_p is an unbiased estimate of the population cumulant κ_p. Thus, for a sample, the ratio statistic k_4/k_2^2 is considered as the coefficient of kurtosis. (Kendall and Stuart [1]).

The asymptotic distribution of k_4/k_2^2 is normal. The exact distribution has not been obtained, and is almost impossible to obtain theoretically. Using computer methods, the exact distributions are obtained here, for sampling from a given finite population of size N, when samples of size n are drawn with or without replacement.

For sample $x_1, x_2, \ldots, x_n$, the relevant k-statistics are

$$k_2 = \frac{\sum_i x_i^2}{n} + \frac{\sum_{i \neq j} x_i x_j}{n(n-1)},$$

$$k_4 = \frac{\sum_i x_i^4}{n} - 4\frac{\sum_{i \neq j} x_i^3 x_j}{n(n-1)} - 3\frac{\sum_{i \neq j} x_i^2 x_j^2}{n(n-1)}$$

$$+ 12\frac{\sum_{i \neq j \neq k} x_i^2 x_j x_k}{n(n-1)(n-2)} - 6\frac{\sum_{i \neq j \neq k \neq \ell} x_i x_j x_k x_\ell}{n(n-1)(n-2)(n-3)}.$$

The given finite population considered here is the discrete uniform, with $f(x) = \frac{1}{N}$, $x = 1, 2, \ldots, N$. For $N = 15(5)60$, we draw samples of size $n = 4, 5, 6$, without replacement. We also draw samples with replacement of size 4 from the discrete uniform populations of sizes $N = 10(5)35$.

Sampling distributions of k_4/k_2^2 were obtained for the above situations. However, the computer technique is general enough to allow one to obtain the sampling distributions in other situations.

2. THE COMPUTER ALGORITHM

The computer program used to obtain a sampling distribution can essentially be divided into two parts. Part one consists of the relevant statistic, the underlying population distribution, the sample size, and the type of sampling to be done (with or without replacement, etc.). The second part consists of the histogram which records the location of each sample statistic value and its corresponding probability.

Nested DO-loops are used to consider all possible samples. About a 10 statement series with IF statements in a loop are required for the histogram. We used FORTRAN IV on an IBM 360-65 computer with a Calcomp plotter attached to the computer. However, any computer language with IF statements, DO-loops and subscripted variables would suffice for obtaining sampling distributions.

The range of the histogram is selected and fixed ahead of time. Then this range is divided up into 8, 16, 32, 64, 128, ..., 2^r intervals, depending on the accuracy required for the study. We chose $r = 7$, and hence 128 intervals. A system of 128 pigeonholes is thus set up, based on comparing the calculated value of k_4/k_2^2 with the seven different midpoints depending on the outcome of the preceding comparison.

The program can be modified to sample from any given finite population. For sampling with replacement, we allow unrestricted selection of samples in the nested DO-loops giving N^n instead of $\binom{N}{n}$ sample points. Extra FORTRAN statements are added to ensure that the correct probability is calculated when a sample consists of repititions. Samples where all x_i are identical are removed, so that k_2 and k_4^1 do not become zero.

3. SAMPLING DISTRIBUTIONS OF k_4/k_2^2

Figure 1 shows some examples of sampling distributions of k_4/k_2^2 for samples of size 4, when sampling without replacement from discrete uniform populations. The population and sample sizes are indicated along the ordinate, e.g., 15-4. Figures 2 and 3 show the same for samples of size 5 and 6 respectively, for sampling without replacement.

Figure 4 shows sampling distributions for size 4 for the case of sampling with replacement.

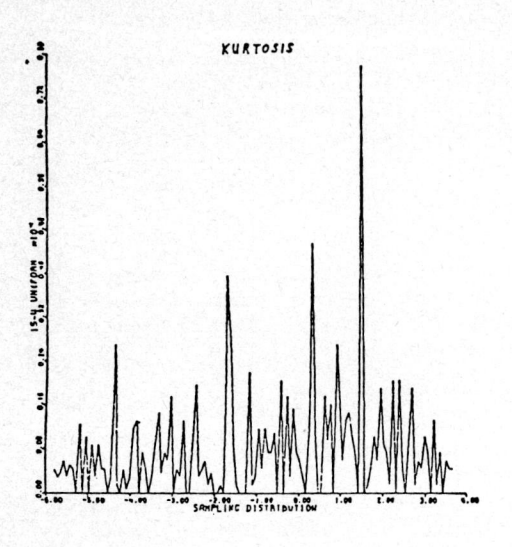

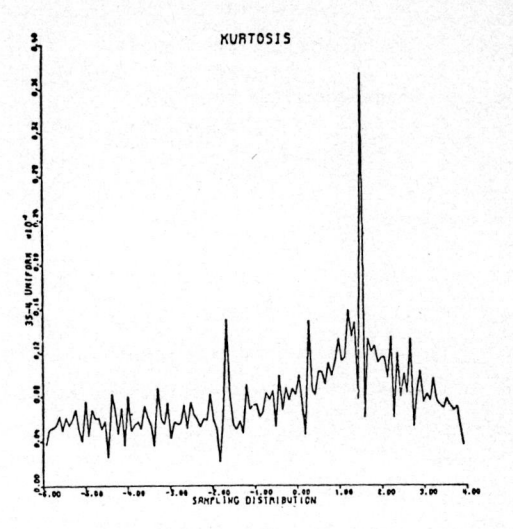

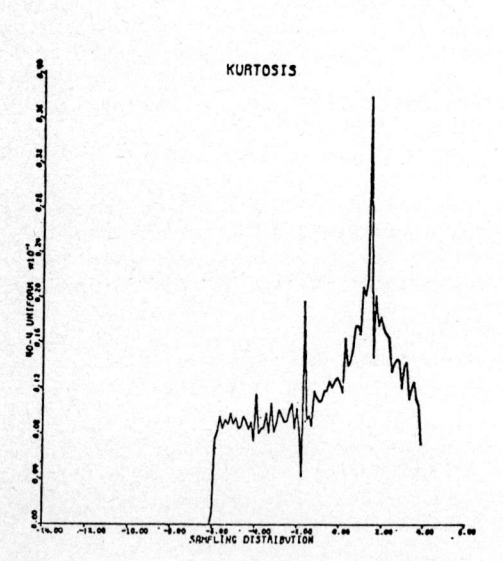

Figure 1 : Sampling distributions for without replacement samples of size 4

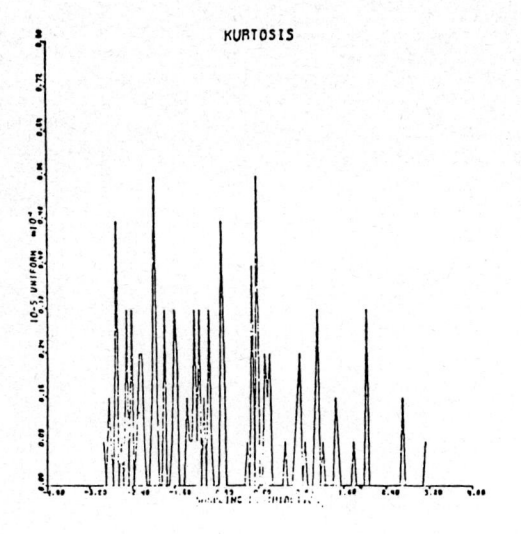

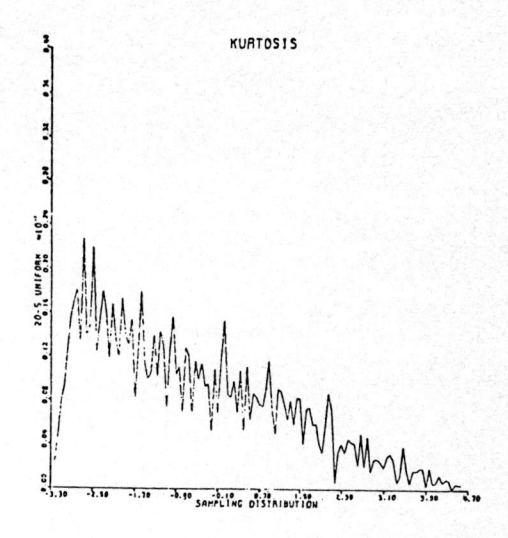

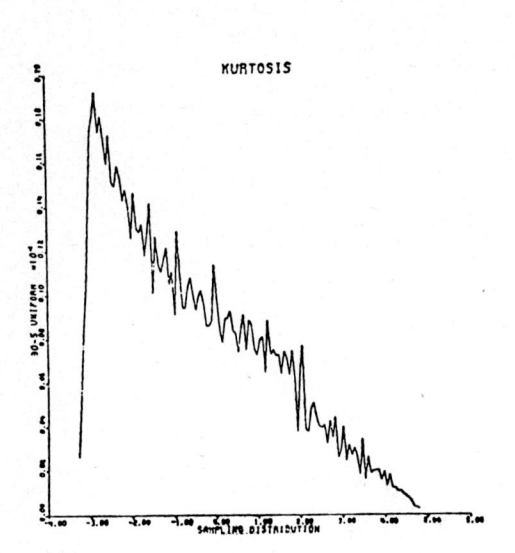

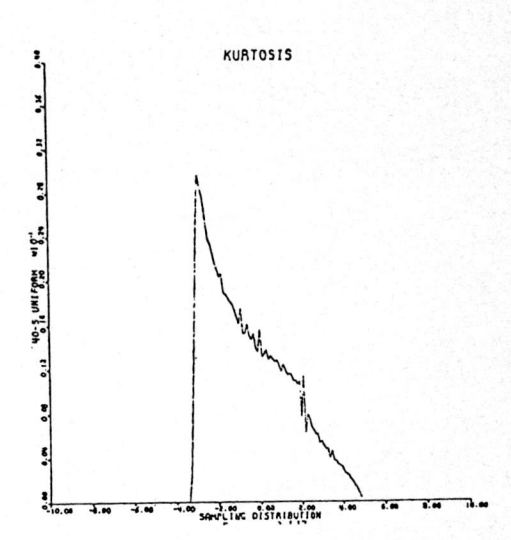

Figure 2 :   Sampling distributions for without replacement samples of size 5

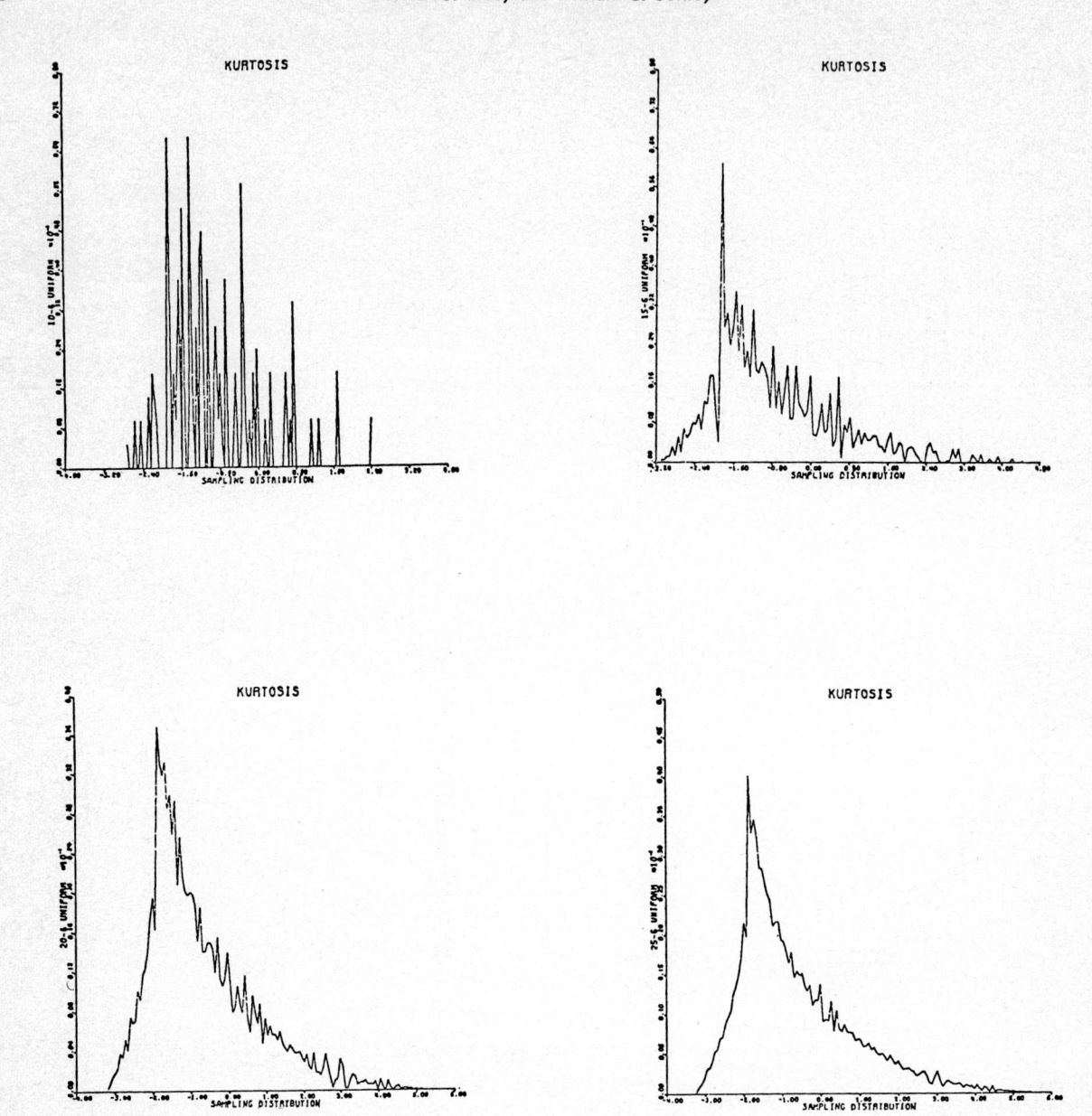

Figure 3 : Sampling distributions for without replacement samples of size 6

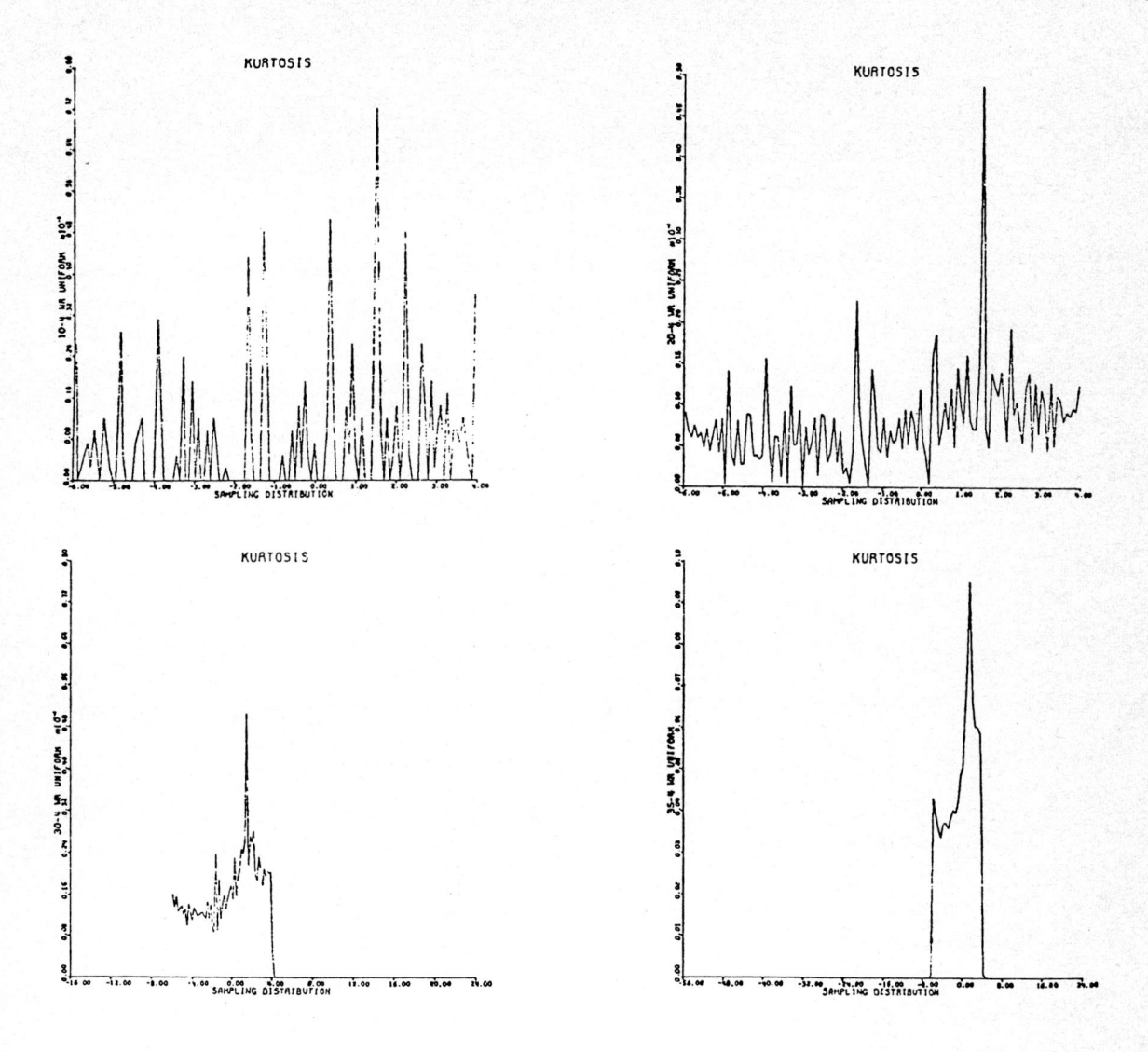

Figure 4 : Sampling distributions for with replacement samples of size 4

4. CONCLUSION

The above sampling distributions are based on all possible samples, unlike the Monte Carlo approach to sampling distributions. Until recently, such an approach to a theoretically difficult sampling distribution would have been impossible. But with recent advances in speed (Ware [2]) and capacity (Houston [3]) of computers, our technique can be useful in a variety of situations where theoretical methods become too involved.

REFERENCES

[1] Kendall, M.G. and Stuart, A., The Advanced Theory of Statistics, 1 (Hafner, New York, 1963).
[2] Ware, W.H., The Ultimate Computer, I.E.E.E. Spectrum (March 1972), 84-91.
[3] Houston, G.B., Trillion Bit Memories, Datamation (Oct. 1973), 52-58.